T0324863

Computer Modelling of Seas and Coastal Regions

INTERNATIONAL CONFERENCE ON COMPUTER MODELLING OF SEAS AND COASTAL REGIONS AND BOUNDARY ELEMENTS AND FLUID DYNAMICS
SOUTHAMPTON, U.K., APRIL 1992

Acknowledgement is made to K.Z. Elahi *et al.* for the use of Fig. 4, page 173, which appears on the front cover of this book.

Computer Modelling of Seas and Coastal Regions

Editor: P.W. Partridge, Wessex Institute of Technology

Computational Mechanics Publications
Southampton Boston

Co-published by

Elsevier Applied Science
London New York

P.W. Partridge
Wessex Institute of Technology
Ashurst Lodge
Southampton
SO4 2AA
U.K.
(also from the Dept. of Civil Engineering
University of Brasilia, Brazil)

C.A. Brebbia
Wessex Institute of Technology
Ashurst Lodge
Southampton
SO4 2AA
U.K.

Co-published by:

Computational Mechanics Publications
Ashurst Lodge, Ashurst, Southampton, UK

Computational Mechanics Inc.
25 Bridge Street, Billerica, MA 01821, USA

and

Elsevier Science Publishers Ltd
Crown House, Linton Road, Barking, Essex IG11 8JU, UK

Elsevier's Sole Distributor in the USA and Canada:

Elsevier Science Publishing Company Inc.
655 Avenue of Americas, New York, NY 10010, USA

British Library Cataloguing in Publication Data
A Catalogue record for this book is available
from the British Library

ISBN 1-85166-779-2 Elsevier Applied Science, London, New York
ISBN 1-85312-164-9 Computational Mechanics Publications, Southampton
ISBN 1-56252-092-X Computational Mechanics Publications, Boston, USA
Set
ISBN 1-85166-799-7 Elsevier Applied Science, London, New York
ISBN 1-85312-193-2 Computational Mechanics Publications, Southampton
ISBN 1-56252-121-7 Computational Mechanics Publications, Boston, USA

Library of Congress Catalog Card Number 91-77632

CONTENTS

SECTION 1: WAVES

SECTION 2: TIDES

SECTION 3: SHALLOW WATER CIRCULATION AND CHANNEL FLOW

SECTION 4: SILTATION AND SEDIMENTATION

SECTION 5: POLLUTION PROBLEMS

SECTION 6: COMPUTATIONAL TECHNIQUES

PREFACE

This book *Computer Modelling of Seas and Coastal Regions* is the first volume of the two volume proceedings of the International Conference on Computer Modelling of Seas and Coastal Regions and Boundary Elements and Fluid Dynamics, held in Southampton, U.K., in April 1992.

The importance of accurate modelling of seas and coastal regions is emphasized by the need for predicting their behaviour under extreme conditions. Problems, such as pollution of these areas, have become a major international concern and the related environmental problems need further study using techniques which can be used to determine the ways in which the water systems respond to different effects and try to minimize the damage. They can also lead to the development of early warning systems in combination with remote sensing equipment and experimental sampling techniques. Furthermore, once a disaster occurs, the model can be used to optimize the use of the available resources.

The conference addresses coastal region modelling both under normal and extreme conditions, with special reference to practical problems, currently being experienced around the world. Many of the delegates are actively involved in the modelling of seas and coastal regions.

This volume includes sections on waves, tides, shallow water circulation and channel flow, siltation and sedimentation, pollution problems, and compututational techniques.

The organizer would like to thank the International Scientific Advisory Committee, the conference delegates and all those who have actively supported the meeting.

P.W. Partridge
April 1992

SECTION 1: WAVES

Modelling Wave Propagation in Large Areas

P. Milbradt, K.P. Holz

Institute of Fluid Mechanics and Computer Application in Civil Engineering, University of Hannover, D-3000 Hannover 1, Germany

1 Introduction

The aim of coastal engineering is to estimate the effects of coastal protection structures. The erosion of coastal sections requires measures to regulate the sedimentbudget. Beach protection works and coastal structures are designed according to the local wave conditions.

During the planning phase it is necessary to have the appropriate instruments to estimate the effects of building measurements. There exist theoretical procedures as well as hydrological and numerical models. In the past the trend goes uniquely to the increase of the use of computers.

The commercial models - corresponding to their desired task - are based on different calculation assumptions and solution algorithms and therefore they are not of universal use.

For their analysis large areas have to be modelled in order to determine the wave characteristics of deep water waves from different origin and direction propagating into shallow water near shore regions. Numerical treatment of nonlinear waves for such areas of some ten kilometers in extension is beyond the capabilities of workstations. A compromise between numerical practicability and physical quality of the results may be based on linear wave theory. This was shown by field measurments and statistical analysis even for a near-shore groinfield [5].

The following presented numerical wave model is a part of the program system TICAD (Tidal Interactive Computation And Design) [12]. It was developed for large-scale areas. The range of application encloses deep and shallow water regions up to breaking zones.

2 Wave model: Theoretical background

The wave model is based on the numerical solution of analytical and empirical approximation functions as they are given for instance in Shore Protectional Manual [1].

On the basis of the linear wave theory of AIRY propagation and changes of a monochromatic wave are calculated by the wave front method. In the case of neglecting external forces (for example wind forces) and of breaking zones the method is based on the conservation of mean energy flux between two wave normals. The mean flux of energy is proportional to the product of the group velocity c_g and the square of the wave height H:

$$F = c_g * \frac{\rho * g * H^2}{8} \tag{1}$$

Under deep water conditions ($d \geq 0.5 * L$) the wave propagates with constant velocity c_0 (index 0 indicates deep water conditions). In this case the wave length L_0, periode T_0 and height H_0 keep their values.

The wave parameters, except the period T, will change if the orbital motion of the wave touches the bottom ($d \leq 0.5 * L$) or if the wave will meet an obstacle like a mole or an end of an island or if it enters into a region with currents. These influences are called shoaling, refraction, diffraction and current-refraction. They are taken into account by the wave model as well as breaking of the waves due to very low water depth and steepness of the wave. In addition the consideration of a windfield is possible. The influences of perculation and reflection are neglected.

The necessary information about the bathymetry of a coastal region are taken from the digital terrain model (DTM). The DTM allows the fitting at every arbitrary terrain bathymetry due to triangular latticenet with variable width of the meshs. On the basis of that latticenet the tidal induced current are calculated by a 2-dimensional hydrodynamic model based on FEM. The result of this tidal calculation (current, water level) serves as input for the wave model.

3 Calculation of the wave parameter

Wave length

The wave length is calculated by the implicitly equation:

$$L = \frac{g * T^2}{2 * \pi} * tanh\left(\frac{2 * \pi * d}{L}\right) \tag{2}$$

Shoaling

Entering more shallow water the wave begins to 'feel bottom', when the water depth is about one half of the wave length. The waves are hereafter slowed, shortened and steepened, as they travel into more shallow water. This process is called shoaling. The group velocity is calculated by

$$c_g = c * n = \frac{c}{2} * \left[1 + \frac{\frac{4*\pi*d}{L}}{sinh(\frac{4*\pi*d}{L})}\right] \tag{3}$$

with

$$c = \frac{L}{T}. \tag{4}$$

The shoaling coefficient k_s is calculated for an arbitrary terrain point by

$$k_s = \frac{H_2}{H_1} = \frac{c_{g1}}{c_{g2}} \tag{5}$$

$$= \sqrt{\frac{L_1 * T_2}{L_2 * T_1} * \frac{\left[1 + \frac{\frac{4*\pi*d_1}{L_1}}{sinh\left(\frac{4*\pi*d_1}{L_1}\right)}\right]}{\left[1 + \frac{\frac{4*\pi*d_2}{L_2}}{sinh\left(\frac{4*\pi*d_2}{L_2}\right)}\right]}}. \tag{6}$$

Depth-Refraction

If the wave front in shallow water meets a bottom contour at an angle, the direction of travel is changed. This process of refraction is due to the fact, that water waves propagate more slowly in shallow than deeper water, and therefore the front tends to get aligned with the contours. This phenomenon is comparable with the refraction of the light described by the *SHNELL*'s refraction law. The change of the wave height due to refraction is calculated using a formula from *WIEGEL* [11]:

$$k_r = \frac{H_2}{H_1} = \sqrt{\frac{b_1}{b_2}} \tag{7}$$

with b =distance between the orthogonals.
Refraction appears always in superposition with shoaling.

Current-Refraction

Changes of phase velocity can also be caused by currents, resulting in current-refraction. In case of coincidence between current and wave propagation direction the wave length will be decreased and the wave height will be increased. Opposite directions cause reverse effects. The influence of the current is calculated using an approach of *JOHNSON* [6]. In Fig.1 the geometrical relations are described for the general case. A wave propagates under the angle α from still to flow water region. The change of the wave parameters are calculable using the geometrical conditions at the discontinuation surface, assuming that the change of the velocity is a jump.

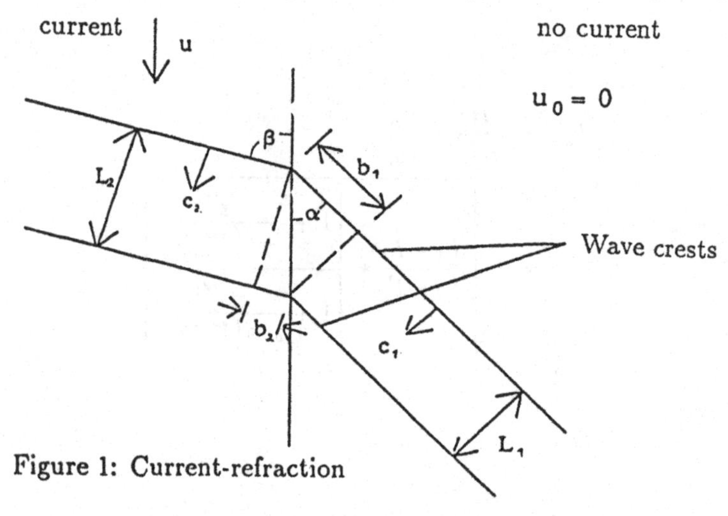

Figure 1: Current-refraction

Hence, the following relations are valid:
wave length

$$\frac{L_1}{sin(\alpha)} = \frac{L_2}{sin(\beta)} \tag{8}$$

propagation velocity

$$\frac{c_1}{sin(\alpha)} = u + \frac{c_2}{sin(\beta)} \tag{9}$$

With the conservation of enery flux and the above equations the change of the wave height is calculated:

$$k_{st} = \frac{H_2}{H_1} = \sqrt{\frac{c_{g_1} * b_1}{(c_{g_2} + u * sin(\beta)) * b_2}} \tag{10}$$

Diffraction

Diffraction is the propagation of a wave behind an obstacle as a mole or an end of an island. In analogy to the geometrical optic the change of wave height is calculated using the equation of *SOMMERFELD*.

Fig.2 shows the relations between the angles and equation (11) is the solution of the *SOMMERFELD*-equation.

$$F(r,\theta) = f(\sigma) * e^{-ikr*cos(\theta-\theta_0)} + f(\sigma') * e^{-ikr*cos(\theta-\theta_0)} \tag{11}$$

with

$$\sigma = 2 * \sqrt{\frac{k*r}{\pi}} * sin\left(\frac{\theta-\theta_0}{2}\right) \tag{12}$$

$$\sigma' = -2 * \sqrt{\frac{k*r}{\pi}} * sin\left(\frac{\theta-\theta_0}{2}\right) \tag{13}$$

$$f(\sigma) = \frac{1+i}{2} * \int_{-\infty}^{\sigma} e^{\frac{-i\pi t^2}{2}} dt \tag{14}$$

$$f(\sigma') = \frac{1+i}{2} * \int_{-\infty}^{\sigma'} e^{\frac{-i\pi t^2}{2}} dt \tag{15}$$

$$k = \frac{2*\pi}{L} \tag{16}$$

$$i = \sqrt{-1} \tag{17}$$

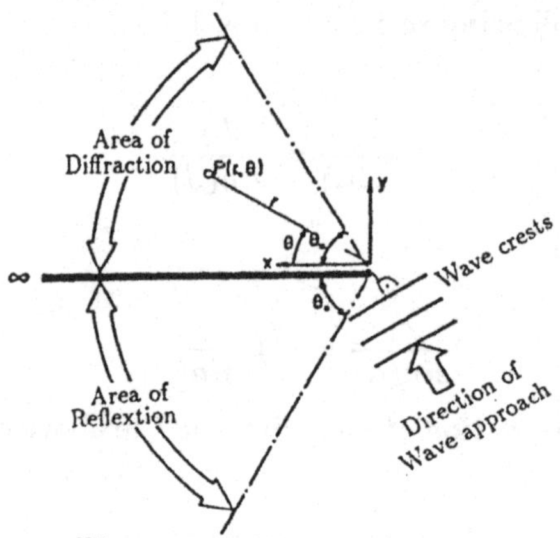

Figure 2: Diffraction

The diffraction coefficient k' is the ratio of local to incoming wave:

$$k' = \frac{H(r,\theta)}{H_0} = \mid F(r,\theta) \mid \tag{18}$$

In the wave model the effect of diffraction with effect of shoaling, refaction and current-refaction are interacting, therefore is valid:

$$H_2 = k_s * k_r * k' * k_{st} * H_1. \tag{19}$$

Breaking of the waves

The breaking of the waves due to very large steepness can be checked using the generalised $MICHE$-criteria in consideration of shallow water conditions [1]:

$$\frac{H_B}{L} = 0.14 * tanh\left(\frac{2 * \pi * d}{L}\right) \tag{20}$$

The index B is considering the breaking conditions.

The assumption from $WEGGEL$ [10], which is derived from the analysis of a lot of labor experiments corresponds to the breaking due to very low water depth independence on the underwater beach slope:

$$\frac{H_B}{d} = b - \frac{a * H_B}{g * T^2} \tag{21}$$

with

$$a = 43.75 * \left(1 - e^{-19*m}\right) \tag{22}$$

$$b = \frac{1.56}{1 + e^{-19*m}} \tag{23}$$

$m =$ is the underwater beach slope

Both breaking criteria are checked by the wave model. The change of the wave height due to energy lost during breaking will be calculated using an approach from *HORIKAWA/KUO* [7] and *ANDERSON/FREDSOE* [4] . In the distance x behind the breaking line one gets the following wave height formula:

$$\frac{H}{H_B} = 0.35 + 0.65 * e^{\frac{-0.12*x}{H_B}} \tag{24}$$

Bottom friction

For waves advancing in still water it is usual to assume that the variation in height with distance may be represented, locally, by

$$H(x_2) = H(x_1) * e^{a_0*(x_2-x1)} \tag{25}$$

where a_0 is a wave attenuation coefficient. The coefficient a_0 is approximated by

$$a_1 = \frac{\left(\frac{2*\pi}{L}\right)^2}{\sqrt{\frac{\pi}{T*\nu}} * \left(\frac{2*\pi*d}{L} + sinh\left(\frac{2*\pi*d}{L}\right)\right)} \tag{26}$$

where ν is the kinematic viscosity and a_1 is the component of a_0 attributable to energy dissipation in the boundary layer at the bed [8]. The bottom friction has the status of lower importance for problems of coastal engineering.

Wind field effects

In most coastal locations in the world no wave records deriving from field measurements are available and both the time scale for the design and financial resources are available make the installation and operation of such devices unfeasible for at least one climatological year.

The dominant wave parameters are generally computed a priori and if possible supported by several measurements. Using the a priori calculation you have the possibility of deterministic and statistic methods. Using the deterministic method the parameters of a "decisive" wave, in generally from engineers point of view the important $H_{1/3}$ value and the accessory period are defined. The statistic methods are describing the totality of waves in a field of waves. The North Sea is a typical case of the "Jonswap-Spectral" based on measures. Both methods have not implemented any information about the effects of bathymetry and the shore for the wind direction. By the numerical modelling there are possibilities given, to calculate the development of wave height also in shallow water areas with complicated depth partitionation, by implementation of refraction, shoaling, diffraction and breaking. Also it is possible to get information about the influence of realistic wind events with differences in surface and time. The program calculates the change of the significant waveheight $H_{1/3}$ and wave priod in an wind field caused by wind appropiated by [1]:

$$\frac{gH}{U_A^2} = 0.283 * tanh\left[0.53 * \left(\frac{gh}{U_A^2}\right)^{\frac{3}{4}}\right] * tanh\left\{\frac{5.65 * 10^{-3} * \left(\frac{gF}{U_A^2}\right)^{\frac{1}{2}}}{tanh\left[0.53 * \left(\frac{gh}{U_A^2}\right)^{\frac{3}{4}}\right]}\right\} \tag{27}$$

$$\frac{gT}{U_A} = 7.54 * tanh\left[0.833 * \left(\frac{gh}{U_A^2}\right)^{\frac{3}{8}}\right] * tanh\left\{\frac{3.79 * 10^{-2} * \left(\frac{gF}{U_A^2}\right)^{\frac{1}{3}}}{tanh\left[0.833 * \left(\frac{gh}{U_A^2}\right)^{\frac{3}{8}}\right]}\right\} \tag{28}$$

with U_A = wind velocity
F = wind fetchlength.

The Jonswap-equations can be calculated alternatively.

In the program, the above equation have been modified. Now the differences of wave height along the running direction are calculated. On this way the influence of bathymetry and the shore course on the development of sea violence can be considered by observing the waves step by step. In the actual program release the uniform wind conditions for the target area can be calculated.

Wave induced streaming

The streaming in nature are superpositon of tide and wave induced streaming which are interacting.

Now one can calculate the induced forces on the background of socalled radiation stresses by $LONGUENT - HIGGINS$ and $STEWART$ [9].

It is convenient to investigate how a progressive wave contributes, through the induced horizontal momentum and pressure components, to the dynamic equilibrium of a water column and to define and formulate the radiation stress magnitudes. The magnitudes will be used in the circulation models.

Independently of the first order wave theory postulation that waves transport no mass in the direction of their propagation due to the periodicity and symmetry of the velocity u magnitude, there is a surplus of momentum flux showing that gradients of induced mean momentum.

Now one can calculate the socalled radiation stresses

$$S_{xx} = \left(n * cos^2(\theta) + n - \frac{1}{2}\right) * E \tag{29}$$

$$S_{xy} = (n * sin(\theta) cos(\theta)) * E \tag{30}$$

$$S_{yy} = \left(n * sin^2(\theta) + n - \frac{1}{2}\right) * E \tag{31}$$

$$\tag{32}$$

where, $E = \frac{1}{8}\rho g h^2$, the wave enery, $n =$ the ratio of group velocity to wave celerity and θ is the angle between the direction of wave propagation and the positive x-axis.

The driving forces are found from the gradients of the wave action, leading to

$$F_x = \frac{1}{\rho * h} * \left(\frac{\delta S_{xx}}{\delta x} + \frac{\delta S_{xy}}{\delta y}\right) \tag{33}$$

$$F_y = \frac{1}{\rho * h} * \left(\frac{\delta S_{xy}}{\delta y} + \frac{\delta S_{yy}}{\delta y}\right) \tag{34}$$

$$\tag{35}$$

With this forces one can calculate the induced velocity by a streaming model.

4 Test cases and applications

The wave program calculates the change of the wave parameters H, L, T, the direction and the possible breaking of a monochromatic wave for the cutting points between wave crests and orthogonals. In this case the quality of this approach is dependent on choosing the step of calculation in dependent on the task. The wave parameters are constant between two orthogonals. The results of the calculations can be written in the following ways:
1. Representation of the wave crests with and without orthogonals in 7 colours indicating the wave height reaches. The boundaries of this reaches can be choosen freely.
2. Output of the wave parameters including the streaming induced forces at determinate points, at the so called pegel points.
3. The same information can output at the points of the digital terrain model.

The calculation of the wavepropagation by the linear wave theory gives good results for a lot of practical cases. In the following for some examples the precise restitution of the model is demonstrated.

As an example the test region is modelised in the same way as published by *DE VRIEND* in [2] and [3]. This is a system with a curved coastline. The distribution of the depth is represented in Fig. 3. For this region the waves simulation is calculated.

The wave distribution is given in Fig. 4, the dotted region shows the breaking of waves. Using in Fig. 6 represented components of forces the wave induced streaming from Fig. 7 is calculated as an example. The correctness of the calculation be examaind in the case of the above presented test region. Of course it is necessary a more complete verification by other natural data.

A large scale area application for this model is the coastal protection investigations in the coastel region near the island of Sylt. Sylt is located in the south-east part of the North Sea near the border between Germany and Danmark. In Fig. 7 the distribution of the depth is represented. Besides the effects according to the bathemetry one can good seen in Fig. 8 and Fig. 9 the effects of tidal streaming at the ends of the island.

The model reproduces well-known results for finite-amplitude waves.

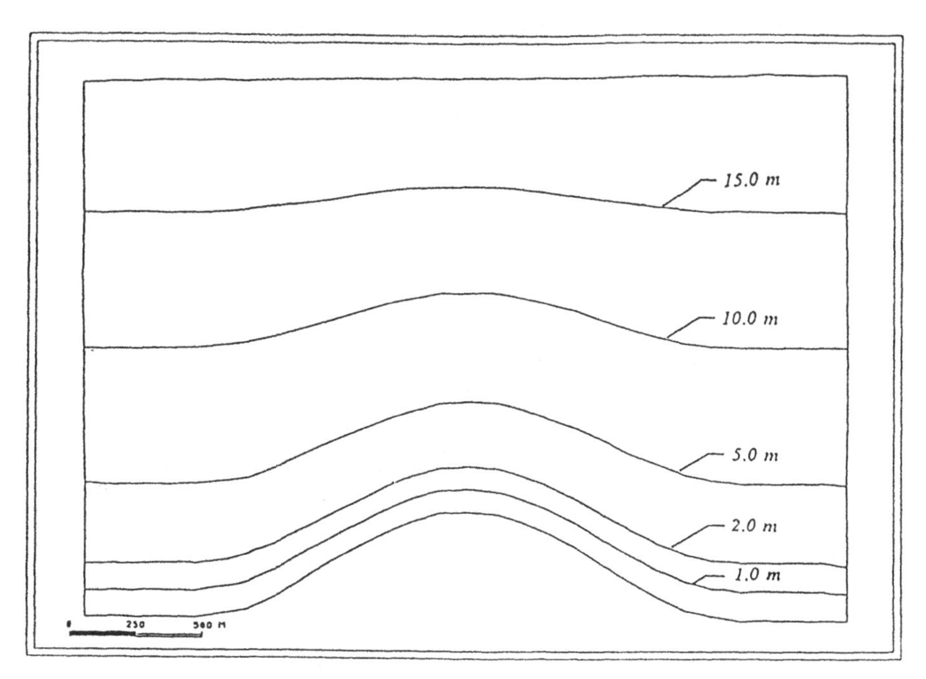

Figure 3: Bathematry of testregion

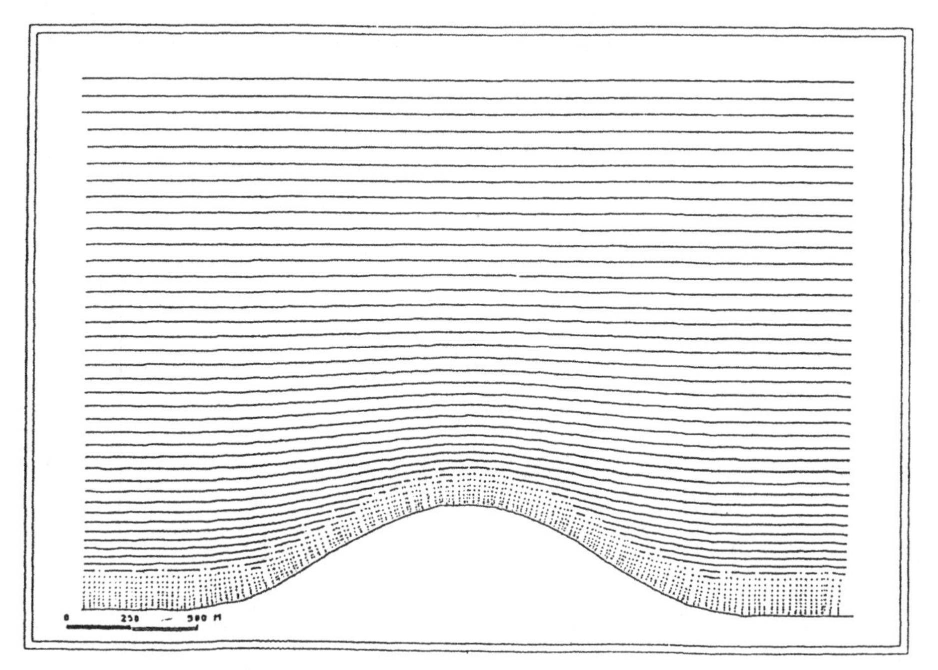

Figure 4: Wave distribution

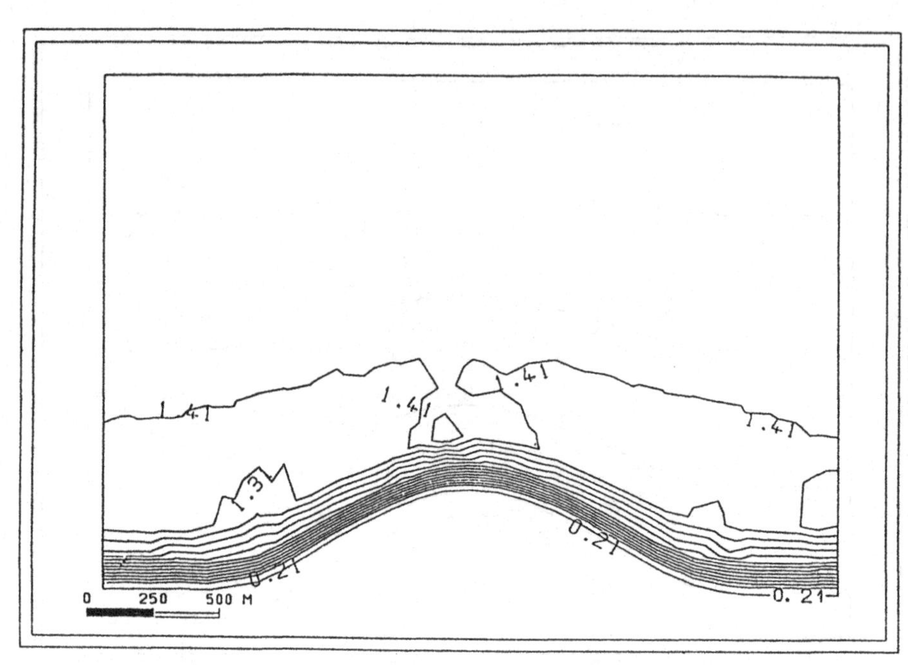

Figure 5: Wave height distribution

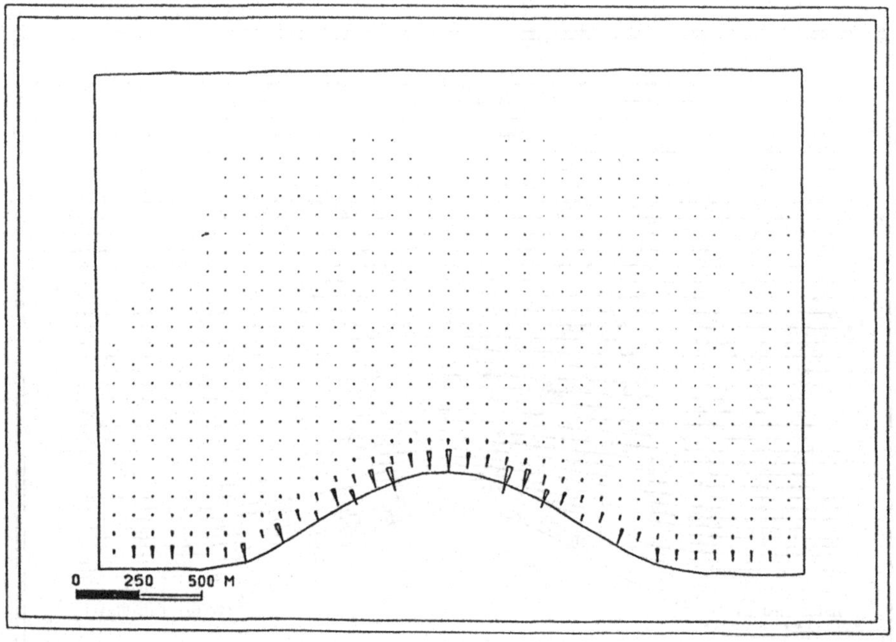

Figure 6: Current induced forces

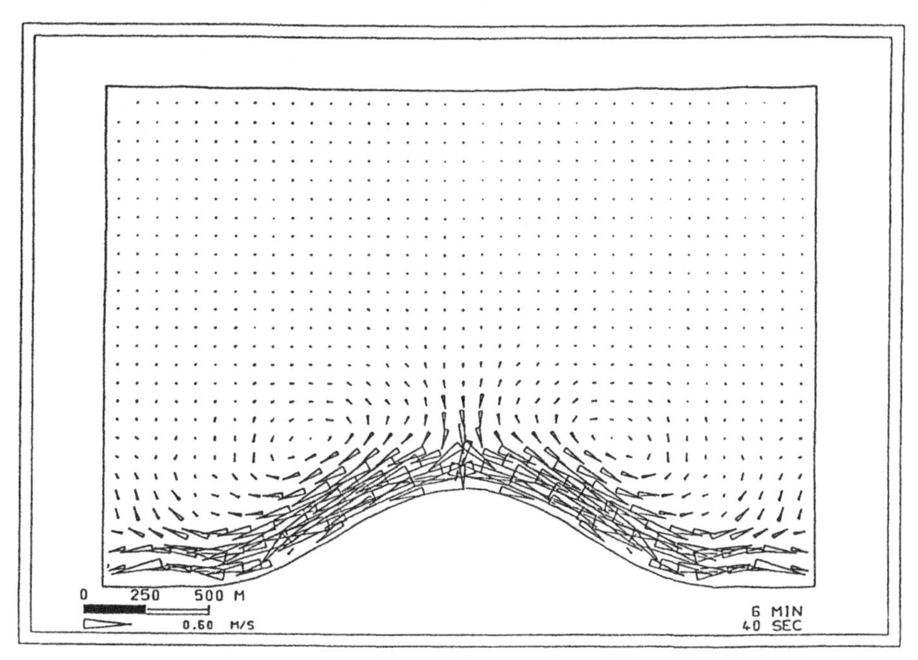

Figure 7: Wave induced currents

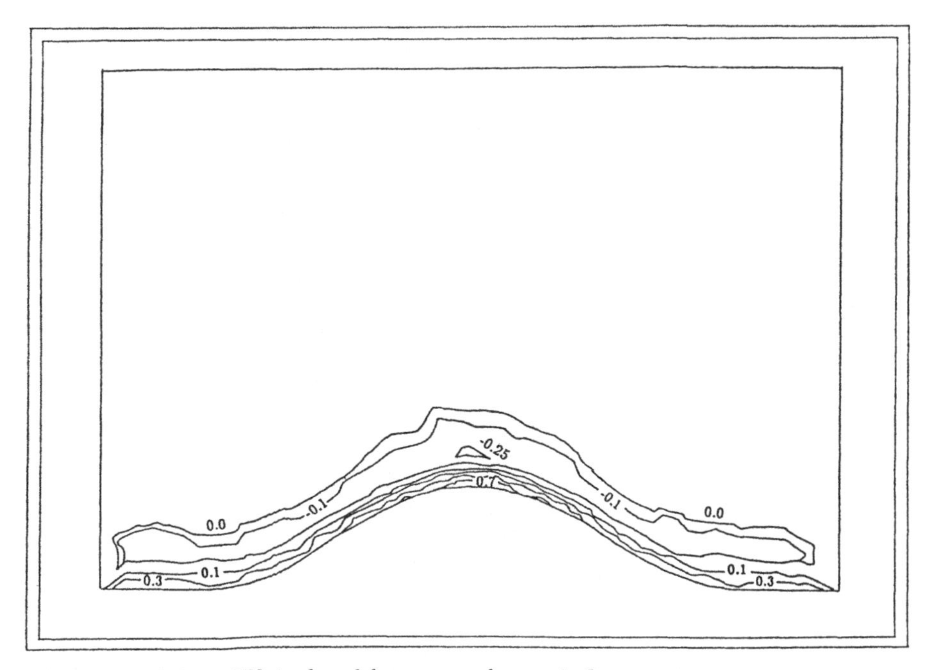

Figure 8: Mean Waterlevel because of waveinfluence

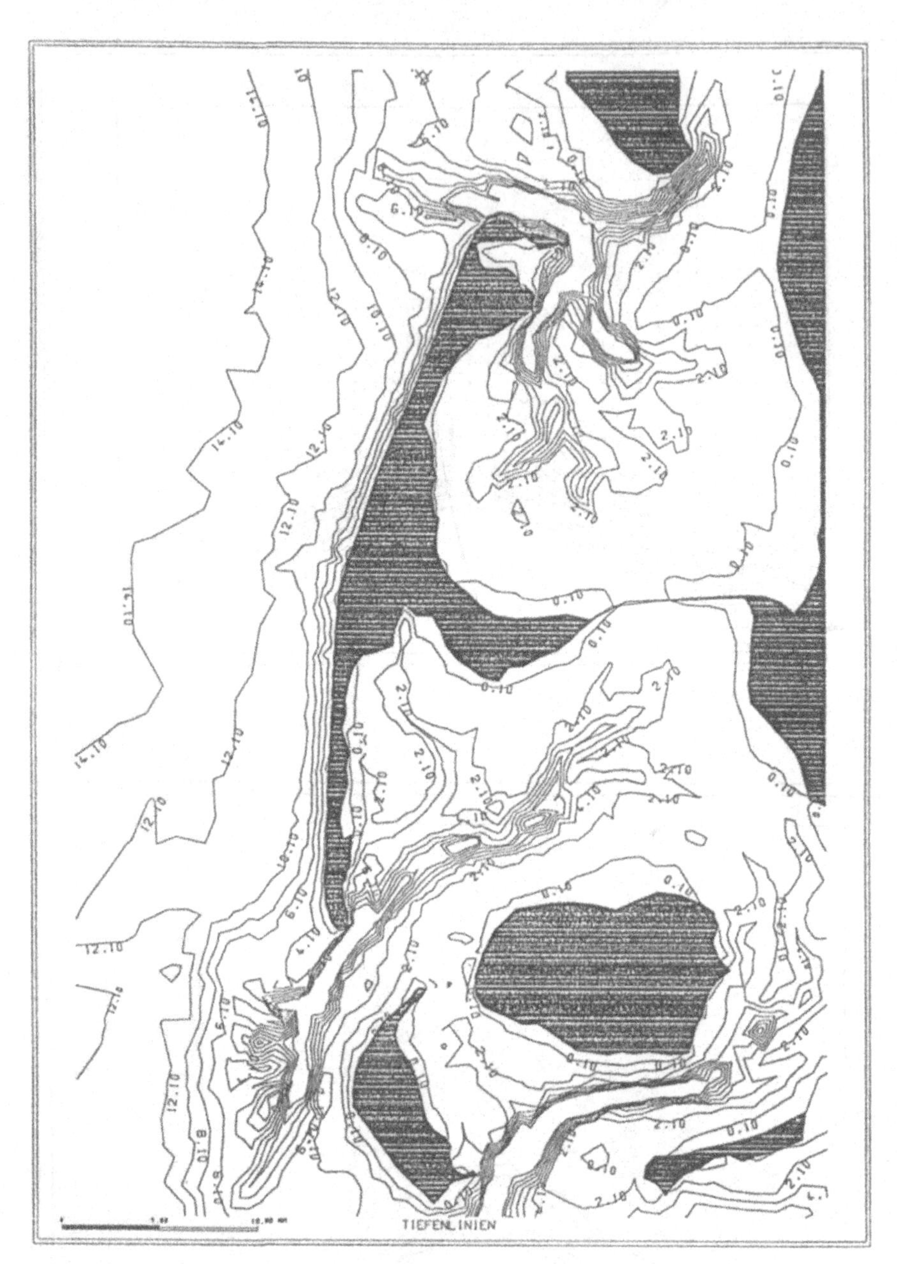

Figure 9: Bathematry arround Sylt

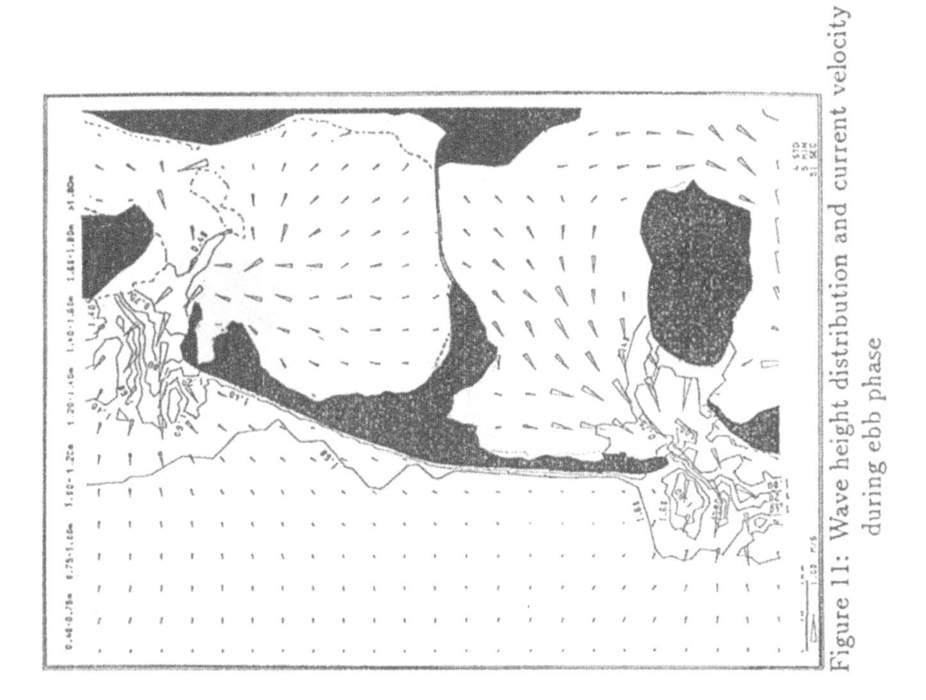

Figure 11: Wave height distribution and current velocity during ebb phase

Figure 10: Wave distribution during ebb phase

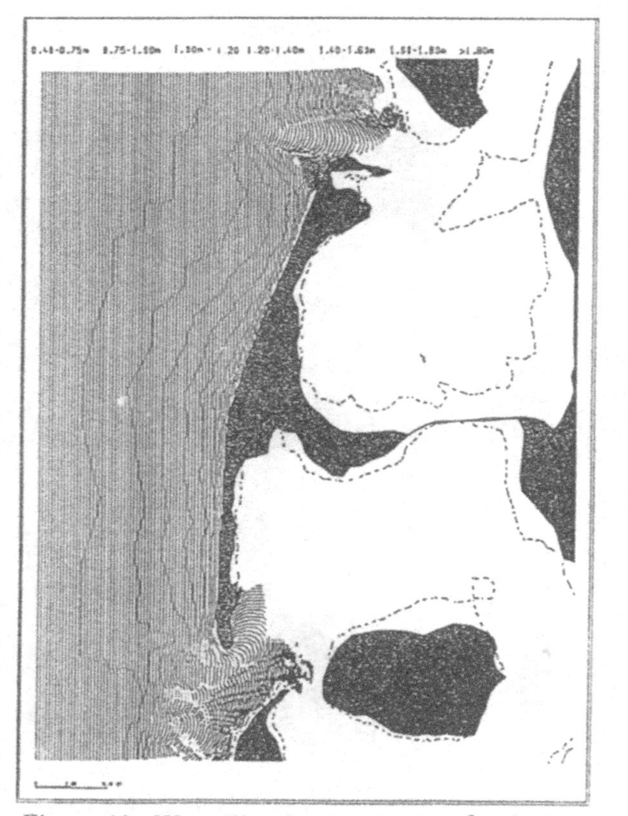

Figure 12: Wave distribution during flood phase

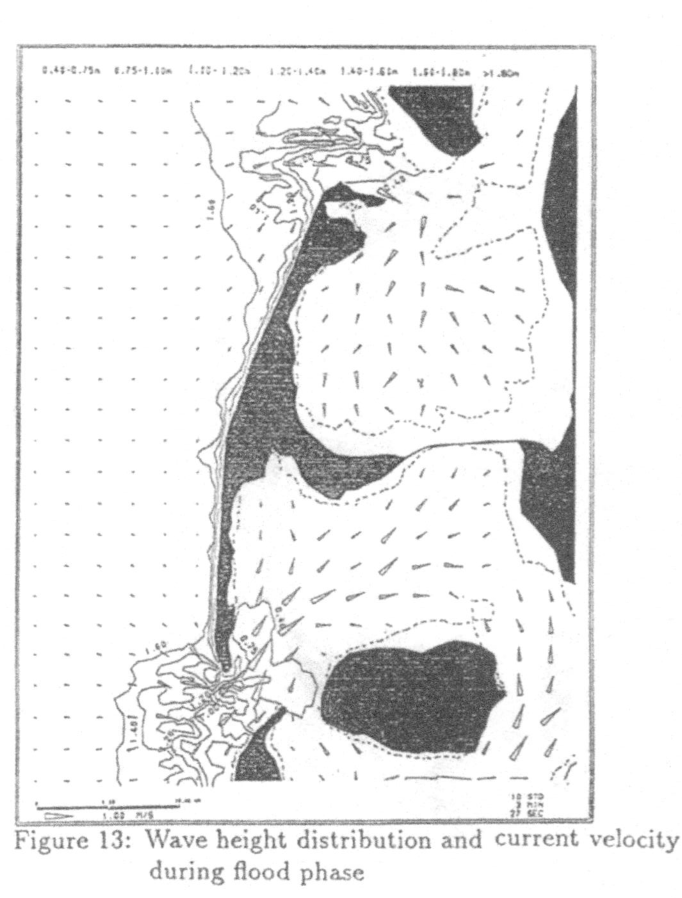

Figure 13: Wave height distribution and current velocity during flood phase

References

[1] *Shore Protection Manual.* U.S. Army Coastal Engineering Research Center, Washington, 1984.

[2] H. J. de Vried. On the applicability of a highly simplified wave propagations model in the computation of wave-driven coastel currents. In *Progress report W 439-3*, Waterloopkundig laboratorium delft hydraulics laboratory, December 1982.

[3] H. J. de Vried. 2 DH mathematical modelling in coastal morphology. In *European Coastal Zones*, Athens, 30.Sep.-04.Oct. 1985.

[4] O.H. Anderson; J. Fredsoe. Transport of Suspended Sediment Along the Coast. Tech. Univ. Danmark, 1983.

[5] V. Sundar; H. Noethel; K. P. Holz. Wave kinematics in a groin Field - Frequency domain analysis. *Coastal Engineering Journal*, 1992. accepted for publication.

[6] J.W. Johnson. The Refraction of Surface Waves by Currents. *Transactions, American Geophysical Union*, 28(6), 1947.

[7] C.T.Kuo K. Horikawa. A study on wave Transformation inside Surfe Zone. *Coastal Engineering Conference*, 1, 1966.

[8] J.F.A. Sleath. *Sea Bed Mechanics.* John Wiley & Sons, 1984.

[9] M. S. Longuet-Higgins; R. W. Stewart. Radiation stress in water waves,"A physical discussion with application". *Deep Sea Res.*, 11(4):225–239, 1977.

[10] J.R. Weggel. Maximum Breaker Height. *Journal of Waterways*, 98, 1972.

[11] R.L. Wiegel. *Oceanographical Engineering.* Prentice Hall International Series in Theoretical and Applied Mechanics. N.J. Englewood Cliffs, 1964.

[12] K.P. Holz; M. Feist; H. Noethel; P. Lehfeldt; A. Pluess; U. Zanke. The TICAD-Toolbox Applied to Coastal Engineering problems. In *Hydraulic Engineering Software Applications in Computational Mechanics Publications*, 1990.

Stokes Drift Effects Computed From Measured Wave Data

G. Piro (*), E. Pugliese Carratelli (*), E. Sansone (**)
(*) Dip. Difesa Suolo, Università della Calabria, Montalto Uffugo (CS), I87040, Italy
(**) Istituto Universitario Navale, Via Acton 38, Naples, I80133, Italy

ABSTRACT

The effect of mean mass transport on the surface of the sea due to wave movement and known as "Stokes Drift" plays an important role in forecasting the movement of floating pollutants and is also of paramount importance when evaluating the boundary conditions for the computation of coastal circulation; this paper presents an attempt to supplement the usual approach based on monochromatic waves or standard spectral simulations with an analysis of measured time series of waves.

Wave heights and periods are computed from records of water height data obtained with a Datawell wave gauge located in the Bay of Naples in different sea states through the usual zero-upcrossing procedure, and wave parameters are estimated; once such parameters are known and a wave theory is assumed the computation of the drift velocity is quite straightforward.

These computations yield a mean drift velocity for all available sea records, which include both calm and very rough sea conditions; correlations are found to relate drift to the sea state.

INTRODUCTION

The presence of waves on the surface of the

water has very important effects on the transport of mass in the sea, and therefore also on the movement of pollutants.

The mean current resulting on the surface of the water as a consequence of the wavy movement is known as the "Stokes' drift" or simply "drift", and its effects are very important when dealing with the movement of floating pollutants (e.g. oil slicks), and particularly so in the Mediterranean, where the tidal effects are often negligible.

The energy transfer between the wind and the upper layer of the sea is mainly based on the wave drift; in fact the waves themselves, in spite of being a basically irrotational phenomenon, act as a sort of intermediate mechanism, or "filter" (Bye, [1]) between the tangential stress and the turbulent structure of the upper sea layer. This effect does of course interact with all the other effects, such as the bottom friction, the tides, the Coriolis forces, and the inertia.

The time scales of the circulation of the whole water body are, of course, much longer than the typical periods of the wind waves; when the global equations of the circulation are considered the drift caused by the waves is therefore taken into account separately and it provides - in mathematical terms - the upper boundary conditions for the underlying currents.

The scientific literature on drift and drift-related phenomena is of course very rich; we shall thus very quickly review only the papers which we consider to be relevant to the results dealt with in this paper.

After Stokes' original work in the last century, the milestone of theoretical research in the field of wave drift is probably Longuet-Higgins' [2] classical paper which is of special interest when dealing with shallow waves, where the effects of viscosity are important. However, the results by Russel and Osorio, as reported by Stolzenbach et al. [3] and by Dyke and Barstow [4] can be used to assess the acceptability of non viscous theories in the context of our work, which only deals with deep water waves.

The paper by Bye [1] quoted above, and another

one by Unulata and Mei [5] are particularly useful in clarifying the concept and nature of the wave drift at sea. On the same line it is also worth mentioning a paper by Kit and Stiassnie [6], and one by Chu [7], together with the discussion of this latter by Darlymple and Svendsen [8] which clarify some conceptual problems about the drift and the possibility of computing it from the existing regular wave models.

A very interesting discussion of the interaction between the waves, their drift and the currents is presented in [9]; the relative importance of Eckmann layer and wave drift in determining surface mass transport, is given in [10]

Two papers by Kenyon [11,12] supply a wave-spectrum based analysis of the drift and an evaluation of the mean surface velocity as a function of the wind speed for random waves in fully developed seas; more detailed numerical results on the same line, as well as experimental data obtained in a water tank for both monochromatic and random waves, are presented by Dyke and Barstow [4].

Bullock and Short [13] offer some useful laboratory experimental results for water particle velocities in regular waves, together with an assessment of the predictive capabilities of various theories.

The procedure we present in the following is based on directly processing the time series of the water height as measured with a wave meter, rather than making use of a simulation based on the spectral analysis as some of the researchers quoted above. Through this approach the experimental data can be more directly employed, and higher order wave theories can be used in order to yield a more reliable estimate of the drift velocity, as compared with the spectrum/linear theory approach.

PROCEDURE AND RESULTS

The water height data records have been obtained with a Datawell accelerometric buoy wave gauge located in the Bay of Naples in a period of time spanning from the summer 1986 to the winter of 1987, at a depth of 90 metres.

Details on the location and the experimental techniques and the pre-processing procedures - which, however, are rather simple by present standards - have been supplied elsewhere (Pugliese Carratelli and Sansone, [14]). It is enough here to say that the sampling rate is 0.39 seconds, and each single record of data is composed by 512 or 1024 samples, thus yielding a record length of about three or six minutes. 69 three-minutes and 34 six-minutes records were considered.

Wave heights and periods have been computed from the water height data through the usual zero-upcrossing procedure, each record thus yielding a number of waves varying around 30 (for the shorter 512 samples data) or around 60 (for the longer ones).

A knowledge of the height and the period of a wave is not sufficient to compute the drift; a mathematical theory with the appropriate parameters must be assumed so that all the hydrodimamic properties are known.

This procedure has been followed twice for all the data records, by computing the displacement due to the drift for each single wave with both Stokes II and Stokes III theories.

Once this is done, the average drift velocity can be simply evaluated for each record by performing this computation for each wave of the record, summing up the displacements and dividing by the total duration of the wave train; this procedure automatically takes into account the different duration of the single waves and thus implicitly supplies correct weighted average.

When the Stokes II theory is used, the computations are quite straightforward, since a classical analytical solution for the drift velocity Ud is available, again from Stokes' own work, (see for instance Stolzenbach et al., [3] or Dyke and Barstow, [4]):

$$U_d = \frac{H^2 \sigma K \cosh[2 K (z+h)]}{2 \sinh^2(K h)} \tag{1a}$$

where H, σ and K are the height, the wave number and the angular frequency respectively; z is the vertical abscissa, measured from the mean free surface, and h is the depth. K is obviously related to h by the dispersion equation (the deep water hypothesis has been held valid in all the cases examined).

To account for mass balance in simple steady-state one dimensional flows a further term must be added to include a current in the direction opposite to the wave velocity, thus yielding:

$$U_d = H^2 \sigma K \left[\frac{\cosh[2K(z+h)]}{2 \sinh^2(K h)} - \frac{\coth(K h)}{2 K h} \right] \qquad (1b)$$

(in order to be consistent with the deep water hypothesis the constant term should be very small)

For more complex wave theories there is no such simple and widely accepted formula for the drift velocity. Even for the relatively simple Stokes III model the equations are quite complicated and cumbersome; the authors have employed the formulas supplied by Scarsi and Stura in [15].

The results are supplied in fig. 1) and fig 2) where the average drift velocity is plotted against the energy E of the sea state, as evaluated by computing the MSR of the n recorded water height values Yi in the relevant record:

$$E = \left[\Sigma_i^n \; (Yi^2 - Ym^2) \; / \; (n-1) \right]^{1/2} \qquad (2)$$

(Ym is of course the average of the Yi values)

The results are plotted in the figures from 1 to 3.

Figure 1 presents the drift velocity Ud, calculated according to equations 1a) and 1b) for each single record, as a function of E; a strong correlation is visible, even though there is a much wider dispersion if compared with the results of previous researchers in the field, who -as stated above - worked with analytical and numerical derivations from theoretical wave spectra. The dispersion of values we found is obviously due to our use of real experimental data rather than of an idealized description of the sea state.

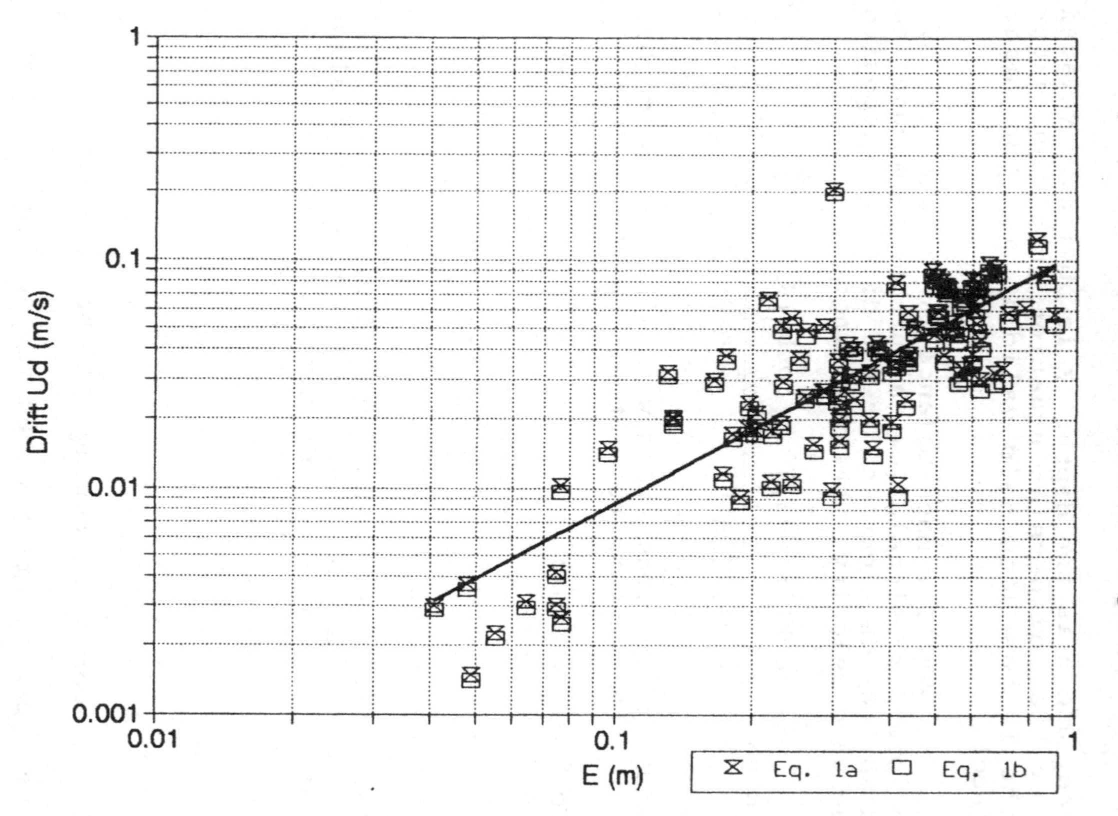

Fig. 1 - Drift velocity evaluated through Stokes'II theory.

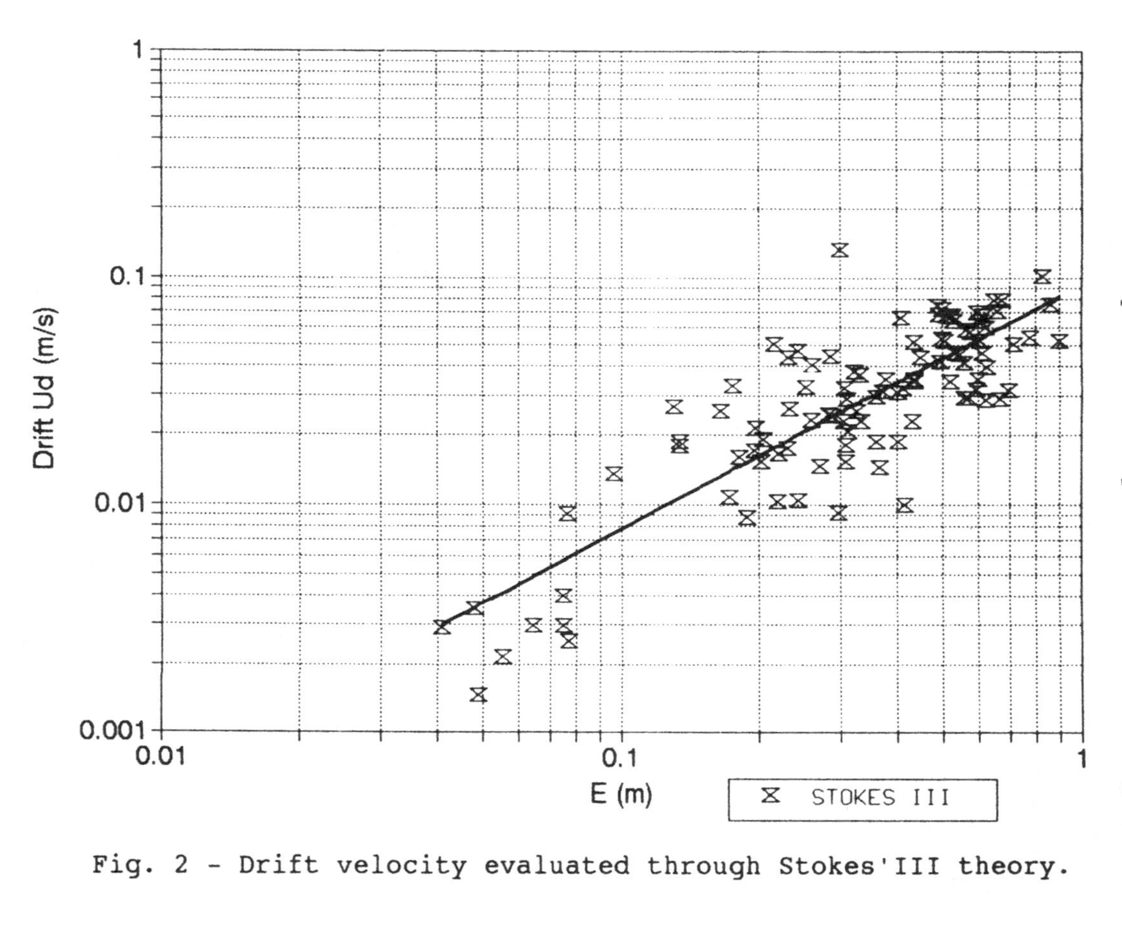

Fig. 2 - Drift velocity evaluated through Stokes'III theory.

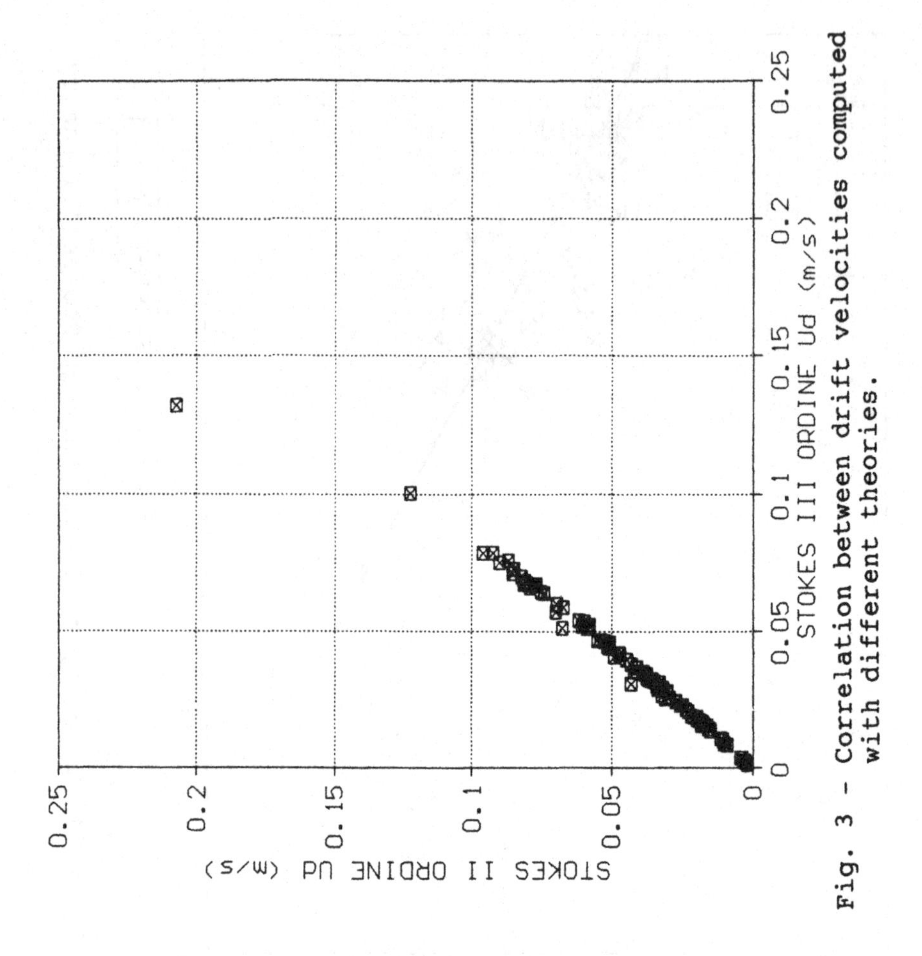

Fig. 3 - Correlation between drift velocities computed with different theories.

Also reported are the best fit straight lines of the correlations; the difference between the results obtained with the two equations is so slight that the two lines practically coincide, as could be expected due to the deep water hypothesis.

In figure 2 the Ud values are plotted as computed with the Stokes III theory, and again the best fit line is calculated and drawn; while the overall trend is the same, there is a difference from the results supplied in the previous picture.

Such a difference is better highlighted in the following picture (figure 3), where the drift obtained with the two different theories are directly compared.

The correlations is strong, but there are enough differences to suggest that perhaps a closer look at higher order theories when evaluating mass transport in real sea would be worth while.

Finally, it might be interesting to estimate the drift velocity as a function of the wind speed; unfortunately this kind of analysis, which is straightforward for a fully developed wave field would lead nowhere in our situation where the sea state is the result of very complex meteorological and geographical situations.

The interested reader can refer to a previous work (Marone et al, [15]), where most of the data employed here have been analyzed and correlated to the measured wind velocity.

CONCLUSIONS AND FUTURE DEVELOPMENTS

Recent research (see for example Benassai G., Rebaudengo Landò and Sansone, [16]) is evaluating the possibility and the reliability of spectrum-based procedures for simulating various sea state parameters. The present work is in fact meant to supply some elements which should eventually lead to a similar assessment for the surface wave drift.

The reader may find it somewhat disconcerting that no experiments on the drift in real sea conditions have been quoted or considered in this context.

It is our opinion however - also on the basis of experience gained during sea measurement campaigns carried out in the past by one of the authors (De Maio et al., [17], Spezie et al.[18]) - that the experimental difficulties involved, as well as the conceptual difficulty of separating the various effects that concur in determining the movement of floaters or tracers, render the analysis of field data a poor tool towards the comprehension of sea wave drift; unless, of course a clear picture is first gained of the interaction between the sea state and the movement of its upper surface itself.

The experimental situation may have changed now that more sophisticated techniques are available, such as microwave measurement water surface velocity, so that it might soon be possible to effectively test different models against real field data. With this work we hope to have supplied some useful elemnts to the future builders of such models.

ACKNOWLEDGMENTS: The work presented here was financially supported by the Italian Ministry of University and Scientific Research (MURST)

REFERENCES

1. Bye A.T.J., The Wave Drift Current, Journal of Marine Research, 25(1), 95-102, 1967

2. Longuet Higgins, M. S., Mass Transport in Water Waves, Phil. Trans. Roy. Soc., pp. 535-581, 1953

3. Stolzenbach K.D., O.S. Madsen, E.E. Adams, A. M. Pollack, C. K. Cooper, A review and Evaluation of Basic Techniques for predicting the Behaviour of Surface Oil Slicks, M.I.T. Report MITSG 77-8, Index 77-308-Cfn.

4. Dyke P.P.G., Barstow S.F., Wave Induced Mass

Transport: Theory and Experiment, Journal of Hydraulic Research 19, N. 2, 1981

5. Unulata U., Mei C.C., Mass Transport in Water Waves, J. Geophys. Res., Vol. 75, n. 36, Dec. 1970

6. Kit E., Stiassnic M., Particle Motion under Stokes Waves, ASCE Journal of Waterway, Port, Coastal and Ocean Ocean Engineering, Vol. 107, n. 3, May 1981.

7. Chu, Yen-hsi, Orbital Motion of Water Particles in Oscillatory Waves, ASCE Journal of Waterway, Port, Coastal and Ocean Ocean Engineering, Vol. 109, n. 2, May 1983

8. Darlymple A.R., I.A. Svendsen, Discussion to [7], ASCE Journal of Waterway, Port, Coastal and Ocean Ocean Engineering, Vol. 110, n. 4, Nov. 1984.

9. Srokosz, M.A., Models of Wave Current Interactions, Chapter 19, Modelling the Offshore Environment, Society for Underwater Technology, London 1987

10. Jenkins, A.D., A Dynamically Consistent Model for Simulating Near-Surface Ocean Current in the Presence of Waves, Chapter 21, Modelling the Offshore Environment, Society for Underwater Technology, London 1987

11. Kenyon, K. E., Stokes drift for Random Gravity Waves, J. Geophys. Res., Vol. 74, N. 28, Dec. 1970.

12. Kenyon, K. E., Stokes Transport, J. Geophys. Res., Vol. 75, N. 6, Feb. 1970

13. Bullock G.N., I. Short, Water Particle Velocities in Regular Waves, ASCE Journal of Waterway, Port, Coastal and Ocean Ocean Engineering, Vol. 111, n. 2, March 1985

14. E. Pugliese Carratelli, E. Sansone, Rilievi Ondametrici nel Golfo di Napoli, Annali Istituto Universitario Navale, Naples 1987, Vol. LV

15. Scarsi G., S. Stura, Trasformazione di onde cilindriche di ampiezza e ripidità finite su fondali a dolce acclività, L' Energia Elettrica, 1970

16. V. Marone, E. Pugliese Carratelli, E. Sansone,

Spinte idrodinamiche verificatesi su condotte sottomarine durante un evento di particolare intensità, Convegno A.I.I. sull' Immissione di acque reflue in mare, Ischia, May 1989

16. Benassai G., L. Rebaudengo Landò, E. Sansone, Reliability of the Linear Numerical Simulation with Respect to the Random Wave Grouping on Deep Water, Computer Modelling in Ocean Engineering, Barcellona, 1991

17. De Maio A., M. Moretti,E. Sansone, G. Spezie, M. Vultaggio, Influenze del vento sulle misure di corrente mediante traccianti, Annali Istituto Universitario Navale, Naples 1976, Vol XLV-XLVI

18. Spezie G., M. Vultaggio, Su l' influenza dell' errore di localizzazione del tracciante nelle misure di corrente, Annali Istituto Universitario Navale, Naples 1974, Vol XLIII-XLIV

The Transfer of the Donelan et al. Spectrum on Shoaling Water

L. Rebaudengo Landò, G. Scarsi, A.C. Taramasso
Institute of Hydraulics, University of Genoa,
Via Montallegro 1, 16145, Genova, Italy

ABSTRACT

Following a procedure adopted to rederive the *TMA* spectrum, a model suitable to transfer the Donelan et al. sea wave frequency spectrum from deep water to shallow water is suggested. Some considerations about the changes of the spectral width, the energy density and the spectral height are carried out and a first approach analysis of the non-linear effects is given.

INTRODUCTION

Starting from an observation by Phillips [1,2], Kitaigorodskii et al. [3,4] introduced a self-similarity hypothesis which stipulates that the form of the relationship giving the sea wave spectrum in the equilibrium range is identical whatever the depth, when the wave number space is considered. With reference to that range [1], Kitaigorodskii et al. took into account, on deep water, the Phillips spatial spectrum

$$F_{ro}(k_o) = (\alpha/2)k_o^{-3} \tag{1}$$

and, according to the above-mentioned hypothesis, introduced on finite depth the spatial spectrum

$$F_r(k,h) = (\alpha/2)k^{-3} \tag{2}$$

where k, k_o are the wave numbers, α is the equilibrium parameter and h is the depth of the bottom [2].

It is worth noting that the spectrum (1), which presupposes a wave

---

1 The equilibrium range of a spectrum is the part just above the wave number peak (wave number space) or the frequency peak (frequency space).

2 Hereafter, the index o indicates deep water conditions and the index p means a quantity referred to the spectral peak.

generation in steady condition, was obtained on purely dimensional grounds and its shape was regarded as governed by the net energy input from wind and wave breaking due to the gravitational instability.
To refer the spectrum (2) to the frequency f space, Kitaigorodskii et al. used the simple algebric relationship

$$S_P(f,h) = F_P(k,h)[dk/df] \tag{3}$$

where $S_P(f,h)$ is the frequency (or temporal) spectrum on finite depth and f_p is the peak frequency. The derivative is performed by adopting the linear isotropic dispersion relationship

$$(2\pi f)^2 = gkTh(kh) = gk/\chi(\sigma_h) \tag{4}$$

where χ is a dimensionless function which is found from

$$\chi Th(\sigma_h^2 \chi) = 1 \tag{5}$$

σ_h being the dimensionless depth parameter

$$\sigma_h = 2\pi f\sqrt{h/g} \tag{6}$$

with g the acceleration of gravity. At the end, the spectrum (3) becomes

$$S_P(f,h) = \alpha g^2 (2\pi)^{-4} f^{-5} \Xi_K(\sigma_h) \tag{7}$$

Ξ_K being the dimensionless depth function (Kitaigorodskii factor)

$$\Xi_K = \chi^{-2}[1 + 2\sigma_h^2 \chi / Sh(2\sigma_h^2 \chi)]^{-1} \tag{8}$$

which correctly exhibits the value one on deep water where the Phillips frequency spectrum is

$$S_{Po}(f) = \alpha g^2 (2\pi)^{-4} f^{5} \tag{9}$$

showing an f^{-5} power law as the principal frequency dependence.
Thus, by Eq. (9), Eq. (7) can be written as

$$S_P(f,h) = S_{Po}(f) \Xi_K(\sigma_h) \tag{10}$$

after assuming f_p equal to f_{po} whatever the depth.

Bouws et al. [5] replaced in Eq. (10) the Phillips frequency spectrum $S_{Po}(f)$ by a *JONSWAP* frequency spectrum $S_{Jo}(f)$ [6] and extended the relationship so obtained to the entire range of frequencies, giving rise to the *TMA* model. This model was tested with an extensive set of field data and was rederived by Scarsi et al. [7] in order to make it suitable to verify the self-similarity hypothesis in the k space.
The modified *TMA* model so deduced differs from the original one in the shape functions ϕ_{PM} and ϕ_J and it gives the frequency spectrum

$$S_T(f,h) = \alpha g^2 (2\pi)^{-4} f^{-5} \phi_{PM}(f, f_p, \chi, \chi_p) \cdot \\ \cdot \phi_J(f, f_p, \chi, \chi_p, \gamma, \omega) \Xi_K(\sigma_h) \tag{11}$$

$$\phi_{PM} = \exp\left[-1.25(f\sqrt{\chi}/f_p\sqrt{\chi_p})^{-4}\right] \tag{12}$$

$$\phi_J = \exp\left[\ln(\gamma)\exp\left[-0.5(f\sqrt{\chi}/f_p\sqrt{\chi_p}-1)^2/\omega^2\right]\right] \quad (13)$$

with γ the peak enhancement factor and ω the peak width parameter.

In the present investigation, the Donelan et al. frequency spectrum $S_{Do}(f)$ on deep water [8] is taken into account and a model suitable to give the relevant spectrum in shoaling water is supplied by following the procedure adopted to obtain the modified *TMA* model. In detail: the spectrum $S_{Do}(f)$ is transferred into the spatial spectrum $F_{Do}(k_o)$; the form of the $F_{Do}(k_o)$ spectrum is kept for the spatial spectrum $F_D(k,h)$ on finite depth; the spectrum $F_D(k,h)$ is transferred into the frequency spectrum $S_D(f,h)$ by introducing the derivative of Eq. (3).
Adopting the spectrum $S_D(f,h)$ so constructed, the following topics were dealt with: the behaviour of the spectral width, the energy density and the spectral height on shoaling water; the comparison of the obtained energy density and spectral height with those deduced by the modified *TMA* model; the analysis of the non-linear effects through a first approach scheme based on an extension and an adjustment of the second order model suggested by Tayfun & Lo [9] on deep water.

THE DONELAN ET AL. SPECTRUM ON FINITE DEPTH

The Donelan et al. frequency spectrum $S_{Do}(f)$ on deep water was determined from very controlled data obtained in field (Lake Ontario) and in a large laboratory tank and it shows an f^{-4} power law as the principal frequency dependence. After some simple manipulations carried out starting from the original form, this spectrum can be given as

$$S_{Do}(f) = \alpha_* g^{1.45}(2\pi)^{-3.45} U^{0.55} f^{-4} f_p^{-0.45} \phi_{PMo}(f, f_p)\phi_{Jo}(f, f_{po}, \gamma, \omega) \quad (14)$$

where the shape functions ϕ_{PMo} and ϕ_{Jo} are supplied by

$$\phi_{PMo} = \exp[-(f/f_{po})^{-4}] \quad (15)$$

$$\phi_{Jo} = \exp[\ln(\gamma)\exp[-0.5(f/f_{po}-1)^2/\omega^2]]. \quad (16)$$

The wind velocity U is considered in the overall mean wave direction and the parameter α_* (related to the equilibrium parameter α), the peak enhancement factor γ and the peak width parameter ω are expressed by

$$\alpha_* = 0.006; \quad 0.83 \le A_o \le 5 \quad (17a)$$

$$\gamma = 1.7; \quad 0.83 \le A_o < 1 \quad (17b)$$

$$\gamma = 1.7 + 6\,lg(A_o); \quad 1 \le A_o \le 5 \quad (17c)$$

$$\omega = 0.08[1 + 4/A_o^3]; \quad 0.83 \le A_o \le 5 \quad (17d)$$

where the dimensionless parameter

$$A_o = 2\pi f_{po} U/g \quad (18)$$

is equal to the ratio U/c_{po}, c_{po} being the phase velocity corresponding to the spectral peak.
The spatial spectrum $F_{Do}(k_o)$ related to the frequency spectrum (14) is deduced from

$$F_{Do}(k_o) = S_{Do}(f)[df/dk_o] \tag{19}$$

where the derivative is performed using Eq. (4) on deep water, i.e.

$$(2\pi f)^2 = gk_o. \tag{20}$$

It turns out

$$F_{Do}(k_o) = (\alpha_*/2)g^{-0.275}U^{0.55}k_o^{-2.5}k_{po}^{-0.225}\phi_{PMo}(k_o, k_{po})\phi_{Jo}(k_o, k_{po}, \gamma, \omega) \tag{21}$$

where the functions ϕ_{PMo} and ϕ_{Jo} are supplied by

$$\phi_{PMo} = \exp\left[-(\sqrt{k_o/k_{po}})^{-4}\right] \tag{22}$$

$$\phi_{Jo} = \exp\left[\ln(\gamma)\exp\left[-0.5(\sqrt{k_o/k_{po}}-1)^2/\omega^2\right]\right]. \tag{23}$$

The parameters α, γ, ω are given by Eqs. (17) after replacing the parameter A_o by

$$B_o = U\sqrt{k_{po}/g} \tag{24}$$

deduced from Eq. (18) taking into account Eq. (20).

To obtain in finite depth h the spectra $F_D(k,h)$ and $S_D(f,h)$ corresponding to the spectra $F_{Do}(k_o)$ and $S_{Do}(f)$ on deep water, the procedure which makes it possible to deduce the spectra (2) and (7) starting from the spectrum (1) is adopted. This procedure leads to the following results.
The spatial spectra $F_D(k,h)$ are expressed by

$$F_D(k,h) = (\alpha_*/2)g^{-0.275}U^{0.55}k^{-2.5}k_p^{-0.225}\phi_{PM}(k, k_p)\phi_J(k, k_p, \gamma, \omega) \tag{25}$$

where the functions ϕ_{PM} and ϕ_J are supplied by

$$\phi_{PM} = \exp\left[-(\sqrt{k/k_p})^{-4}\right] \tag{26}$$

$$\phi_J = \exp\left[\ln(\gamma)\exp\left[-0.5(\sqrt{k/k_p}-1)^2\right]\right]. \tag{27}$$

The parameters α_*, γ, ω are given by Eq.(17) after replacing the parameter A_o by

$$B = U\sqrt{k_p/g} \tag{28}$$

deduced from Eq.(24), taking into account the self-similarity hypothesis.
Thus, the frequency spectrum $S_D(f,h)$ is

$$S_D(f,h) = \alpha_* g^{1.45}(2\pi)^{-3.45}U^{0.55}f^{-4}f_p^{-0.45}\phi_{PM}(f, f_p, \chi, \chi_p)\cdot$$
$$\cdot\phi_J(f, f_p, \chi, \chi_p, \gamma, \omega)\Xi_D(\sigma_h, \sigma_{hp}) \tag{29}$$

where the functions ϕ_{PM} and ϕ_J are supplied by

$$\phi_{PM} = \exp\left[-(f\sqrt{\chi}/f_p\sqrt{\chi_p})^{-4}\right] \quad (30)$$

$$\phi_J = \exp\left[\ln(\gamma)\exp\left[-0.5(f\sqrt{\chi}/f_p\sqrt{\chi_p}-1)^2/\omega^2\right]\right] \quad (31)$$

which show that the function ϕ_J is identical to that relevant to the modified *TMA* model (See Eq.13).

The function χ is found from Eq.(5); the parameters σ_h and σ_{hp} are defined by Eq.(6); the depth function Ξ_D is expressed by

$$\Xi_D = \chi^{-1.5}\chi_p^{-0.225}[1+2\sigma_h^2\chi/Sh(2\sigma_h^2\chi)]^{-1} \quad (32)$$

and it differs from the Kitaigorodskii factor (See Eq.8).

Besides, the parameters α_*, γ, ω are given by Eq. (17) after replacing the parameter A_o by the parameter

$$A = 2\pi U f_p\sqrt{\chi_p}/g = A_o\sqrt{\chi_p} \quad (33)$$

deduced from Eq. (28) taking into account Eq. (4).

On deep water, the functions χ and Ξ_D assume the value one and Eqs. (29), (30), (31), (33) correctly become Eqs. (14), (15), (16), (18).

THE BEHAVIOUR OF THE DONELAN et Al. SPECTRUM ON SHOALING WATER

Figures 1 and 2, suggested as an example, show the spatial spectra $F_D(k,h)$ and the corresponding frequency spectra $S_D(f,h)$ for different depths ranging from deep water to shallow water conditions. The values of the selected depths h and those of the parameters A_o and B_o, peak frequency f_{po}, peak wave number k_{po} are indicated in the same figures.

Table 1 gives the values of the following quantities: the zero-*th* moments m_{0FD} and the maxima F_{MD} for the spectra $F_D(k,h)$; the zero-*th* moments m_{0D}, the maxima S_{MD}, the spectral width ϵ_{2D} as defined by Longuet-Higgins [10] and the energy densities E_D for the spectra $S_D(f,h)$, taking into account that ϵ_{2D} and E_D are expressed by

$$\epsilon_{2D} = [m_{0D}m_{2D}/m_{1D}^2 - 1]^{1/2} \quad (34)$$

$$E_D = \rho g m_{0D} \quad (35)$$

where ρ is the density of the water and m_{1D}, m_{2D} are the 1*st* and 2*nd* moments of the spectra $S_D(f,h)$ considered in the range of frequencies from 0.4 f_p to 6.5 f_p.

Obviously, the energy densities of the spectra $F_D(k,h)$ are equal to the ones of the spectra $S_D(f,h)$.

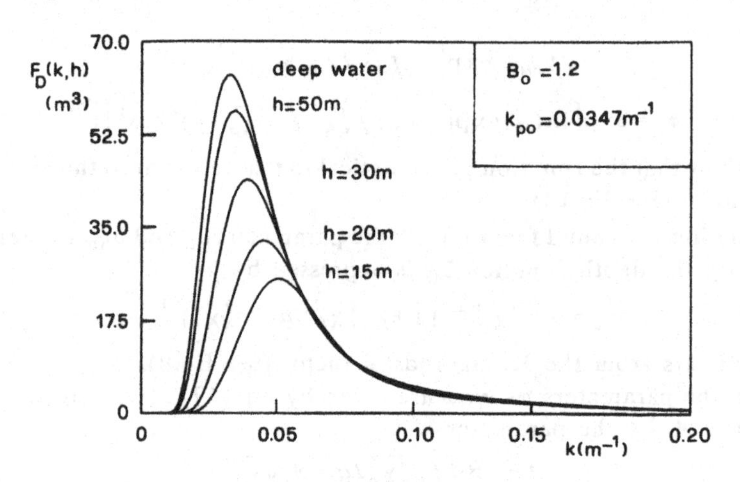

Figure 1 – Spatial spectra $F_D(k,h)$.

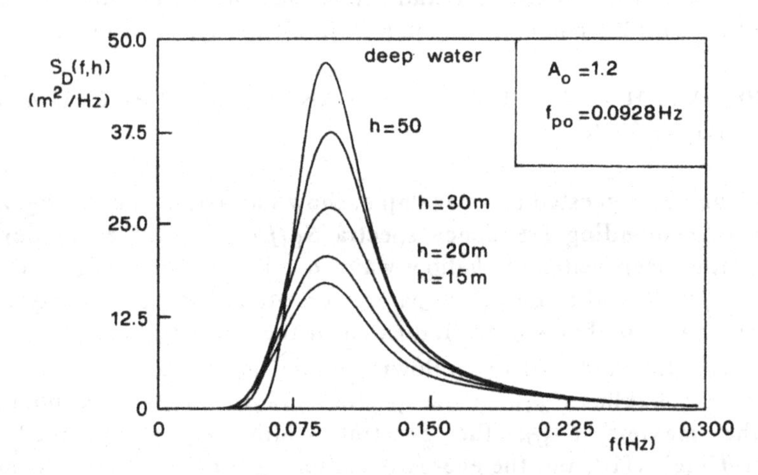

Figure 2 – Frequency spectra $S_D(f,h)$.

h (m)	m_{0FD} (m^2)	F_{MD} (m^3)	m_{0D} (m^2)	S_{MD} (m^2/Hz)	ϵ_{2D}	E_D (N/m)
∞	2.66	63.7	2.66	46.8	0.44	26090
50	2.45	57.1	2.45	37.4	0.46	24030
30	2.03	44.1	2.03	27.2	0.48	19910
20	1.62	32.6	1.62	20.6	0.51	15890
15	1.35	25.4	1.35	17.0	0.53	13240

Table 1 – Quantities referred to the spectra of Figures 1 and 2.

An examination of the figures and table leads to the following observations.
The zero-*th* moments of the frequency spectra are correctly equal to those of the spatial spectra.
The zero-*th* moments, the energy densities and the maxima of the spectra decrease as the depths decrease; in particular, for the conditions taken into account, the decreases of the zero-*th* moments and energy densities can reach ~50% and that of the maxima ~60%.
The maxima of the spatial spectra shift towards higher wave numbers as the depths decrease whereas the maxima of the frequency spectra occur for frequencies very close to $f_p = f_{po}$ despite the presence of the Ξ_D and χ functions in Eq.(29); thus, the frequency f_p actually keeps the meaning of peak frequency whatever the depth.
The spectral widths increase as the depths decrease; in particular, for the conditions taken into account the increases of the spectral widths can reach ~20%.

Bearing in mind the form of Eq.(29), the ratio $\tilde{E}_{SD} = E_{SD} / E_{SDo}$ between the energy densities on finite depth and on deep water, can be written 'a priori' as

$$\tilde{E}_D = \psi'_{ED}(\sigma_{hp}, A_o) \tag{36}$$

involving the dimensionless parameters σ_{hp} and A_o.
A numerical investigation carried out for values of σ_{hp} and A_o respectively ranging from 0.5 to 2.5 and from 0.83 to 5 showed that the dependence of $\tilde{E}_D$ on A_o is much less important than the one on σ_{hp} and, consequently, it can be ignored without appreciably modifying the behaviour of that ratio. In such a way, Eq. (36) becomes

$$\tilde{E}_D = \psi_{ED}(\sigma_{hp}) \tag{37}$$

which, taking into account Eq. (5), can be written in the very simple form

$$\tilde{E}_D = \chi_p^{-1.65} \tag{38}$$

obtained by a least-square regression method (goodness coefficient ~0.98).
Besides, Eqs. (35) and (38) lead to

$$\tilde{H}_D = \chi_p^{-0.825} \tag{39}$$

$\tilde{H}_D = H_D / H_{Do}$ being the ratio between the spectral heights on finite depth and on deep water, defined as

$$H_D = 4\sqrt{m_{0D}}; \quad H_{Do} = 4\sqrt{m_{0Do}}. \tag{40}$$

In the considered range of σ_{hp} the decrease of the energy densities and spectral heights from deep water to the lowest adopted shallow water condition (σ_{hp}=0.5, χ=2.1) reaches respectively ~70% and 45%; thus, the spectral heights are substantially reduced to a half.

In order to test the predictive capability of Eq. (39), the field data taken into account by Hughes & Miller [11] are considered. These data were obtained by the Coastal Engineering Center's Field Research Facility located on the Outer Banks at Duck in North Caroline (USA) and they are given in Figure 3 by Hughes & Miller in terms of the dimensional parameter $H_D/L_p^{0.75}$ determined from measurements of wave conditions at a depth h_1=18 m versus its counterpart determined from analogous measurements at a depth h_2=8 m, L_p being the peak length. In the same figure the dashed area, which contains the representative points of the results deduced from Eq.(39) in the form

$$H_{D1}/L_{p1}^{0.75} = (H_{D2}/L_{p2}^{0.75})(\chi_{p2}/\chi_{p1})^{0.075} \tag{41}$$

is plotted; the indexes 1 and 2 indicate quantities evaluated with respect to h_1 and h_2 depths. That area is bounded by the straight lines corresponding to the adopted extreme peak frequencies f_p=0.05 Hz, f_p=0.5 Hz and it is located in the central belt of the region of the experimental data, where they are in large amount.

The figure clearly shows a very satisfactory mean behaviour of the aforementioned results, which confirms the global validity of Eqs.(38) and (39).

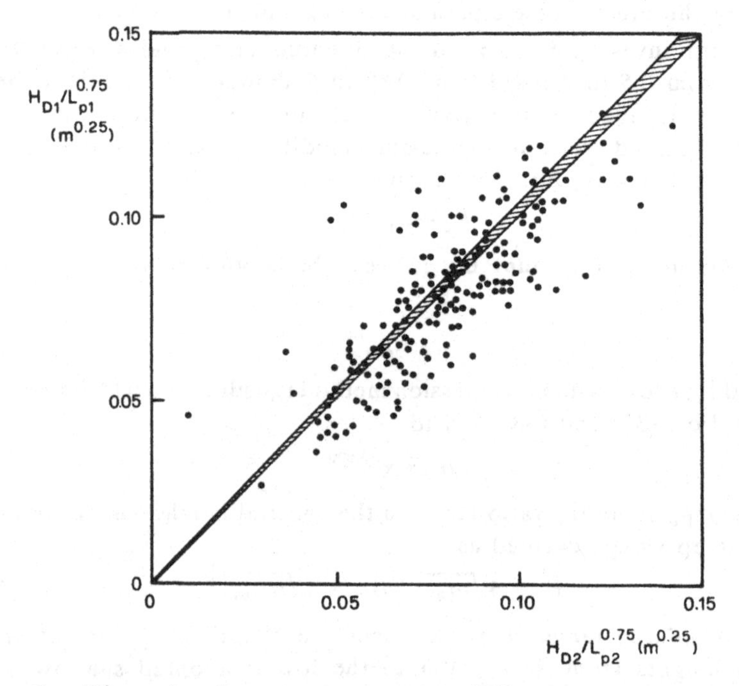

Figure 3 – Dimensional parameter $H_{D1} / L_{p1}^{0.75}$ relevant to the depth h_1 versus the parameter $H_{D2} / L_{p2}^{0.75}$ relevant to the depth h_2 (dark circles: field data; dashed area: suggested model).

With reference to the modified *TMA* model schematically described in the Introduction, the ratio $\tilde{E}_T = E_T / E_{To}$ between the energy densities on finite depth and on deep water formally keeps the dependence (37), i.e.

$$\tilde{E}_T = \psi_{ET}(\sigma_{hp}). \tag{42}$$

The function ψ_{ET}, like the function ψ_{ED} in Eq.(37), can be written in terms of χ_p, as already shown by Scarsi et al., and it turns out

$$\tilde{E}_T = \chi_p^{-2.00} \tag{43}$$

which leads to

$$\tilde{H}_T = \chi_p^{-1.00} \tag{44}$$

$\tilde{H}_T = H_T / H_{To}$ being the ratio between the corresponding spectral heights on finite depth and on deep water.
Eqs. (38),(43) and Eqs. (39),(44) supply the following relationships

$$\tilde{E}_D / \tilde{E}_T = \chi_p^{0.35}; \qquad \tilde{H}_D / \tilde{H}_T = \chi_p^{0.175} \tag{45}$$

which show that the energy densities and the spectral heights obtained by the Donelan et al. spectrum in shoaling water diminish, with respect to the values on deep water, less than the ones obtained by the modified *TMA* model, taking into account that $\chi_p \geq 1$. In particular, the ratios (45) increase as the dimensionless depth σ_{hp} decreases reaching the values $\tilde{E}_D / \tilde{E}_T \sim 1.30$ and $\tilde{H}_D / \tilde{H}_T \sim 1.14$ in the lowest adopted shallow water condition.

SOME CONSIDERATIONS ABOUT THE NON-LINEARITIES

The Donelan et al. spectrum $S_D(f,h)$ in shoaling water keeps a single peak whatever the depth, as happens for the original and modified *TMA* spectra.
In general, observations and theories support the occurrence of a secondary peak in the region close to $2f_p$ but its presence becomes energetically important only when the non-linearities play a significant role, essentially because of the decrease in depth.
To analyse the non-linearities which arise from the $S_D(f,h)$ spectra, an investigation to the second order was carried out. Attention was focused on the ratio $\tilde{\sigma}^2_D = \sigma^2_{\eta 2D} / \sigma^2_{\eta 1D}$, where $\sigma^2_{\eta 1D}$ is the variance of the 1*st* order component η_{1D} (Gaussian component) of the surface elevation η_D and $\sigma^2_{\eta 2D}$ is the variance of the relevant 2*nd* order component η_{2D}, noting that the aforementioned ratio corresponds to the one between the energy densities.

The evaluation of γ^2_D was made by following this procedure: the model suggested by Tayfun & Lo for deep water, unidirectional waves, narrow-band spectra was adjusted for its use on finite depths too; the model so obtained was specified starting from the Gaussian components η_{1D} constructed, in the time t domain, through linear numerical simulations of sea states corresponding to the $S_D(f,h)$ spectra on different depths; after this specification, the non-linear components η_{2D} became available and the variances $\sigma^2_{\eta 1D}$ and $\sigma^2_{\eta 2D}$ could be calculated.
It is worth noting that the numerical simulations were carried out by a single summation model already adopted and appropriately tested by the authors (Rebaudengo Landò et al. [12]).

Following the Tayfun & Lo model, the *2nd* order components η_{2Do} of the surface elevations η_{Do} on deep water can be expressed by

$$\eta_{2Do} = (a_o^2 k_{mo}/2)\cos(2\Theta_o) \tag{46}$$

which becomes operative when the corresponding Gaussian components η_{1Do} are known. In fact, the wave amplitudes a_o and the wave phases Θ_o are given by

$$a_o = (\eta_{1Do}^2 + \hat{\eta}_{1Do}^2)^{1/2}; \quad \Theta_o = \tan^{-1}(\hat{\eta}_{1Do}/\eta_{1Do}) \tag{47}$$

where $\hat{\eta}_{1Do}$ is the Hilbert transform of η_{1Do}. Besides, the spectral mean wave numbers k_{mo} are provided by Eq.(20) specified with the spectral mean frequency f_{mo}, which is supplied by

$$f_{mo} = m_{1Do}/m_{0Do}. \tag{48}$$

Taking into account Eqs.(47), η_{1Do} can also be written in the form

$$\eta_{1Do} = a_o\cos(\Theta_o) \tag{49}$$

which associated with Eq. (46), allows η_{Do} to be expressed as

$$\eta_{Do} = a_o\cos(\Theta_o) + (a_o^2 k_{mo}/2)\cos(2\Theta_o) \tag{50}$$

which is in line with the Rice–Dugundji representation of the wave envelope, giving extremal values that differ from those of the envelope for quantities of the $\epsilon_{2Do}{}^2$ order.
Eq.(50) is consistent with the second order Stokes expansion on deep water and it can be extended to the finite depths h after an appropriate adjustment, according to the form of that expansion on these depths. The adjustment allows the surface elevations η_D on the depths h to be expressed as

$$\eta_D = a\cos(\Theta) + Ch(k_m h)[[2 + Ch(2k_m h)]/4Sh^3(k_m h)]a^2 k_m\cos(2\Theta) \tag{51}$$

where the wave amplitudes a, the wave phases Θ and the spectral mean wave numbers k_m are obtained starting from Eq.(4) and Eqs.(47),(48) specified by the η_{1D} components and the m_{0D} and m_{1D} moments relevant to the finite depths.

The investigation performed to evaluate the $\tilde{\sigma}^2_D$ ratio was based on Eq.(51) after constructing, by numerical simulations, the discrete time histories of the Gaussian components η_{1D} of sea states related to the $S_D(f,h)$ spectra. These where characterized by the value of the dimensionless parameters σ_{hp} and A_o ranging respectively from 0.5 to 2.5 and from 0.83 to 5, as already assumed.
The obtained results lie in the region of Figure 4 bounded by the curves corresponding to A_o=0.83 and to A_o=5. The figure shows that the $\tilde{\sigma}^2_D$ ratio increases as σ_{hp} decreases and A_o increases, the dependence on these parameters being of the same order of importance. The drawn behaviour means that the energy of the *2nd* order components η_{2D}, and thus the non-linearities, are not negligible on shallow water for σ_{hp} which become lower and lower by increasing A_o and consequently by increasing the wind velocity U in comparison with the spectral phase velocity c_{po} on deep water. For example, assuming $\tilde{\sigma}^2_D$=0.05 as threshold which not to be exceeded in order to keep the linear assumption appropriate, the range of σ_{hp} is reduced by increasing A_o, like the dashed straight line in the figure shows. In particular, for A_o=0.83 that range starts at σ_{hp}~0.65 whereas for A_o=5 it starts at σ_{hp}~0.90.
Obviously, the curves plotted in the figure and the quantitative indications given above must be considered within the limits of both the hypotheses of the Tayfun & Lo model and the simplified way followed to obtain Eq.(51).
Figure 5 shows partial time histories of the Gaussian component η_{1D} and *2nd* order component η_{2D} of the surface elevations η_D relevant to a depth h=14.8 m and a $S_D(f,h)$ spectrum characterized by a peak frequency f_p=0.0648 Hz and a wind velocity U=20 m/s, which give A_o=0.83 and σ_{hp}=0.5. The figure indicates that the second component, which is very important in the situation considered, vertically skews the wave profile giving rise to sharper higher crests and more rounded flatter troughs, as the crosses clearly evidence. This fact leads to a probability density function of the surface elevation which can appreciably deviate from the Gaussian one.

CONCLUSIONS

Following the procedure already adopted by the authors to rederive the *TMA* spectrum, a model able to give the Donelan et al. frequency spectrum on shoaling water is suggested.
This spectrum (Eq.29) correctly verifies (Eqs.21,25) the self-similarity hypothesis introduced by Kitaigoroskii et al. in the space of the wave numbers and, in decreasing depths, it leads to the following behaviours:

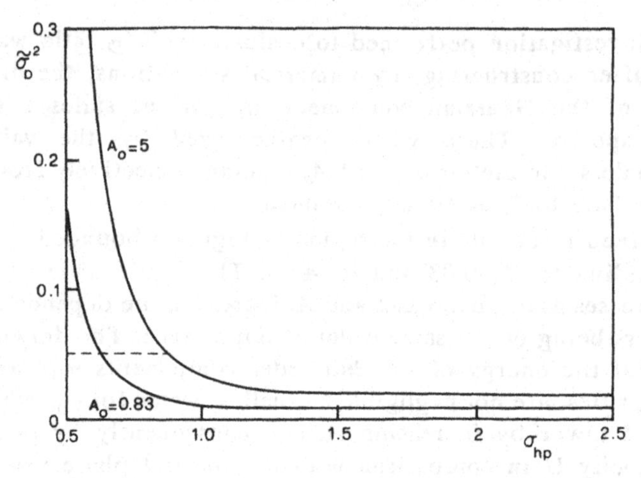

Figure 4 – Ratio $\tilde{\sigma}^2_D = \sigma^2_{\eta 2D}/\sigma^2_{\eta 1D}$ versus the dimensionless depth σ_{hp}.

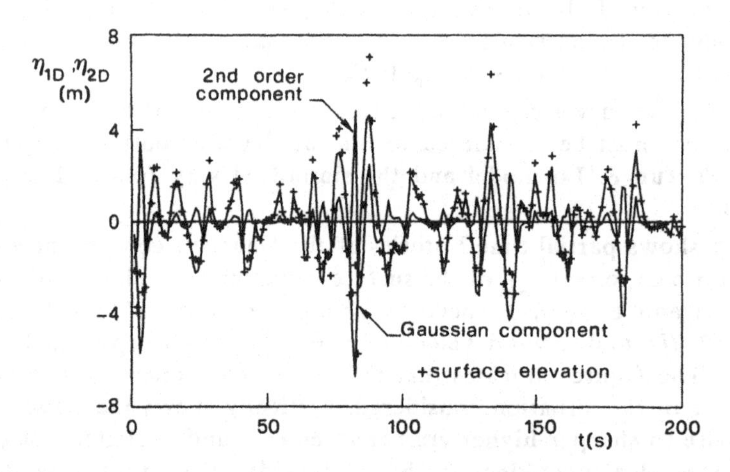

Figure 5 – Time histories of the Gaussian and *2nd* order components (continuous lines) relevant to the surface elevation (crosses).

the spectral maximum, the energy density, the spectral height decrease and the spectral width increases (Table 1); the ratio between the local energy density and the one on deep water, and the ratio between the corresponding spectral heights can be expressed in very simple forms (Eqs.38,39) – satisfactorily tested with field data (Figure 3) – involving a function that depends on a dimensionless depth parameter only (Eq.6); the local energy density and the spectral height are greater (Eqs.45) than those obtained from the rederived *TMA* spectrum (Eq.11).

The suggested spectrum keeps a single peak whatever the depth and it can be adopted until the non-linearities, which give rise to a secondary peak, play a significant role. To understand when this occurs, a simplified scheme based on the Tayfun & Lo model was adopted, focusing the attention on the ratio between the variances of the *2nd* order component and the Gaussian component relevant to the surface elevation (Eq.51) and carrying out numerical simulations of several sea states in order to calculate that ratio. The results deduced allow some quantitative evaluations to be supplied, especially for the ranges of the dimensionless depth where the proposed linear model can indicatively be used.

REFERENCES

1. Phillips, O.M. The Equilibrium Range in the Spectrum of Wind-Generated Waves. J.Fluid Mech., Vol.4, pp.426–434, 1958.
2. Phillips, O.M. The Dynamics of the Upper Ocean. Chapter 4, Ocean Surface Waves, pp.99–189, Cambridge University Press, Cambridge, 1977.
3. Kitaigorodskii, S.A., Krasitskii, V.P. and Zaslavskii, M.M. On Phillips' Theory of Equilibrium Range in the Spectra of Wind Generated Gravity Waves. Phys. Oceanog., Vol. 5, pp.410–420, 1975.
4. Kitaigorodskii, S.A. On the Theory of the Equilibrium Range in the Spectrum of Wind-Generated Gravity Waves. Phys. Oceanog., Vol.13, pp.816–827, 1983.
5. Bouws, E., Gunther, H., Rosenthal, W. and Vincent, C.L. Similarity of the Wind Wave Spectrum in Finite Depth Water, Part I- Spectral Form. J.Geophys.Res., Vol.90, pp.975–986, 1985.
6. Hasselmann, K. et al. Measurements of Wind-Wave Growth and Swell Decay During the Joint North Sea Wave Project (JONSWAP). Deutsch. Hydrogr. Z., Vol.12, pp.1–95, 1973.
7. Scarsi, G., Taramasso, A.C., Rebaudengo Landò L. and Benassai, G. Rederivation of the TMA Model for Wind Wave Spectra on Finite Depth. IM1/91, Institute of Hydraulics, Genoa, Italy, pp.1–12, 1991.
8. Donelan, M.A., Hamilton, J. and Hui, W.H. Directional Spectra of Wind-Generated Waves. Phil. Trans. R. Soc. Lond., Vol. 315, pp.509–562, 1985.
9. Tayfun, M.A. and Lo, J.M. Envelope, Phase and Narrow-Band Models of Sea Waves. J.Waterway, Port, Coastal and Ocean Eng., ASCE, Vol.115, pp.594–613, 1989.
10. Longuet-Higgins, M.S. On the Joint Distribution of the Periods and Amplitudes of Sea Waves. J. Geophys. Res., Vol.80, pp.2688–2694, 1975.

11. Hughes, S.A. and Miller, H.C. Transformation of Significant Wave Heights. J.Waterway, Port, Coastal and Ocean Eng., ASCE, Vol.113, pp.588–605, 1987.

12. Rebaudengo Landò, L., Scarsi, G. and Taramasso, A.C. Statistical Properties of Wave Fronts in Directional Seas, pp.3–47 to 3–54, Proc. of the 1st. Int. Offshore and Polar Eng. Conf., Edinburgh, U.K., 1991.

A Numerical Solution for the Breakwater Gap Wave Diffraction Problem

P.F.C. Matsoukis

Department of Hydraulics, School of Engineering, Democritus University of Thrace, Xanthi 67100, Greece

ABSTRACT

The breakwater gap wave diffraction problem is investigated by employing a system of first order partial differential equations equivalent to Berkhoff's mild slope equation which describes the combined diffraction-refraction phenomena in shoaling waters. This system of equations has been previously solved by a finite difference technique (see Copeland [5]) but in this work, an explicit characteristics scheme is employed for the solution running over a non-staggered grid of points. The scheme is compared with a number of analytical and numerical solutions and through this comparison, it is proven to be a reliable means for practical calculations given that its average error remains at a 5% level.

INTRODUCTION

Harbour entrances are very commonly constructed by the protrusion into the sea of two breakwater arms allowing the formation between themselves of a gap of certain width. In such a case, the shelter afforded by the harbour depends on the wave diffraction patterns inside the harbour introduced by the impinging waves as these are passing through the gap.

The mathematical description of these breakwater gap wave diffraction patterns inside a harbour is a difficult problem and numerous attempts have been made in the past at its solution. Some date as far back as 1932 like, for example, the one by Lamb [9] for the case of a small gap width B in comparison with the wavelength L. However, the solutions most often used in civil engineering practice are those provided by Penny and Price [13] and Carr and Stelzriede [3]. The former have applied Sommerfeld's theory on the diffraction of light waves in the case of water waves, while the latter a solution by Morse and Rubenstein [12] for the diffraction of sound and electromagnetic waves by a slit in an infinite plane. Johnson [8] has incorporated these two solutions in generalised wave diffraction diagrams which are the well-known diagrams included in CERC [4].

Berkhoff [1] has developed the so-called "mild slope" equation for the description of this kind of problem which has found wide practical use. Bettess and Zienkiewicz [2] have provided a numerical solution to this equation by using finite and infinite elements ,while Copeland [5] has given an explicit finite difference solution to a hyperbolic approximation of the equation.

The "mild-slope" equation is an elliptic-type equation posing a boundary value problem which requires for its solution a large amount of computational work, even for coastal areas of a limited extent. This shortcoming has prompted the development in recent years, of parabolic approximations to the equation which have the advantage of drastically reducing the necessary computations by excluding the reflected waves from the solution (see e.g. Radder[14], Southgate[15] etc.). According to Copeland[5], the number of operations required for the solution of the elliptic version of the mild-slope equation is M^4 where M is the number of the grid points describing the wave field, the number of operations required for the parabolic version M^2 and the equivalent number for the hyperbolic one equal to M^3. It can be seen, therefore, that the hyperbolic approximation to the equation constitutes an intermediate solution whereby the number of calculations is reduced without it being necessary to exclude the presence of the reflected wave.

In this paper, we introduce an explicit numerical scheme for the solution of the hyperbolic form of the equation which is based on the method of characteristics and uses a non-staggered grid of solution points.

THE MATHEMATICAL SOLUTION

According to Ito and Tanimoto[7] and Copeland[5], the mild slope equation decribing wave propagation in the horizontal plane may be replaced by a hyperbolic system of first order equations in the following way:

$$\frac{\partial \bar{\zeta}}{\partial t} + \frac{1}{n}\frac{\partial Q}{\partial x} + \frac{1}{n}\frac{\partial P}{\partial y} = 0 \qquad \text{(Continuity Equation)}$$

$$\frac{\partial Q}{\partial t} + C\,C_g\frac{\partial \bar{\zeta}}{\partial x} = 0 \qquad \text{(Momentum Equations)} \qquad (1)$$

$$\frac{\partial P}{\partial t} + C\,C_g\frac{\partial \bar{\zeta}}{\partial y} = 0$$

where $C_g = n.C$, $P = n.p$, $Q = n.q$ and n = shoaling number, ie

$$n = \frac{1}{2}\left[1 + \frac{2kh}{\sinh (2kh)} \right] \tag{2}$$

The quantities p,q are the velocity integrals over the depth, ie

$$p = \int_{-h}^{o} u.\, dz \text{ and } q = \int_{-h}^{o} v.\, dz \tag{3}$$

These may be further written as

$$p = n.\, \bar{U}.h \text{ and } q = n.\, \bar{V}.h \tag{4}$$

where

$$\bar{U} = \frac{1}{h} \int_{-h}^{o} u.\, dz \text{ and } \bar{V} = \frac{1}{h} \int_{-h}^{o} v.\, dz \tag{5}$$

are the mean velocities in the vertical.

Under these conditions and accounting only for linear waves propagating over mild slopes of the sea bed, the system of Equations (1) takes the form

$$\frac{\partial \zeta}{\partial t} + U \frac{\partial h}{\partial x} + V \frac{\partial h}{\partial y} = -h \left[\frac{\partial U}{\partial x} + \frac{\partial V}{\partial y} \right]$$

$$\frac{\partial U}{\partial t} = - \frac{C^2}{h} \frac{\partial \zeta}{\partial x} \tag{6}$$

$$\frac{\partial V}{\partial t} = - \frac{C^2}{h} \frac{\partial \zeta}{\partial y}$$

where $\zeta = n.\bar{\zeta}$, $U = n.\bar{U}$ and $V = n.\bar{V}$.

For shallow water waves, it will be $C^2 = gh$ and therefore, Equations (6) become identical with the equations describing the propagation of tidal waves under linear conditions. A multitude of numerical techniques have been applied for the integration in time and space of these equations (e.g. explicit or implicit finite difference, ADI, finite element techniques etc.) and it is obvious, therefore, that these can be equally applied for the integration of Equations (6) describing the propagation of wind-generated waves. The application of a characteristics technique for the solution of system (6) is demonstrated in the next section. The technique has been already tested and successfully validated in a previous paper where tidal wave propagation is simulated (Matsoukis [11]).

THE METHOD OF CHARACTERISTICS

The method of characteristics has the advantage of transforming the initial system of partial differential equations to an equivalent set of equations (the "characteristic conditions) which contain only total derivatives and are valid only along specific lines, the so-called "characteristic lines" or "bicharacteristics". In the (x,y,t) domain, these can be proven to be the generators of a conoid defined locally by the following equations (see Dauber and Graffe [6]) :

$$\frac{dx}{dt} = U + C\cos\varphi \text{ and } \frac{dy}{dt} = V + C\sin\varphi \tag{7}$$

where φ = parametric angle which is measured anticlockwise and defines the space-time direction of each generator (Fig. 1).

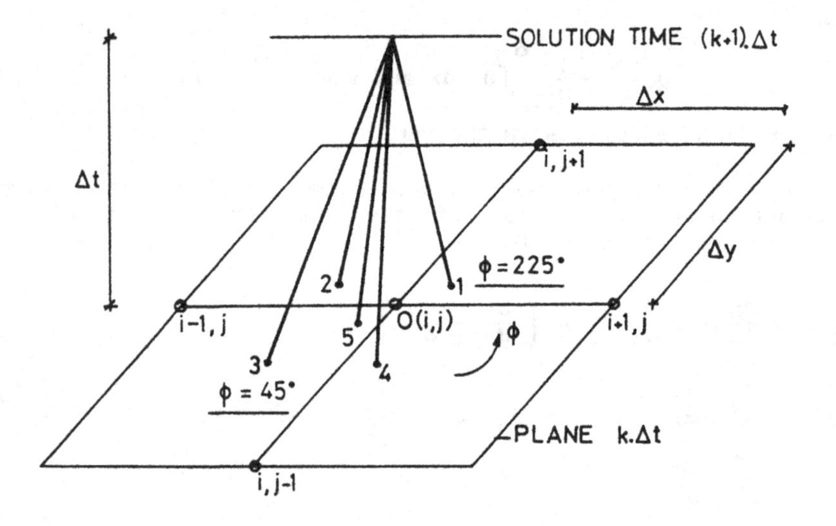

Figure 1. Characteristic lines and intersection points 1 to 5.

It may be proven that along these generators, the following "characteristic" condition holds true:

$$g'\frac{dZ}{dt} + C\cos\varphi\frac{dU}{dt} + C\sin\varphi\frac{dV}{dt} = f \tag{8}$$

where $g' = C^2/h$ and

$$f = g'C\cos\varphi\frac{\partial h}{\partial x} + g'C\sin\varphi\frac{\partial h}{\partial y}$$

$$- C^2\left[\frac{\partial U}{\partial x}\sin^2\varphi - \left(\frac{\partial U}{\partial y} + \frac{\partial V}{\partial x}\right)\sin\varphi\cos\varphi + \frac{\partial V}{\partial y}\cos^2\varphi\right]$$

Also, the trajectory followed by an individual water particle is proven to be a characteristic line (the so-called "particle path" line) defined by

$$\frac{dx}{dt} = U \text{ and } \frac{dy}{dt} = V \tag{9}$$

The corresponding characteristic condition is as follows:

$$\frac{dZ}{dt} = -h\left[\frac{\partial U}{\partial x} + \frac{\partial V}{\partial y}\right] \tag{10}$$

It can be seen that this is in fact identical with the first of Equations (7) (ie the continuity equation), if its terms are re-arranged and therefore, it is not a characteritic condition in the strict sense.

THE NUMERICAL SCHEME

The field of application of the equations is divided to a network of solution points by considering straight lines running parallel to the two axes x and y at distances Δx and Δy ,respectively. The unknown values of ζ, U and V at a certain time $(k+1).\Delta t$ are all calculated simultaneously at every point of the grid by using known values of the variables at the previous time level $k.\Delta t$. In this way, an explicit technique is introduced for the solution running over a non-staggered grid.

Accuracy and numerical stability considerations have shown (Matsoukis[10]) that the computational algorithm must be established by considering the characteristic conditions (8) along four bicharacteristics corresponding at values of φ equal to $5\pi/4, 7\pi/4$, $\pi/4$ and $3\pi/4$ and the characteristic condition (10) along the particle path line (9). Accordingly, it is (see Fig. 1):

Characteristic No 1 ($\varphi = 5\pi/4$)

$$\frac{dZ}{dt} - \frac{c}{g'}\frac{\sqrt{2}}{2}\frac{dU}{dt} \quad \frac{c}{g'}\frac{\sqrt{2}}{2}\frac{dV}{dt} = -c\frac{\sqrt{2}}{2}\frac{\partial h}{\partial x} - c\frac{\sqrt{2}}{2}\frac{\partial h}{\partial y} - \frac{1}{2}\frac{c^2}{g'}\left[\frac{\partial U}{\partial x} - \frac{\partial U}{\partial y} - \frac{\partial V}{\partial x} + \frac{\partial V}{\partial y}\right]$$

Characteristic No 2 ($\varphi = 7\pi/4$)

$$\frac{dZ}{dt} + \frac{c}{g'}\frac{\sqrt{2}}{2}\frac{dU}{dt} - \frac{c}{g'}\frac{\sqrt{2}}{2}\frac{dV}{dt} = c\frac{\sqrt{2}}{2}\frac{\partial h}{\partial x} - c\frac{\sqrt{2}}{2}\frac{\partial h}{\partial y} - \frac{1}{2}\frac{c^2}{g'}\left[\frac{\partial U}{\partial x} + \frac{\partial U}{\partial y} + \frac{\partial V}{\partial x} + \frac{\partial V}{\partial y}\right]$$

Characteristic No 3 ($\varphi = \pi/4$) (11)

$$\frac{dZ}{dt} + \frac{c}{g'}\frac{\sqrt{2}}{2}\frac{dU}{dt} + \frac{c}{g'}\frac{\sqrt{2}}{2}\frac{dV}{dt} = c\frac{\sqrt{2}}{2}\frac{\partial h}{\partial x} + c\frac{\sqrt{2}}{2}\frac{\partial h}{\partial y} - \frac{1}{2}\frac{c^2}{g'}\left[\frac{\partial U}{\partial x} - \frac{\partial U}{\partial y} - \frac{\partial V}{\partial x} + \frac{\partial V}{\partial y}\right]$$

Characteristic No 4 ($\varphi = 3\pi/4$)

$$\frac{dZ}{dt} - \frac{c}{g'}\frac{\sqrt{2}}{2}\frac{dU}{dt} + \frac{c}{g'}\frac{\sqrt{2}}{2}\frac{dV}{dt} = -c\frac{\sqrt{2}}{2}\frac{\partial h}{\partial x} + c\frac{\sqrt{2}}{2}\frac{\partial h}{\partial y} - \frac{1}{2}\frac{c^2}{g'}\left[\frac{\partial U}{\partial x} + \frac{\partial U}{\partial y} + \frac{\partial V}{\partial x} + \frac{\partial V}{\partial y}\right]$$

The numerical solution is established by approximating the total derivatives on the left hand side by forward in time finite differences and, therefore, the conditions above may be written as follows:

$$\frac{Z-Z1}{\Delta t} - \frac{c}{g'}\frac{\sqrt{2}}{2}\frac{U-U1}{\Delta t} - \frac{c}{g'}\frac{\sqrt{2}}{2}\frac{V-V1}{\Delta t} = -c\frac{\sqrt{2}}{2}\frac{\partial h}{\partial x} - c\frac{\sqrt{2}}{2}\frac{\partial h}{\partial y} - \frac{1}{2}\frac{c^2}{g'}\left[\frac{\partial U}{\partial x} - \frac{\partial U}{\partial y} - \frac{\partial V}{\partial x} + \frac{\partial V}{\partial y}\right]$$

$$\frac{Z-Z2}{\Delta t} + \frac{c}{g'}\frac{\sqrt{2}}{2}\frac{U-U2}{\Delta t} - \frac{c}{g'}\frac{\sqrt{2}}{2}\frac{V-V2}{\Delta t} = c\frac{\sqrt{2}}{2}\frac{\partial h}{\partial x} - c\frac{\sqrt{2}}{2}\frac{\partial h}{\partial y} - \frac{1}{2}\frac{c^2}{g'}\left[\frac{\partial U}{\partial x} + \frac{\partial U}{\partial y} + \frac{\partial V}{\partial x} + \frac{\partial V}{\partial y}\right]$$

$$\frac{Z-Z3}{\Delta t} + \frac{c}{g'}\frac{\sqrt{2}}{2}\frac{U-U3}{\Delta t} + \frac{c}{g'}\frac{\sqrt{2}}{2}\frac{V-V3}{\Delta t} = c\frac{\sqrt{2}}{2}\frac{\partial h}{\partial x} + c\frac{\sqrt{2}}{2}\frac{\partial h}{\partial y} - \frac{1}{2}\frac{c^2}{g'}\left[\frac{\partial U}{\partial x} - \frac{\partial U}{\partial y} - \frac{\partial V}{\partial x} + \frac{\partial V}{\partial y}\right] \qquad (1$$

$$\frac{Z-Z4}{\Delta t} - \frac{c}{g'}\frac{\sqrt{2}}{2}\frac{U-U4}{\Delta t} + \frac{c}{g'}\frac{\sqrt{2}}{2}\frac{V-V4}{\Delta t} = -c\frac{\sqrt{2}}{2}\frac{\partial h}{\partial x} + c\frac{\sqrt{2}}{2}\frac{\partial h}{\partial y} - \frac{1}{2}\frac{c^2}{g'}\left[\frac{\partial U}{\partial x} + \frac{\partial U}{\partial y} + \frac{\partial V}{\partial x} + \frac{\partial V}{\partial y}\right]$$

$$\frac{Z-Z5}{\Delta t} = -h\left[\frac{\partial U}{\partial x} + \frac{\partial V}{\partial y}\right] \qquad \text{(particle path line)}$$

where Z, U and V the values of the unknown variables at the apex of the conoid ie at the time level of solution $(k+1).\Delta t$ and Z1,U1,V1,Z2,U2,V2 etc. the values of Z,U and V at the intermediate points 1 to 5 at the previous time level $k.\Delta t$. These are defined as the intersection of the bicharacteristics with the time plane $k.\Delta t$ (Fig. 1) and, therefore, their coordinates x_1, y_1, x_2, y_2 etc. are as follows:

$x_5 = -U.\Delta t$, $y_5 = -V.\Delta t$ and

$$x_{1,2} = x_5 \pm c\frac{\sqrt{2}}{2}\Delta t \qquad y_{1,3} = y_5 \pm c\frac{\sqrt{2}}{2}\Delta t \qquad (13)$$

$$x_{3,4} = x_5 \mp c \frac{\sqrt{2}}{2} \Delta t \qquad y_{2,4} = y_5 \pm c \frac{\sqrt{2}}{2} \Delta t$$

The quantities U,V and C in Equations (13) above and also the partial derivatives on the right hand side of Equations (12) are all calculated at k.Δt at the point of solution O(iΔx,jΔy).

By combining Equations (13) , we have

$$Z = \frac{R1+R2+R3+R4}{2} - Z5$$

$$U = \sqrt{2}\, g \frac{R2+R3-R1-R4}{4C} - \frac{c^2}{h} \frac{\partial h}{\partial x} \Delta t \tag{14}$$

$$V = \sqrt{2}\, g \frac{R3+R4-R1-R2}{4C} - \frac{c^2}{h} \frac{\partial h}{\partial y} \Delta t$$

where
$$R1 = Z1 - \frac{c}{g'} \frac{\sqrt{2}}{2} U1 - \frac{c}{g'} \frac{\sqrt{2}}{2} V1$$

$$R2 = Z2 + \frac{c}{g'} \frac{\sqrt{2}}{2} U2 - \frac{c}{g'} \frac{\sqrt{2}}{2} V2$$

$$R3 = Z3 + \frac{c}{g'} \frac{\sqrt{2}}{2} U3 + \frac{c}{g'} \frac{\sqrt{2}}{2} V3 \tag{15}$$

$$R4 = Z4 - \frac{c}{g'} \frac{\sqrt{2}}{2} U4 + \frac{c}{g'} \frac{\sqrt{2}}{2} V4$$

are the so-called Riemann invariants.

The value of any variable at the intermediate points 1 to 5 is approximated by means of an interpolating scheme which takes the form of a polynomial following Taylor's formula up to a second order of accuracy, ie

$$Q_{x,y} = Q_{i,j} + x\, L_x(Q) + 0.5\, x^2\, L_{xx}(Q) + y\, L_y(Q) + 0.5\, y^2\, L_{yy}(Q) + xy\, L_{xy}(Q) \tag{16}$$

where Q stands for any quantity involved in the solution and L_x, L_y, L_{xx}, L_{yy} etc. are finite difference operators in the following way:

$$L_x(Q) = \frac{Q_{i+1,j} - Q_{i-1,j}}{2\,\Delta x}, \quad L_{xx}(Q) = \frac{Q_{i+1,j} + Q_{i-1,j} - 2Q_{i,j}}{\Delta x^2} \text{ etc.}$$

As a result, the final solution given by (14) above is finally expressed in terms of the values of the variables not at the intermediate points 1 to 5 but only at the grid points. In practical applications, it is usually taken $\Delta x = \Delta y = \Delta s$ and so, the numerical solution of elevation ζ and mean velocities U and V at a grid point $O(i\Delta s, j\Delta s)$ and at time level $(k+1).\Delta t$,ie ζ^{k+1}, U^{k+1} and V^{k+1}, becomes

$$\zeta^{k+1}_{i,j} = \zeta^{k}_{i,j} + 0.5\ c^2 \Delta t^2 \left[L_{xx}(\zeta^{k}_{i,j}) + L_{yy}(\zeta^{k}_{i,j})\right]$$
$$- U^{k}_{i,j}\ \Delta t\ L_x(h^{k}_{i,j}) - V^{k}_{i,j}\ \Delta t\ L_y(h^{k}_{i,j})$$
$$- h_{i,j}\ \Delta t\ L_x(U^{k}_{i,j}) - h_{i,j}\ \Delta t\ L_x(U^{k}_{i,j})$$

$$U^{k+1}_{i,j} = U^{k}_{i,j} + 0.5\ c^2 \Delta t^2 \left[L_{xx}(U^{k}_{i,j}) + L_{yy}(U^{k}_{i,j})\right] \qquad (17)$$
$$- \frac{c^2}{h_{i,j}}\ \Delta t\ L_x(\zeta^{k}_{i,j})$$

$$V^{k+1}_{i,j} = V^{k}_{i,j} + 0.5\ c^2 \Delta t^2 \left[L_{xx}(V^{k}_{i,j}) + L_{yy}(V^{k}_{i,j})\right]$$
$$- \frac{c^2}{h_{i,j}}\ \Delta t\ L_y(\zeta^{k}_{i,j})$$

This solution scheme is restricted by the following CFL condition for stability

$$\frac{c\ \Delta t}{\Delta s} \leq \frac{\sqrt{2}}{2} \qquad (18)$$

Two kinds of boundary points may be recognized within the model area:those lying along an impermeable barrier (e.g. quay wall), ie the so-called "closed" boundary points and those lying along a line connecting the model with the open sea, ie the so-called "open" boundary points. At closed boundary points, the velocity directed at right angles to the boundary line is taken as zero. At the open boundary points, the amplitude of the incident wave is usually prescribed as a sinusoidal function of time , while the amplitude of the outgoing wave is calculated by the model itself using the bicharacteristics lying only in the interior of the model.

MODEL APPLICATIONS

To test the validity of the model, four different cases of breakwater gap width to wavelength are examined , ie B/L = 0.5, 1.0, 1.41 and 1.64 and the resulting diffraction coefficients isolines are compared with those provided by the generalised diffraction diagrams in CERC.

Waves travelling over a water depth of h = 15m are considered with a tidal period of T = 16secs, an amplitude of 0.5m and a 90° angle of incidence to the breakwater gap axial line. Preliminary tests have shown that the method of solution developed above (Equations (17)), depends strongly on the number of grid points per wavelength and in this respect, around 40 points per wavelength has been found to be a more than adequate number to ensure a high level of accuracy. Due to the symmetry of solution , only half of the wave field needs to be considered in practical applications (see Fig. 2, 3, 4 and 5). A suitable number of iterations is also selected, so that the wave front does not reach the limit of the internal boundaries. In the opposite case, unwanted reflections will take place along the

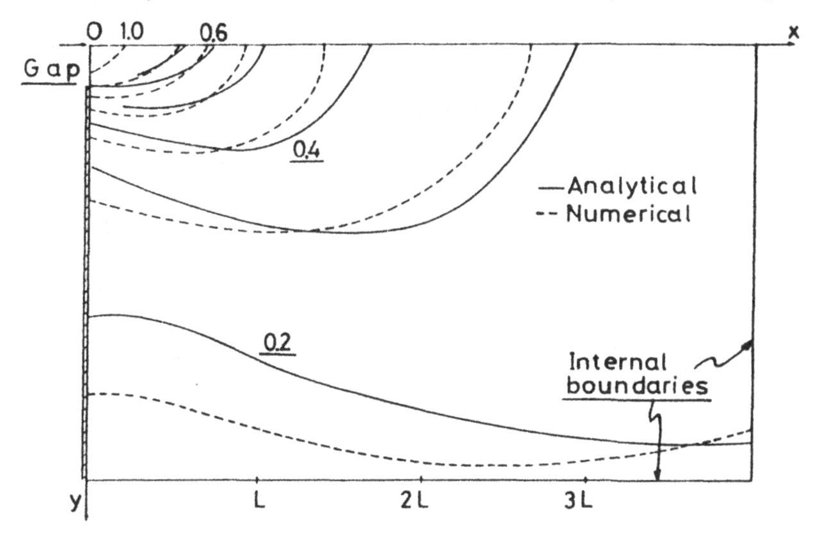

Figure 2. Isolines of diffraction coefficients. Ratio B/L=0.5

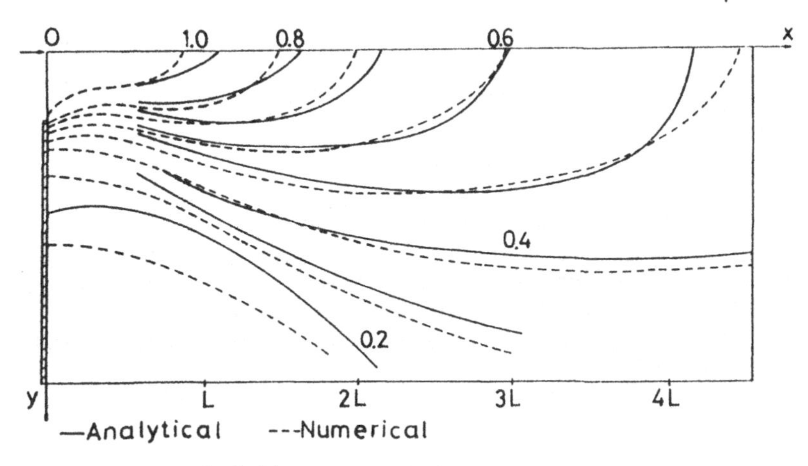

Figure 3. Isolines of diffraction coefficients. Ratio B/L = 1.0

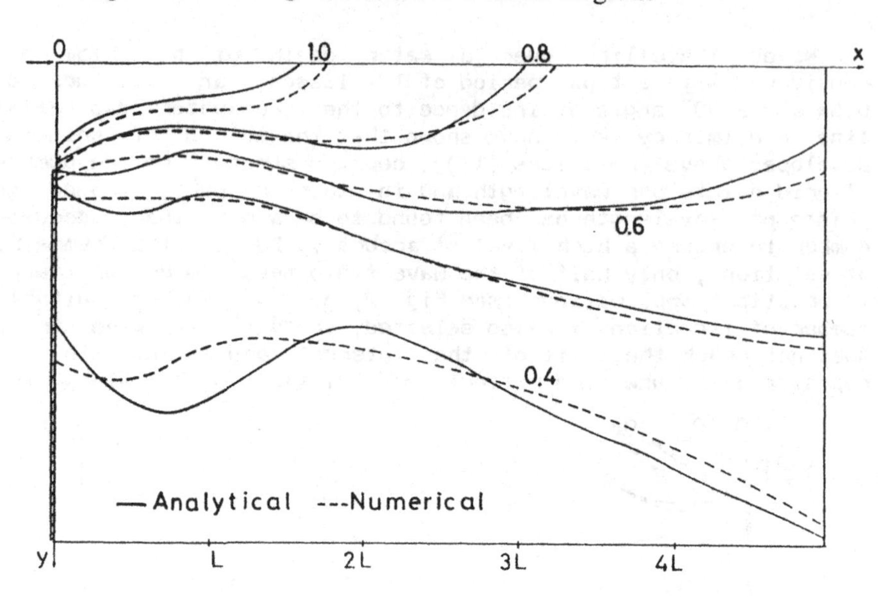

Figure 4. Isolines of diffraction coefficients.Ratio B/L = 1.41

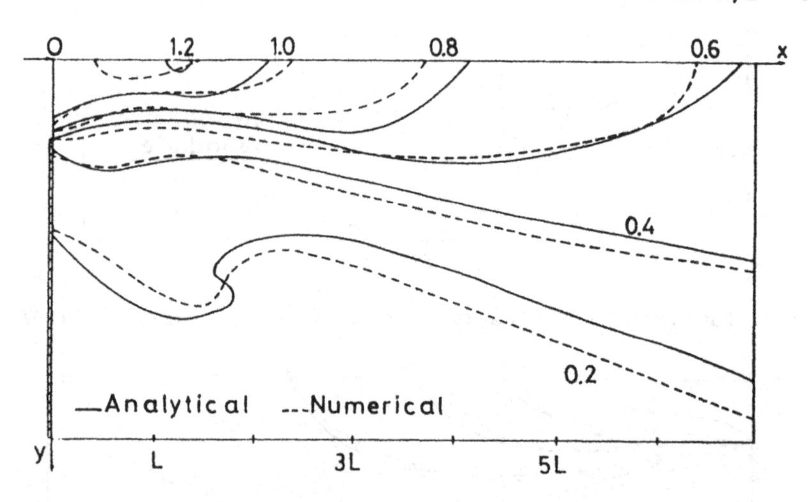

Figure 5. Isolines of diffraction coefficients.Ratio B/L = 1.64

boundaries which are bound to contaminate the solution in interior of the model. During the running of the explicit scheme, wave amplitude maxima and minima are stored at each grid point which are then used to calculate wave heights and also, wave diffraction coefficients.Depending on the size of the grid, the required CPU time is of the level of 1 to 2 minutes on a micro VAXII computing system.

By comparing the isolines of diffraction coefficients produced by the model inside the harbour with those provided by the generalised diffraction diagrams for the selected ratios of B/L, the following table can be formed.

Table 1. Average percent(%) error per isoline

Ratio B/L	Isoline							
	1.0	0.8	0.7	0.6	0.5	0.4	0.3	0.2
0.5			4.9	5.1	4.2	4.1		
1.0		3.1	2.6	3.1	2.2	1.7	5.2	
1.41	7.3	2.7		3.4		1.5		1.5
1.64	2.3	3.8		2.6		7.0		

Under these conditions, it becomes obvious that the average error of the numerical solution lies at a 5% level and, therefore, we can safely conclude that the characteristics technique is indeed a reliable alternative for practical applications of civil engineering importance.

REFERENCES

1. Berkhoff, J.C.W., Computation of combined refraction - diffraction, pp. 471-490, Proc. 13th Inter. Conf. Coastal Engrg., Vancouver, Canada, 1972.
2. Bettess, P. and Zienkiewicz, O.C. Diffraction and refraction of surface waves using finite and infinite elements, Int. J. Numer. Meth. Engrg. ,11, pp. 1271-1290, 1977.
3. Carr ,J.H. and Stelzriede, M.E. Diffraction of water waves passing through a breakwater gap , Gravity Waves, Circular No 521, U.S. Nat. Bur. Standards, 109-125, 1952.
4. CERC Shore Protection Manual, 1, 4th ed. U.S. Army Coastal Engineering Research Center (cerc), 1984.
5. Copeland, G.J.M. A practical alternative to the mild slope wave equation , Coastal Eng., 9:125-149, 1985.
6. Daubert, A. and Mlle Graffe, D. Quelques aspects des ecoulements presque horizontaux a deux dimensions en plan et non permanents. Application aux estuaires, pp. 847, La Houille Blanche, No 8 ,1967.
7. Ito .Y. and Tanimoto K. A method of numerical analysis of wave propagation - application to wave diffraction and refraction, pp. 503 - 522 , Proc. Conf. Coastal Eng., 134th Ch. 26, 1972.
8. Johnson, J.W., Generalized wave diffraction diagrams, pp. 6-23 ,Proc. 2nd Inter. Conf. Coast. Engrg., 1952.
9. Lamb, H. Hydrodynamics, 6th ed., Cambridge University Press, Cambridge, England, 1932.

10. Matsoukis, P.F.C. The analysis and application of the method of characteristics for simulating long wave propagation in the two-dimensiobnal space, Ph.D. Thesis, Strathclyde University of Glasgow, UK, 1980.
11. Matsoukis, P.F.C. A tidal model using the method of characteristics , a paper to appear in the J. of Ports, Waterway, Coastal and Ocean Engineering, ASCE, May 1992.
12. Morse, P.M. and Rubenstein, P.J. The diffraction of waves by ribbons and by slits, Phys. Rev., 54, ppp. 895-898, 1938.
13. Penny, W.G. and Price A.T. The diffraction theory of sea waves and the shelter afforded by breakwaters, Philosoph. Trans. Royal Soc. London, Ser. A, 244, 236-253, 1952.
14. Radder, A.C. On the parabolic equation method for water wave propagation, pp. 159-176, J. Fluid Mech., 95(1), 1979.
15. Southgate, H.N. Ray methods for combined refraction and diffraction problems, Hydraulics Research Report IT 214, Wallingford, UK, July 1981.

NOTATION

The following symbols are used in this paper:

C = wave celerity

C_g = wave group celerity

Δx, Δy = space increments along x- and y- axis

Δt = time increment

h = mean water depth relative to a certain datum

ζ = water elevation relative to a certain datum

i,j,k = integers

n = shoaling number

U,V = depth mean velocities along x- and y- axis

t = time variable

x,y = cartesian coordinates

Z = $h+\zeta$ = total water depth

Boundary Integral Formulation of Wave Energy Dissipation in Porous Media

G.Z. Gu (*), H. Wang (**)

(*) Dallas E & P Engineering, Mobil R & D Corp., 13777 Midway, Dallas TX 75244, U.S.A.

(**) Coastal and Oceanographic Engineering Dept., University of Florida, Gainesville, FL, 32611, U.S.A.

ABSTRACT

When gravity waves interact with porous media, a significant portion of the kinetic energy is dissipated inside the pores due to turbulence and friction. It is important to estimate such energy loss when designing porous ocean structures, such as rubble-mound breakwaters in which the wave energy dissipation is usually the major goal. In this paper, the energy dissipation process is numerically modeled with the Boundary Integral Element Method (BIEM). To apply the BIEM efficiently, a boundary integral formulation for the wave energy dissipation has been developed, based on energy-flux deficit concept, to replace the commonly used volumetric expression (Sulisz [7], Madsen [5], and Sollitt et al.[6]).

The numerical model, which combines BIEM, the nonlinear-unsteady porous flow model and the boundary integral formula of wave energy dissipation, is capable of simulating the complex interaction process of waves with porous media of complex geometry for the full range of permeability. Compared to the conventional volumetric formulation, the boundary integral formulation for the energy dissipation is proved to be a very efficient when applied with the BIEM.

INTRODUCTION

When surface gravity waves interact with porous media, such as rubble-mound breakwaters or gravel islands, a significant amount of the wave energy is being dissipated within the media due to turbulence and friction. The computation of this energy dissipation is important in the design of rubble-mound breakwaters.

Due to high permeability of rubble-mound structures, the dissipation process is highly nonlinear; also the inertial resistance becomes significant

---

[1] Formerly with Coastal and Oceanographic Engineering Dept. University of Florida

when the flow is unsteady. The porous flow of such kind must be modeled by a nonlinear-unsteady percolation model such as the one proposed by Sollitt and Cross [6]. Due to the difficulty of solving the nonlinear problem directly, the nonlinear formulation is usually linearized to obtain an equivalent linear resistance coefficient, based on the principle of equal energy dissipation. Such a technique has been successfully employed by many investigators such as Sollitt, et al.[6], Madsen [5], Sulisz [7] and others when modeling porous ocean structures. However, in their models, the expression of the energy dissipation is in the form of a volumetric integration which is awkward and leads to tedious and needless computations when boundary element method is used. Since the linearization process accounts for a large portion of the total computational effort in the solution of a nonlinear problem, a more efficient method for evaluating the wave energy dissipation is highly desirable.

NONLINEAR-UNSTEADY PERCOLATION MODEL

For porous media made of rubble stones, the porous flow can be described by (see Wang and Gu [8], Gu and Wang [3], Sollitt and Cross [6])

$$-\frac{1}{\rho}\nabla P(x,y,z,t) = \left(\frac{\nu}{K_p} - i\sigma\beta + \frac{C_f}{\sqrt{K_p}} \mid \vec{q}(x,y,z,t) \mid\right) \vec{q}(x,y,z,t) \quad (1)$$

$$= \sigma(f_1 + f_2 \, |\vec{q}|) \, \vec{q} \quad (2)$$

where

$P(x,y,z,t)$ is the pore pressure function;

ν **and** ρ are the kinematic viscosity and fluid density, respectively;

K_p is the intrinsic permeability of porous media; it is related to particle diameter by (Engelund [4])

$$K_p = \frac{n^2 d_s^2}{a_0\,(1-n)^3} \quad (3)$$

where

a_0 is an empirical constant and

n is volumetric porosity;

σ is wave frequency;

β is the inertial resistance parameter defined as

$$\beta = \frac{n + C_a(1-n)}{n^2}$$

C_a is the virtual mass coefficient;

C_f is a non-dimensional constant which characterizes the nonlinear resistance and

f_1 and f_2 are two complex coefficients introduced for simplicity. The definitions of them are straightforward from the equation;

$\vec{\mathrm{q}}(x, y, z, t)$ is the complex vector of discharge velocity in the porous medium, the real velocity vector is

$$\vec{q}(x, y, z, t) = Re(\vec{\mathrm{q}}(x, y, z, t))$$

Introducing a non-dimensional parameter called the intrinsic permeability parameter, defined as $R = \sigma K_p/\nu$, Eq.(1) becomes

$$-\frac{1}{\rho}\nabla P(x, y, z, t) = \sigma(\frac{1}{R} - i\beta + \frac{C_f}{\sqrt{\sigma\nu R}} \mid \vec{\mathrm{q}}(x, y, z, t) \mid) \vec{\mathrm{q}}(x, y, z, t) \quad (4)$$

BOUNDARY INTEGRAL FORMULATION OF THE ENERGY DISSIPATION

The conventional formulation for the wave energy dissipation e_D within a volume V of a porous medium during the time period T is (Sollitt et al.[6], Madsen [5], and Sulisz [7])

$$e_D = \int_V \int_t^{t+T} \vec{F} \cdot \rho\vec{q}\, dt\, dv \quad (5)$$

where $\vec{F}$, which is a function of spatial coordinates and time, is the dissipative stress in the medium and $\vec{q}$ is the discharge velocity of the porous flow. Here both $\vec{F}$ and $\vec{q}$ are real quantities.

With the nonlinear percolation model Eq.(1), the dissipative stress is defined as the real part of the quantity in the brackets:

$$\vec{F} = (\frac{\nu}{K_p} + \frac{C_f}{\sqrt{K_p}} \mid \vec{q} \mid)\vec{q} \quad (6)$$

The inertial term in Eq.(1), $i\sigma\beta\vec{q}$, which is a pure imaginary number in this case, is a non-dissipative stress and therefore will not contribute to the energy dissipation process.

The volumetric integration in Eq.(5) is usually difficult to carry out and in many cases, approximations must be made to simplify the integrand. However, if the wave energy dissipation can be expressed as a contour integration along the boundary of the computational domain, the calculation of energy dissipation will be considerably simplified, especially for a boundary element solution where P or $\vec{q}$ are usually well specified along the boundaries. As a matter of fact, such an expression can be easily obtained by the use of a control volume and the Green's theorem.

For simplicity, we consider only two dimensional problems in the $x - z$ plane. We define U and W to be the discharge velocities (real quantities) in the x and z directions, respectively, and $\bar{e}_D$ to be the rate of energy dissipation per unit volume (also a real quantity, considered as a positive value). The rate of total energy dissipation in an arbitrarily small cube of $dx \cdot 1 \cdot dz$ can be expressed as

$$\bar{e}_D = -(P_r\frac{\partial U}{\partial x} + P_r\frac{\partial W}{\partial x} + U\frac{\partial P_r}{\partial x} + W\frac{\partial P_r}{\partial z}) \tag{7}$$

$$= -[\frac{\partial}{\partial x}(UP_r) + \frac{\partial}{\partial z}(WP_r)] \tag{8}$$

with P_r being the real part of the complex pore pressure function.

By applying the continuity condition of pore fluid and the nonlinear percolation model given in Eq.(1) to Eq.(7), it can be readily proven that Eq.(8) is just an alternate expression of the integrand in Eq.(5) (see Gu [2]).

The total energy dissipation within the entire computational domain during a time period T is then

$$e_D = \iint_A \int_t^{t+T} \bar{e}_D \, dt \, dx \, dz$$

$$= -\int_t^{t+T} \iint_A [\frac{\partial}{\partial x}(UP_r) + \frac{\partial}{\partial z}(WP_r)] \, dx \, dz \, dt \tag{9}$$

where A is the total area of a computational domain, such as the cross sectional area of a submerged breakwater.

Eq.(9) is an equivalent expression to Eq.(5) for the energy dissipation in a porous medium. The only difference between the two is that the non-dissipative resistance $i\sigma\beta\vec{q}$ is included in Eq.(9) but not in Eq.(5). The inclusion of this term should not affect the value of e_D because of the non-dissipative nature of this resistance.

Applying the Green's theorem to Eq.(9), which converts an aerial integration into a contour integration, the energy dissipation within the area bounded by S during a time period T becomes

$$e_D = -\int_t^{t+T} \oint_S P_r U_n ds \, dt \tag{10}$$

The above equation simply states that the volumetric energy dissipation in the time period T is equal to the net energy flux across the boundary enclosing the volume in the same period; i.e., the physical principle of energy conservation.

Expressed in terms of complex variables, Eq. (10) becomes

$$E_D = -\frac{1}{2}\int_t^{t+T} \oint (pu_n e^{2i\sigma t} + u_n p^*) \, ds \, dt \tag{11}$$

with

$$u_{nr} = Re(u_n) \tag{12}$$

$$p_r = Re(p) \tag{13}$$

$$e_D = Re(E_D) \tag{14}$$

where u_{nr}, p_r and e_D are real velocity, pressure and energy dissipation, respectively.

For linear wave problems, it is convenient to chose T as the wave period. The integration of the first term of Eq.(11) with respect to time vanishes, and the complex energy dissipation is reduced to

$$E_D = -\frac{T}{2} \oint u_n p^* \, ds \tag{15}$$

LINEARIZATION

The linearization of the nonlinear formulation of Eq.(1) is accomplished by equating the energy dissipation by the linearized system to that by the true nonlinear system.

For the nonlinear system,

$$(u_n)_{nl} = -\frac{1}{\rho\sigma(f_1 + f_2 \,|\vec{q}|)} p_n \tag{16}$$

and

$$(E_D)_{nl} = \frac{i}{2} \int_C \frac{p_n p^*}{f_1 + f_2 \,|\vec{q}|} \, ds \tag{17}$$

where C is the portion of the closed boundary of porous domain where u_n is nonzero.

For the linearized system,

$$(u_n)_l = -\frac{1}{\rho\sigma f_0} \frac{\partial p}{\partial n} = -\frac{1}{\rho\sigma f_0} p_n \tag{18}$$

and

$$(E_D)_l = \frac{i}{2 f_0} \int_C p_n p^* \, ds \tag{19}$$

with f_0 being the linearized (or equivalent) resistance coefficient, an unknown for the problem.

Equating $(E_D)_l$ to $(E_D)_{nl}$ and taking approximately $|\vec{q}| \simeq |p_n/\rho\sigma f_0|$, the linearized coefficient f_0 is then

$$f_0 = \frac{\displaystyle\int_C p_n p^* \, ds}{\displaystyle\int_C \frac{p_n p^*}{f_1 + f_2 \,|\, p_n/\rho\sigma f_0 \,|} \, ds} \tag{20}$$

This is an implicit equation of f_0, the solution of the problem is achieved by iteration, starting with an initial guess of f_0. Since it involves contour integrals instead of volumetric integrals, Eq.(20) is much more efficient than its conventional counter part.

BOUNDARY INTEGRAL ELEMENT FORMULATION AND NUMERICAL MODEL

In the BIEM numerical model, the velocity potential function Φ in the fluid domain D_1 bounded by a closed contour C_1 and the pore pressure function P in a porous domain D_2 bounded by C_2 can be expressed by the following two equations, respectively

$$\alpha\Phi(\mathbf{x}_0) = \oint_{C_1}[\Phi(\mathbf{x})\frac{\partial G}{\partial n}(\mathbf{x}_0,\mathbf{x}) - \frac{\partial \Phi}{\partial n}(\mathbf{x})G(\mathbf{x}_0,\mathbf{x})]ds \tag{21}$$

$$\alpha P(\mathbf{x}_0) = \oint_{C_2}[P(\mathbf{x})\frac{\partial G}{\partial n}(\mathbf{x}_0,\mathbf{x}) - \frac{\partial P}{\partial n}(\mathbf{x})G(\mathbf{x}_0,\mathbf{x})]ds \tag{22}$$

where $G(\mathbf{x}_0, \mathbf{x})$ is a free space Green's function and α is a coefficient depending on the position of the point $\mathbf{x}_0$, (it is 2π when $\mathbf{x}_0$ is an interior point, and the inner angle of the boundary when it is a boundary point); $\mathbf{x}_0$ is a point in the domain of $D_1 \cap C_1$ or $D_2 \cap C_2$ and $\mathbf{x}$ is a boundary point on C_1 or C_2.

The free space Green's function is

$$G(\mathbf{x}_0, \mathbf{x}) = \ln r \tag{23}$$

where r is the distance between $\mathbf{x}$ and $\mathbf{x}_0$ and it is

$$r = \sqrt{(x - x_0)^2 + (z - z_0)^2}$$

on the $x - z$ plane.

The boundary conditions used for the problem are: the linearized free surface condition on the free surface, the radiation condition on the two vertical lateral boundaries and the non-flux condition on the seabed. The boundary conditions on the common portion of C_1 and C_2, i.e., the interface of the fluid and porous domains, are established based on the continuity of velocity and pressure.

Discretizing the boundaries C_1 and C_2 and carrying out the contour integration over each boundary segment, Eqs.(21) and (22) give rise to two sets of linear algebraic equations. Solving these equations simultaneously together with the boundary conditions, the unknowns Φ, $\frac{\partial \Phi}{\partial n}$, P and $\frac{\partial P}{\partial n}$ can be obtained along the boundaries C_1 and C_2 on a discretized basis (Gu [2]), if the linearized resistance coefficient f_0, which is introduced into the computation through the matching condition along the breakwater surface, is known.

The equation for f_0 is given by eq.(20) and is solved by iterating the solution process with a guessed initial value. For details of the numerical model, readers are referred to Gu [2].

NUMERICAL EXAMPLES

The numerical model described above is applied to a porous submerged breakwater of model scale. The geometric parameters of the breakwater are: 15 cm high, 60 cm wide at the crest with slope 1:1.5. The stone size d = 0.93 cm and the porosity n = 0.349. The mean water level is 23 cm above the bottom and 8 cm above the breakwater crest. The parameters adopted in the numerical model are: $a_0 = 570$, $C_f = 1.0$ and $C_a = 0.46$ (Gu and Wang [3]).

The wave transmission and reflection coefficients K_T and K_R calculated by the numerical model for different wave periods due to percolation are plotted in Fig. 1(a) and (b) against the intrinsic permeability parameter R ($log_{10}R$). The energy dissipation e_D can be obtained by

$$e_D = (1 - K_T^2 - K_R^2)\frac{\gamma H_i^2}{8}$$

It is interesting to note the existence of the minima of K_T (therefore the maxima of e_D). Such maximum energy dissipation was found to occur at a permeability where the dissipative resistance (velocity related) equals to the non-dissipative resistance (acceleration related). The magnitude of the maximum energy dissipation rate is slightly affected by wave period. Utilizing the relationship between permeability and particle diameter described in Eq.(3), the range of stone size where maximum e_D's will occur can be established to aid design of structures such as breakwaters and gravel islands.

Experiments were conducted in a wave tank at University of Florida on a model of same configuration to verify the numerical model. Figure 2 and 3 show the transmitted and reflected wave heights versus the incident wave heights. The continuous curves are the numerical results. The incident wave height at which breaking waves were observed is marked in the figure as the breaking point. For wave heights smaller than the breaking height, the energy dissipation is solely due to percolation and the numerical model and physical model agree well in this range. After breaking point, the energy dissipation by the physical model is significantly greater than that calculated by the numerical model. This is because the energy dissipation of the physical model contains the portion due to breaking which is not considered in the numerical model. The difference between the numerical results and the experimental data, however, can be viewed as the dissipation attributed to breaking.

CONCLUDING REMARKS

1. With the formulation of boundary integration, the computation of wave energy dissipation due to nonlinear percolation can be considerably simplified. The linearization process using this expression becomes much simpler and much more efficient, as compared to that of using the conventional volumetric integration.
2. The rate of wave energy dissipation due to nonlinear percolation, when plotted against the intrinsic permeability parameter R, has a well defined peak for each wave period. These maxima occur when the dissipative resistance (velocity related) in the porous structure is equal to the non-dissipative resistance (acceleration related). The magnitudes of the peak energy damping are slightly different for different periods.
3. The numerical model results compare well with the laboratory data for non-breaking waves. Work is being continued to include the energy dissipation due to wave breaking into the model.

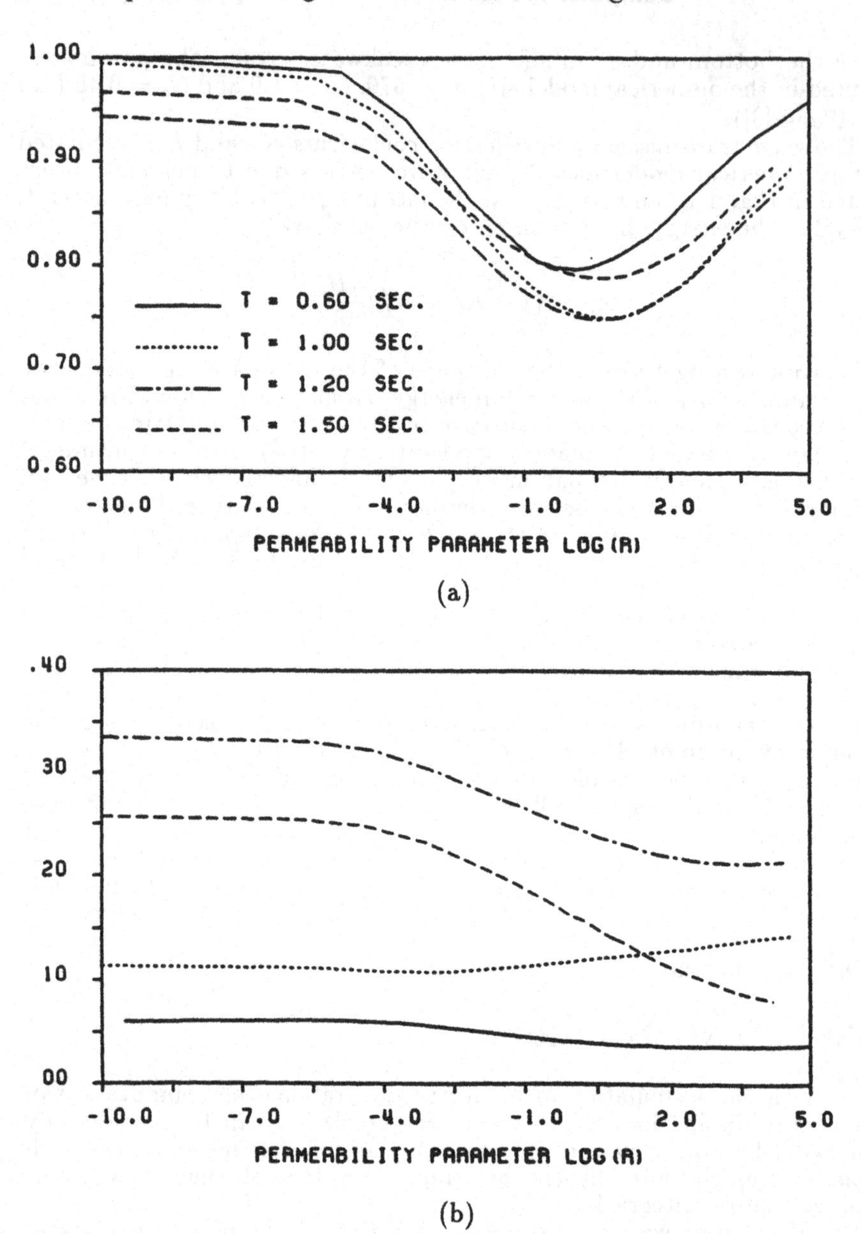

Figure 1: Predicted transmission and reflection coefficients vs. R for different wave periods. (a) Transmission coefficient K_T; (b) Reflection coefficient K_R

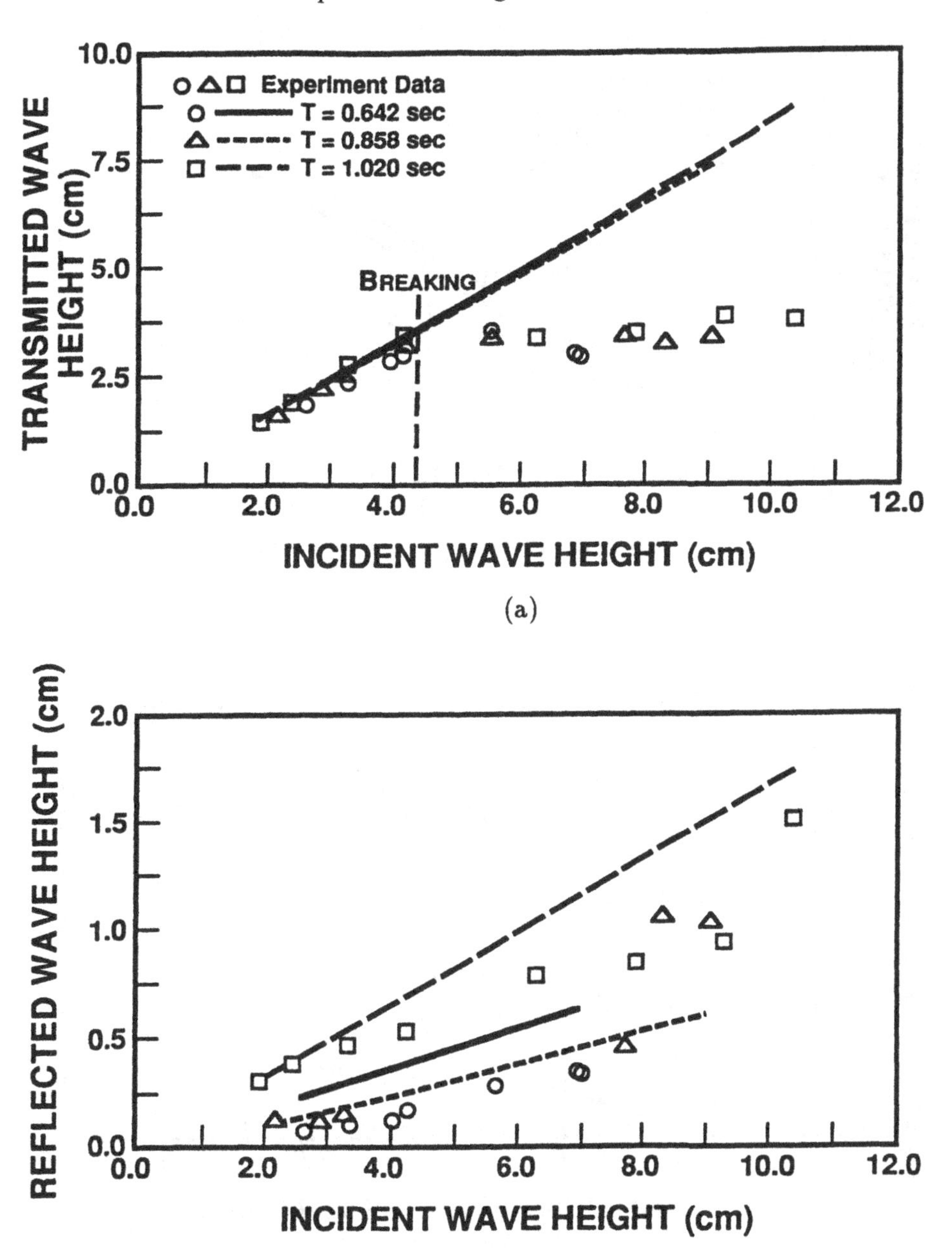

Figure 2: Comparison of predictions and measurements of transmitted and reflected wave heights vs. incident wave heights. (a) Transmitted wave heights vs. incident wave heights; (b) Reflected wave heights vs. incident wave heights

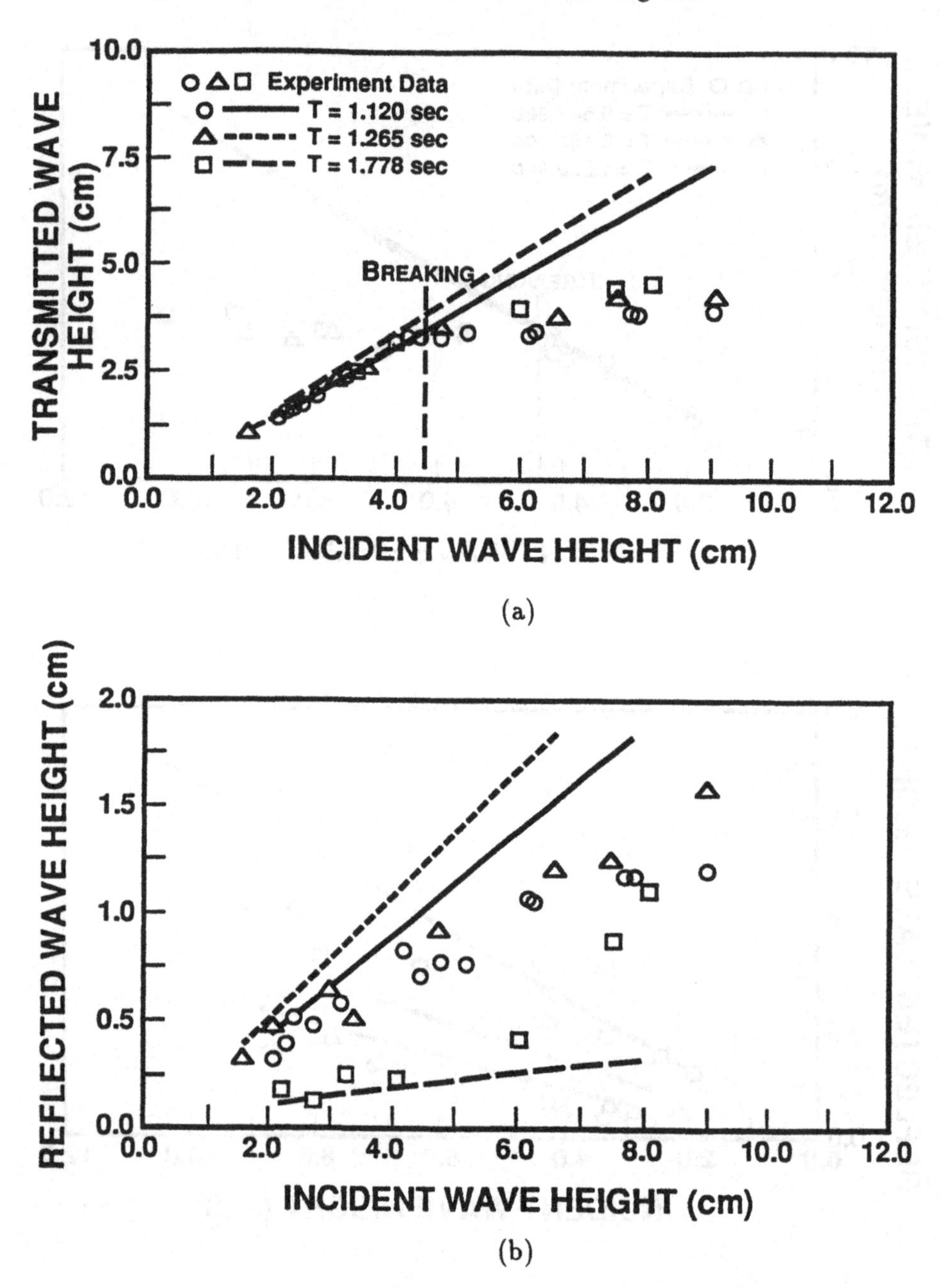

Figure 3: Comparison of predictions and measurements of transmitted and reflected wave heights vs. incident wave heights. (a) Transmitted wave heights vs. incident wave heights; (b) Reflected wave heights vs. incident wave heights

References

[1] Engelund, F., 1953, On the Laminar and Turbulent Flows of Ground Water Through Homogeneous Sand. *Trans. of the Danish Academy of Technical Sciences* **3**, No. 4.

[2] Gu, Z., 1990, Water Wave Interaction with Porous Structures with Irregular Cross Sections. *PhD Dissertation, University of Florida.*

[3] Gu, Z. and Wang, H, 1991, Gravity Waves over Porous Bottoms. *Coastal Engineering***21**, 497-524.

[4] Ligget., J.A., and Liu, P.L-F., 1983, *The Boundary Integral Equation Method for Porous Media Flow*, George Allen & Unwin Ltd.

[5] Madsen, O.S., 1974, Wave Transmission through Porous Structures, *J. of the Waterways Harbors and Coastal Engineering Div.*, ASCE, Vol. 100, No.WW3.

[6] Sollitt, C.K., and Cross, R.H., 1972, Wave Transmission through Permeable Breakwaters. *Proc. 13th Coastal Eng. Conf., ASCE* **III**, 1827-1846.

[7] Sulisz, W., 1985, Wave Reflection and Transmission at Permeable Breakwaters of Arbitrary Cross Section, *Coastal Engineering* Vol. 9, 371-386.

[8] Wang, H. and Gu, Z., 1988. Gravity Waves over Porous Bottom. 2nd International Symp. on Wave Research and Coastal Engineering, pp. 1-21.

Numerical and Experimental Studies on Diffraction of Water Waves Around Three-Cylinder Group

T.V. Gopalakrishnan (*), C.P. Vendhan (**), H. Raman (***)

(*) Centre for Water Resources, College of Engineering, Anna University, Madras -25, India
(**) Ocean Engg. Centre, I.I.T., Madras-36, India
(***) Hydraulic Engg. Dept, I.I.T., Madras-36, India

ABSTRACT

The present study is concerned with numerical and experimental investigations on wave diffraction around bottom-fixed, surface-piercing, rigid, vertical, three-cylinder group of large diameter situated in constant water depth. The numerical analysis is based on the first order diffraction theory. Numerical solutions of the complex velocity potential and associated wave hydro-dynamic forces are obtained using two-dimensional finite elements in the inner domain in combination with infinite elements in the outer domain which directly satisfy the radiation condition. The assembly and solution of the finite element equations are based on the frontal scheme due to Irons (1). Extensive numerical results of wave forces have been obtained for the three cylinder group for various scattering parameters, wave angular approaches and spacing parameters. Experiments have been conducted on model cylinders to verify the validity of the numerical solutions. The present FEM results and experimental results have been compared with the other published analytical solutions. Due to the interference effect the increase in force on one cylinder due to the presence of the other cylinders is also presented.

INTRODUCTION

Offshore engineers are concerned with the effects of diffraction of water waves by large fixed bodies in the ocean and the resulting diffraction forces and structural loadings exerted by incident waves on groups of pilings supporting various types of offshore drilling platforms. Spring and Monkmeyer(2) obtained a solution for the interaction of plane waves with three arbitrary rigid, vertical, circular cylinders at arbitrary wave angles under the conditions in which the inertial forces on the cylinders dominate over the drag forces. In this method a direct matrix solution as well as multiple scattering are used to obtain the velocity potential in the vicinity of the cylinders. The resulting potential function is then applied to calculate force components in the direction of wave advance and orthogonal to it. Chakrabarti(3) extended the work by spring and Monkmeyer and presented analytical results on the wave forces on a three-legged and a four-legged platform. Ohkusu(4) used the method of multiple scattering in which the full scattered wave field is determined by considering separately each scattering event within the cylinder group. McIver and Evans(5) presented an approximate method for the estimation of wave forces on groups of fixed vertical cylinders. The method is based upon a large spacing approximation and involves replacing scattered diverging waves by plane waves. In this paper, we apply the finite element coupled with infinite element method for the solution of the diffraction of waterwaves by three-cylinder group.

MATHEMATICAL FORMULATION OF DIFFRACTION PROBLEM

The diffraction of water waves around solid obstructions such as large offshore structures may be studied using a linear diffraction theory governing the irrotational motion of an incompressible fluid, wherein the wave amplitudes are assumed to be small. Denoting the velocity potential by $\Phi(x,y,z,t)$ and assuming harmonic waves with frequency ω, complex potential $\bar{\phi}$ may be written for the case of constant water depth as :

$$\Phi(x,y,z,t) = \mathrm{Re}\{\bar{\phi}(x,y,z)e^{-i\omega t}\} \tag{1}$$

in which $i = \sqrt{-1}$ and Re{ } implies that only the real part of the quantities inside the bracket has physical meaning. Then the governing equation in terms of $\bar{\phi}$ may be shown to be:

$$\nabla^2 \bar{\phi} = 0 \tag{2}$$

where ∇ denotes the three dimensional laplace operator in the Cartesian (x,y,z) system. At the fluid-structure interface as well as at the seabed, which are assumed to be rigid and impermeable, one has:

$$\frac{\partial \bar{\phi}}{\partial n} = 0 \text{ on } S \tag{3}$$

where S is the rigid surface which is assumed to be stationary and n is normal to the surface S which includes all solid surfaces. Therefore, only the scattering problems are discussed herein. In addition, the free surface boundary condition can be written as :

$$\frac{\partial \bar{\phi}}{\partial z} - \frac{\omega^2 \bar{\phi}}{g} = 0 \text{ at } z = 0 \tag{4}$$

where g is the acceleration due to gravity. Since the fluid domain is unbounded in the plan dimensions, the Sommerfeld radiation boundary condition, which requires the scattered waves to be outgoing, can be expressed as :

$$\lim_{r \to \infty} r^m \left[\frac{\partial \bar{\phi}_s}{\partial r} - ik\bar{\phi}_s \right] = 0 \tag{5}$$

where $m = (P-1)/2$, P being the number of dimensions, and $\bar{\phi}_s$ is the scattered wave potential. By definition, the superposition of scattered waves and incident waves yields the total wave field. Thus for linear waves :

$$\bar{\phi} = \bar{\phi}_i + \bar{\phi}_s \tag{6}$$

For the case of bottom fixed, surface piercing, prismatic cylinders it can be shown that the solution of Eq.(2) is of the form :

$$\bar{\phi} = Z(z)\ \phi(x,y) \tag{7}$$

$$\eta(x,y) = \frac{i\omega}{g} \phi(x,y) \tag{7a}$$

Where ϕ is a two dimensional complex potential and the depth transfer function Z is given by :

$$Z(z) = \text{Cosh } k(h+z)/\text{Cosh } kh \tag{8}$$

where h is the water depth and k is the wave number which is related to ω by the well known linear dispersion relation :

$$\omega^2 = gk \tanh (kh) \tag{9}$$

In view of the solution in Eq.(7), the diffraction formulation in Eqs. (2 to 5) may be replaced by a 2-D formulation. The Laplace equation in (2) is replaced by the classical Helmholtz equation:

$$\frac{\partial^2\phi}{\partial x^2} + \frac{\partial^2\phi}{\partial y^2} + k^2\,\phi = 0 \qquad (10)$$

The boundary condition in Eq.(3) reduces to

$$\frac{\partial\phi}{\partial n} = 0 \quad \text{on } \Gamma \qquad (11)$$

where the curve Γ denotes the intersection of the structure and the still water surface. The general radiation condition in Eq.(5) reduces to

$$\lim_{r\to\infty} \sqrt{r}\left[\frac{\partial\phi_s}{\partial r} - ik\phi_s\right] = 0 \qquad (12)$$

where ϕ_s is scattered part of ϕ [See Eq.(6)]. The reduced governing Eqs.(10 to 12) involving only two dimensions (x,y) are analytically simpler to handle.

VARIATIONAL FORMULATION AND DISCRETIZATION

The solution of the 2-D diffraction problem in Eqs.(10 to 12) along with the free surface boundary condition may be solved in closed form for simpler geometries. However, for cases with multiple structures and complex geometries only numerical solutions are possible. A numerical method for the solution of the diffraction problem using finite and infinite elements has been proposed by Bettess and Zienkiewicz (6). In this approach the problem domain is divided into two regions. The first one is the inner domain encompassing the structure which is essentially the near field region. The remaining far field is treated as the outer domain. The inner domain is discretized using finite elements and the outer domain with infinite elements proposed by Bettess (7). The infinite elements are formulated such that they satisfy the radiation condition in eq.(12). These elements in effect represent the influence of the far field on the near field diffracted pattern in a compact manner. In the finite element analysis over the inner domain the total potential may be used as the field variable, in which case the functional for variational formulation turns out to be :

$$\pi = \iint_{\Omega^i} \tfrac{1}{2}\,\{(\nabla\phi)^2 - k^2\phi^2\}\,d\Omega - \int_\Gamma \phi\left(\frac{\partial\phi}{\partial n}\right) d\Gamma \qquad (13)$$

where Ω^i denotes a domain in the inner region and Γ is the bounding curve. In the outer domain it is convenient to employ the diffracted potential as the variable so that the radiation condition may easily be imposed on the infinite element formulation. Then the appropriate functional for this domain turns out to be:

$$\pi = \iint_{\Omega^o} \tfrac{1}{2} \{ (\nabla \phi_s)^2 - k^2 \phi_s{}^2 \} d\Omega + \oint \left(\frac{\partial \phi_i}{\partial x} \phi_s \, dy - \frac{\partial \phi_i}{\partial y} \phi_s \, dx \right) \quad (14)$$

where Ω^o denotes the domain of an infinite element.

Upon minimising the functionals in Eqs.(13) & (14) defined over a finite or infinite element, as the case may be, using the Rayleigh Ritz technique the element equations for both the domains are obtained. The element property matrices may be assembled following the standard procedure to get system equations in the form

$$\left[\begin{array}{c|c} K_{11} & K_{12} \\ \hline K^T{}_{12} & K_{22} \end{array} \right] \left\{ \begin{array}{c} b \\ \hline d \end{array} \right\} = \{f\} \quad (15)$$

where {b} denotes the nodal variables in the inner domain (excluding those at the interface of the inner and outer domains) and {d} denotes the remaining nodal variables. The load vector {f} vanishes except at the interface nodes. The global equations in (15) are complex and hence complex arithmetic is used for solution. A computer programme for the above 2-D diffraction analysis has been developed which computes wave induced inline and lateral forces on the individual cylinders. The programme has been implemented on IBM 370/155 computer system at Indian Institute of Technology, Madras. The programme uses 6-noded isoparametric triangular and 8-noded isoparametric quadrilateral elements in the inner domain and 9-noded parametric infinite elements over the outer domain. A typical element mesh used in the numerical solution is shown in Fig.1.

EXPERIMENTAL INVESTIGATIONS

Experiments have been conducted on three-cylinder model group (Fig.2) in a laboratory wave flume at the Ocean Engineering Centre, Indian Institute of Technology, Madras, to verify the validity of the numerical solutions. In the present analysis the

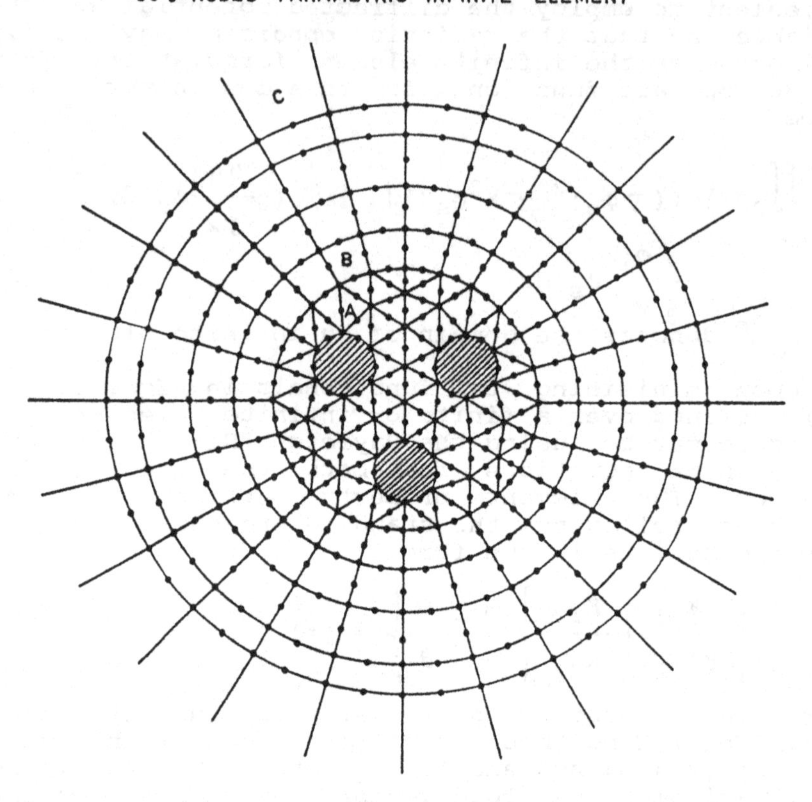

NUMBER OF NODES = 508
NUMBER OF FINITE ELEMENTS = 150
NUMBER OF INFINITE ELEMENTS = 24

FIG.1 THREE- CYLINDER DISCRETIZATION.

following cases (Table 1) with various geometric and wave parameters have been considered for experimental and numerical studies on model cylinders. The water depth, spacing parameter, scattering parameter and wave angular approach are varied as shown in table 1 both for experimental and numerical studies where S is the centre to centre distance between adjacent cylinders, a is the radius of the cylinder and k is the wave number.

Table 1. Parameters for Experimental and Numerical Investigation on Model Cylinders.

Case	Diameter of the cylinder D (cms)	Water depth h(cm)	Spacing parameter S/a	Range of scattering parameter ka	Direction of wave propagation θ (deg)
3 CYL. group	20	50&100	4,5	0.1to 1.6	0,90 & 180

Inline forces on the individual cylinders and incident wave heights are measured for the various parameters. The force measurement is repeated by varying the space between the cylinders and the orientation of the cylinder group with respect to the wave propogation direction.

RESULTS AND DISCUSSION

FEM results for the variation of the force ratio R on the leading cylinder (cyl.3) of the three-cylinder group for $\theta=0^o$ and $S/a = 4$ is compared with the experimental results in Fig.3, where R is the ratio of the force on the leading cylinder of the three-cylinder group to the corresponding force on the single -cylinder case. Out of the three wave angular approaches ($\theta = 0^o, 90^o$ & 180°) and two spacing parameters ($S/a = 4,5$) considered the case with $\theta=0^o$ and $S/a=4$ gives the largest increase in force ratio $R = 1.51$ and exhibits marked fluctuations between maximum and minimum values of R with ka. Agreement between the numerical predictions and experiment is satisfactory for all the cases considered. The variation of force ratio R with scattering parameter ka for the leading cylinder of the three-cylinder group is shown in Fig.4 for $\theta=0^o$ and $S/a = 5$, which is typical of some gravity structures. This figure compares the present FEM results and experimental results with the approximate analytical solution due to Spring and Monkmeyer (2) and also with the results due to McIver and Evans (5). The correlation obtained

FIG. 2 THREE-CYLINDER MODEL ($\theta=0°$, $S/a=5$)

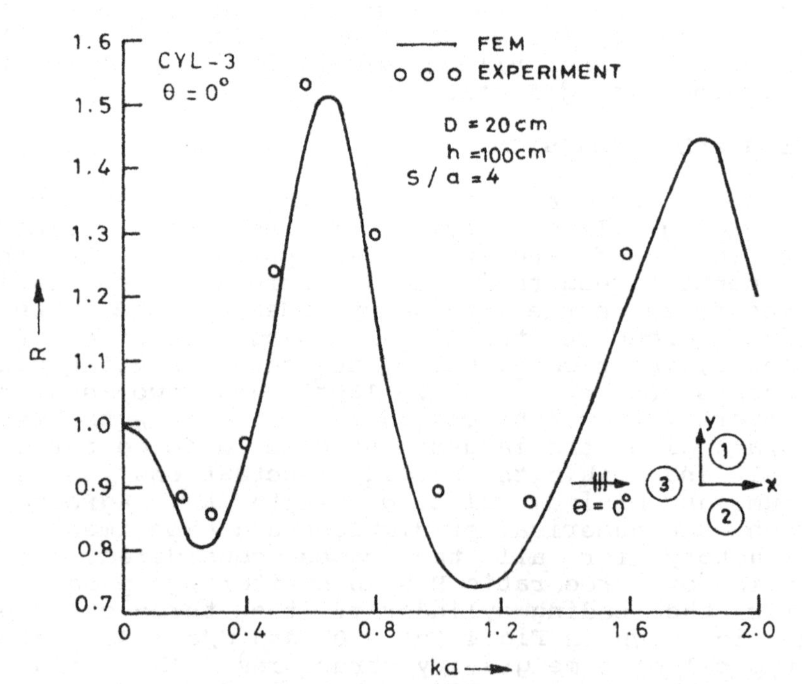

FIG. 3 COMPARISON OF NUMERICAL PREDICTIONS WITH EXPERIMENT SHOWING THE VARIATION OF FORCE RATIO R WITH ka FOR $S/a=4$ (3-CYL. CASE)

among the four methods is generally good. The results obviously exhibit an oscillating trend of the force ratio R with respect to ka, R being maximum at ka = 0.5.

Numerically evaluated maximum inline and lateral non-dimensional wave forces on large diameter three-cylinder group are graphically presented in Figs. 5 to 7 for $\theta=0^{o}$, 90° and 180° respectively for S/a = 4. In all these cases the results are compared with the non-dimensional wave force on a 20m diameter isolated vertical circular cylinder. In Fig.5 it is observed that the peak inline force on the leading cylinder for $\theta=0^{o}$ and S/a=4 is about 42 percent larger than the peak force experienced by an isolated cylinder which occurs at different ka values, thus showing the importance of interference effect brought about by two trailing cylinders. On the other hand the two trailing cylinders experiences only 6 percent increase of force over the isolated cylinder value. Considering the case with $\theta=0^{o}$ and S/a = 5 the increase of the peak inline force on the leading cylinder is about 34 percent over the isolated cylinder peak value, the increase being smaller compared to the case with S/a = 4. Due to the symmetry of the geometry about x-axis the inline and lateral forces on the two trailing cylinders are identical.

For $\theta=90^{o}$ and S/a = 4 (Fig.6) the leading cylinder experiences the highest increase in the peak inline force of about 31 percent for this case. For the same case with S/a = 5 the increase in the peak inline force on the leading cylinder is of the order of only 21 percent. For $\theta=180^{o}$ and S/a = 4 (Fig.7) it is observed that the increase of peak force on the two leading cylinders over an isolated cylinder peak force is only about 10 percent. Apparently the single trailing cylinder has less of interference effect than in the previous cases with $\theta=0^{o}$ and 90°. The peak force on the trailing cylinder is almost of the same magnitude as that of an isolated cylinder.

CONCLUSIONS

From the results of this study the following conclusions are drawn :

i) It is generally observed that the peak horizontal wave force on an isolated cylinder and on any one member of a group of cylinders occur at different ka values over a given range of scattering parameter ka. For all the cases studied the peak

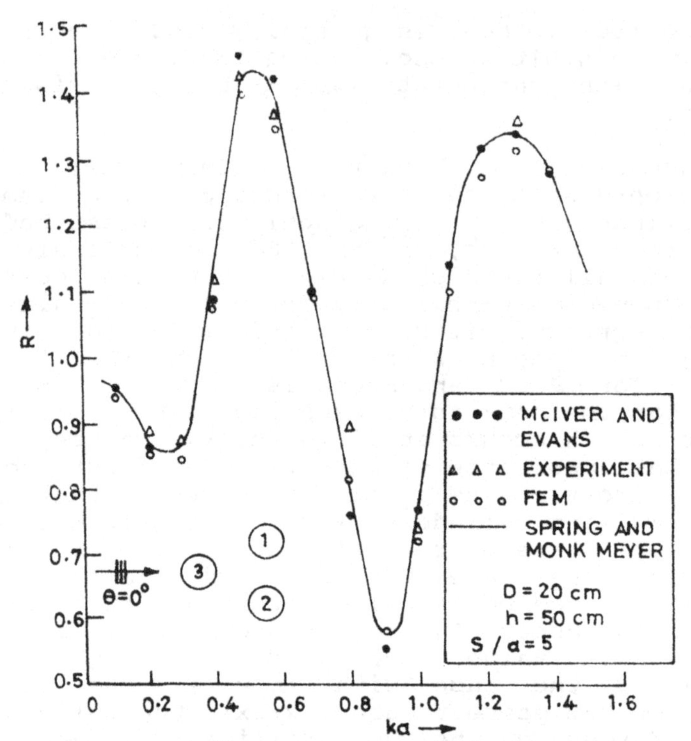

FIG. 4 VARIATION OF INLINE FORCE RATIO R FOR S/a= 5 (FOR CYL-3 OF THREE-CYLINDER CASE)

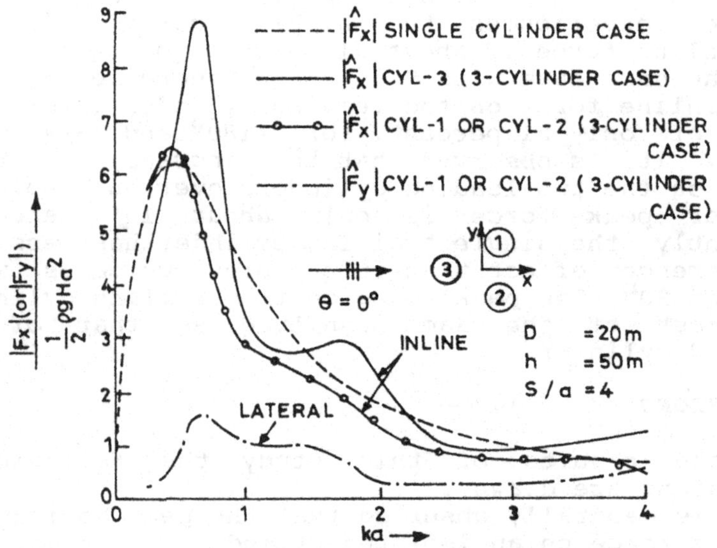

FIG.5 NON-DIMENSIONAL FORCE FOR THREE-CYLINDER CASE (S/a=4, θ=0°) COMPARED WITH SINGLE CYLINDER CASE

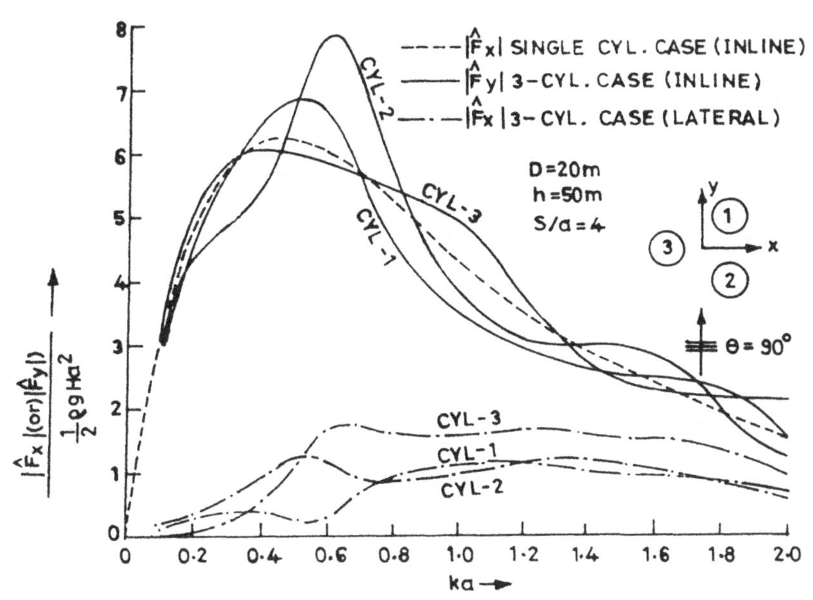

FIG. 6 NON-DIMENSIONAL FORCE FOR THREE-CYLINDER CASE (S/a=4, θ=90°) COMPARED WITH SINGLE-CYLINDER CASE.

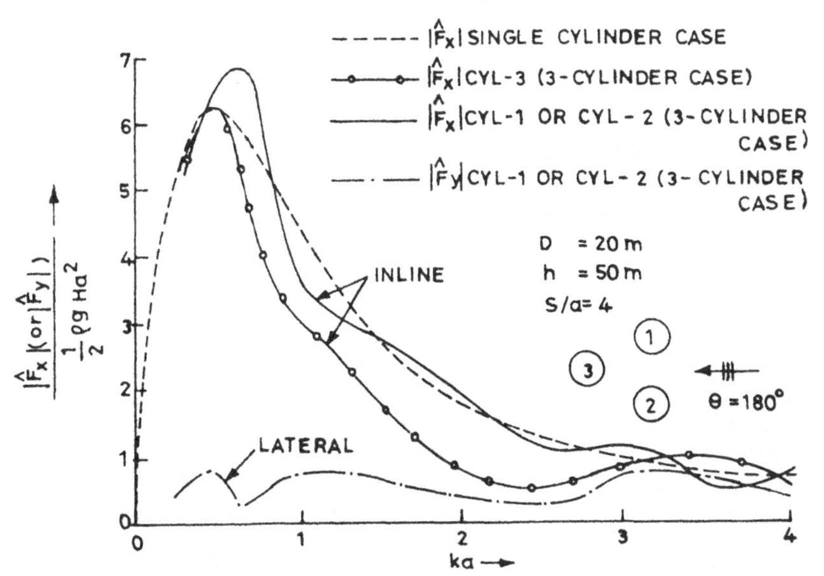

FIG.7 NON-DIMENSIONAL FORCE.THREE-CYLINDER CASE (S/a=4, θ=180°) COMPARED WITH SINGLE-CYLINDER CASE

force on the multiple cylinder shifts towards higher ka value when compared to an isolated cylinder, in some cases the shift being very pronounced.

ii) When the spacing parameter S/a for the group of cylinders is decreased the peak horizontal force on the leading cylinder increases significantly, whereas the force on the trailing cylinder is less sensitive, although marginally larger than the single cylinder value.

iii) As for the numerical technique used, it has been found that, for better accuracy, atleast 3 to 4 elements must span half the wave length.

iv) For the three-cylinder case the increase in force is most significant when one cylinder is directly in front of the remaining two cylinders for $\theta=0^o$. Out of the three wave angular approaches ($\theta=0^o$, 90° & 180°) and two spacing parameters (S/a = 4 and 5) considered, the interference effect is severe on the leading cylinder of the three-cylinder case for $\theta=0^o$ and S/a = 4. For this case there is 51% increase in force on the leading cylinder compared to an isolated cylinder at ka=0.65 (Fig.3), whereas the increase in peak force on the leading cylinder when compared to the peak force on an isolated cylinder is 42% (Fig.5). This obviously brings out the predominant interference effect on the three-cylinder case.

REFERENCES

1. Irons, B. A frontal solution program for finite element analysis. Int. J. num. Meth. Eng. Vol.2, pp.5-32, 1970.
2. Spring, B.H. and Monkmeyer, P.L. Interaction of plane wave with vertical cylinders. Proc. 14th Int. Conf. on Coastal Eng. Copenhagen, Denmark, ASCE, pp.1828-47, 1974.
3. Chakrabarti,S.K. Wave forces on multiple vertical cylinders. J. Waterway, Port, Coastal and Ocean Division, ASCE, 104 (WW2), pp.147-161, 1978.
4. Ohkusu, M. Wave action on groups of vertical circular cylinders. J. Soc. Nav. Arch. Japan., Vol.11, 37, 1973.
5. McIver, P. and Evans, D.V. Approximation of wave forces on cylinder arrays. Applied Ocean Research, Vol.6, No.2, pp.101-107, 1984.
6. Bettess, P. and Zienkiewicz, O.C. Diffraction and refraction of surface waves using finite and infinite elements. Int. J. Numer. Methods in Eng. Vol.2, 1271, 1977.
7. Bettess, P. Infinite elements. Int. J. Numer. Methods in Eng. Vol.11, pp.53-64, 1977.

Wave Breaking over a Submerged Plate - A Numerical Study

C.M. Lemos

Department of Coastal and Estuarine Dynamics, Instituto Hidrografico, Rua das Trinas 49, Lisbon, Portugal

ABSTRACT

In this work, the performance of a submerged breakwater consisting of a submerged horizontal plate was investigated, using a numerical model capable of treating arbitrary free-surface configurations, including breaking waves. It was found that for the given conditions of incident waves, breakwater dimensions, and placement, the device has a hydrodynamic efficiency of 0.75. The mechanisms responsible for this high efficiency were clearly identified. The computed vertical wave loading on the structure was much higher than linear estimates predicted. The results of the present study are part of a research program on the coastal protection works for the Barcelona 92 Olympic Marina. Some future developments are also briefly discussed.

1. Introduction.

The Olympic Games of Barcelona 92 will be a major world event. This fact is reflected in the extensive coastal protection works presently being built at the Olympic marina (Barcelona) in the coast of Catalonia (Spain). Since the coast of Catalonia is plagued with erosion problems, the organizers of Barcelona 92 turned their attention to the possibility of using alternative shore protection structures which lower the wave height significantly without disrupting the sediment balance.

A submerged horizontal plate is an example of these alternative devices, which for certain conditions of incident waves, and placement, can be very effective (Guevel et al., [2]). However, the use of submerged breakwaters in coastal protection has been precluded by several difficulties. First, the performance of the device is strongly dependent on the ratio of the wave length to the plate length, relative immersion of the plate, and even the thickness of the device. Secondly, the hydrodynamic loading on the structure may be very strong, requiring expensive construction to avoid collapse.

The purpose of this work is to evaluate the transmission coefficient, the hydrodynamic forces, and the qualitative characteristics of the flow around a submerged breakwater with a scale of 1/20 relative to the device proposed for the Barcelona Olympic marina, using a numerical model capable of treating strong wave deformations before and after breaking. Wave flume experiments of the reduced model were available for comparison. The mechanisms responsible for the efficiency of the submerged breakwater were clearly identified. It was found that the transmitted waves were irregular, even though the boundary conditions for the incident waves were periodic. The numerically computed transmission coefficient was very close to the experimental value. The numerically computed wave loadings were much higher than suggested by simple linear estimates, with significant implications for the structural design of the device.

2. Problem definition.

The problem is sketched in Figure 1, where $\underline{H_i}$ is the height of the incident waves; $\underline{H_r}$ is the height of the reflected waves; $\underline{H_t}$ is the height of the transmitted waves; $\underline{h}$ is the still water depth; $\underline{z_p}$ is the immersion of the top face of the plate; $\underline{e}$ is the thickness of the plate; and $\underline{L_p}$ is the length of the plate. The efficiency of the device as a wave attenuator is due to three mechanisms (Kojima et al. , [3]). The first is the interaction between the waves passing over the plate (region II) and the flow underneath (region IV), which results in more or less strong reflections (region I). The second is wave breaking and turbulence production, as the incident waves are suddenly intercepted by the plate. The third is wave disintegration due to nonlinear effects (region III). The relative importance of these factors is strongly dependent on the conditions controlling wave propagation (height, period, and depth) for given breakwater dimensions and placement. Also, strong hydrodynamic loadings are to be expected, since the disruption of wave orbital motion by the plate is accompanied by strong accelerations in the fluid, and by significant frictional forces (causing decay of wave momentum and energy). A qualitative and quantitative determination of the hydrodynamic behaviour of the plate requires a numerical model which is able to treat strongly distorted waves, during and after breaking.

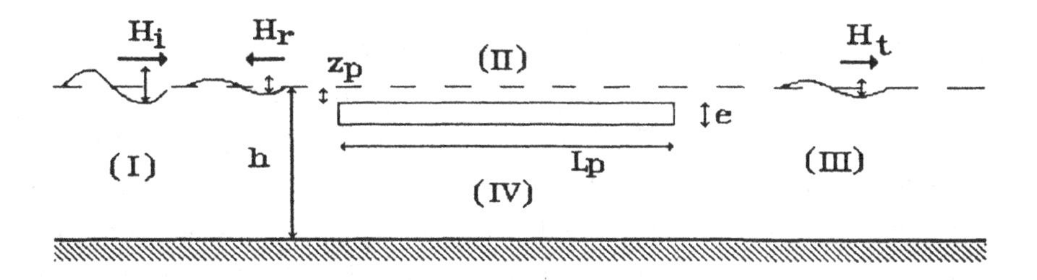

Figure 1. Definition sketch for the submerged breakwater problem.

3. Theoretical formulation.

The numerical model used in the present study is based on the two dimensional Navier-Stokes equations for incompressible flow, together with an additional equation which defines the fluid configuration (Nichols & Hirt, [7]; Nichols et al. , [8]):

$$\partial \vec{u} / \partial t + \vec{u} \cdot \nabla \vec{u} = -\nabla \phi + \nu \nabla^2 \vec{u} + \vec{g} \tag{1}$$

$$\nabla \cdot \vec{u} = 0 \tag{2}$$

$$\partial F / \partial t + \nabla \cdot (\vec{u} F) = 0 \tag{3}$$

where $\vec{\underline{u}}=(u, w)$ is the velocity vector, $\phi=p/\rho$ is the kinematic pressure (pressure divided by constant density), ν is the kinematic viscosity, $\vec{\underline{g}}=(0,-g)$ is the acceleration due to gravity, $\nabla\equiv(\partial/\partial x, \partial/\partial z)$, and $\underline{F(x,z,t)}$ is a volume-of-fluid function whose value is 1 for a point inside the fluid and 0 elsewhere. Hence, the average value of F in a computational cell is equal to the fractional volume of the cell which is "wet". Cells with F=1 are full of fluid, and cells with F=0 are empty. Cells with values of F between 0 and 1 contain a free surface. In this way, free surfaces with arbitrary time-dependent configurations can be represented efficiently in numerical algorithms for incompressible fluid dynamics. Equation (3) states the Lagrangian invariance of F, in an Eulerian representation.

The governing equations are solved by a finite-difference method, using a stretched, and staggered, cartesian grid. The momentum equations are advanced in time using an explicit scheme with third-order spatial accuracy for the convection term. The continuity equation is solved using a pressure-velocity iteration method. The volume-of-fluid advection equation is advanced in time using the donor-acceptor flux approximation introduced by Nichols and Hirt [7], which avoids the smearing of the interface resulting from the computation of convective fluxes in Eulerian difference schemes, while maintaining stability and overall volume conservation.

The free surface is identified as the transition between fluid-occupied and void regions. In cells containing a free surface, the appropriate boundary conditions are imposed. The numerical algorithm is able to treat any surface configuration, including overturning and broken waves, without restrictive approximations. Internal obstacles can be defined by blocking out any desired combination of mesh cells. Further details of the numerical model can be found in Lemos [4, 5].

4. Numerical results.

The physical parameters used in the numerical simulations were chosen to match a 1/20 scale model of the breakwater for the Barcelona 92

Olympic marina, which was tested in a wave flume by Océanide, France (unpublished report). The water depth was specified h=50 cm. The dimensions of the plate were length L_p=80 cm, and thickness e=5 cm. The immersion of the top face of the slab was z_p=7.5 cm below the still water level. The plate was positioned in the middle of a computational domain 13 m long and 0.75 m high, leaving a distance of approximately 3 wave lengths free on either side of the plate for allowing the reflected and transmitted waves to stabilize. The computational domain was discretized using a mesh of 250 by 30 cells. The molecular viscosity was set to $\nu=10^{-4}$ m^2s^{-1} to ensure stability, the acceleration due to gravity was g=9.81 ms^{-2}, and the time step was $\Delta t=10^{-3}$ s. At the left boundary, periodic waves with period 1.4 seconds and height 12.5 cm were generated by specifying the velocity components and the water level according to second-order Stokes theory. Around the slab, free slip boundary conditions were imposed, due to the impossibility of solving the details of the boundary layer with the finite difference mesh used. This choice rules out the calculation of the frictional drag originated in the boundary layer around the slab. However, since in prototype conditions the flow will be turbulent, the drag forces will not follow the Froude similarity, and thus the drag computed in the reduced model cannot be extrapolated to prototype conditions.

Figure 2 shows two “snapshots” of the velocity field, separated by one half of the wave period, for fully developed flow conditions. The first of these (top) shows vigorous breaking of an incident wave coming from the left, after being intercepted by the plate. There was a massive plunging jet, striking the thin layer of fluid above the plate left by the passage of the previous wave. Near the end of the plate, a strong vortex was observed. Associated with the flow around this vortex, a small wave propagating to the left was formed near the end of the structure. The collision of this reflected wave with the incident wave greatly enhances dissipation of the incident wave energy. This feature was clearly observed in video recordings (unpublished) of the physical experiments. The second picture (bottom) shows the same wave after passing by the structure. Nonlinear deformation was violent, and the vortex system was replaced by a massive surface shear layer with very high fluid velocities near its "bump". Although the height of the transmitted wave is still relatively high, the flow in the transmission region was highly rotational (Figure 3). Hence the transmitted waves are subject to strong dissipation, since the rate of dissipation is proportional to the viscosity times vorticity squared.

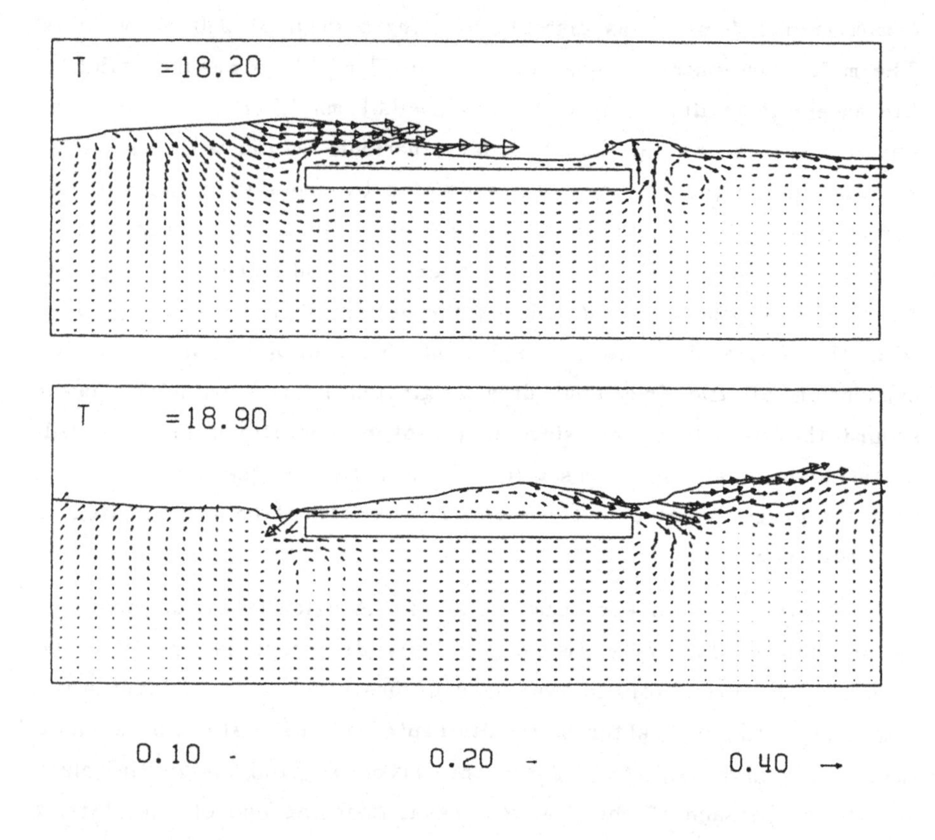

Figure 2. Velocity field near the submerged plate.

The flow under the trough of the transmitted wave generated a suction jet at the front of the plate, and a clockwise vortex underneath. These localized jets and vortices also contribute to dissipation.

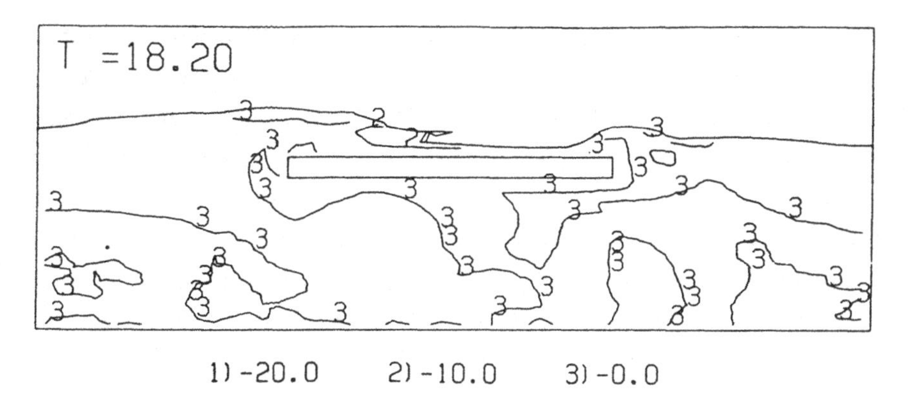

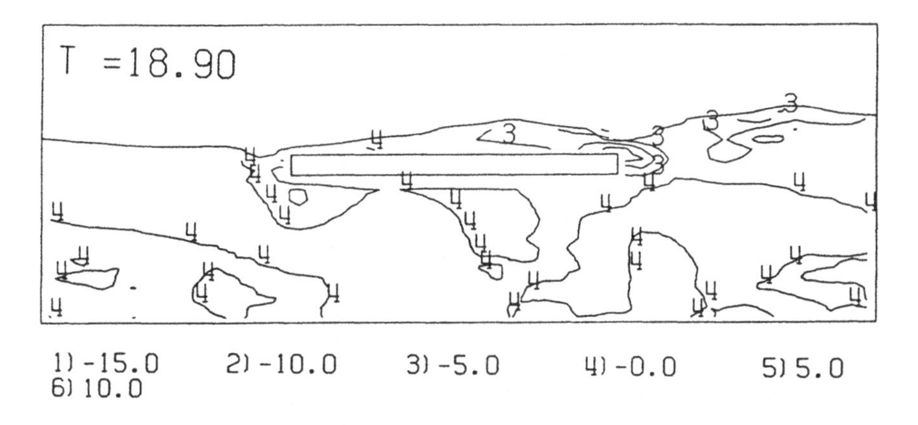

Figure 3. Vorticity contours around the submerged plate.

Figure 3 shows the corresponding vorticity contour plots, i.e. $\vec{\omega}=\nabla\times\vec{u}$ for t=18.2 s (top) and t=18.9 s (bottom). The highest values of $|\vec{\omega}|$ were concentrated in the regions of high shear, as expected. Other concentrations of vorticity, either positive or negative, showed up in regions where counterclockwise or clockwise vortices were observed in the velocity plots. In regions where the flow was nearly irrotational, the vorticity was small and spatially uncorrelated. The vorticity plots also reveal that free-slip conditions were effectively used around the plate, otherwise vortex sheets would have been observed. Thus, for the purpose of realistic calculation of the horizontal drag force, it is necessary to refine the mesh near the plate and specify a no-slip or partial-slip (law-of-the-wall) condition. But in this work the interest was mainly concentrated on the vertical forces (see § 6). The numerical results indicate that breaking and breaker-induced vorticity and dissipation are the main agents responsible for the hydrodynamic efficiency of the breakwater. There is some reflection at the end of the plate, but this appears to be originated by the pulsating vortex structure in the transmission region, rather than by interference between the flow above and under the plate. The calculation here reported was performed in an IBM-PC 386 computer running at 33 MHz.

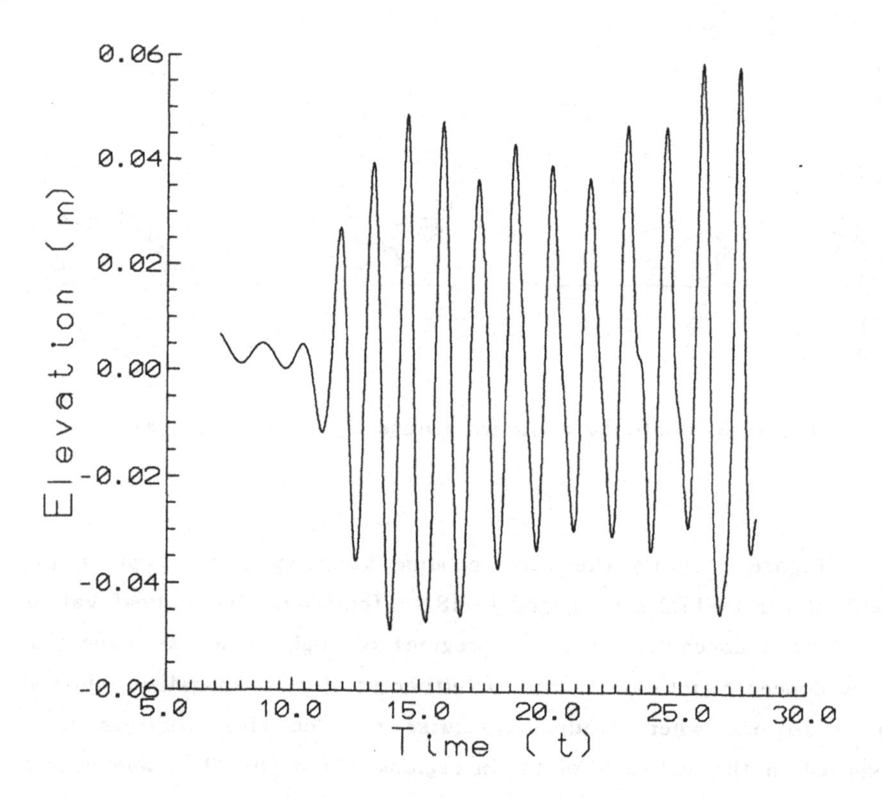

Figure 4. Time history of the water surface elevation near the right boundary.

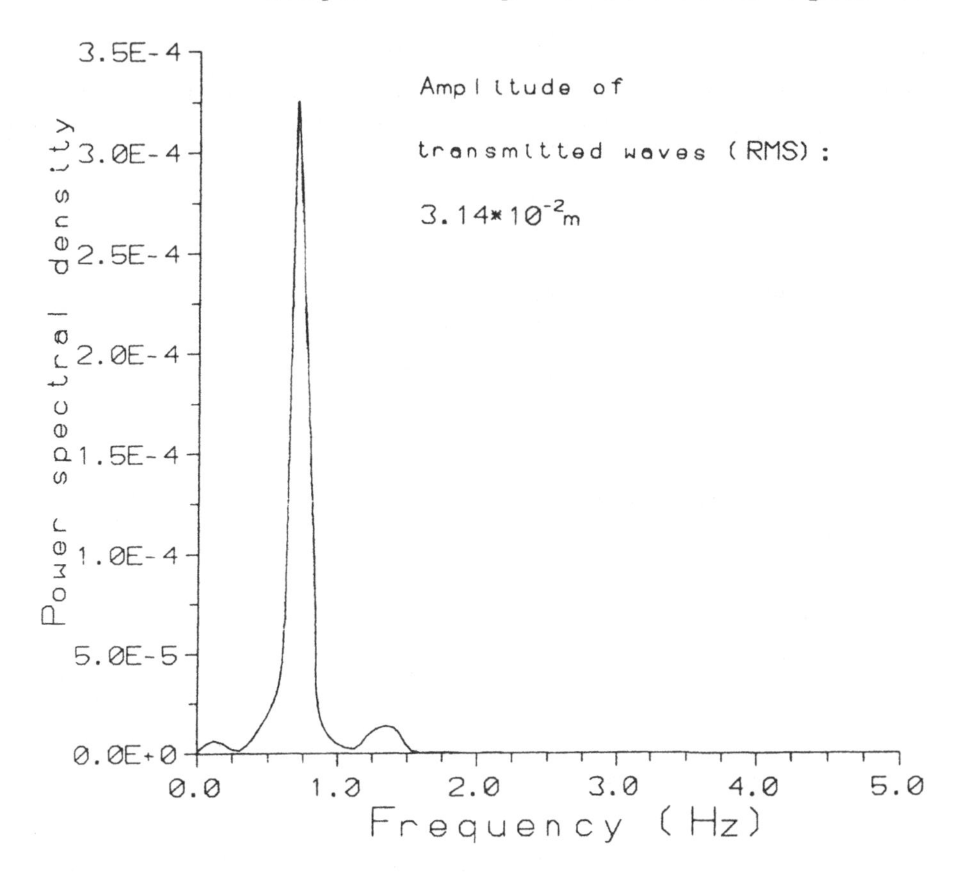

Figure 5. Power spectral density (PSD) estimator and RMS amplitude of the transmitted waves.

5. Transmission coefficient and hydrodynamic efficiency.

Figure 4 shows the time history of the water surface elevation, taken at the center of the column of computational cells near the right (radiative) boundary. It is observed that, due to nonlinear effects, the transmitted waves were irregular, even though the boundary conditions for the incident waves were periodic. The (RMS) amplitude was determined by taking the periodogram of the data in figure 4 using standard FFT routines, and using Welch's window to minimize leakage

(Press et al. , [10]). Figure 5 shows the power spectral density (PSD) estimator and RMS amplitude of the transmitted waves. The transmission coefficient was

$$C_t = \left(\frac{0.0314}{0.0625}\right)^2 \simeq 0.25$$

and the efficiency of the device was

$$E = 1 - C_t \simeq 0.75$$

This efficiency could be expected in view of the strength of the breaking. The experimentally observed value (Océanide, unpublished) was 0.79 ($C_t \simeq 0.21$).

6. Wave loading.

In this work, the interest was mainly concentrated on the vertical force exerted by the flow on the plate. The vertical force was calculated using the following formula :

$$F_z = \int_0^{L_p} \{(p - \rho g z)_{bottom} - (p - \rho g z)_{top}\} dx \tag{4}$$

i.e. only the dynamic pressure is taken into account. Due to the structure of the staggered mesh, the integral was evaluated most simply using the midpoint rule. Figure 6 shows the variation of the total vertical force, which is the relevant parameter for the design of the submerged breakwater, was found to be nearly 400 N/m, and typical peaks had a value ~280 N/m. These values are 38 and 27 times higher than the linear estimate (e. g. Dean & Dalrymple, [1]):

$$F_z = \rho \frac{4}{k^2} \sinh(k \frac{e}{2}) \sin(k \frac{L_p}{2}) \frac{\partial w}{\partial t} \tag{5}$$

where ρ is the water density, k is the wave number, and $\partial w / \partial t$ is evaluated using linear theory. This is not surprising, for the linear estimate does not take into account the added-mass inertia force and the vortex drag force (Lighthill, [6]), which on a physical basis are seen to be dominant in this situation. The large vertical wave loadings have a significant impact on the structural design of the device.

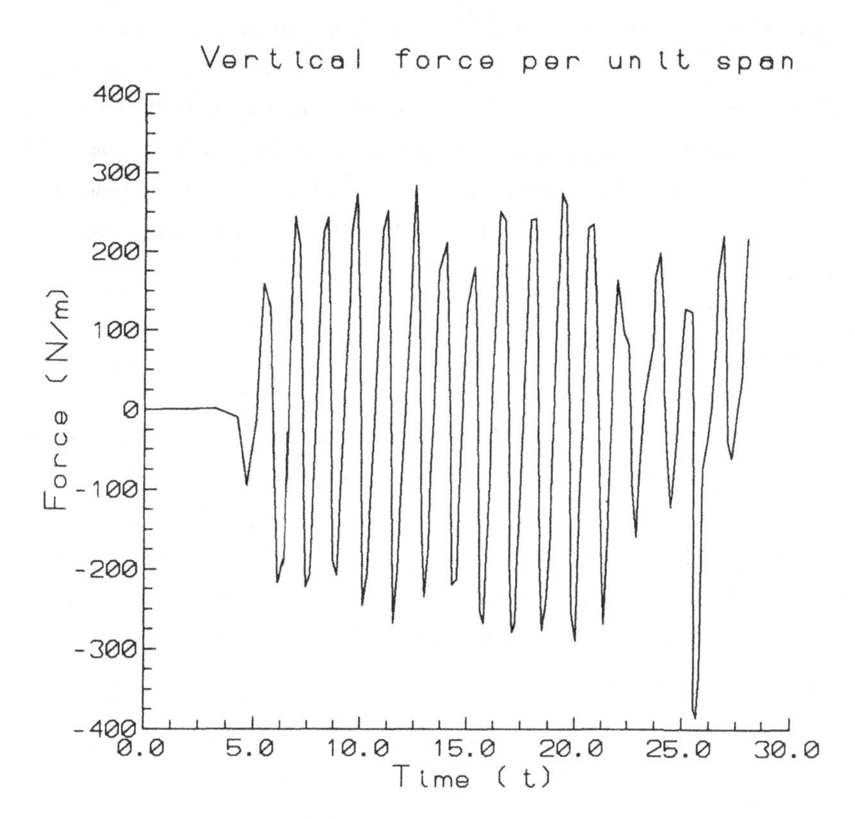

Figure 6. Time history of the vertical force on the plate.

7. Conclusions and future developments.

The hydrodynamic efficiency of a breakwater, consisting of a submerged horizontal plate, was investigated using a numerical model capable of treating breaking waves. It was shown that dissipation of wave energy by breaking and by vortex motion is the key factor responsible for the efficiency of the device, and not other mechanisms that have been proposed in previous studies. The computed transmission coefficient was 0.25, in fair agreement with the value found in physical experiments. The computed vertical forces were rather large, with peaks of up to 40 times higher than predicted by an approximate linear theory.

Some improvements now under way are: implementation of a new boundary condition around the plate, possibly using the capability of the model for representing turbulence effects (Lemos, [4, 5]); estimation of the reflected component using a special version of Orlanski's [9] technique; irregular wave input, to determine more realistic peaks of wave heights and loading; and a parametric study of the transmission coefficient and wave loading for different periods and relative immersions of the plate.

Acknowledgements

I am grateful to Mr. J. R. Clascá and to Dr. J. L. Monsó of EUROPROJECT (Barcelona, Spain) for suggesting this study, and for their help and interest during all stages of this work. This work is part of an EUREKA research program on coastal protection structures. Thanks are also due to Dr. Daniel A. Rodrigues of the University of Lisbon (Portugal), for his constant help and advice in the elaboration of the numerical model.

References

1. Dean, R. G. and Dalrymple, R. A. (1984). *Water wave mechanics for scientists and engineers*. Prentice Hall, Englewood Cliffs.

2. Guevel, P. , Landel, E. Bouchet, R. and Manzone, J. (1985). *Le phénomène d'un mur d'eau oscillant et son application pur portéger un site côtier soumis à l'action de la houle*. ATMA 1985, Principia Recherche Développment S. A. , 18 p.

3. Kojima, H. , Ijima, T. and Yoshida, A. (1990). *Decomposition and interception of long waves by a submerged horizontal plate*. 22th Int. Conf. on Coastal Eng. , ASCE, Delft, paper No. 263.

4. Lemos, C. M. (1992a). *Wave breaking—A numerical study*. Lecture

Notes in Engineering, vol. 71, Springer Verlag (to appear).

5. Lemos, C. M. (1992b). *A simple numerical technique for turbulent flows with free surfaces.* Int. Jour. for Num. Meth. in Fluids, to appear.

6. Lighthill, M. J. (1979). *Waves and hydrodynamic loading.* Proc. 2nd. Int. Conf. Behaviour of Off-shore Structures, vol. 1 , pp 1-40.

7. Nichols, B. D. and Hirt, C. W. (1975). *Methods for calculating multi-dimensional, transient free-surface flows past bodies.* Proc. 1st Int. Conf. Num. Ship Hydrodynamics, Gaithersburg, Maryland, October 1975.

8. Nichols, B. D. , Hirt, C. W. and Hotchkiss, R. S. (1980). *SOLA-VOF: A solution algorithm for transient fluid flow with multiple free boundaries.* Los Alamos Scientific Laboratory report LA-8355, 1980.

9. Orlanski, I. (1976). *A simple boundary condition for unbounded hyperbolic flows.* Jour. Comp. Phy. , vol. 21, pp 251-269.

10. Press, W. H. , Flannery, B. P. , Teukolsky, S. A. and Vetterling, W. T. (1989). *Numerical Recipes—The Art of Scientific Computing* (FORTRAN Version). Cambridge University Press.

Numerical Modelling of Wave Field in Harbours Using Boundary Element Method

M.N. Anwar, M. Sayed

Department of Engineering Mathematics, Faculty of Engineering, Alexandria, Egypt

ABSTRACT

The boundary element approach is implemented to investigate the nature of wave field inside harbours. The problem is mathematically formulated within the context of linear theory. Numerical results are obtained using constant elements for the case of an idealized circular harbour and for a more general practical situation.

INTRODUCTION

The adequate engineering planning and development of harbours is a subject of great importance. The environmental impact assessment of suggested coastal structures or other engineering development activities in harbour facilities is one of the major issues that affect the final decision making regarding approval of such designs. A first step toward achieving this goal is to develop an efficient and flexible tool to analyse the wave pattern within the study area. Although physical modelling has its own merits but for financial reasons such an approach may not be implemented without severe limitations. Numerical modelling, on the other hand,has been well established as a powerful approach to investigate several coastal engineering problems. Among the numerical methods that has attracted the interests of many authors is the boundary element method. This is due to its evident numerical efficiency, flexibility and convenience for the investigation of engineering problems. The method has been successfully implemented to analyse a number of coastal engineering problems, among these studies, we mention the work of Ijima and Youshida [4] , Grilli and Lejeune [3], Dermiral [2], and Mansur and Brebbia [6] .

The present paper is concerned with determining the waves response and the amplification nature of surface gravity waves entering a harbour of almost uniform depth but with general geometrical configuration. The problem is mathematically formulated assuming that the motion is irrotational and the fluid is taken as an inviscid incompressible one. The solution is developed based on boundary element formulation using constant elements for an idealized circular harbour and for a practical case representing an existing harbour in Alexandria Egypt.

MATHEMATICAL FORMULATION OF THE PROBLEM

The fluid is assumed to be inviscid and incompressible. Cartesian coordinates are used with x, y axes are horizontal and z axis vertically upwards. The governing equations are given by

$$\nabla^2\phi = 0 \quad \text{in the fluid} \tag{1}$$

$$\frac{\partial \tau}{\partial t} = \left.\frac{\partial \phi}{\partial z}\right|_{z=0} \quad \text{on the free surface} \tag{2}$$

$$\tau = -\frac{1}{g}\left.\frac{\partial \phi}{\partial t}\right|_{z=0} \tag{3}$$

The last two equations can be combined to give

$$\frac{\partial^2\phi}{\partial t^2} + g\frac{\partial \phi}{\partial z} = 0 \quad \text{at } z = 0 \tag{4}$$

The boundary conditions to be satisfied at the bottom assuming the water to be of uniform depth h are :

$$\left.\frac{\partial \phi}{\partial z}\right|_{z=-h} = 0 \tag{5}$$

$$\frac{\partial \phi}{\partial n} = 0 \quad \text{on the fixed solid boundary} \tag{6}$$

where $\tau(x,y;t)$ is the free surface elevation and g is the gravitational acceleration.

Since water depth is uniform, we can introduce the velocity potential ϕ in the form

$$\phi(x,y,z;t) = \frac{1}{iw}\ \psi\ (x,y)\ Z(z)\ e^{-iwt} \tag{7}$$

The boundary condition at the bottom given by equation (5) becomes

$$\left.\frac{dZ}{dz}\right|_{z=-h} = 0 \tag{8}$$

Substituting the above expression of ϕ into Laplace's equation (2), we get

$$\frac{\partial^2\psi}{\partial x^2} + \frac{\partial^2\psi}{\partial y^2} + k^2\psi = 0 \tag{9}$$

Two dimension helmholtz equation, and

$$\frac{dZ}{dz} - K^2\psi = 0 \tag{10}$$

where K is the wave number. The solution of equation (10) satisfying the boundary condition (8) can be written as

$$Z(z) = B \cosh (K(z+h)) \tag{11}$$

where B constant to be determined.

After substituting from (11) into (7) the velocity potential $\phi(x,y,z;t)$ will have the form

$$\phi(x,y,z,t) = \frac{B}{iw} \psi(x,y) \cosh K(z+h) e^{-iwt} \tag{12}$$

Since the solid surface is parallel to z axis, then by using equation (12), the boundary condition (5) can be written as

$$\frac{\partial \psi}{\partial n} = 0 \qquad \text{on the fixed solid} \tag{13}$$

We can assume that at infinity, the influence of the harbour is minimal and that $\psi(x,y)$ is equal to $\psi_0(x,y)$ which is given by

$$\psi_0(x,y) = \cos (Kx \cos \beta) e^{-iky \sin \beta} \qquad 0<\beta<\pi \tag{14}$$

where ψ_0 is a straight crested standing wave with the crest inclined by an angle β to the shore line.

The problem now is to find the solution of the equation (9) that satisfies the boundary conditions (13) and (14).

INTEGRAL EQUATION FORMULATION

The scattered wave caused by the presence of the boundary of the harbour can be written in the form of contour integral as

$$\psi(x,y) = \int_S f(x_1,y_1) \; G \; (x,y;x_1,y_1) ds \tag{15}$$

where f(x,y) is source function that depends on the boundary condition given by equation (13), Carrier and Pearsm [1] . The Green's function is chosen to be such that

$$G(x,y;x_1,y_1) = - 1/4 \; i \; H_0^1(KR) \tag{16}$$

$$R = \sqrt{(x-x_1)^2 + (y-y_1)^2} \tag{17}$$

and H_0^1 (KR) is a Hankel function of the first kind.

Since disturbance due to the harbour at infinity is vanishing, thus the solution of equation (9) can be written as

$$\psi \; (x,y) = \psi_0 \; (x,y) + \int_S f(x_1,y_1) \; G \; (x,y;x_1,y_1) ds \tag{18}$$

At any point (x',y') on the boundary

$$\lim_{x,y \to x',y'} \left[\frac{\partial \psi_0(x,y)}{\partial n} + \frac{\partial}{\partial n} \int f(x_1,y_1)\, G\,(x,y,x_1,y_1) ds \right] = 0 \quad (19)$$

Since $H_0^1(KR)$ is singular for a small value of the argument, the following limiting relation is used

$$\frac{1}{4}\, i\, H_0^1(KR) \longrightarrow \frac{1}{2\pi} \ln \left(\frac{KR}{2} \right) \quad (20)$$
$$\text{as } KR \longrightarrow 0$$

We consider the path of the integral to be along the boundary except at the point (x',y') where the contour is deformed into a small circle of radius ϵ.

Hence the equation (19) can be written as

$$\frac{\partial \psi_0}{\partial n}(x',y') + \lim_{x,y \to x',y'} \frac{\partial}{\partial n} \left[f(x_1,y_1)\, G\,(x,y;x_1,y_1) ds \right] = 0$$

$$\frac{\partial \psi_0}{\partial n}(x',y') + \int_S f(x_1,y_1)\, G_n(x',y';x_1,y_1) ds +$$

$$\lim_{x,y \to x',y'} \int_S f(x_1,y_1)\, Gn(x,y;x_1,y_1) ds = 0 \quad (21)$$

From equation (20) as $R \longrightarrow 0$, it follows that,

$$\lim_{R \to 0} \frac{1}{2\pi} f(x',y') \int_0^{\pi} \frac{\partial}{\partial R} \ln \left(\frac{KR}{2}\right) R\, d\theta = 1/2\, f(x',y')$$

Thus equation (21) can be written as Liggett [5]

$$\frac{\partial \psi_0}{\partial n}(x',y') + \int_S f(x_1,y_1) G_n(x',y';x_1,y_1) + \frac{1}{2} f(x',y') = 0 \quad (22)$$

where

$$G_n\,(KR) = -\frac{1}{4} \frac{\partial}{\partial n} H_0^1(KR) \quad (23)$$

$$R = \sqrt{(x'-x_1)^2 + (y'-y_1)^2} \quad (24)$$

NUMERICAL SOLUTION

It is clear that the solution $\psi(x,y)$ for equation (9) is contingent on generating a solution for f(x,y) in equation (22). This is achieved numerically by dividing the boundary S into finite number of elements N defined by a fixed number of nodes. The source function in each element is approximated by

$$f(x,y) = \sum N_i(x,y)\ f_i \tag{25}$$

Where $N_i(x,y)$ are the shape element functions and f_i are the unknown nodal value of $f(x,y)$.

Substituting equation (25) into equation (22) we get

$$\frac{\partial \psi_0}{\partial n}(x_i,y_i) + \sum_e \sum_i f_i \int G_n(KR_i) N_i(x,y)\ ds + \frac{1}{2} f_i = 0 \tag{26}$$

$$R_i = \sqrt{(x_j - x_i)^2 + (y_j - y_i)^2} \tag{27}$$

Equation (26) may be written in the form

$$\sum_{j=1}^{N} a_{ij}\ f_j = b_i \qquad i = 1 \text{ to } N \tag{28}$$

for constant elements case, we have

$$a_{ij} = \frac{1}{2}\delta_{ij} + \int_e G_n(KR)\ ds \tag{29}$$

$$b_i = -\frac{\partial \psi_o}{\partial n}(x_i,y_i) \tag{30}$$

δ_{ij} is the kronecker delta. Equation (28) is the set of N linear algebraic equations in N unknown f_i, $i = 1.....N$.

These N algebraic equations can be solved to give the source distribution function $f(x,y)$ at the N points of the boundary S.

We can now define the wave amplification factor A at pint one the plane Z=0 to be equal to the ratio of the maximum wave height at that point to the maximum wave height at infinity.

From equation (4) and (12) the surface elevation $\tau(x,y,t)$ can be written as

$$\tau(x,y;t) = (1/g)\ B \cosh(Kh)\ \psi(x,y)\ e^{-iwt} \tag{31}$$

$$|\tau|_{max} = (1/g) B \cosh(Kh) |\psi(x,y)| \tag{32}$$

From equations (14) and (31) it follows that

$$\tau_0 = (1/g) \cosh(Kh) \cos(Kx \cos\beta)\ e^{-i(wt+k \sin\beta)} \tag{33}$$

$$|\tau_0|_{max} = (1/g) B \cosh(Kh) \tag{34}$$

and hence

$$A = \frac{|\tau|_{max}}{|\tau_0|_{max}} = \left|\psi\ (x,y)\right| \tag{35}$$

NUMERICAL RESULTS

To illustrate the implementation of the boundary element solution discussed above the following two cases are considered. The procedure was implemented on a VAX system using a double precision pascal.

Case (1) : Circular harbour

In these case the idealized circular harbour shown in figure (1) is considered. The boundary has been subdivided into 20 elements figures (2)-(3) give the graphical representation of the variation of amplification factor A against certain point on the boundary. Maximum amplification occurs at the points in between G and H and at points in their neighbourhood. When γ is increased from $\pi/3$ to $\pi/4$ maximum amplification occurs in the same region. But the corresponding value of A are higher. Figure (4) shows the value of A at a specific point G for a large range of values of K. Figure (5) shows the calculated values of A inside the harbour, there is an increase in the value of A as X increases, and no change in the value of A as Y is changed. Outside the harbour A is approximately constant.

Case (2) : A practical existing harbour

In this case we consider the Eastern harbour in Alexandria city, Egypt. The geometric configuration of this harbour as shown in figure (6). The boundary element approach discussed in this paper has been implemented to obtain the nature of waves amplification within the area. Results obtained are graphically displayed in figure (7) some point inside the harbour.

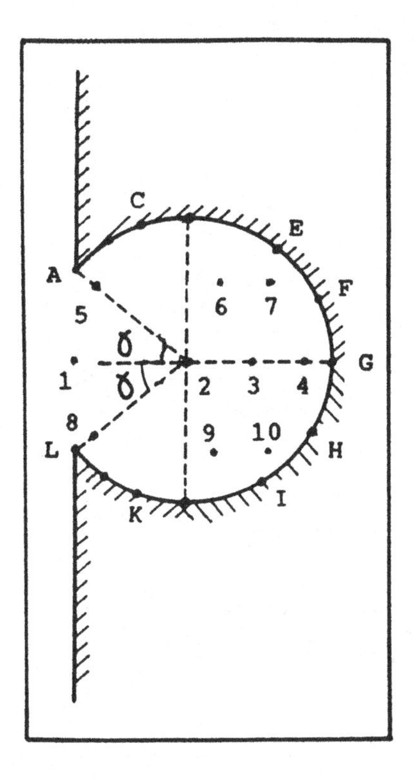

Figure 1. Geometric configuration of a circular harbour.

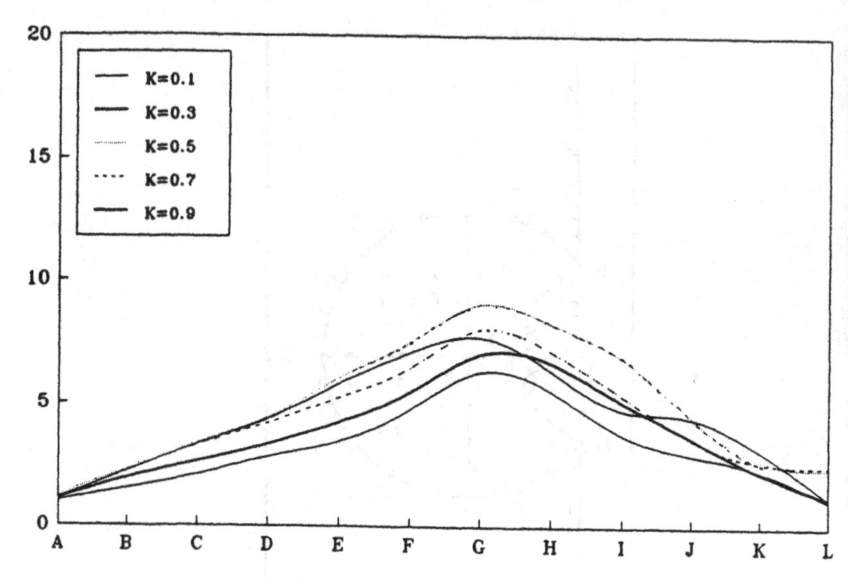

Figure 2. Frequency response of different point of the harbour for different values of k $\gamma = \pi/3$ and $\beta = 0$.

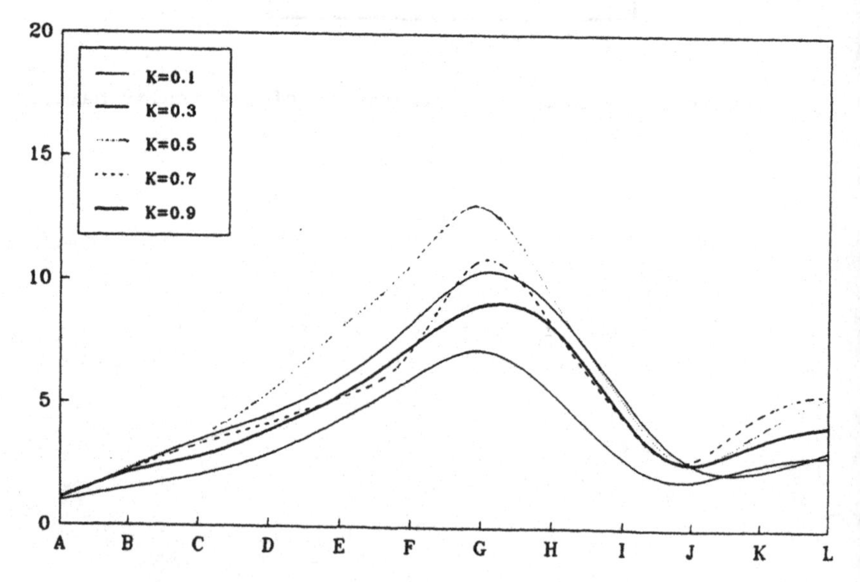

Figure 3. Frequency response of different points of the harbour for different values of k $\gamma = \pi/4$ and $\beta = 0$.

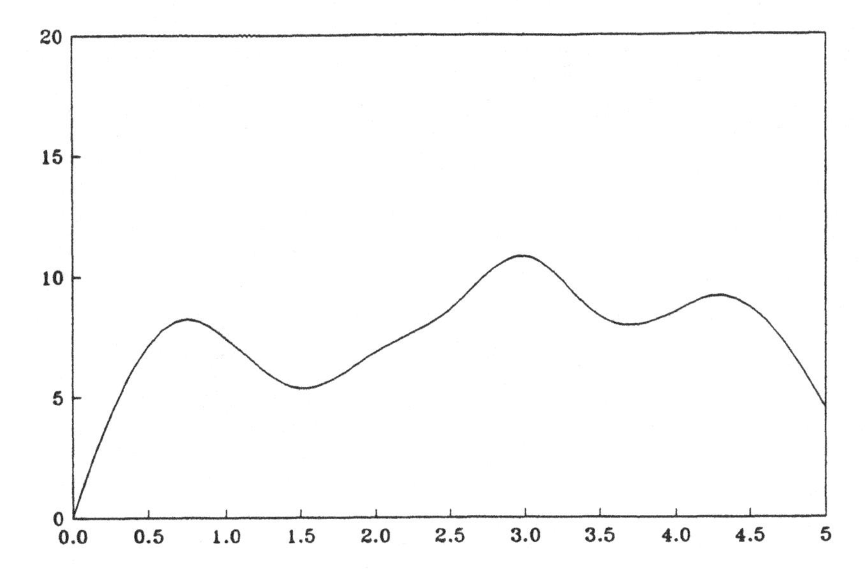

Figure 4. Frequency response at midpoint G .

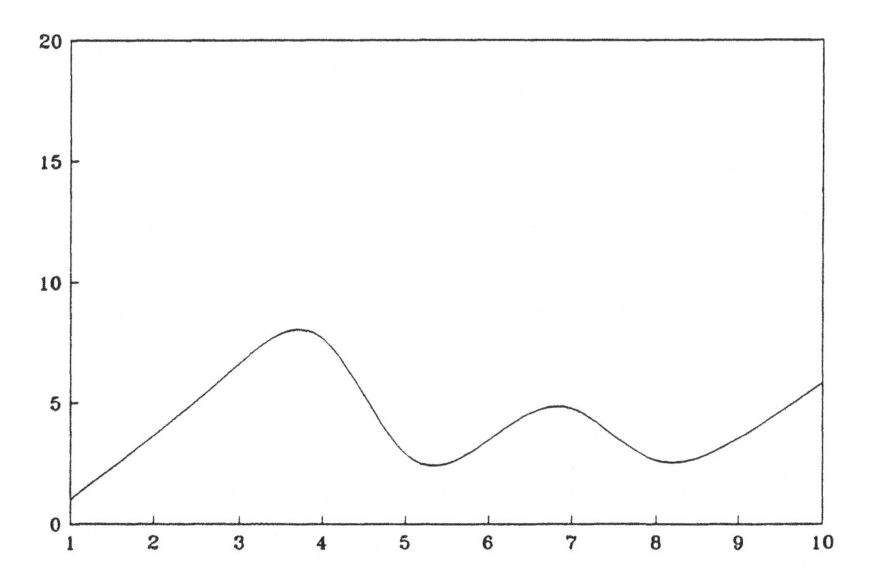

Figure 5. Frequency response for selected points inside the harbour .

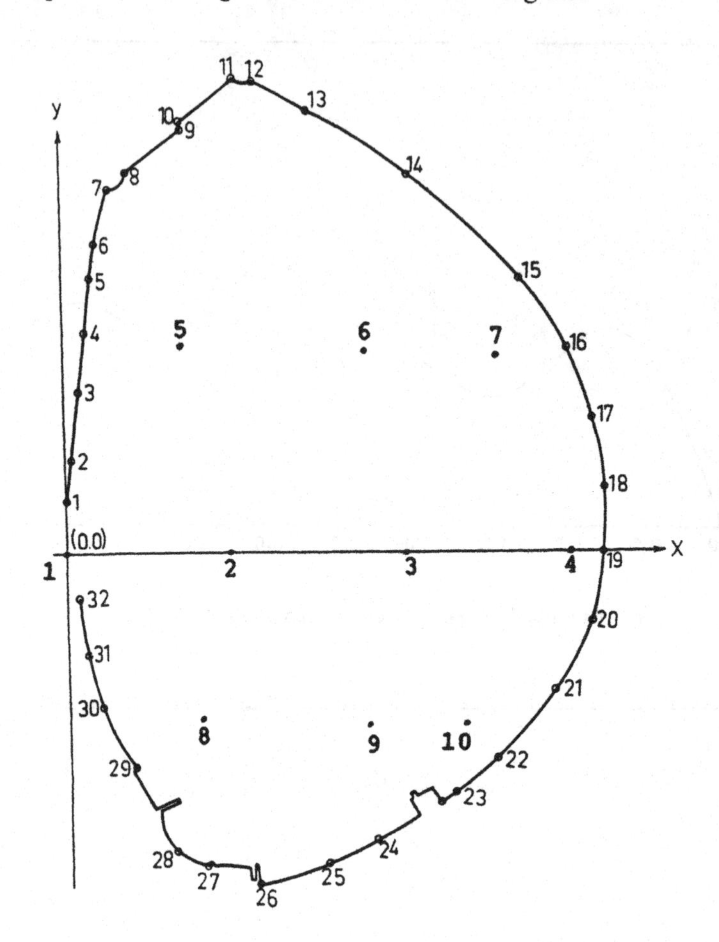

Figure 6. Geomertric configuration of Alexandria harbour.

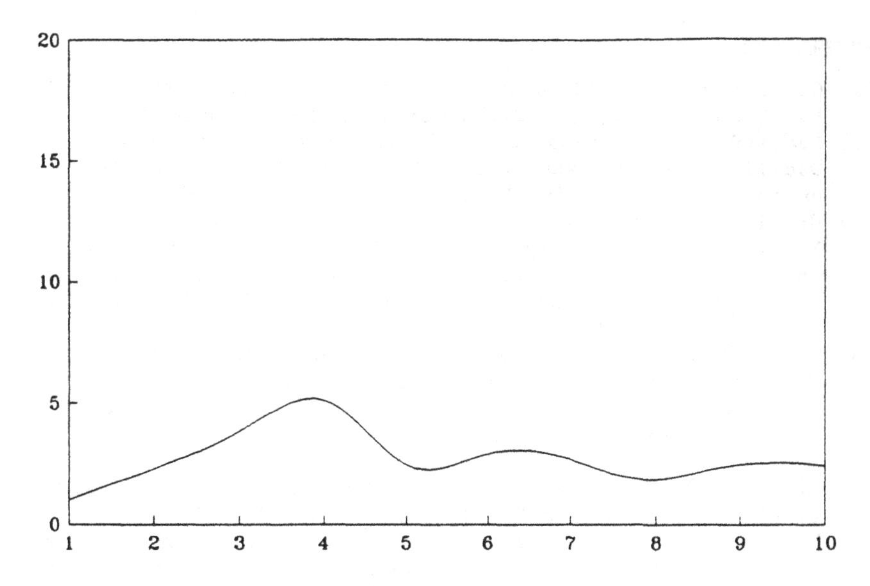

Figure 7. Frequency response for selected points inside the harbour .

REFERENCES

[1] Carrier, G.F. and pearson, C.E., Partial Differential Equations. Theory and Techniques, Academic Press, (1976).
[2] Demired, V. Transient Linear Long Wave Induced Oscillations Arbitary Shaped Harbours: A Boundary Element approach. M.S. Thesis, University of Miami. (1986).
[3] Grilli, S. Leijeune, A. study of the Wave Action on Flooting Structures by Boundary Elements. Comparison with the pp.227 to 237, Proceeding of the 5th International Conference on Boundary Elements, Hiroshima, Springer Verlag, Berlin (1983).
[4] Ijima, T., Yoshida, A. Resonance in Harbours of Arbitrary Topography, in BEM/83 (Ed. Brebbia, C. A.), pp. 217 to 226, proceedings of the 5th International Conference on Boundary Elements, Hiroshima, Springer Verlag, Berlin (1983).
[5] Liggett, J. A. Location of Free Surface in Porous Media, J. Hydraulics Div., ASCE, 103 (HY4), 353-65 (1977).
[6] Mansur, W. J. and Brebbia, C.A. Formulation of the Boundary Element Method for Transient Problems Governed by the Scalar wave Equation. Applied Mathematical Modelling, vol. 6, pp 307-311. (1982)

Numerical Hindcasting and Forecasting of Wind Waves in La Plata River, Argentina

R. Días (*), E.D. Kreimer (**), R.M. Cecotti (**)
(*) Area Hidráulica Marítima, Facultad de Ingeniería, Universidad Nacional de La Plata, Argentina
(**) C.I.C. Provincia de Buenos Aires, Area Hidráulica Marítima, Facultad de Ingeniería, U.N.L.P., Calle 47 No.200, 1900 La Plata, Argentina

ABSTRACT

The aim of this research was the forecasting of wave spectra under storm conditions in La Plata River. Since this phenomenon is heavily dependent upon the water levels where the waves propagate the computation of the storm surge levels throughout the river basin had to be made prior to the wave forecasting. For this purpose a two- dimensional modelling of wind-induced currents and levels was carried out. Calibration with field data to adjust the friction coefficient was done. Later on, wave spectra generation and propagation for different locations within the river was computed using an uncoupled numerical model. This model was also calibrated and checked against field data. Despite of the simplicity of the model itself, the results agree very well with records obtained under storm conditions.

INTRODUCTION

The La Plata river is a large water body connected to the Atlantic Ocean. It is *290 km* long and has a width ranging from *40 km* upstream to *220 km* at the mouth, as seen in Figure 1. In addition to these large dimensions a particular feature of this river is its shallowness: on *75%* of the basin surface depths smaller than *7 m* can be found.

This study started in 1988 with the aim of investigating the wave characteristics at selected points of the river caused by severe storm surges that are regularly caused by E and SE winds blowing from the ocean.

Since the characteristics of the waves reaching a certain zone do not only depend upon the initial conditions of the wave field but also on the water depths along the paths of the incoming wave rays, a previous estimation of these depths had to be made. This is particularly important in La Plata river due to the tremendous variation of the depth field in the middle and upper

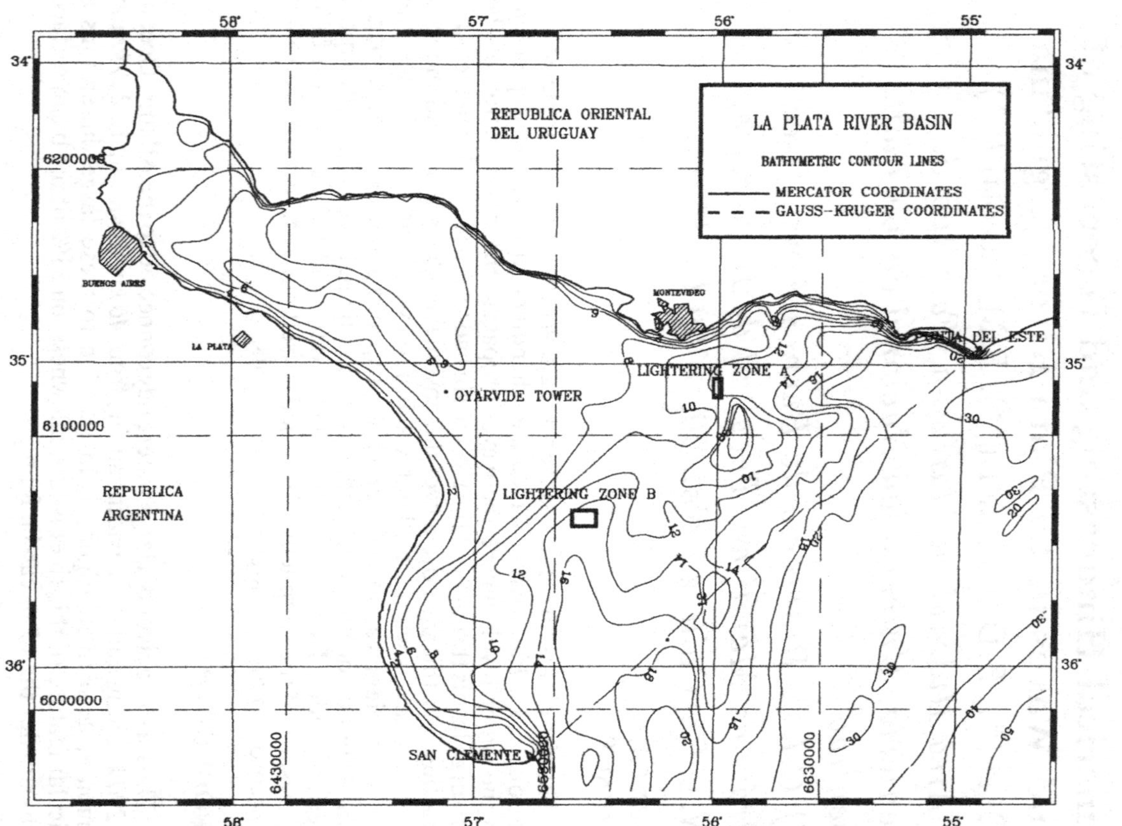

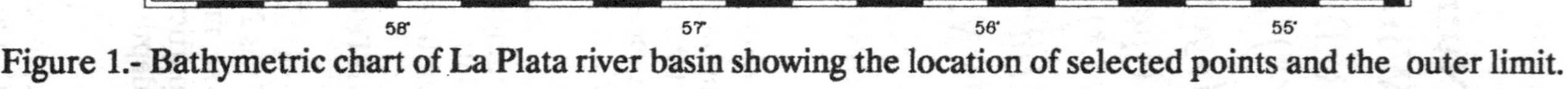

Figure 1.- Bathymetric chart of La Plata river basin showing the location of selected points and the outer limit.

zones of the river produced by intense wind action as said in the previous paragraph. To provide just an idea of the influence of this phenomenon, it is possible to find storm surge records of up to *4 m* in certain cases, in addition to the astronomical tide levels. Taking into account the mean water depth of the basin this effect cannot be disregarded. In fact, a thorough modelling of the level field is as important as the the wave modelling itself due to the causes mentioned above.

Therefore, a two-dimensional modelling to reproduce the current and level field was developed; by this means the astronomical tide effect could be studied and also the storm surge effect could be simulated for different wind directions and intensities. This model was calibrated with field data so that to insure a reasonable accuracy of the simulations. These data were obtained from tide level meters at coastal stations along the river basin in both Argentina and Uruguay. This was a particularly useful step with regard to the level field reproduction due to the large horizontal dimensions and shallow depths of the river basin. In such a situation, the turbulent bottom friction plays a dominant role in both the current-level and wave computations, for this reason the friction coefficient must be very carefully evaluated if a reasonable reproduction is desired.

SIMULATION OF THE ASTRONOMICAL TIDE

This step of the research was made using a numerical model that solves the depth-integrated, time-averaged Navier-Stokes equations:

$$\frac{DU}{Dt} + f_c \, . \, V = -\frac{1}{\rho}\frac{\partial p_n}{\partial x} - g\frac{\partial \eta}{\partial x} + \frac{1}{\rho}\frac{\tau_{nx} - \tau_{bx}}{h + \eta} + A_u\left[\frac{\partial^2 U}{\partial x^2} + \frac{\partial^2 U}{\partial y^2}\right] \quad (1)$$

$$\frac{DV}{Dt} - f_c \, . \, U = -\frac{1}{\rho}\frac{\partial p_n}{\partial y} - g\frac{\partial \eta}{\partial y} + \frac{1}{\rho}\frac{\tau_{ny} - \tau_{by}}{h + \eta} + A_v\left[\frac{\partial^2 V}{\partial x^2} + \frac{\partial^2 V}{\partial y^2}\right] \quad (2)$$

$$\frac{\partial \eta}{\partial t} + \frac{\partial [U(h+\eta]}{\partial x} + \frac{\partial [V(h+\eta]}{\partial y} = 0 \quad (3)$$

where U and V are the depth-averaged current velocities, η is the free surface elevation, h is the mean local depth, f_c is the Coriolis force and τ_{ni} and τ_{bi} are the wind and bottom stresses respectively. This equation system was discretized by an *ADI* numerical scheme. To carry out the model calibration, the following steps were undertaken:

1.- A line passing through the cities of San Clemente (Argentina) and Punta del Este (Uruguay) was adopted as the river's outer limit, as shown in Figure 1. This was done so because the storm surge effect beyond this limit can be considered negligible as compared with the astronomical tide effect.

2.- This line represents an outgoing open boundary of the numerical model where it is neccesary to define the water levels in the course of time. This condition was therefore provided by the astronomical tide levels calculated by means of the tidal constituents at both locations. A linear interpolation of water levels along this line was then assumed to complete the formulation of this boundary condition.

3.- Three ingoing open boundaries were defined in coincidence with the major tributaries of La Plata river, namely: the Paraná de las Palmas (*3630* m^3/s, direction *20º*), Paraná Guazú (*12,870* m^3/s, direction *-56º*) and Uruguay (*6,900* m^3/s, direction *225º*) rivers for which a mean annual discharge and incidence direction were adopted.

4.- A water depth grid was then prepared based on existing surveys of the river basin.

The calibration process consisted basically in modelling the astronomical tide adding the incoming flow of the tributaries and carrying out an adjustment of the bottom friction (Chezy coefficient) in a way such as to get a good correspondence between the computed and predicted levels. To simplify the computations, a constant friction coefficient was adopted for the whole river basin. A careful analysis on the real value of this coefficient was made, values currently cited in the literature could not be used because of the special features of this river. Values provided by the river engineering experience shown dissimilar results due to the fine sediments of the bottom, on one side, and the lack of knowledgde of the bedforms on the other side. Based on this analysis and preliminary runs of the model it was concluded that a constant value of $C = 115\ m^{1/2}/s$ would provide the best agreement.

Some results of the calibration procedure can be seen in Figure 2 where the predicted (by harmonic analysis) and computed astronomical tide levels at the Montevideo and Buenos Aires stations have been included. As seen there, there is a good coincidence in the amplitude and phases of both levels.

Once the friction coefficient could be determined, the modelling of astronomical tide and flow discharge of the tributaries was made, later on, wind action was added. All the driving forces were gradually included in the modelling process during an initial, short, time of the process to avoid the generation of gravity waves caused by an instantaneous driving force.

A special consideration deserves the available wind data for this study: at first, the Hydrographic Survey of the Argentine Navy provided an estimation of the wind fields on the whole region for the selected storms based on coastal measurements which proved to be of very low intensity to generate the recorded storm surges and waves. Therefore, the authors decided to use a wind field characterized by a constant speed and direction. Even so, an analysis based on synoptic weather charts had to be made in order to select storm situations during which such a constancy could be reasonable. This was especially important for the wind-wave forecasting carried out later.

A wind speed of *10 m/s* was used in coincidence with the E and SE directions and resulting current fields corresponding to the final time of the simulation can be seen in Figures 3 and 4; the isolines of water levels can be

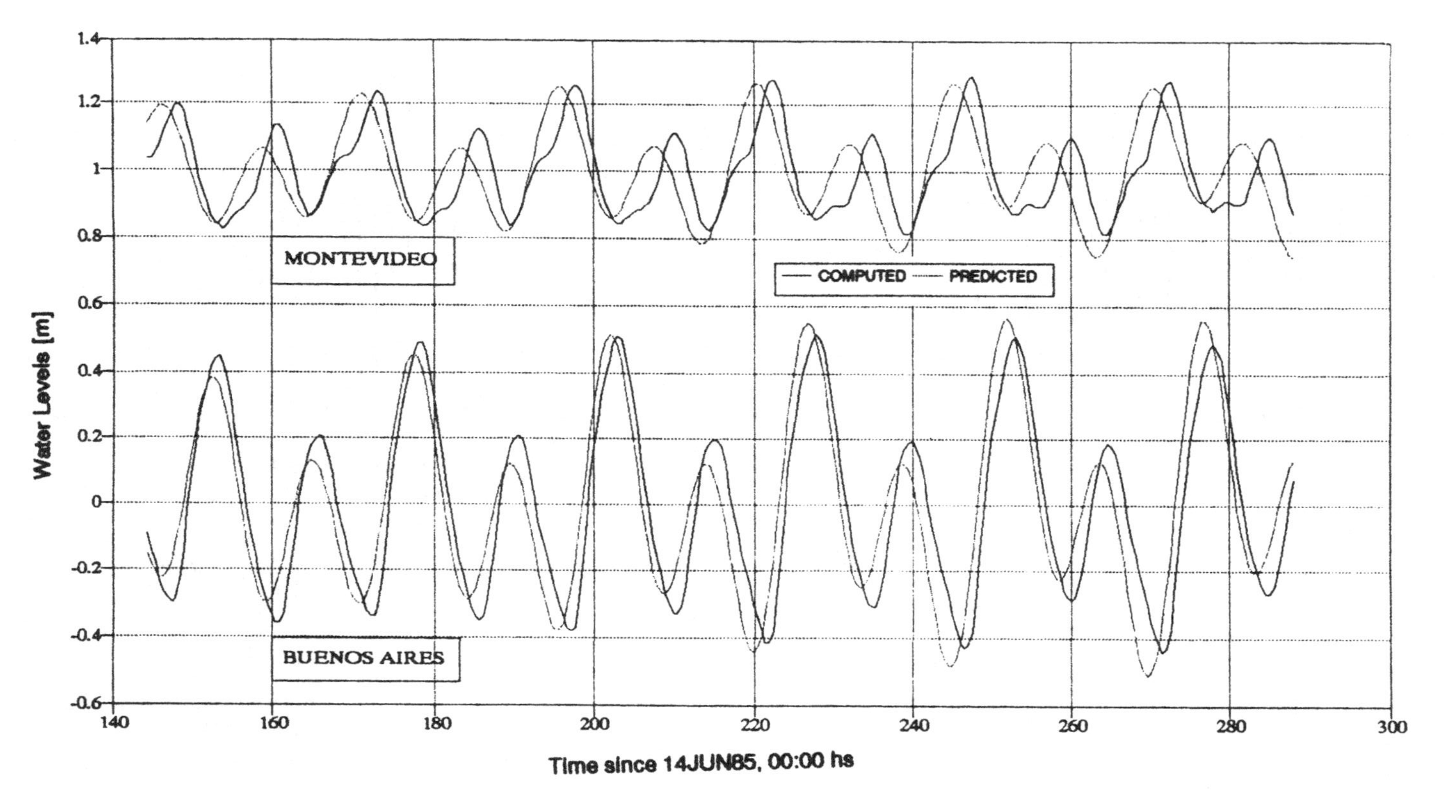

Figure 2.- Computed and predicted astronomical tide levels at the ports of Montevideo and Buenos Aires from 20JUN85 to 26JUN85 (Note: mean levels were changed to avoid superposition).

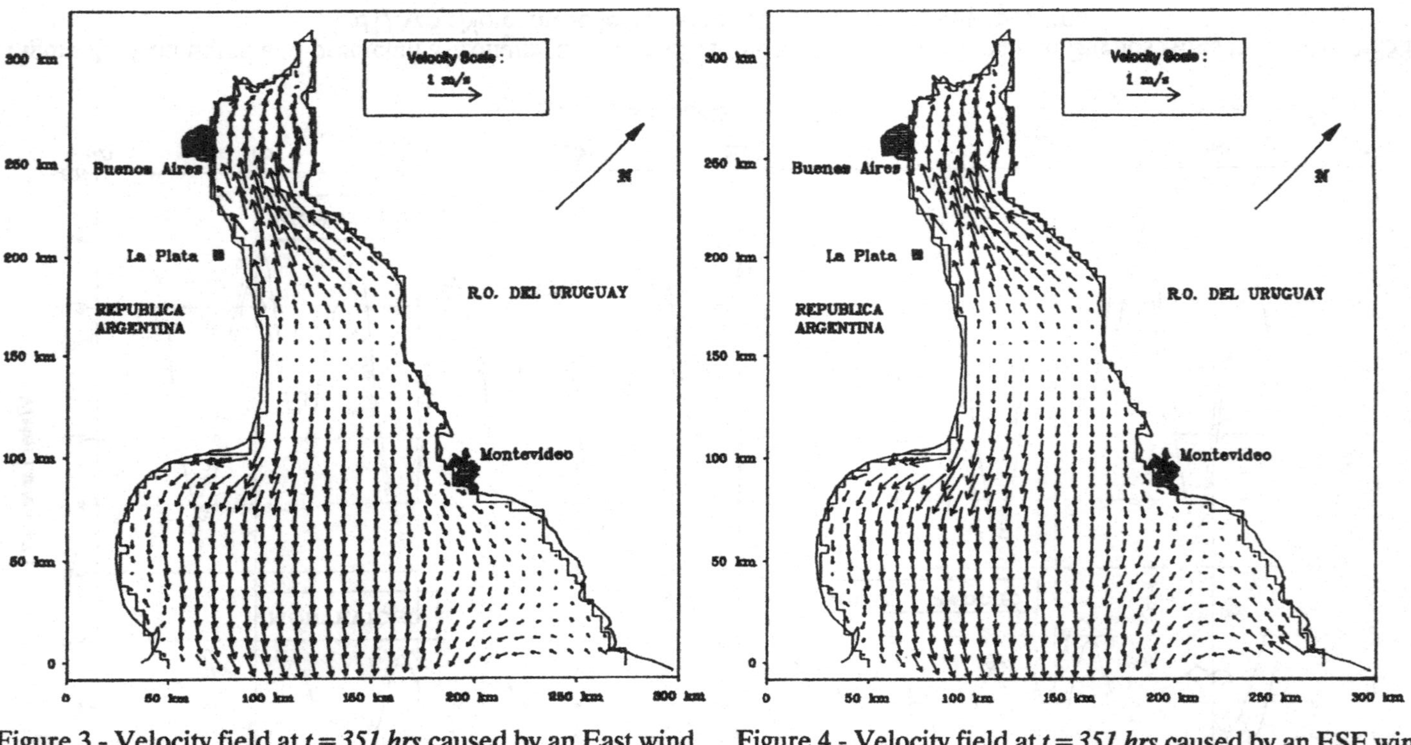

Figure 3.- Velocity field at $t = 351$ *hrs* caused by an East wind blowing at *10 m/s*.

Figure 4.- Velocity field at $t = 351$ *hrs* caused by an ESE wind blowing at *10 m/s*.

observed in Figures 5 and 6. These level fields were then used in the study of wave generation and propagation throughout the river basin.

FORECASTING OF WIND-WAVE SPECTRA

The method used to compute the generation, dissipation and propagation of wind-wave spectra has been extensively been studied by Collins [3] and Cavaleri et al.[2]. It consists essentially of solving a steady- state equation such as:

$$\vec{c_g} . \nabla S = S_g + S_{fr} \tag{4}$$

where S_g and S_{fr} represent the wind generation and turbulent bottom friction terms respectively, S is the spectral density and c_g is the wave group velocity. The wind generation term can be, as usual, be represented by means of the Philips and Miles generation mechanisms so that, if k is the wave number:

$$S_g = \alpha + \beta . S(k) \tag{5}$$

The bottom dissipation term, according to Collins [3], is:

$$\frac{c_f . g . k . c_g}{2\pi . \sigma^2 . cosh(kh)} . S(f) . D(\theta) \tag{6}$$

where c_f is the friction coefficient, is the mean wave velocity at the bottom, S(f) is the spectral density at frequency f and $D(\theta)$ is the directional spreading function adopted as a function of $\cos^4\theta$ in this case. The friction coefficient was evaluated with a formulation given by Swart(1978). $S(f)$ was represented by means of the *JONSWAP* formulation while energy saturation was modelled as suggested by Philips (1957) in the following way:

$$S_{sat} = 0.073\, g^2 . \sigma^{-5} . S(\delta) \tag{7}$$

$$S(\delta) = \frac{8}{3}\pi . \cos^4(\delta) \tag{8}$$

where $S(\delta)$ is due to Barnett [1]. This saturation limit is then the maximum which a particular wave frequency can attain. For this reason a truncation factor, μ, for the spectral growth mechanism is defined as follows:

$$\mu = \left[1 - \left(\frac{S}{S_{sat}}\right)\right] \tag{9}$$

In practice, this is a simple model which carries out the accounting of all the energy packets transported by the wave travelling along the different wave rays arriving at the point under consideration. Summation of all the wave packets for each frequency gives then the spectral density contribution to the whole spectrum. A further summation of the remaining frequencies gives finally the final spectrum. This is an uncoupled modelling technique due to the fact that each wave ray corresponding to each frequency behaves

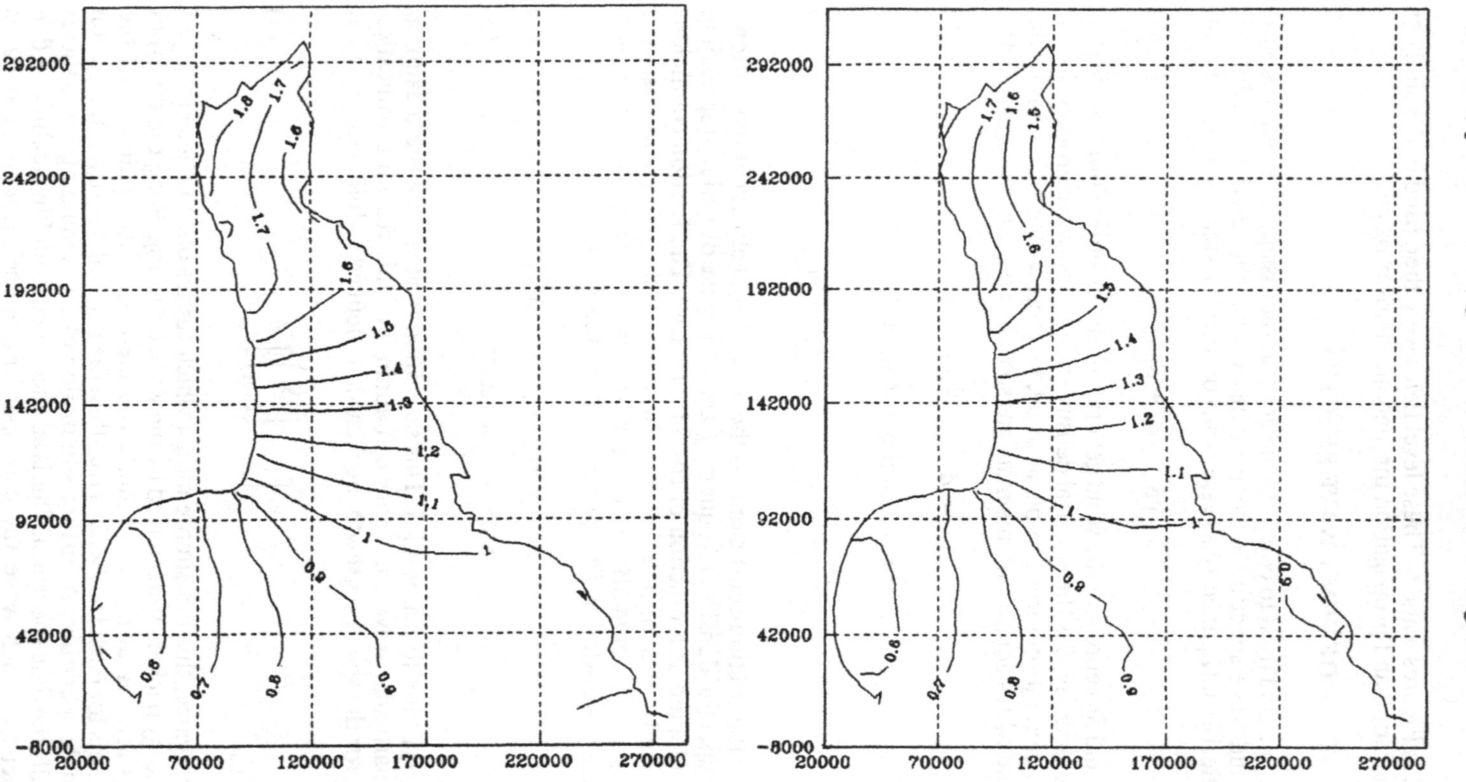

Figure 5.- Isolines of water levels at $t = 351$ *hrs* (28JUN85, 15:00 hrs) caused by an East wind blowing at *10 m/s.*

Figure 6.- Isolines of water levels at $t = 351$ *hrs* (28JUN85, 15:00 hrs) caused by an ESE wind blowing at *10 m/s.*

independently of each other; therefore, non-linear wave interactions cannot be modelled. It simplicity however makes it very attractive for practical use since it can be even used on a personal computer.

Due to the large distances involved in the computations, the phenomenon become friction-dominated, therefore a calibration stage had to be made in order to find an optimum value of the friction coefficient. Two values of c_f were tested, *0.005* and *0.01*, the latter being currently cited in the literature. The results showed however that a $c_f = 0.005$ provided better results due to the large distances travelled by the wave rays, shallowness of the river basin and the fine-grained bottom sediments. Wave spectra measured by a buoy at the port of La Plata in 1985 were used to check the validity of the computations.

A previously developed wave refraction model was used to compute the wave rays for five different locations within the river, namely: the ports of Buenos Aires and La Plata, the Oyarvide Tower and Lightering Zones A and B as shown in Figure 1. Examples of waves irradiated from the port of La Plata and Lightering Zone A can be observed in Figures 7 and 8. Results of the forecasting have been included in Figure 9 where one may see the good coincidence between the computed and measured spectra at the port of La Plata. It is worthy to be noted that, despite of the selection of the appropriate friction coefficient, this acts only as a sort of "scaling" of the whole spectrum because the peak frequency remains practically constant for a much different value of this coefficient. The shape of the spectrum can be also reproduced fairly well. The difference in the high frequency range are due to the lack of modelling of these frequencies since they are not important for engineering purposes. The low frequency range of the measured spectra instead can be mainly attributed to transmission noise.

It must be pointed out that a steady state of the water levels in the river was considered in this modelling. This simplifying assumption arises from the fact that the duration of the selected storms was longer than the travelling time of the main spectral constituents.

CONCLUSIONS

As one would expect, the numerical forecasting of the wave spectra at five different locations within the La Plata river basin showed increasing wave heights and decreasing peak frequencies with an increase of the wind speed. This general result holds for each separate location.

Comparing the results from different locations, it can be observed a strong filtering effect caused by the bottom topography of the river basin on the wave rays which are able to arrive at each one of them. Therefore, the upstream locations exhibit much smaller wave heights and peak frequencies than those located at the outer part of the river. Here, of course, there are no restrictions imposed by either the bottom friction effect or the coastlines as found in the inner part of the river.

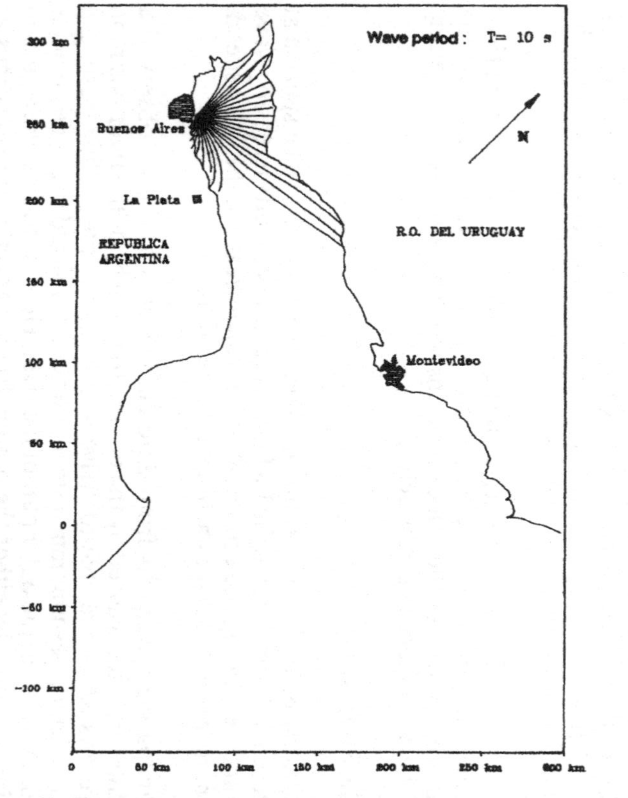

Figure 7.- Wave rays (T = 10 s) irradiated from the Buenos Aires port on a tidal stage caused by a SE wind blowing at 15 m/s.

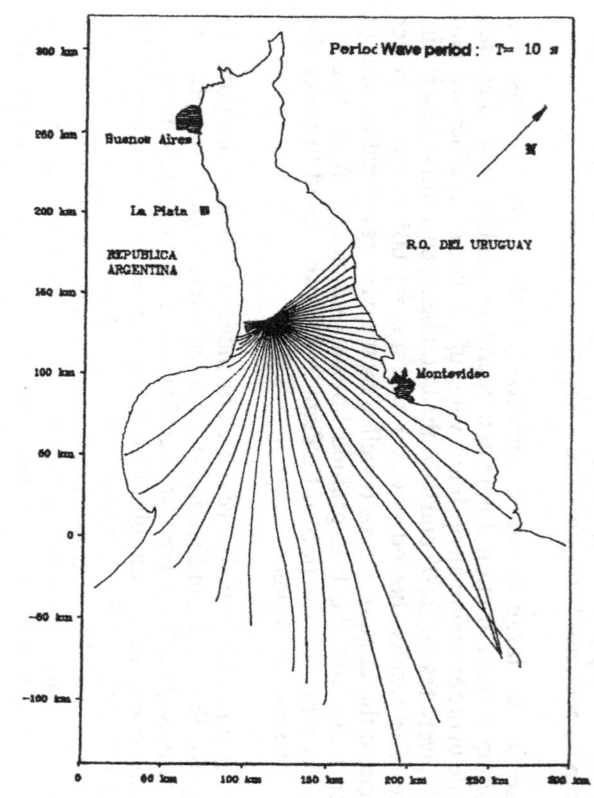

Figure 8.- Wave rays (T = 10 s) irradiated from the Oyarvide Tower on a tidal stage caused by a SE wind blowing at 15 m/s.

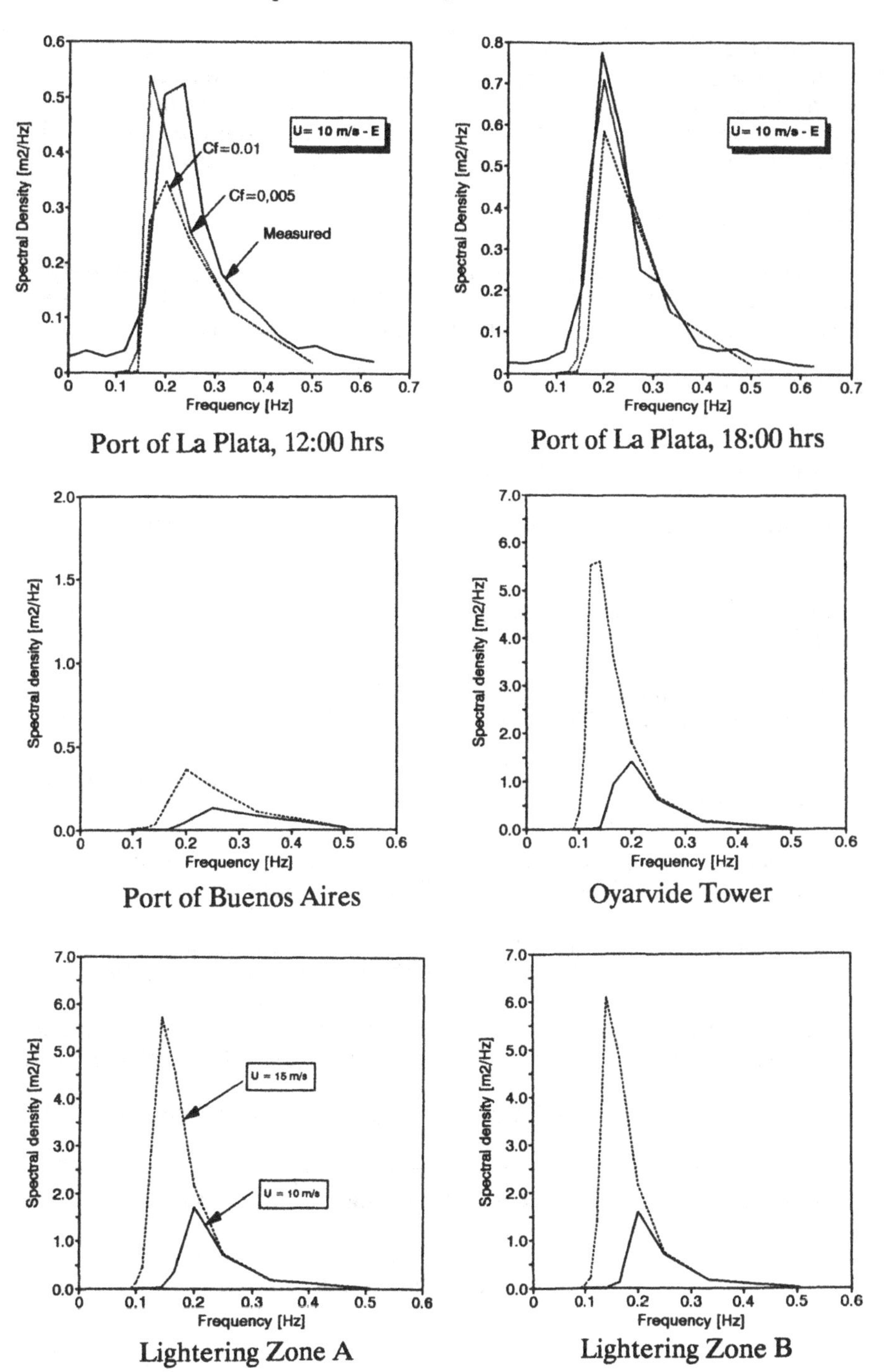

Figure 9.- Results of the wave spectra forecasting for the five selected locations within the river basin.

An interesting result found in this study is that the most severe wave conditions are not those coincident with SE winds but they are produced instead by E winds. This phenomenon is surely due to a slightly more efficient "piling up" effect of the water body of the river when winds blow from the E direction that provide also somewhat deeper depths than SE winds do.

ACKNOWLEDGMENTS

This research was carried out under the sponsorship of the Administration Commission of the La Plata River (C.A.R.P.) and the authors want to express their gratitude for the permission to publish this paper.

BIBLIOGRAPHY

1.- Barnett, T.P.: *On the generation, dissipation and prediction of ocean wind waves*, J.of G.Res., Vol.73, Nº2, 1968.

2.- Cavaleri, L. and M.Rizzoli, P.: *Wind wave prediction in shallow water: theory and applications*, J.of G.Res., Vol.86, NºC11, 1981.

3.- Collins, J.: *Prediction of shallow water spectra*, J.of G.Res., Vol.77, Nº15, 1972.

4.- Horie, T.: *Numerical models on flow-dispersion*, P.H.R.I., Ministry of Transport, Japan, 1979.

5.- Holthouysen, L.: *Ocean wave theory*, Int.Inst.for Hyd.and Env.Engg., Delft, 1982.

6.- Karlsson, T.: *Refraction of a continuous wave spectra*, J.of the Waterways, Harbors & Coastal Engg.Div., NºWW4, Proc.Paper Nº6881, 1969.

7.- Kreimer, E. and Cecotti, R.: *Predicción de espectros de energía de olas en aguas poco profundas*, XII Latin American Congress of Hydraulics, 1986.

8.- Lendeertsee, J.: *Aspects of a computational model for long-period water-wave propagation*, U.S.Air Force Proj., Rand Memo.RM 5294-PR, 1967.

9.- Philips, O.: *The dynamics of the upper ocean*, Cambridge Univ.Press, 1977.

10.- Swart, D.: *Offshore sediment transport and equilibrium beach profiles*, Doctorate Dissertation, Dept. of Civil Engg., Delft Univ.of Technology, 1978.

11.- Vanoni, V.: *Sedimentation Engineering*, A.S.C.E., 1975.

On the Calculation of Extreme Waves and Design Waves for Designing Coastal Structures

R. Bilgin, T. Cam

Department of Civil Engineering, Karadeniz, Technical University, 61080 Trabzon, Turkey

ABSTRACT

The hindcast study of extreme wave conditions at the sites of Ordu - Giresun, Trabzon and Hopa of the Eastern Black Sea coast of Turkey is described. The average hourly wind speed values observed at the meteorological stations and extracted from the synoptic weather charts prepared periodically for weather forecasting are used to predict waves by the method of SMB (CERC [1]). Then, the Extreme Value Type I (Gumbel) distribution has been applied to the annual maximum waves, consequently the extreme waves with various return periods are estimated. Among the results the estimates made by using the synoptic data showed higher values about 1.5 to 3 times, than those of the meteorological stations.

INTRODUCTION

The growing importance of the coastal and offshore engineering structures requires estimates of extreme wave climates and in particular reliable design criteria and of the environmental information. This task is ideally realized by analysing long time series of the parameters of interest, but, where these are not available, other means have to be derived.

Around the coasts of Turkey wave measurements, one of the most critical parameters for the design of a sea structure are sparse, especially at the coasts of the Black Sea are almost absent. The availability in the same region of reliable meteoroligal data during the last 50 years allowed us to make use of the hindcasting technique.

In obtaining the reliable estimates of extreme waves there are various methods developed among which to choose the most appropriate one and the distribution is the main topic of interest. Besides, there are not any strict principles or standards to be obeyed by the design engineer for coastal structures. The risk or encounter probability that during its lifetime a structure has to face conditions beyond some predetermined limits is closely related to return period of the design wave as follows :

$$E_p = 1 - \left(1 - \frac{1}{R_p}\right)^L \qquad (1)$$

where E_p indicates risk, L lifetime(year), R_p return period(year). The common principle is that return period of design wave varies with the functions and the coasts of the structure. For example, a design wave with return period of 25 years is reasonable for designing a rubble-mound coastal protective structure but for a breakwater of a large port, a design wave with return period of at least 100 years has to be projected (Muir and El-Shaarawi[2]).

METHODS OF WAVE PREDICTION AND THE DESIGN WAVE

Sea waves have the most complicated processes in nature so that it is extremely important to define characteristics of the waves and the functions of the structure to be built. There are two main approaches to define the mathematical structure of complex wind waves : Determination of probability distributions of wave parameters and of the power spectrum. As stated above, because of the scarcity of wave measurements in Turkish coasts it has not been examined and tested for observed probability distribution of wave parameters whether they are adequately fitted to the theoretical distribution or of the power spectrums to the theoretical models spectrums (Ozhan[3]).Therefore a hindcast study of extreme wave conditions is done to predict the design waves.

The methods of wave prediction in coastal engineering practice may be grouped as follows :

a) The methods of prediction of statistical wave height and period; namely the significant wave height ($H_{1/3}$) and significant wave period ($T_{1/3}$) which are determined from the observational figures (Breitschneider[4], Favey[5]).

b) The methods of wave power spectrum : A relevant model of a power spectrum is chosen first, then by applying to the wave data, power spectrums of the waves, significant wave heights and periods are calculated (Silvester[6]).

In this study the significant wave concept is considered.

Significant wave height and period

Significant wave height is defined as the average

height of the one - third highest waves and it was about equal to the average height of the waves as estimated by an experienced observer. The significant wave period is similarly defined as the significant wave height.

Statistics of the extreme waves

In this study, Extreme Value Type I distribution is used for estimating the design waves. The distribution function is as follows:

$$F(H_{1/3}) = \exp\left[-e^{-A(H_{1/3}-B)}\right] \tag{2}$$

$$A = \frac{\pi}{\sigma_{H_{1/3}}\sqrt{6}}, \quad B = \mu_{H_{1/3}} - \frac{\gamma}{A}, \quad \gamma = 0.5772 \tag{3}$$

$$\hat{\mu}_{H_{1/3}} = \bar{H}_{1/3} = \frac{1}{N}\sum_{i=1}^{n}(H_{1/3})_i \tag{4}$$

$$\hat{\sigma}_{H_{1/3}} = s_{H_{1/3}} = \frac{1}{N-1}\sum_{i=1}^{n}\left[(H_{1/3})_i - \bar{H}_{1/3}\right]^2 \tag{5}$$

where $H_{1/3}$ is the maximum significant wave height observed or estimated during the year; $F(H_{1/3})$ is the total probability that the maximum wave height is less than or equal to the significant wave height; A and B are the parameters of the distribution; $\mu_{H_{1/3}}$, $\sigma_{H1/3}$, $\bar{H}_{1/3}$ and $S_{H1/3}$ are the statistical parameters of the distribution belonging to the population and the sample respectively.

To estimate the significant wave height the following equation is used :

$$H_{1/3} = B - \frac{1}{A}\,\ell n\ell n\left[F(H_{1/3})\right] \tag{6}$$

Supposing the highest significant wave heights for the referenced time period are ordered in the following way:

$$(H_{1/3})_1 < (H_{1/3})_2 < (_{1/3})_3 \ldots < (H_{1/3})_m \ldots < (H_{1/3})_n$$

where n indicates the no. of values, m = 1,2,..,n indicates ordered no. of wave height. The probability corresponding to each $H_{1/3}$ in the array is obtained by the following equation :

$$F(H_{1/3}) = \frac{m}{N+1} \qquad (7)$$

If each value of $H_{1/3}$ in the list plotted against the corresponding probability computed from the eq.(7) on the Gumbel probability paper it is expected that all the plotted points are clustered around a straight line. The equation of this line is given by the Eq.(6).

Choosing the design wave

As a general approach; first of all, depending on lifetime of the structure a return period is defined and the corresponding wave height is found from the prediction line on the Gumbel probability paper.

The equation of the relation between R_p and $F(H_{1/3})$ can be written as follows:

$$R_p = \frac{\tau}{1-F(H_{1/3})} \qquad (8)$$

where R_p is the return period , $1-F(H_{1/3})$ is the exceedance probability of highest wave in comparison with the design wave, τ is the avarage time period between events, generally taken as one year.

APPLICATION

Wind data

Since the estimation of the design waves with various return periods are aimed at the coasts of the Eastern Black Sea (from Ordu to the Russian border) the historical wind measurements observed at the meteorological stations in the region are used in the analysis. Wind data are derived from two different sources; namely the meteorological stations situated in the coastal area and the synoptic weather charts drawn periodically each day by the Meteorological Department in Ankara.

The wave prediction methods explained in detail in CERC use wind data which are supposed to have been observed above sea surface but the available wind data are observed by the land stations. This creates some problems although the necessary adjustments are made, it is well known that it is difficult to quantify the difference between the wind speed observed on the land and of the sea.

In the investigations done by Hsu[7], following equation has been developed :

$$U_{deniz} = 3.0\ (U_{kara})^{0.67} \tag{9}$$

where U_{sea} and U_{land} represent wind speed(m/sec) on the sea and the land respectively. The conclusions reached in this study gave greater estimates than that of Hsu.

The wind speeds extracted from the synoptic weather charts are known as "geostrophic winds" which are about % 25-40 less than the wind speeds above the sea. Thousands of the synoptic charts had to be examined to obtain candidate storms producing the highest waves in each year. The wind speed of at least 5 m/sec is considered as the base limit for the synoptic data. The values below this limit are not included in the analysis. There has to be at least 3.62 cm (winter) or 4.80 cm (summer) between two adjacent isobars to have this wind speed on the latitude of 45°. Therefore, the charts displaying higher values are eliminated at the preliminary examination.

Synoptic wind speeds are calculated for three coastal sub-areas indicated below

coast 1 : 37° E - 39° E Ordu-Giresun

coast 2 : 39° E - 40.4° E Trabzon

coast 3 : 40.4° E - 41.7° E Hopa

The positions of these coasts are shown in Fig. 1.

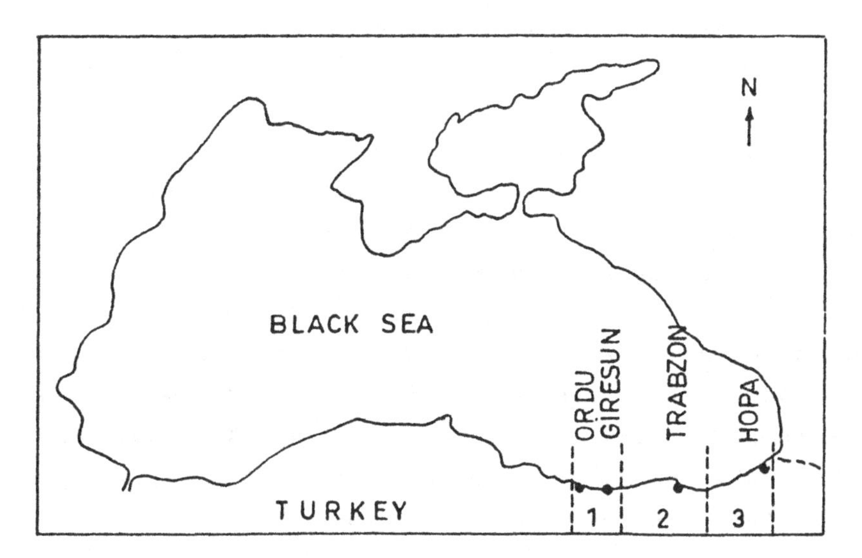

Fig. 1. The coastal bands and the meteorological stations where the wind speeds are observed or estimated

Computation of fetch lengths

Any wind speed recorded at a coastal station with a direction that can create waves is supposed to be effective along the fetch lengths. Effective fetch lengths for the directions that create waves in the region are computed and summarized in Table 1.

Table 1. The effective fetch lengths (km)

Direction	Ordu	Giresun	Trabzon	Hopa
W	-	-	-	698.00
WNW	-	484.90	307.60	658.70
NW	352.20	629.50	676.90	519.10
NNW	485.80	450.60	419.00	257.20
N	390.20	366.90	308.10	130.80
NNE	321.70	299.90	231.00	84.60
NE	269.40	281.40	201.20	-
ENE	274.20	282.80	143.80	-

Computation of storm waves from average wind speeds

Storm waves are computed by using both recorded wind speed and the wind speeds computed from the synoptic weather charts for the Eastern Black Sea Region where is divided into three sub-areas. Transformations of wind characteristics into waves is done by the method of SMB. The maximum annual storm wave heights are determined from these wave values. The period of observations taken into consideration to evaluate the wind waves are summarized in Tablo 2.

Extreme wave statistics

Annual maximum significant wave heights estimated for each sub-region are arranged in increasing order with corresponding significant wave period and nonexceedance probabilities based on the Weibull plotting positions formula(Eq.7) as shown in Table 3. These data are plotted using an Extreme Value Type I distribution and the fitted distributions obtained as in Fig.2 for Hopa.

Table 2. The period of observations

Synoptic chart		Meteorological stations	
Coast No.	period	City	period
I	1977-1985	Ordu	1969-1985
II	1977-1985	Giresun	1969-1985
III	1977-1989	Trabzon	1969-1985
		Hopa	1972-1988

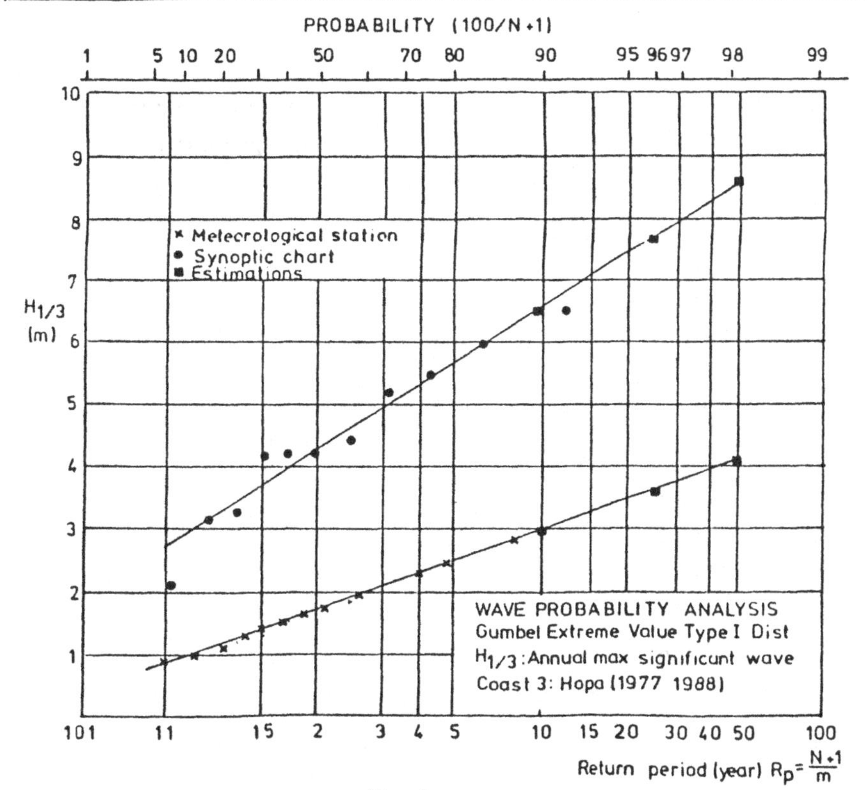

Fig. 2.

Using the fitted line on the Gumbel probability paper design wave heights and associated wave period for the return periods of 10, 25 and 50 years are estimated for the sub-regions(Tables 4, 5, 6, and 7). Computational of the wave period is done by using the observed average wave steepness coefficient (wave height/wave length) that is known as varying around 0.04 from the work of Bilgin et. al.[8]. Wave period is computed from the following equation :

$$(H_{1/3})_o = \frac{0.04\ g}{2\pi}\ T^2_{1/3} \qquad (10)$$

where $(H_{1/3})_o$ is deep sea wave height.

Table 3. Ordered wave data estimated from the synoptic weather charts(Coast 3-Hopa)

order No. m	year	significant wave $H_{1/3}$(m)	significant wave period $T_{1/3}$ (sec)	Non-exceedance probability $F(H_{1/3})$X100
1	1981	2.10	5.80	8
2	1984	3.17	7.10	17
3	1982	3.23	7.20	25
4	1979	4.18	8.25	33
5	1983	4.20	8.20	42
6	1977	4.22	8.20	50
7	1980	4.40	8.40	58
8	1978	5.20	9.10	67
9	1985	5.50	9.40	75
10	1987	6.00	9.80	83
11	1988	6.50	10.20	92

Table 4. Estimated design waves (Coast 1 - Ordu)

return period	synoptic chart		meteorological station	
R_p (year)	$H_{1/3}$ (m)	$T_{1/3}$ (sec)	$H_{1/3}$ (m)	$T_{1/3}$ (sec)
10	4.00	8.00	1.59	5.05
25	4.65	8.63	1.73	5.26
50	5.15	9.08	1.83	5.41

Table 5. Estimated design waves (Coast 1 - Giresun)

return period	synoptic chart		meteorological station	
R_p (year)	$H_{1/3}$ (m)	$T_{1/3}$ (sec)	$H_{1/3}$ (m)	$T_{1/3}$ (sec)
10	4.00	8.00	2.30	6.07
25	4.65	8.63	2.67	6.54
50	5.15	9.08	2.91	6.83

Table 6. Estimated design waves (Coast 2 - Trabzon)

return period	synoptic chart		meteorological station	
R_p (year)	$H_{1/3}$ (m)	$T_{1/3}$ (sec)	$H_{1/3}$ (m)	$T_{1/3}$ (sec)
10	5.36	9.26	2.30	6.07
25	6.09	9.88	2.67	6.54
50	6.63	10.30	2.91	6.83

Table 7. Estimated design waves (Coast 3 - Hopa)

return period	synoptic chart		meteorological station	
R_p (year)	$H_{1/3}$ (m)	$T_{1/3}$ (sec)	$H_{1/3}$ (m)	$T_{1/3}$ (sec)
10	6.55	10.24	3.00	6.93
25	7.75	11.13	3.65	7.65
50	8.60	11.73	4.10	8.10

CONCLUSIONS

In this study, with the aim of providing reliable data of deep sea wave characteristics, in particular, design waves for coastal and harbour engineering structeres, following conclusions have been reached:

1) The design waves of various return periods estimated by the wind speeds observed at the meteorological stations in the Eastern Black Sea region showed smaller values than those of the wind speeds extracted from the synoptic weather charts (Tables 4, 5, 6 and 7). This significant difference is mainly due to the erroneous wind measurements at some of the meteorological stations surrounded by high buildings. Therefore, the design waves of synoptic origin are relatively reliable and should be preferred.

2) The estimations have shown that design wave characteristics in the Eastern Black Sea are increasing from West to East. This is due to the prevailing wind of North - West direction so that the fetch lengths increasing towards East. Therefore,

Hopa sub-region has longer fetch lengths and higher design waves.

3) In the Eastern Black Sea from West to East, design wave heights for 10-year return period vary between 4.0 to 6.55m, for 25-year return period, from 4.65 to 7.75m, for 50-year return period, from 5.15 to 8.60 m.

REFERENCES

1. CERC (U.S. Army Coastal Engineering Center),'Shore Protection Manual,'Corps of Engineers, Washington, 1984
2. Muir, L. R. and El Shaarawi, A. H.' On the Calculation of Extreme Wave Heights: A Rewiew,' Ocean Engng., Vol 13, No. 1, pp. 93-118, 1986
3. Özhan, E. 'Estimation of Wind Waves by Computers,' TÜBİTAK VII. Science Congress, İzmir, pp. 797-813, 1977 (in Turkish)
4. Breitschneider, C.L.'Revisions in Wave Forecasting Deep and Shallow Water,' Proc. 6th. Coastal Eng. Conf., ASCE, Florida, pp. 30-67, 1975
5. Favey, H.T.'Prediction of Wind Wave Heights', ASCE, Journ. Waterways, Harbours and Coastal Eng. Div., Vol. 100, No: WWI, pp.1-3, 1974
6. Silvester, R. and Vongvisessomjai, S. 'Computation of Storm Wave and Swell,'Proc.Inst. Civil Engngs., Vol. 48, pp.259-283, 1971
7. Hsu, A.'On the Correction of Landbased Wind Measurements for Oceanographic Application,' Proc. of 17 th Coastal Eng. Conf., Sydney, Vol. 1, pp. 709-724, 1980
8. Bilgin, R., Ertaş, B. and Günbak, A.R. 'Estimation of Wind Waves and Designing the Rubble - Mound Coastal Protective Structures in the Eastern Black Sea Region,' Proc. Symp. on "The role of Engineering on the Development of Turkey," Yıldız University, İstanbul, 1988 (in Turkish)

Refraction and Diffraction of Surface Water Waves Using a Coupled FEM-BEM Model

A. Dello Russo, A. Del Carmen

C.I.C. Provincia de Buenos Aires, Area Hidráulica Marítima, Departamento de Hidráulica, Facultad de Ingeniería, Universidad Nacional de La Plata, Calle 47 No 200, 1900 La Plata, Argentina

ABSTRACT

The aim of this research work was the calculation of the wave calmness in arbitrary-shaped port basins. The governing differential equation was firstly devised by Berkhoff (1972) and Schoenfeld (1972). This is the so-called mild slope equation and can be used only in case of gently bottomed or gradually-varied depths. So far several methods have been proposed to find out solutions to the aforementioned equation. The domain of the equation is the fundamental difficulty of this problem since this extends to infinity. In this paper an adaptation of a coupled FEM-BEM model developed by He Yinnian and Li Katai (1987) is presented that leads to the obtention of useful solutions when very simple elements are used in the domain discretization and allows an a-priori estimation of the error involved in the approximation. Results obtained with this method are checked against theoretical solutions and other numerical methods and the order of convergence is empirically established.

INTRODUCTION

He Yinnian and Li Katai [1] have studied the use of a finite element method combined with a boundary element method to radiation problems that lead to the well-known Helmholtz differential equation:

$$\nabla^2 \varphi + k^2 \varphi = f \qquad in\ R^2 \tag{1}$$

In their work, these researchers show the theoretical correction of the formulation of the coupled method and obtain an a-priori estimation of the error in the approximate solution. It is possible to adapt this method to determine the propagation of water waves on a horizontal plane.

The phenomenon is governed by the mild slope differential equation:

$$\nabla (c \,.\, c_g .\nabla \varphi) \; + \; \omega^2 . \frac{c_g}{c} . \varphi \; = \; 0 \qquad in \; R^2 \tag{2}$$

by the boundary conditions at infinity:

$$\begin{aligned} \varphi &= O\,(r^{-1}) \\ \frac{\partial \varphi}{\partial r} - i\,.\,k\,.\,\varphi &= O\,(r^{-1/2}) \qquad r = |x| \rightarrow \infty \end{aligned} \tag{3}$$

which are known as the Sommerfeld radiation conditions and by appropriate reflection conditions on solid boundaries, provided that they existed.

Wave agitation in port basins with reflecting boundaries can be computed by means of Equation (2).

ANALYTICAL FORMULATION

A schematization of a typical geometry to be studied can be seen in Figure 1, this is a port basin. The basin has an irregular shape together with

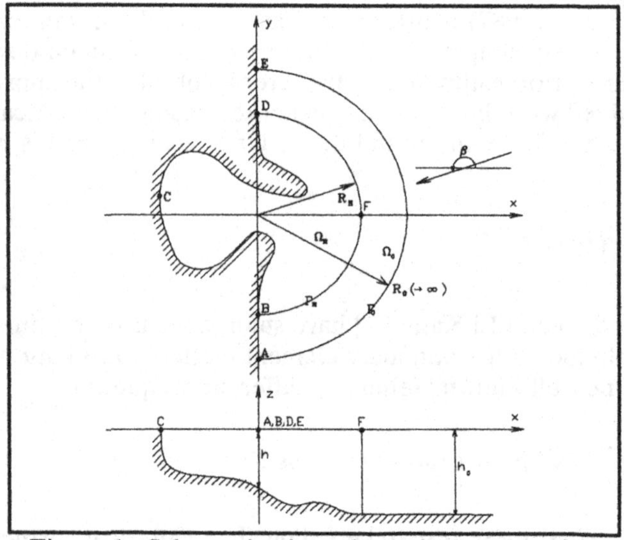

Figure 1.- Schematization of a typical port basin geometry.

varying depth.

The solution will be calculated on a domain Ω made up by joining two non-overlapping regions, that is to say:

$$\Omega = \Omega_H + \Omega_O \tag{4}$$

Region Ω_H corresponds to the port basin itself and is limited by the BCD and Γ_H curves. In the figure mentioned above, this boundary is shown as a semicircle of radius R_H but, in general, Γ_H can take on an arbitrary shape. The water depth may vary on this region.

The port will be able to be limited by vertical walls or by a combination of vertical and inclined walls; then, the energy can be fully reflected at the BCD boundary or not. The outer region, Ω_O is limited by Γ_H, the physical boundaries AB and DE, and extends towards infinity. This region is used to model the open ocean or water body surrounding the basin.

We shall asume that the boundaries AB and DE are vertical and perfectly reflecting ones and that the water depth is constant within Ω_O. In this way, the solution within this region can be linearly decomposed in a system of incident and reflected waves and another system of radiated waves. This first system is defined as the one which would be obtained if the depth were constant throughout the whole region $x > 0$ and the port basin would not exist, with a vertical wall extending from point A to point E. This wave system is determined once the incident wave system is specified.

The radiated wave system is then simply defined as the difference between the real wave system within Ω_O and the incident- reflected wave system and represents the radiated energy from the mouth of the port basin towards infinity. This wave system satisfies asymptotically the Sommerfeld radiation condition. Therefore, the problem stated within Ω is decomposed in to two problems on different domains: Ω_H and Ω_O.

The reduced small-amplitude wave Equation (2) will have to be solved within the domain Ω_H specifying along the boundary BCD a partially-reflecting condition as follows:

$$\bar{n} \,.\, \nabla\varphi - i \,.\, k \,.\, \alpha \,.\, \varphi = 0 \tag{5}$$

In Equation (5) α is a real coefficient in a way such that $\alpha = 0$ means a complete reflection and $\alpha = 1$ a complete absorption. Within the domain

Ω_0, whose depth is considered to be constant, the Helmholtz differential equation shown below will have to be solved:

$$\nabla^2 \varphi + k^2 \varphi = 0 \tag{6}$$

If, in addition to that, φ is linearly decomposed in a system of incident and reflected waves, $\tilde{\varphi}$ and a system of radiated waves, φ_R, Equation (6) will then be transformed to:

$$(\nabla^2 \tilde{\varphi} + k^2 \tilde{\varphi}) + (\nabla^2 \varphi_R + k^2 \varphi_R) = 0 \tag{7}$$

Since the radiated wave system is assumed to be known, the reflected wave system originated by the existence of an impervious vertical wall that is limited by points A and E of the sketch of Figure 1 is not difficult to find out. As an example, let us suppose that the incident perturbation has the form:

$$\varphi_i = \exp\,[\,i\,.\,k\,(x\,.\,\cos\beta + y\,.\,\sin\beta\,] \tag{8}$$

where β is the incidence angle, as shown in Figure 1. The reflected wave will be then as follows:

$$\varphi_r = \exp\,[\,i\,.\,k\,(-x\,.\,\cos\beta + y\,.\,\sin\beta\,] \tag{9}$$

therefore, for the incident-reflected wave system it follows that:

$$\tilde{\varphi} = R\,.\,\cos\,(k\,.\,x\,.\cos\beta)\,.\,e^{i.k.y.sen\beta} \tag{10}$$

Carrying out a direct substitution it can be shown that $\tilde{\varphi}$ satisfies the Helmholtz differential equation and that along the AE boundary the following condition is verified:

$$\bar{n}\,.\,\nabla\tilde{\varphi} = 0 \tag{11}$$

The radiated wave system will then be obtained as a solution of:

$$\nabla^2 \varphi_R + k^2 \varphi_R = 0 \tag{12}$$

within Ω_0 together with the boundary conditions:

$$\bar{n}\,.\,\nabla\varphi_R = 0 \tag{13}$$

along the physical boundaries AB and DE, and also

$$\lim_{r\to\infty} \sqrt{r}\,.\left(\frac{\partial\varphi_R}{\partial r} - i\,.\,k\,.\,.\varphi_R\right) = 0 \tag{14}$$

along the boundary at infinity.

Both regions, Ω_H and Ω_0, will have to be coupled by means of appropriate continuity conditions along the common boundary Γ_H. Therefore:

$$\varphi = \tilde{\varphi} + \varphi_R \tag{15}$$

$$\bar{n}\,.\,\nabla\varphi = \bar{n}\,.\,\nabla(\tilde{\varphi} + \varphi_R) \tag{16}$$

represent the continuity of the free surface and the velocity in the normal direction to Γ_H respectively.

VARIATIONAL FORMULATION

Using the Green integral formulae and bearing in mind that throughout Ω_H Equation (2) is the governing equation, the variational problem associated with the differential problem raised in the previous section can be established as finding out φ in a way such that:

$$-\int_{\Omega_.} c.c_g.\nabla\varphi.\nabla v.d\Omega + \int_{\Omega_.} \omega^2.\frac{c_g}{c}.\varphi.v.d\Omega + \int_{\partial\Omega_.} c.c_g.\bar{n}.\nabla\varphi.v.dS = 0 \tag{17}$$

holds for all the v members of the $H^1(\Omega_H)$ class of functions. The boundary $\partial\Omega_H$ has been made joining curves BCD and Γ_H. If we say that

$$\lambda = \bar{n}\,.\,\nabla\varphi|_{\Gamma_.} \tag{18}$$

$$-\int_{\Omega_.} c.c_g.\nabla\varphi.\nabla v.d\Omega + \int_{\Omega_.} \omega^2.\frac{c_g}{c}.\varphi.v.d\Omega + \int_{BDF} c.c_g.\bar{n}.\nabla\varphi.v.dS + \int_{\Gamma_.} c.c_g.\lambda.v.dS = 0 \tag{19}$$

Now, Equation (6) governs within region Ω_0. Making use again of the Green integral formulae together with the boundary conditions expressed by Equations (13) and (14), one may obtain:

$$\frac{1}{2} \cdot \varphi_R(P) = \int_{\Gamma_H} \varphi_R(Q) . \bar{n} . \nabla\omega(P,Q) . dS_Q - \int_{\Gamma_H} \bar{n} . \nabla\varphi_R(Q) . \omega(P,Q) . dS_Q \tag{20}$$

for every P that belongs to Γ_H. In Equation (20) the function $\omega(P,Q)$ is the fundamental solution associated with the bidimensional Helmholtz problem:

$$\omega(P,Q) = \frac{1}{4i} . H_0^1(k . r_{PQ}) \tag{21}$$

H_0^1 being the Hankel function of the first class and zero order.

If P lies within Γ_H however, the coupling conditions establish the following relations:

$$\varphi_R(P) = \varphi(P) - \tilde{\varphi}(P) \tag{22}$$

$$\bar{n} . \nabla\varphi_R(P) = \lambda - \bar{n} . \nabla\tilde{\varphi}(P) \tag{23}$$

In Equation (23) the unit vector $\bar{n}$ represents the outer normal along Γ_H from the region Ω_H; in Equation (20) instead, it represents the outer normal along Γ_H from Ω_0. The relation between both of them is given by:

$$\bar{n}_{\Omega_H} = -\bar{n}_{\Omega_0} \tag{24}$$

Bearing this in mind and substituting the Equations (22) and (23) in Equation (20), it can be obtained:

$$\begin{aligned} &\frac{1}{2} \cdot \varphi(P) + \int_{\Gamma_H} \varphi(Q) . \bar{n} . \nabla\omega(P,Q) . dS_Q - \int_{\Gamma_H} \lambda . \omega(P,Q) . dS_Q = \\ &\frac{1}{2} \cdot \tilde{\varphi}(P) + \int_{\Gamma_H} \tilde{\varphi}(P) . \bar{n} . \nabla\omega(P,Q) . dS_Q - \int_{\Gamma_H} \bar{n} . \nabla\tilde{\varphi}(Q) . \omega(P,Q) . dS_Q \end{aligned} \tag{25}$$

for every point P on Γ_H.

Let $\mu(P)$ a member of the $H^{-1/2}(\Gamma_H)$ class of functions; multiplying formally Equation (25) by $\mu(P)$ and carrying out an integration over Γ_H the following can be obtained:

$$\frac{1}{2}\int_{\Gamma_*} \varphi(P) \,.\, \mu(P) \,.\, dS_P + \int_{\Gamma_*} \mu(P) \times$$
$$\times \int_{\Gamma_*} \varphi(Q) \,.\, \bar{n} \,.\, \nabla\omega(P,Q) \,.\, dS_Q\, dS_P - \int_{\Gamma_*} \mu(P) \int_{\Gamma_*} \lambda \,.\, \omega(P,Q)\, dS_Q\, dS_P =$$
$$= \frac{1}{2}\int_{\Gamma_*} \tilde{\varphi}(P) \,.\, \mu(P) \,.\, dS_P + \int_{\Gamma_*} \mu(P) \times \tag{26}$$
$$\times \int_{\Gamma_*} \tilde{\varphi}(Q) \,.\, \bar{n} \,.\, \nabla\omega(P,Q) \,.\, dS_Q\, dS_P - \int_{\Gamma_*} \mu(P) \int_{\Gamma_*} \bar{n} \,.\, \nabla\tilde{\varphi}(Q)\, \omega(P,Q)\, dS_Q\, dS_P$$

for every P on Γ_H.

Combining Equation (19) with Equation (26) the problem can be formulated in the following way: to find out (φ,λ) that belongs to $H^1(\Omega_H) \times H^{-1/2}(\Gamma_H)$ such that Equation (19) can be satisfied for any function v of $H^1(\Omega_H)$ and Equation (26) for every function μ of $H^{-1/2}(\Gamma_H)$. These equations are the starting point for the formulation of the coupled finite-element - boundary element method.

APPROXIMATE SOLUTION

Let us consider a regular triangulation of Ω_H : $T_h = \{T\}$. If h represents the diameter of the circle that circumscribes the greatest triangle, the discretization process partitions Γ_H in segments whose length is at most h. Let us consider then the partition $S_h = \{S\}$ of the boundary Γ_H.

Let V_h be the set of all the continuous functions φ_i which are polynomials of degree $m \geq 1$ in each T that belongs to T_h. If M represents the number of elements of the triangulation network, an approximation φ_h of φ can be developed by means of:

$$\varphi_h = \sum_{i}^{M} \alpha_i \,.\, \varphi_i \tag{27}$$

with α_i, $i = 1, \dots, M$ being constants.

Let H_h be the set of all the continuous functions λ_i which are polinomials of a degree smaller than m in each S that belongs to S_h. Provided that N is the number of segments of the partition, an approximation λ_h of λ can be developed in the following way:

$$\lambda_h = \sum_{i}^{N} \beta_i \,.\, \lambda_i \tag{28}$$

with β_i, $i = 1, \dots, N$ being constants.

We can establish the following approximate problem: to find out (φ_h, λ_h) that belongs to $V_h \times H_h$ such that

:

$$-\int_{\Omega} c \,.\, c_g \,.\, \nabla\varphi_h \,.\, \nabla v \,.\, d\Omega + \int_{\Omega} \omega^2 \,.\, \frac{c_g}{c} \,.\, \varphi_h \,.\, v \,.\, d\Omega + \\ + \int_{BCD} c \,.\, c_g \,.\, \bar{n} \,.\, \nabla\varphi_h \,.\, v \,.\, dS + \int_{\Gamma} c \,.\, c_g \,.\, \lambda_h \,.\, v \,.\, dS = 0 \qquad (29)$$

holds for every v that belongs to V_h and

$$\frac{1}{2}\int_{\Gamma} \varphi_h(P) \,.\, \mu(P) \,.\, dS_P + \int_{\Gamma} \mu(P) \times \\ \times \int_{\Gamma} \varphi_h(P) \,.\, \bar{n} \,.\, \nabla\omega(P,Q) \,.\, dS_Q \, dS_P - \int_{\Gamma} \mu(P) \int_{\Gamma} \lambda_h \,.\, \omega(P,Q) \, dS_Q \, dS_P = \\ = \frac{1}{2}\int_{\Gamma} \tilde{\varphi}(P) \,.\, \mu(P) \,.\, dS_P + \int_{\Gamma} \mu(P) \times \\ \times \int_{\Gamma} \tilde{\varphi}(Q) \,.\, \bar{n} \,.\, \nabla\omega(P,Q) \,.\, dS_Q \, dS_P - \int_{\Gamma} \mu(P) \int_{\Gamma} \bar{n} \,.\, \nabla\tilde{\varphi}(Q) \, \omega(P,Q) \, dS_Q \, dS_P \qquad (30)$$

holds for every μ that belongs to H_h.

In this approximate problem, let us take: $v = \varphi_j$ and $\mu = \lambda_j$. Then, an algebraic equation system from Equations (29) and (30) can be obtained making it possible to obtain the coefficients α_i,, $i = 1, ..., M$ together with β_i, $i = 1, ..., N$ which lead to the solution of the problem. It must be denoted that this formulation leads to non-symmetrical matrices.

RESULTS OBTAINED

The method described above was applied to situations where it is possible to get an analytical or semi-analytical solution to the problem, always within the limit of what is knwon as the long wave approximation.

Figure 2 shows the response of a rectangular port located on a straight infinite coast subjected to incoming waves of different lengths. This problem has been studied by several researchers from both the theoretical as well as from the experimental viewpoint [4, 5, 6]. As seen in that figure, the wave amplitude exhibits a maximum for some of the modelled wave lengths with corresponding frequencies which are denominated resonant frequencies of the port basin.

It was also investigated the effect of the water depth on the response of the basin when it is excited by long-period waves.

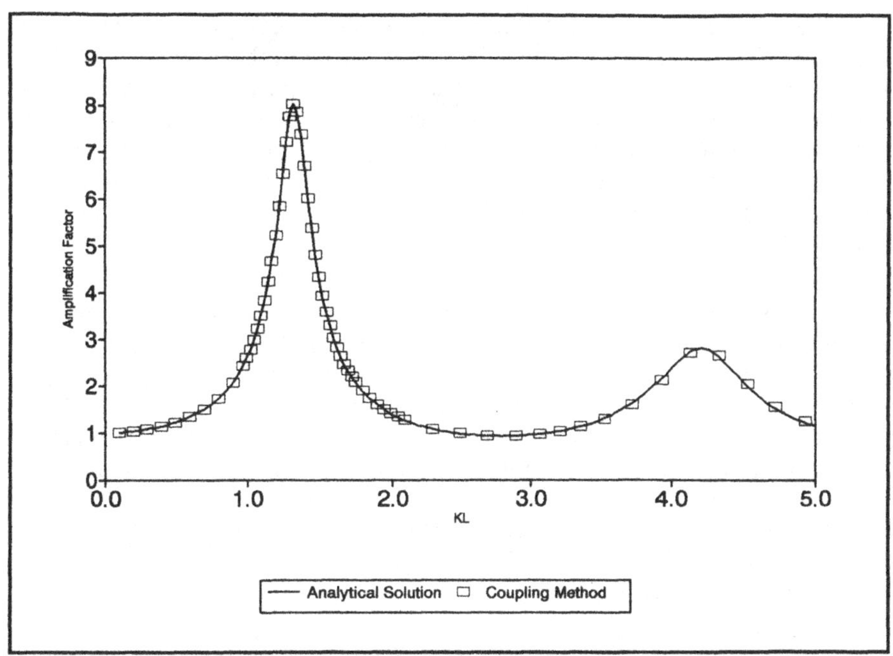

Figure 2.- Reponse of a rectangular basin with constant water depth located on a semi-infinite coast.

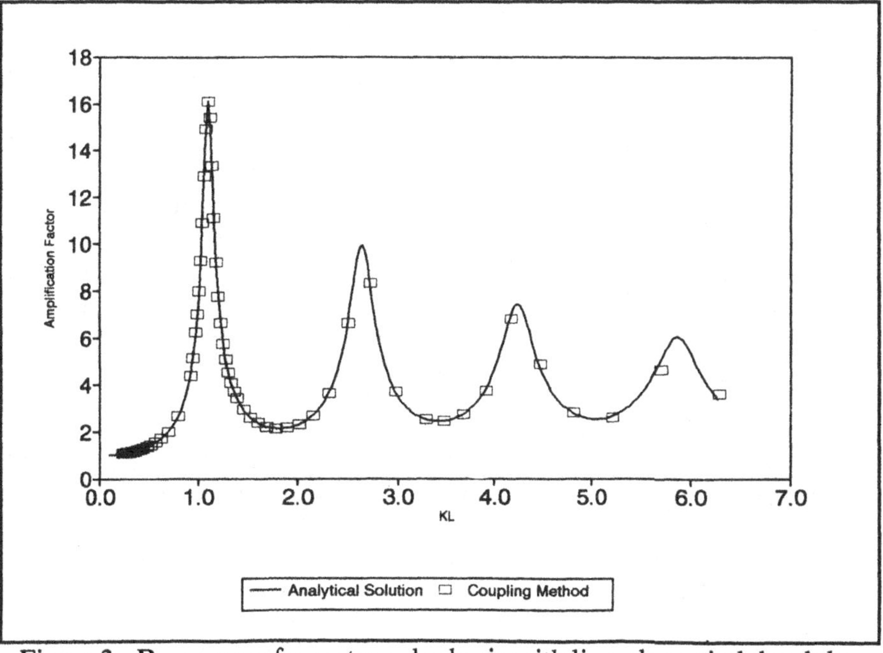

Figure 3.- Response of a rectangular basin with linearly-varied depth located on a semi-infinite coast.

Figure 3 corresponds to the response of a rectangular basin with a linearly-varied water depth. It must be denoted that in these two case studies the basins have the same length, width and water depth at the entrance; the resonant response curves are different however. Not only the peak magnitude is much different but their frequency location is different as well. Therefore, in order to estimate the resonant frequencies of a port it is not enough to define the basin geometry, its bathymetric features must also be carefully defined.

Looking at Figures 2 and 3, a simple comparison leads to conclude that the agreement between the numerical and the analytical solutions is very satisfactory. Comparisons with results obtained by other methods were also made. Solutions obtained by a BEM model with indirect formulation, by a FEM model, where the calculation domain must be compulsory truncated using some criterion and the coupled method, where the boundary condition at infinity is implicitly imposed were also carried out. These results, shown in Figure 4, were obtained for the same partition of the basin perimeter and for the same discretization of the enclosing region, something that highlights the aptitude of the coupled method.

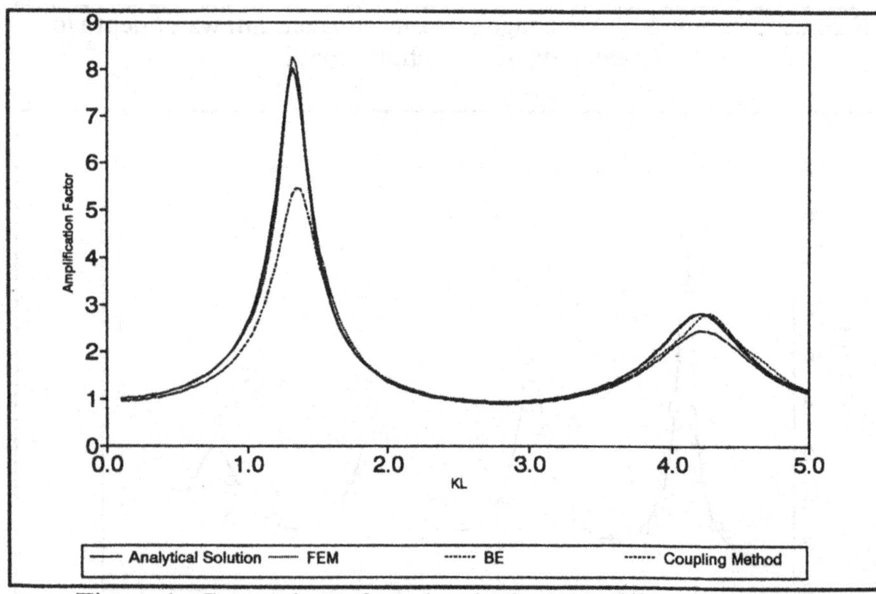

Figure 4.- Comparison of results obtained by different methods.

CONVERGENCE

Numerical experiments that were carried out with three-node triangular elements in order to interpolate the solution inside them and with C^0 continuity elements to interpolate the normal derivative of the solution on the transmitting boundary let us conclude that the method has a $O(h)$ convergence, h being the diameter of the circumscribing circle of the largest element used in the discretization.

The variation of the computed error in the norm L_2 is presented in Figure 5 as a function of the size of the elements used to determine the response of the rectangular port basin with constant depth subjected to intermediate-length incident waves. It can be clearly seen a linear decrease of the error with an increase of the number of elements used in the discretization. This result agrees well with theoretical results obtained by He Yinnian and Li Katai [1].

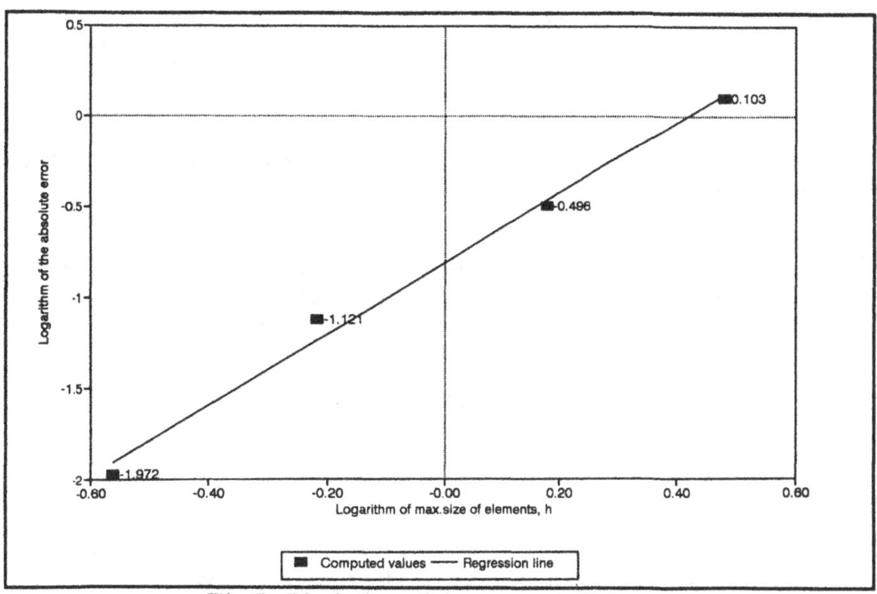

Fig.5.- Variation of the computed error.

CONCLUSIONS

Despite of the fact that the variational formulation, that is the starting point of the coupled method, makes it impossible to obtain symmetrical matrices, the method used in this research work gives particularly useful solutions when very simple elements are used in the domain discretization.

On other respect, as it includes into the formulation the boundary condition at infinity, there is no need to truncate the region using ambiguous criteria.

A fundamental advantage of the method shown in the preceding sections lies on the possibility of making an a-priori estimation of the error involved in the approximation.

BIBLIOGRAPHY

1.- He Yinnian and and Li Katai: *The coupling method of finite elements and boundary elements for radiation problems*, IMA Preprint Series #296, 1987.

2.- Berkhoff, J.C.W.: *Mathematical models for simple harmonic water waves; wave difraction and refraction*, Delft Hydraulics Laboratory, Publ.Nº163, 1976.

3.- Schönfeld, J.Ch.: *Propagation of two- dimensional short waves*, Delft Hydraulics Laboratory, Manuscript,, 1972.

4.- Lee, J.J.: *Wave induced oscillations in harbours of arbitrary shape*, W.M.Keck Laboratory of Hydrualics and Water Resources, California Institute of Technology, Report NºKH-R-20, 1969.

5.- Bettes, P. and Zienkewicz, O.: *Difraction and refraction of surface waves using finite and infinite elements*, Int.J.for Num.Meth.Engg., Vol.11, pp.1271-1290, 1976.

6.- Zelt, J.A.: *The response of harbors with sloping boundaries to long wave excitation*, W.M.Keck Laboratory of Hydraulics and Water Resources, California Institute of Technology, Report NºKH-R-47, 1986.

SECTION 2: TIDES

On Quality Assurance for Numerical Tidal Models

S.G. Wallis (*), D.W. Knight (**)
(*) Department of Civil & Offshore Engineering, Heriot-Watt University, Edinburgh, Scotland, U.K.
(**) School of Civil Engineering, The University of Birmingham, Birmingham, England, U.K.

ABSTRACT

One-dimensional numerical tidal models are frequently used by civil engineers. Whether the engineer is developing in-house software or running a commercially available package, some form of quality assurance testing needs to be undertaken so that he/she can be confident that there are no errors in the computer code. The practical operating characteristics of the numerical scheme employed in the model also need to be assessed.

The problem addressed in this paper is how should the engineer check the veracity of the model output, bearing in mind that exact solutions to the non-linear equations of motion are not available? Four possible strategies are considered involving the use of: field data, simplified analytical solutions, laboratory data and alternative numerical solutions. The advantages and disadvantages of these strategies are discussed with reference to a set of numerical solutions for the case of tidal propagation in a simple idealised estuary.

It is concluded that laboratory data offers the best source of objective validation data, but difficulties remain with the representation of frictional resistance.

INTRODUCTION

The propagation of tides in estuaries is frequently modelled using a one-dimensional scenario in which solutions are sought to the spatially integrated Navier-Stokes and mass conservation equations. Since the one-dimensional equations are non-linear, and also because natural estuaries are inherently non-uniform, numerical solution methods are required to give tidal elevations and flows throughout an estuary. The successful application of the numerical model is critically dependent on the effort expended at the calibration stage. This should include a study of: (1) the degree to which the governing equations actually represent the physical processes; (2) the spatial schematisation used; (3) the characteristics of the numerical scheme adopted; (4) the degree to which the resistance and/or other coefficients are consistent with field measurements; and (5) the application of the model to as wide a range of tidal and fluvial conditions as possible.

During the development of the model it is important to carry out some form of quality assurance testing in order to eliminate errors in the computer code and to assess the practical operating characteristics of the numerical method. Also, it is a prudent measure to test the model over the full range of tidal frequencies found to exist in the tidal response of the estuary under study.

This paper considers the problem of how the modeller can check the validity of the model output. The model results are, after all, approximate in the sense that they are numerical solutions of a pair of non-linear partial differential equations for which exact solutions are not available. In practical applications, of course, models can be calibrated by tuning the friction coefficient such that the model output "agrees" with observations. But this does not constitute an objective check of the model.

What the modeller needs is an independent check to ensure that the model is working correctly. One way of furnishing this need is to run the model for idealised conditions for which analytical solutions to the appropriately simplified equations of motion are available.

Unfortunately, however, these analytical solutions are of limited use in this regard because the idealised conditions under which they apply nearly always include a linearisation of the friction and/or advective acceleration terms in the equations of motion. Thus the numerical treatment of, and the computer code relating to, the non-linear terms cannot be checked in this way.

An alternative source of independent data for model checking is laboratory data. In theory, there is a good chance of friction coefficients being known in the laboratory, but for oscillatory flows the frictional behaviour of the boundary layer depends on the frequency of oscillation, so that in practice an adequate description of frictional losses may be difficult to achieve using the same resistance equations as those used for real estuaries.

One further way of checking a numerical model is to compare its output with that of another numerical model. Although this is not ideal, because both outputs are only approximate solutions, agreement would strengthen the belief that both models were working correctly.

This paper addresses some of the ideas introduced above. Firstly, we present one set of numerical solutions to the fully non-linear equations of motion describing the tidal hydraulics of a closed, horizontal rectangular estuary. The solutions were obtained using the Preissmann finite difference scheme. Secondly, we discuss the use of field data, analytical solutions, laboratory data and alternative numerical solutions to validate these results.

THEORETICAL BACKGROUND

The movement of water in the one-dimensional estuary under consideration is described by the St. Venant equations (Cunge et al [3]:

$$\frac{1\partial Q}{g\partial t} + \frac{1\partial}{g\partial x}\left[\frac{\beta Q^2}{A}\right] + \frac{A\partial H}{\partial x} + \frac{A\tau}{\rho g R} = 0 \tag{1}$$

$$\frac{W\partial H}{\partial t} + \frac{\partial Q}{\partial x} = 0 \tag{2}$$

where W=width of the estuary, H=water surface elevation above the horizontal bed, Q=discharge, A=cross-sectional area of flow, g=gravitational acceleration, τ=mean bed shear stress, ρ=water density, R=hydraulic radius, x=longitudinal co-ordinate direction and t=time. The origin of the co-ordinate system is taken at the closed end (head) of the estuary and the mouth of the estuary is at x=l, where l=estuary length. Ebb flows are taken as positive. The momentum coefficient (β) in equation (1) is set to unity everywhere, which in the absence of better information is normal practice.

Solutions to equations (1) and (2) are sought under the following boundary conditions: at the estuary mouth there is a cosinusoidal variation of water surface elevation (of amplitude a and period T) about the mean depth, h, and at the estuary head there is no flow, so that the tidal oscillation is reflected. The tidal response of the estuary is that of a damped standing wave with high water at the mouth occurring at a phase of 0° and low water there occurring at a phase of 180°.

NUMERICAL SOLUTIONS

Equations (1) and (2) were solved using the Preissmann finite difference scheme (Abbott & Basco [1]; Cunge et al [3]; Liggett & Cunge [10]; Samuels & Skeels [17]). The numerical characteristics of this scheme are well documented, and it has an excellent track record in computational hydraulics. The estuary case defined in Table 1 was chosen as a "benchmark" problem and numerical solutions were obtained over the ranges of frequency and frictional resistance indicated by $\sigma l/c$ and f/α, respectively. Here, σ is the tidal frequency ($=2\pi/T$), c is the shallow water small amplitude wave celerity ($=\surd(gh)$), f is a resistance parameter ($=2\tau A^2/(\rho Q^2)$ = Darcy-Weisbach coefficient/4) and α is a geometrical parameter accommodating the width to mean depth ratio of the estuary ($=R/h$, evaluated at mean depth). $\sigma l/c$ was changed by varying the length of the estuary and f/α was changed by varying the value of f: other parameters took the constant values shown in Table 1. ψ is a non-dimensional group ($=(a/h)[c/(h\sigma)](f/\alpha)$), see Knight [6].

Parameter	Dimensionless Group
Mean depth, h=10m Width, W=200m	$a/h=0.05$
Geometrical parameter, $\alpha=0.909$	$c/(h\sigma)=7092$
Amplitude at mouth, a=0.5m	$\sigma l/c=0.226$ - 1.410
Time period, T=12.5hr Length, l=16 - 100km	$f/\alpha=0.00033$ - 0.0330
Resistance parameter, f=0.0003 - 0.0300	$\psi=0.117$ - 11.7

Table 1. Details of the estuary benchmark case under study.

A substantial number of preliminary model runs were undertaken in order to define the conditions under which the numerical solutions were to be obtained. Based on these results the following guidelines were adopted:

(1) A distance step of 4km was used in all cases. Hence the number of space steps in the model increased as $\sigma l/c$ was increased by lengthening the estuary. However, the spatial representation of the undistorted tidal oscillation remained constant and approximately equal to 111 parts per wavelength.
(2) The temporal weighting parameter in the finite difference scheme was set at 0.6.
(3) One iteration of a double sweep elimination procedure (i.e. two passes) was used to solve the non-linear difference equations at each time step.
(4) All runs were started at a high water condition of no flow everywhere and a water surface elevation given by linear theory ($=h + a/\cos(\sigma(l-x)/c)$).
(5) The convergence of any run was taken to have occurred when the elevations of both high and low water at the estuary head from successive tidal cycles changed by less than 2mm.
(6) The same time step was used for all simulations at a particular value of $\sigma l/c$. The size of the time step was

determined by monitoring the convergence of the numerical solution as the time step was successively halved from an initial value of 450s (100 per tidal cycle). The same numerical criterion as in (5) was adopted in order to determine the largest time step that could be used, and at all values of $\sigma l/c$ this was determined by the minimum friction case.

RESULTS

The results from the final runs are presented in Tables 2 & 3 and Figures 1 & 2. The error in the calculated levels is estimated to be less than ±5mm, however, the results have been rounded to the nearest 1cm since this is certainly the highest order of accuracy expected in any practical application of an estuary model. Similarly the peak velocities have been rounded to the nearest 1cm/s and all the phases are given to the nearest degree.

$\sigma l/c$	f	ψ	LEVEL		AMP	PHASE		VELOCITY		PHASE		N
			HW (m)	LW (m)		HW (o)	LW (o)	EBB (m/s)	FLO (m/s)	EBB (o)	FLO (o)	
0.226	0.0003	0.12	10.51	9.49	1.03	0	180	0.09	0.09	94	266	100
0.226	0.0030	1.17	10.51	9.49	1.03	0	180	0.09	0.09	94	266	100
0.226	0.0300	11.7	10.51	9.49	1.03	0	180	0.09	0.09	97	270	100
0.451	0.0003	0.12	10.55	9.44	1.11	0	180	0.21	0.21	97	266	100
0.451	0.0030	1.17	10.55	9.44	1.12	0	180	0.21	0.21	101	270	100
0.451	0.0300	11.7	10.58	9.42	1.16	7	191	0.20	0.20	115	288	100
0.620	0.0003	0.12	10.59	9.37	1.22	0	180	0.33	0.34	97	263	100
0.620	0.0030	1.17	10.60	9.36	1.24	4	187	0.31	0.34	97	259	100
0.620	0.0300	11.7	10.62	9.40	1.21	29	212	0.27	0.28	112	281	100
0.733	0.0003	0.12	10.56	9.23	1.34	345	184	0.46	0.49	114	252	800
0.733	0.0030	1.17	10.66	9.30	1.37	357	198	0.36	0.48	106	267	800
0.733	0.0300	11.7	10.61	9.42	1.19	45	231	0.32	0.32	107	231	800
0.846	0.0003	0.12	10.91	9.38	1.54	358	194	0.56	0.61	64	292	800
0.846	0.0030	1.17	10.80	9.27	1.52	7	203	0.48	0.57	83	280	800
0.846	0.0300	11.7	10.57	9.46	1.11	61	247	0.34	0.34	123	307	800
0.959	0.0003	0.12	10.96	9.21	1.75	2	185	0.68	0.70	79	283	200
0.959	0.0030	1.17	10.89	9.20	1.70	18	208	0.63	0.67	95	288	200
0.959	0.0030	11.7	10.52	9.51	1.00	74	261	0.35	0.34	137	319	200
1.072	0.0003	0.12	11.14	9.04	2.10	4	185	0.88	0.89	85	283	200
1.072	0.0030	1.17	10.99	9.11	1.89	32	221	0.77	0.78	106	299	200
1.072	0.0300	11.7	10.47	9.56	0.90	86	272	0.35	0.33	146	326	200
1.184	0.0003	0.12	11.44	8.79	2.66	8	190	1.19	1.19	87	286	400
1.184	0.0030	1.17	11.06	9.05	2.02	49	238	0.87	0.86	121	313	400
1.184	0.0300	11.7	10.43	9.61	0.82	96	284	0.34	0.32	151	331	400
1.297	0.0003	0.12	11.99	8.35	3.63	19	202	1.71	1.68	95	298	800
1.297	0.0030	1.17	11.07	9.03	2.03	68	256	0.94	0.89	138	327	800
1.297	0.0300	11.7	10.39	9.64	0.75	106	294	0.32	0.31	155	334	800
1.410	0.0003	0.12	12.78	7.72	5.06	49	235	2.50	2.34	122	331	1600
1.410	0.0030	1.17	11.02	9.06	1.96	85	274	0.94	0.87	153	340	1600
1.410	0.0300	11.7	10.36	9.67	0.69	115	304	0.31	0.30	158	336	1600

Table 2. Numerical results at the head of the estuary.

Table 2 shows a selection of computed water levels and phases of high and low water at the estuary head together with the amplification of the tidal range (AMP) which is evaluated as the range at the head divided by that at the mouth (1m). The table also shows peak ebb and flood velocity magnitudes (ebb positive, flood negative) and their phases, at a location 4km inland from the mouth. We choose to present velocity data one distance step inland from the mouth to facilitate the use of the data with numerical schemes based on either a staggered or a non-staggered spatial discretisation. The values of N are the number of time steps per tidal cycle used in the simulations.

For two cases from the higher frequency range (i.e. dimensionless frequency, $\sigma l/c > 1$), Table 3 shows the longitudinal variation of the level and phase of high and low water, and the longitudinal variation of the magnitude and phase of the peak flood and ebb velocity. This data will enable model testing to be carried out at a number of locations along the estuary. In the remainder of this paper, however, we concentrate on the results at the estuary head.

$\sigma l/c$	f	l (km)	$\sigma x/c$	LEVEL HW (m)	LEVEL LW (m)	PHASE HW (o)	PHASE LW (o)	VELOCITY EBB (m/s)	VELOCITY FLO (m/s)	PHASE EBB (o)	PHASE FLO (o)
1.072	0.0030	0	0	10.5	9.5	0	180	0.77	0.78	106	299
1.072	0.0030	20	0.282	10.69	9.36	22	202	0.67	0.79	106	299
1.072	0.0030	40	0.564	10.86	9.22	29	214	0.51	0.55	104	295
1.072	0.0030	60	0.846	10.97	9.13	32	220	0.29	0.32	101	288
1.072	0.0030	76	1.072	10.99	9.11	32	221	0.05	0.05	99	284
1.410	0.0030	0	0	10.5	9.5	0	180	0.94	0.87	153	340
1.410	0.0030	20	0.282	10.53	9.49	46	220	0.87	0.84	157	344
1.410	0.0030	40	0.564	10.70	9.37	64	242	0.75	0.77	156	344
1.410	0.0030	60	0.846	10.86	9.23	76	259	0.56	0.62	150	338
1.410	0.0030	80	1.128	10.98	9.11	83	270	0.35	0.41	143	328
1.410	0.0030	100	1.410	11.02	9.06	86	274	0.11	0.13	139	322

Table 3. Numerical results along the estuary.

Figure 1 shows the tidal range amplification as a function of $\sigma l/c$ and ψ, and Figure 2 shows the corresponding average phase difference plotted in the same way. This average phase difference is the average of the high water and low water phase differences between head and mouth. Greater detail of the tidal hydraulics can be gained by examining the tidal amplitude and phase for high and low water individually, see Wallis & Knight [20], but for our purposes here the main features are adequately depicted by Figures 1 & 2 which show how the tidal hydraulics are controlled by $\sigma l/c$ and ψ.

All three of the non-dimensional groups which contribute to ψ may influence the tidal response. However, because a/h and c/(hσ) were kept constant the variations in ψ here are solely due to changes in the frictional resistance. These results show that when σl/c < 1 the frictional resistance has little effect on tide levels, but when σl/c > 1 the resistance influences the tide levels significantly. It is also noticeable that the levels of high and low water are approximately independent of σl/c when ψ > 4: phase differences, however, tend to increase with increasing σl/c and ψ. The behaviour of peak velocities is broadly similar to that of tide levels.

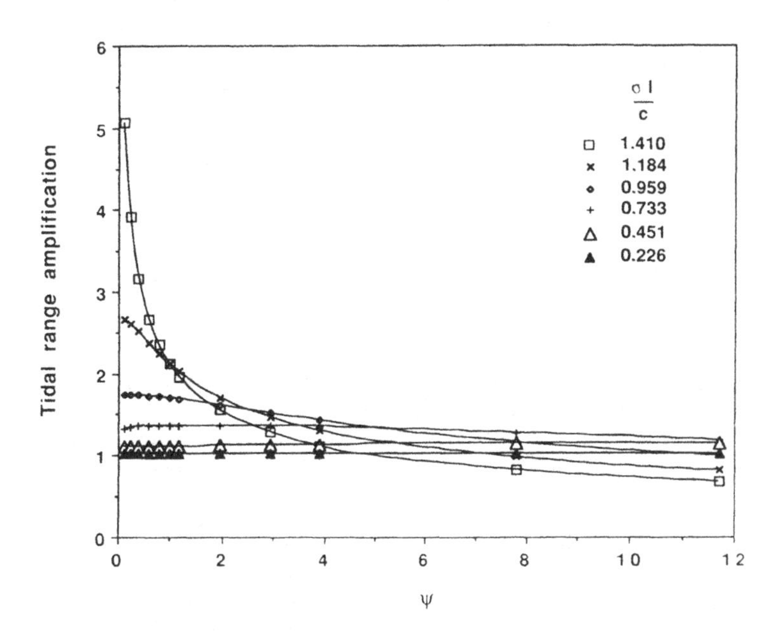

Figure 1. Tidal range amplification as a function of σl/c and ψ (for clarity, only 6 values of σl/c are shown).

An interesting feature of the results, which is more noticeable in the phase data than in the levels, is the occurrence of a weak resonance mode when the frequency is close to $\pi/4$ (see also Knight & Ridgway [8]). The distortion in the otherwise smooth trends in the results is damped out as the resistance increases but is evident in the lower resistance cases for σl/c = 0.733 & 0.846. From the values of N in Table 2, it is clear that the convergence characteristics of the numerical scheme are also affected by this phenomenon.

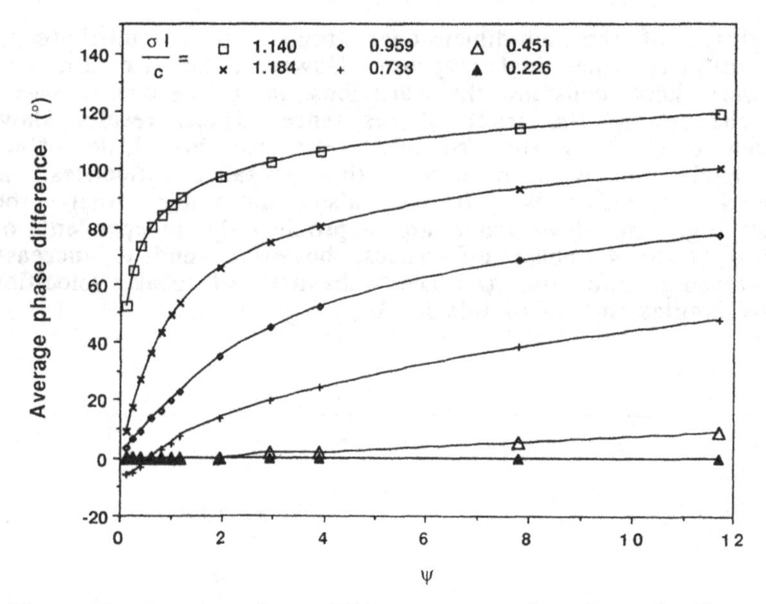

Figure 2. Average phase difference as a function of $\sigma l/c$ and ψ (for clarity, only 6 values of $\sigma l/c$ are shown).

DISCUSSION

The numerical solutions presented here are believed to be the most accurate ones available for the case study. They are superior to analytical solutions because they retain the full non-linear effects of the quadratic resistance term and the advective acceleration term. The problem introduced earlier remains however: how do we know that these results really are solutions to the governing equations of motion?

Although a lot of care has been taken to ensure that in practical terms the solutions have converged, only a check against an independently obtained data set would be entirely convincing. Four sources for such independent data have been considered, namely: field data, analytical solutions, laboratory data and other numerical solutions.

Field data is not particularly useful to us here for a number of reasons. Applying a one-dimensional numerical tidal model to a natural estuary requires the modeller to estimate the resistance coefficient in the dynamic equation at every location included in the model. Since resistance coefficients cannot be directly measured, their values are usually unknown. Hence they have to be estimated, and when model results are compared with measurements of, for example, water levels, the modeller does not know whether discrepancies are due to errors in the resistance coefficients, errors in the field data, errors introduced by a poor representation of the physical processes in the model or errors in the model itself. Modellers should also be aware that in estuaries with a large tidal range to mean depth ratio, resistance coefficients show a dependence on stage (Knight [7]; Wallis & Knight [19]) as well as varying with location.

Analytical solutions are not ideal for model checking because they only apply for special cases involving some simplification of the governing equations of motion. This usually takes the form of a linearisation of the friction and/or advective acceleration terms in the dynamic equation. And, indeed, in many cases one or both of them is deleted altogether (Bowers & Lennon [2]; Ippen [4]; Lynch & Gray [11]; Needham [12]; Ostendorf [13]; Prandle [14]). Thus errors associated with these terms in the fully non-linear model cannot be identified by comparing model results with the analytical solutions. One exception to this is the solution presented by Proudman [15] which does retain the correct non-linear nature of the non-linear terms. Inevitably, however, the solutions are only valid when the non-linear terms are "small". Proudman's solution is discussed at length in Knight [5], Knight [6] and Wallis & Knight [20].

Laboratory data appears at first sight to offer a good way of checking numerical solutions, and the authors have access to such a data set from a previous study (Ridgway [16]; Knight & Ridgway [8]; Knight & Ridgway [9]). The experiments are described in detail in Ridgway [16].

Unfortunately, direct comparisons with the numerical results is not possible because there are doubts over the appropriate values of the Darcy-Weisbach resistance coefficient and the estuary length, i.e the values of ψ and $\sigma l/c$ are not known precisely for the experimental data. The former doubts are related to the effects of the tidal frequency on the vertical and lateral velocity profiles in the estuary channel. In particular this caused the frictional behaviour of the channel to be somewhat different to that normally described by steady flow resistance equations. The latter doubts concerned effects due to a transition section in the apparatus.

Because of these inconsistencies, a comparison of the numerical and experimental data is shown in the form of tidal range amplification plotted against average phase difference (Figure 3) because explicit evaluation of ψ and $\sigma l/c$ is then unneccesary. Clearly, even allowing for the uncertainties in the experimental data, they lend considerable support to the numerical results.

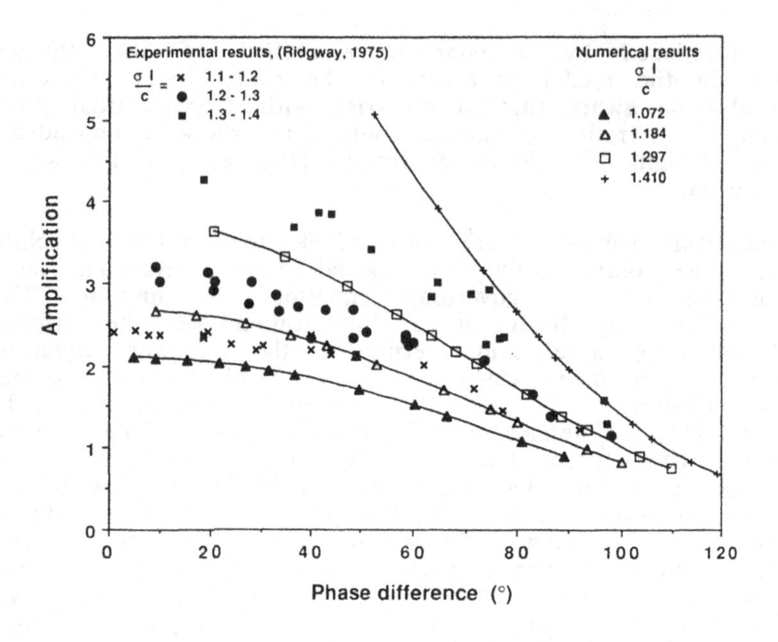

Figure 3. Comparison of numerical and experimental data.

It is apparent that when plotted in this way the data form a family of lines of constant $\sigma l/c$. Indeed, the laboratory data show that data of this type fall on a unique curve for each value of $\sigma l/c$ regardless of mean depth, tidal amplitude or roughness (Wallis [18]). This is discussed at greater length in Wallis & Knight [20] but it is clear that such a graphical representation offers a novel and objective method of comparing tidal data.

Numerical results obtained previously for similar cases (Wallis [18]) also lend strong support to the numerical results reported here. A fully implicit and a central implicit staggered finite difference scheme were used. Direct comparison with the results presented here is not possible, however, because the discretisation grids are incompatible. Nevertheless, allowing for this and the differing quantities of numerical damping in the schemes, all the results are entirely consistent.

CONCLUSIONS

Numerical solutions have been presented for the case of tidal propagation in a parallel sided, uniform, horizontal estuary of rectangular cross-section.

Four methods of objectively assessing the veracity of these results have been considered. Laboratory data have confirmed the numerical solutions to a large extent, but primarily because of inconsistencies between the representation of frictional resistance in the numerical and physical estuaries, the comparison, though good, is rather imprecise. By plotting the data in the form of tidal range amplification against phase difference, however, the agreement is evident.

Analytical solutions, field data and alternative numerical solutions are not as useful as laboratory data for validating numerical model results. Analytical solutions, by their nature, require certain simplifying assumptions which affect the non-linear terms and hence reduce the usefulness of the solutions. Applying a model to a natural estuary and using field data does not actually help to check a model because it is not possible to identify the source of discrepancies between measured and computed values: are the friction coefficients wrong, are there errors in the field data or are there errors in the model? Alternative numerical solutions can offer support, but since these too are approximate, the modeller is limited in his ability to favour one set of results in place of another.

The numerical solutions given in Tables 2 & 3 should be of particular use to workers employing one-dimensional tidal models, since they enable an objective quality assurance procedure to be carried out. It is hoped that modellers will compare their own solutions with those presented here.

REFERENCES

1. Abbott, M.B. and Basco, D.R. Computational Fluid Dynamics: An Introduction for Engineers, Longman, Harlow, 1989.
2. Bowers, D.G. and Lennon, G.W. Tidal Progression in a Near-Resonant System - A Case Study from South Australia, Estuarine, Coastal & Shelf Science, Vol.30, pp 17-34, 1990.
3. Cunge, J.A., Holly, F.M. and Verwey, A. Practical Aspects of Computational River Hydraulics, Pitman, London, 1980.
4. Ippen, A.T. (Ed.). Estuary and Coastline Hydrodynamics. McGraw-Hill, New York, 1966.
5. Knight, D.W. Theoretical Studies of Long Wave Propagation in Estuaries of Finite Length, Proceedings of the International Symposium on River Mechanics, Bangkok, Thailand, 1973, Vol.3, pp. 327-338, IAHR, 1973.
6. Knight, D.W. Long Wave Propagation in an Idealised Estuary, Journal of the Hydraulics Division, ASCE, Vol.99, HY7, July, pp. 993-1007, 1973.
7. Knight, D.W. Some Field Measurements Concerned with the Behaviour of Resistance Coefficients in a Tidal Channel, Estuarine, Coastal & Shelf Science, Vol.12, pp 303-322, 1981.

8. Knight, D.W. and Ridgway, M.A. An Experimental Investigation of Tidal Phenomena in a Rectangular Estuary, Proceedings of the International Symposium on Unsteady Flow in Open Channels, Cranfield, pp 25-40, BHRA, 1976.

9. Knight, D.W. and Ridgway, M.A. Velocity Distribution in Unsteady Open Channel Flow with Different Boundary Roughnesses, Proceedings of the 17th IAHR Congress, Baden-Baden, Germany, Vol.2, pp. 437-444, IAHR, 1977.

10. Liggett J.A. and Cunge J.A. Numerical Methods of Solution of the Unsteady Flow Equations. Chapter 4, Unsteady Flow in Open Channels (Eds. Mahmood K. & Yevjevich V.), Vol.1, pp. 89-182, Water Resources Publications, Fort Collins, Colorado, 1975.

11. Lynch, D.R. and Gray, W.G. Analytical Solutions for Computer Flow Model Testing, Journal of the Hydraulic Division, ASCE, Vol.104, HY10, Oct, pp. 1409-1428, 1978.

12. Needham, D.J. A Simple Model Describing the Tidal Flow in an Alluvial River, Geophysical and Astronomical Fluid Dynamics, Vol.41, pp. 129-140, 1988.

13. Ostendorf, D.W. Linearised Tidal Friction in Uniform Channels, Journal of Hydraulic Engineering, ASCE, Vol.110, HY7, July, pp. 867-885, 1984.

14. Prandle, D. Generalised Theory of Estuarine Dynamics. Part III, Physics of Shallow Estuaries and Bays, (Ed. van de Kreeke, J.), Lecture Notes on Coastal and Estuarine Studies, Vol.16, pp.42-57, Springer, Berlin, 1986.

15. Proudman, J. Oscillations of Tide and Surge in an Estuary of Finite Length, Journal of Fluid Mechanics, Vol.2, pp. 371-382, 1957.

16. Ridgway, M.A. An Experimental Study of Tidal Propagation in a Rectangular Estuary, PhD Thesis, University of Birmingham, 1975.

17. Samuels, P.G. and Skeels, C.P. Stability Limits for Preissmann's Scheme, Journal of Hydraulic Engineering, ASCE, Vol.116, No.8, August, pp. 997-1012, 1990.

18. Wallis, S.G. (1982). The Simulation of Tidal Flows in Natural and Idealised Estuaries, PhD Thesis, University of Birmingham, 1982.

19. Wallis, S.G. and Knight, D.W. Calibration Studies Concerning a One-Dimensional Numerical Tidal Model with Particular Reference to Resistance Coefficients, Estuarine, Coastal & Shelf Science, Vol.19, pp. 541-562, 1983.

20. Wallis, S.G. and Knight, D.W. Tidal Propagation in Rectangular Estuaries, In preparation.

Computer Modelling of the Tides of the Arabian Gulf

K.Z. Elahi, R.A. Ashrafi

Department of Mathematics, King Saud University, Riyadh, Saudi Arabia

ABSTRACT

A depth-averaged hydrodynamical numerical model is developed to investigate the tidal dynamics of the Arabian Gulf. Observed tidal constituents are compared with the computed results and are presented in tabular form. Chi square goodness of fit test was utilized to compare the computated values with the observed values. The existence of amphidromic points for all the major partial tides are also noticed and reproduced by the model.

INTRODUCTION

A computational algorithm for the solution of the shallow water equations based on explicit finite difference scheme, is used to reproduce the surface elevation and depth-averaged velocity components in the Arabian Gulf (from 24°N to 30°N and 48°E to 56°E). It covers nearly all the major coastal areas of the region. The bathometry of the gulf is reported in Lehr [1], Koske [2], Hughes and Hunter [3], and Defence Mapping Agency [4].

Some of the major studies on the tides of the Arabian Gulf are reported by Lardner et al [5], Evans-Roberts [6], Von Trepka [7], and many others. Hunter [8] has reviewed the literature on circulation and mixing process in the Gulf. Lehr [9] has surveyed the literature on oceanographic modelling and oil spill in the Gulf. Le Provost [10] has also reviewed three Gulf tidal models.

Surface elevations are used to predict amplitudes and phases of the partial tides. Computer model is used to study four partial tides M_2, S_2, K_1 and O_1 as these sufficiently describe tidal dynamics of the area. The model is based on the shallow-water equations, obtained from the three-dimensional hydrodynamical equations by averaging over the depth.

OBSERVATIONAL TIDAL DATA

The observational data on the tidal constituents along the coastal area of the Arabian Gulf is obtained from the special publication of the International Hydrographic Bureau in Monaco. There are quite a good number of the coastal gauges on the Southern Coast of the Gulf. The northern coast of the Gulf that is the Iranian coast and the interior of the Gulf is less observed. The available data is given in Table [1].

Tides in the Gulf are mixed in nature. They are classified by using the ratio of the amplitudes of the two main diurnal constituents K_1 and O_1 to that of the two main semi-diurnal amplitudes M_2 and S_2. The ratio is known as "form ratio", Pond and Pickard [11].

$$F = \frac{K_1 + O_1}{M_2 + S_2}$$

$F = 0$ to 0.25 — Semi-diurnal tides, mean spring range = $2(M_2 + S_2)$.

$F = 0.25$ to 1.5 — Mixed, mainly semi-diurnal tides.

$F = 1.5$ to 3.0 — Mixed, mainly diurnal tides, mean spring range = $2(K_1 + O_1)$.

$F > 3.0$ — Diurnal tide.

MATHEMATICAL MODEL

Mathematically the tidal propagation is described by the quasi-linear hyperbolic partial differential equation, with necessary initial and boundary conditions. Since the water movements are mainly horizontal and the horizontal dimensions are much greater than the depth, the system of equations has been simplified to a two-dimensional system. Vertical structure is included as the system is averaged over depth. Effect of the Earth's curvature is taken into account by considering the depth averaged equations of motion and continuity in spherical polar coordinates as follows:

$$\frac{\partial \zeta}{\partial t} + \frac{1}{R \cos\phi} \left[\frac{\partial (HU)}{\partial \lambda} + \frac{\partial}{\partial \phi} (HV \cos\phi)\right] = 0 \tag{2}$$

$$\frac{\partial U}{\partial t} - \Omega V + \frac{1}{H} \tau_b^{\lambda} - A_h \nabla^2 U + \frac{g}{R \cos\phi} \frac{\partial \zeta}{\partial \lambda} = 0 \tag{3}$$

$$\frac{\partial V}{\partial t} - \Omega U + \frac{1}{H} \tau_b^{\phi} - A_h \nabla^2 V + \frac{1}{R} \frac{\partial \zeta}{\partial \lambda} = 0 \tag{4}$$

No.	PLACE	Position N	Position E	M_2 a	M_2 κ	S_2 a	S_2 κ	K_1 a	K_1 κ	O_1 a	O_1 κ	F
		° '	° '									
1	Zabut	24.00	52.27	17.1	237.0	7.6	259.0	46.9	161.0	19.2	108.0	1.9
2	Jabal Dhanna	24.11	52.38	10.0	271.8	2.5	256.2	45.9	163.0	22.0	108.3	5.44
3	Umm-al-Hatab	24.13	51.52	23.5	224.0	9.4	244.0	5.2	156.0	24.4	114.0	0.84
4	Abu Al-Abyad	24.15	53.49	45.0	20.0	14.0	89.0	42.0	164.0	22.0	102.0	1.08
5	Yas	24.17	52.37	9.9	271.2	1.9	239.8	48.6	159.4	24.7	107.0	6.2
6	Khor Zubaya	24.18	54.09	44.8	48.0	17.1	94.0	24.7	178.0	18.9	114.0	0.7
7	Khawr Zubayyah	24.20	54.10	37.0	45.0	15.0	103.0	30.0	180.0	19.0	137.0	0.94
8	Jazar-Ghagha	24.24	51.33	33.2	221.0	13.1	251.0	50.9	158.0	20.4	117.0	1.54
9	Ghasha Tower	24.25	52.34	12.7	245.3	5.9	237.9	45.5	161.3	23.0	107.1	0.27
10	Ummal Nar	24.26	54.30	25.0	79.0	8.0	146.0	22.0	211.0	12.0	170.0	1.03
11	Halat Al-Mubaras	24.27	53.22	28.3	17.0	12.5	75.0	42.7	162.0	23.2	108.0	1.6
12	Dalma, Jazirat	24.28	52.19	14.2	231.2	5.8	241.9	47.2	161.4	24.2	122.4	3.57
13	Abu Zabi	24.29	54.21	41.0	20.0	15.0	78.0	32.0	158.0	17.0	105.0	0.88
14	Mina Zayed	24.31	54.23	35.0	27.0	13.0	86.0	27.0	166.0	18.0	139.0	0.93
15	Umm Addalkh	24.36	54.05	44.0	13.0	17.0	70.0	34.0	157.0	18.0	18.0	0.85
16	Khor Udeid	24.42	51.27	30.0	179.0	13.0	209.0	50.0	137.0	27.0	90.0	1.79
17	Ardhana	24.46	52.34	6.3	210.5	4.0	215.1	40.4	150.5	18.5	118.1	5.7
18	Ghar Al Buraid	24.48	50.52	23.0	280.0	4.0	344.0	2.0	199.0	5.0	70.0	0.26
19	Khor Ghanada	24.50	54.46	41.5	19.0	14.9	78.0	26.5	165.0	15.2	112.0	0.74
20	Mubarras Approaches	24.51	53.31	26.0	16.0	12.0	67.0	34.0	153.0	18.0	100.0	1.37
21	Zarqa,	24.53	53.05	11.2	21.8	4.0	97.1	35.8	155.6	16.3	117.4	3.42
22	Umm Said	24.57	51.35	26.5	202.0	11.4	233.4	45.8	152.7	20.3	101.9	1.74
23	Al Khraij	25.00	50.48	21.0	259.0	8.0	337.0	2.0	177.0	4.0	20.0	0.21
24	Ras Al Alaj	25.01	51.39	26.5	180.9	10.9	213.9	42.7	139.5	19.2	85.8	1.65
25	Fujairah	25.08	56.21	67.0	302.0	25.0	313.0	44.0	46.0	1.8	62.0	0.5
26	Jazirat Das	25.09	52.53	7.9	70.0	5.2	129.0	35.6	146.0	17.1	107.0	4.0
27	Wakrah	25.10	51.37	26.8	182.0	6.1	222.0	36.9	141.0	15.2	82.0	1.6
28	Port Rashid	25.15	55.16	44.2	353.4	15.9	41.4	22.6	151.0	16.0	97.3	0.64
29	Dubai Al Matoum Bridge	25.15	55.19	47.0	8.0	16.0	60.0	19.0	167.0	19.0	107.0	0.6
30	Dabei Khor	25.16	55.17	26.8	1.0	8.8	60.0	13.7	174.0	11.6	161.0	0.71
31	Dubayy	25.16	55.17	32.0	10.0	11.0	64.0	19.0	170.0	15.0	127.0	0.79
32	Ad Dawhah	25.18	51.31	32.1	137.6	10.8	171.5	36.2	117.2	16.4	69.4	1.23
33	Khor Al Fakkan	25.21	56.22	66.0	278.0	27.0	312.0	35.0	40.0	19.0	42.0	0.58
34	Sharjah, Aliya	25.22	55.23	48.2	341.0	19.8	26.0	21.0	146.0	13.4	101.0	0.51
35	Ash Shariqah	25.22	55.23	44.0	353.0	17.0	39.0	23.0	144.0	16.0	102.0	0.64
36	Ajman	25.25	55.26	43.0	4.0	13.0	61.0	17.0	163.0	15.0	106.0	0.57
37	Zekrit,	25.28	50.51	16.2	13.0	4.3	44.0	8.2	144.0	2.4	60.0	0.52
38	Sumaismah	25.34	51.30	34.0	124.0	11.0	157.0	34.0	109.0	15.0	59.0	1.1
39	Umm Al Qaywayn	25.35	55.35	45.0	350.0	15.0	27.0	20.0	134.0	17.0	89.0	0.61
40	Jazirat Halul	25.40	52.24	20.0	139.0	6.0	174.0	38.0	145.0	17.0	92.0	2.1
41	Halul	25.40	52.25	20.0	136.0	7.0	165.0	29.0	140.0	15.0	90.0	1.63
42	Ras Al Khaimah	25.48	55.57	54.0	326.0	20.0	11.0	13.0	111.0	16.0	88.0	0.39
43	Hadd Al Yamal	25.51	50.37	11.3	267.0	4.0	339.0	3.7	146.0	2.7	29.0	0.42
44	Qurayah	25.53	50.07	16.0	256.0	4.0	313.0	7.0	149.0	5.0	51.0	0.6
45	Ras Laffan	25.54	51.35	38.0	124.0	11.0	166.0	25.0	98.0	12.0	42.0	0.75
46	Jazirat Sirri	25.54	54.33	39.0	350.8	14.5	39.1	24.5	137.1	17.6	87.0	0.79
47	Khor Khwair	25.58	56.03	62.0	326.0	20.0	10.0	21.0	89.0	16.0	84.0	0.45
48	Ras Ashairiq	25.59	51.00	41.7	188.0	11.3	248.0	7.0	76.0	3.7	357.0	0.2

Table 1: Continued

No.	PLACE	Position		M_2		S_2		K_1		O_1		F
		N	E	a	κ	a	κ	a	κ	a	κ	
		° '	° '									
49	Zellaq	26.03	50.29	4.9	312.0	2.1	33.0	1.2	176.0	1.8	92.0	0.42
50	Jabal Fuwairat	26.03	51.22	42.4	160.0	13.1	208.0	20.1	114.0	9.4	56.0	0.53
51	Ruwais Inshore	26.09	51.13	51.0	144.0	18.0	192.0	15.0	65.0	8.0	350.0	0.33
52	Sitra	26.10	50.40	66.0	174.0	18.9	240.0	9.8	47.0	6.1	320.0	0.19
53	Ruwais Offshore	26.10	51.11	54.0	144.0	19.0	196.0	15.0	64.0	7.0	343.0	0.3
54	Khasab Bay	26.12	56.15	67.0	312.0	22.0	359.0	22.0	68.0	16.0	71.0	0.43
55	Sibi Isthmus	26.12	56.24	67.0	316.0	25.0	4.0	22.0	78.0	16.0	70.0	0.41
56	Habalain Ghubbat Al Gahazirah,	26.12	56.24	73.0	288.0	27.0	335.0	34.0	43.0	18.0	41.0	0.52
57	Manamah Anchorage	26.14	50.35	66.0	152.0	22.0	195.0	10.0	22.0	12.0	318.0	0.25
58	Mina Salman	26.14	50.36	66.1	152.2	21.8	213.4	7.1	43.6	4.8	332.1	0.13
59	Jazirat Farur	26.15	54.31	45.1	3.0	15.2	43.0	37.5	140.0	22.3	92.0	0.99
60	Jezirat Tunbh	26.16	55.18	59.1	336.0	20.1	17.0	29.3	120.0	18.9	89.0	0.61
61	Bahrain Approach Beacon	26.22	50.47	63.0	170.0	20.0	222.0	9.0	70.0	7.0	324.0	0.19
62	Khor Kawi	26.22	56.22	68.9	309.0	25.3	347.0	26.2	70.0	15.8	67.0	0.45
63	Little Quoin Island	26.28	56.33	76.8	301.0	27.4	341.0	28.7	60.0	20.4	51.0	0.47
64	Bander Lingeh	26.33	54.53	59.7	346.0	22.6	27.0	32.6	127.0	21.9	89.0	0.66
65	Tarut Bay	26.39	50.22	54.0	182.0	17.0	249.0	12.0	10.0	9.0	323.0	0.3
66	Ras At Tannura	26.39	50.10	59.5	129.4	19.8	187.9	14.4	339.3	11.6	280.3	0.33
67	Henjam	26.41	55.54	73.8	320.0	25.0	7.0	29.0	88.0	20.4	70.0	0.5
68	Jazirah Ye Lavan	26.48	52.23	33.0	73.0	12.0	111.0	30.0	145.0	16.0	114.0	1.0
69	Jazirat Shaikh Shuaib	26.48	53.23	30.0	76.0	12.0	115.0	29.0	147.0	15.0	104.0	1.0
70	Ras Al Qulayah	26.51	49.54	48.0	123.0	16.0	183.0	18.0	319.0	10.0	260.0	0.44
71	Bandar Abbas	27.11	56.17	100.0	298.0	36.0	334.0	33.8	64.0	20.7	52.0	0.4
72	Berri Dawhat Abu Ali	27.13	49.43	44.0	124.0	16.0	197.0	17.0	318.0	14.0	269.0	0.5
73	Asalu	27.28	52.37	51.2	120.0	17.1	162.0	23.8	168.0	11.9	138.0	0.5
74	Ras Al Mishaab	28.07	48.38	25.0	3.0	8.0	65.0	38.0	304.0	21.0	263.0	1.79
75	Lavar, IRAN	28.15	51.16	49.7	169.0	17.7	222.0	25.3	262.0	18.0	220.0	0.64
76	Ras-Al-Khafji	28.25	48.31	32.7	342.5	11.6	50.8	42.3	301.0	21.6	262.4	1.44
77	Mina Saud	28.44	48.24	42.0	336.0	14.0	34.0	43.0	305.0	27.0	259.0	1.25
78	Bushire	28.54	50.45	33.7	8.1	12.3	54.5	30.7	174.6	20.4	140.4	1.1
79	Mina-Al-Ahmadi	29.04	48.10	62.7	335.0	16.8	42.0	42.9	308.0	28.7	257.0	0.9
80	Kharg Island	29.16	50.20	36.4	250.2	12.8	301.4	38.8	285.6	25.7	241.2	1.3
81	Shat Al Arab Bar	29.50	48.43	84.1	308.4	28.6	8.6	49.7	295.4	29.8	247.1	0.7
82	Fao	29.58	48.30	82.4	337.1	25.3	39.2	43.8	315.5	25.4	267.8	0.64
83	Warba Spit	29.59	48.09	126.2	343.3	43.0	57.4	66.1	306.3	31.4	263.8	0.58
84	Khor Musa Bar	30.00	49.00	86.9	313.2	31.3	13.1	49.4	300.9	31.7	252.7	0.69

Table 1: Amplitude a (cm), Phase κ (degree) of the Major Tidal Constituents, and Form Ratio F.

If U(z) and V(z) denote the horizontal velocity components at depth z below the undisturbed sea surface, then

$$U = \frac{1}{h + \zeta} \int_{-h}^{\zeta} u(z)dz \quad (5)$$

$$V = \frac{1}{h + \xi} \int_{-h}^{\zeta} v(z)dz \quad (6)$$

$(\tau_b^\lambda, \tau_b^\phi)$ are the components of the bottom frictional stresses and are parametrized emprically by a quardratic law (G.I. Taylor [12]) relating bottom stresses to the depth mean velocity

$$\tau_b^\lambda = r\ U(U^2 + V^2)^{1/2} \quad (7)$$

$$\tau_b^\phi = r\ V(U^2 + V^2)^{1/2} \quad (8)$$

where r is a non-dimensional friction coefficient of the order 3×10^{-2}.

The coefficient of horizontal eddy viscosity A_h is directly proportional to grid size and inversely proportional to water depth and timestep. Considering the Arabian Gulf as shallow water body, the empirical value of the horizontal eddy viscosity is $.1 \times 10^3\ m^2s^{-1}$.

The tide in the model is generated by prescribing amplitudes and phases of tidal constituents at the open boundary. Water levels as a function of time for each partial tide are computed by

$$\zeta(t) = A \cos(\sigma t - \kappa) \quad (9)$$

where A is amplitude, κ is phase of incoming tide and σ is frequency.

The Coriolis force Ω results from the fact that our reference system is fixed to the earth, but the earth itself is moving through the space. The effects in the horizontal plane are considered only.

$$\Omega = 2\omega \sin\phi \quad (10)$$

INITIAL AND BOUNDARY CONDITIONS

The solution of the system of equation (2-4) requires the knowledge of initial and boundary conditions. There are two types of boundary conditions:

1. Solid boundary (coastal line)

   no-slip condition

   $$U = V = 0$$

2. Open boundary (supposed joining line of the Gulf with the Arabian Sea).

Waterlevels are prescribed at every time step using Eq.(9). Moreover, the velocity gradients in the normal direction are zero.

$$\frac{\partial V}{\partial n} = 0$$

As initial condition the values of waterelevation and velocity components are taken to be zero at t=0.

NUMERICAL MODEL

The system of hyperbolic partial differential equations (2-4) along with initial and boundary conditions can be transformed into a system of explicit finite difference equations by replacing the space derivatives by the central differences and time derivatives by forward differences. The system of explicit finite difference equations thus obtained is given in Elahi [13], and is solved numerically by using the hydrodynomical-numerical method. The area is covered by the spatial mesh of 39x54 computational points. The grid size is .167° = 18 Km. Programs were written in FORTRAN IV. Computations were done on personal computer PC-AT 368 with RM/FORTRAN compiler. The personal computer includes the Math Coprocesser 80387 and a hard disk of 40 Mg. Graphic work was done by using GEOGRAF VER 4.0.

RESULTS AND ANALYSIS

Tidal elevations are plotted in the form of the tidal charts for M_2, S_2, K_1 and O_1 and are given in Fig. 1,2,3 and 4 respectively. Special feature of the tide in the Arabian Gulf is the existance of amphidromic points for all the major tides. Two amphidromic points appears in case of the semi-diurnal tides M_2 and S_2. One amphidromic point appears for each diurnal tide K_1 and O_1. Location of the amphidromic points is the position of zero amplitudes

Tide	Location of Amphidromic Points	
M_2	28°.15' N ,	49°.39' E
	25°.00' N ,	53°.10' E
S_2	28°.25' N ,	49°.10' E
	25°.00' N ,	53°.02' E
K_1	26°.57' N ,	50°.40' E
O_1	26°.50' N ,	51°.10' E

The locations of the amphidromic points for M_2, S_2 and K_1 tides are compared with the charts constructed from experimental observations of tidal heights [14] and these are in reasonable agreement with each other. Patterns of the co-tidal and co-phase lines are also quite similar. Results are also compared with the observational data of IHO Tidal Constituents Bank [15] at 18 different tidal gauges around the coast of the Arabian Gulf (Table 2). Around the coast of Bahrain the degree of accuracy of the results is not very encouraging, whereas in the rest of the Arabian Gulf, amplitude and phases are reproduced with maximum error of ± 10 cm, and ± 5 deg. respectively.

The Chi-square goodness-of-fit test has been applied to asses the accuracy of the computed values with the observed values. The results of this test are shown in Table 3. It is evident that all the models fitted very well and the computed results are not statistically significant from the observed values.

	P-value (d.f.)	Significance
M_2 tide		
Amplitude	0.99 (32)	not
Phase	0.99 (32)	not
S_2 tide		
Amplitude	0.59 (33)	not
Phase	0.47 (34)	not
K_1 tide		
Amplitude	0.72 (34)	not
Phase	0.99 (34)	not
O_1 tide		
Amplitude	0.94 (36)	not
Phase	0.99 (36)	not

Table 3 Chi-square goodness-of-fit test results.

No.	PLACE	Position		M_2		S_2		K_1		O_1	
		N	E	O	C	O	C	O	C	O	C
		° '	° '								
1	FAO	29.58	48.30	82.4	85.6	25.3	27.01	43.8	47.4	25.4	28.9
				337.1	333.6	39.2	39.6	315.5	313.8	267.8	263.0
2	Shatal Arab Bar	29.50	48.43	84.1	74.7	28.6	27.5	49.7	49.3	29.8	28.2
				308.4	319.4	8.6	12.8	295.4	301.7	247.1	255.9
3	Kharg Island	29.16	50.20	36.4	37.9	12.8	16.5	38.8	43.0	25.7	21.7
				250.2	253.7	301.4	305.2	285.6	276.8	241.2	232.1
4	Mina Saud	28.44	48.24	42.0	45.2	14.0	15.6	43.0	39.8	27.0	23.8
				336.0	337.6	34.0	26.1	305.0	310.1	259.0	261.3
5	Berri	27.13	49.43	44.0	34.0	16.0	13.8	17.0	16.9	14.0	11.9
				124.0	127.9	197.0	191.8	318.0	313.8	269.0	271.4
6	Munifa	27.35	48.54	22.0	17.3	8.0	6.4	31.0	24.3	21.0	15.4
				100.0	105.0	167.0	172.2	316.0	313.7	267.0	270.5
7	Asalu	27.28	52.37	51.2	53.1	17.0	21.5	23.8	33.1	11.9	13.6
				120.0	115.5	162.0	162.5	168.0	170.5	138.0	134.3
8	Ras Al-Tannora	26.39	50.10	59.5	55.7	19.8	17.0	14.4	11.9	11.6	9.9
				129.4	132.0	187.9	193.0	339.3	337.6	280.3	285.6
9	Henjam	26.41	55.54	74.0	73.0	25.0	28.6	29.0	31.3	20.4	19.5
				320.0	323.4	7.0	5.4	88.0	94.7	70.0	71.8
10	Bander	26.33	54.53	59.7	58.7	22.6	24.1	32.6	35.9	21.9	19.8
				346.0	351.5	27.0	34.5	127.0	123.8	89.0	88.3
11	Sibi Isthmas	26.12	56.24	67.0	67.7	25.0	26.9	22.0	18.5	16.0	15.6
				316.0	313.7	364.0	355.5	78.0	77.0	70.0	68.1
12	Bandar Abbas	27.11	56.17	100.0	97.0	36.0	35.1	33.8	34.7	20.7	22.6
				298.0	310.0	334.0	340.1	64.0	70.6	52.0	57.7
13	Jazirat Sirri	25.54	54.33	39.0	43.8	14.5	18.6	24.5	32.5	17.6	17.8
				350.8	356.5	39.1	43.4	137.1	138.8	87.0	94.9
14	Khor Khwair	25.58	56.03	54.0	58.0	20.0	24.4	13.0	16.8	16.0	15.6
				326.0	324.8	11.0	4.9	111.0	110.5	88.0	76.5
15	Ras Al Khajmah	25.48	55.57	62.0	62.4	20.0	23.03	21.0	21.7	16.0	14.9
				326.0	319.1	10.0	8.2	89.0	94.4	84.0	80.4
16	Umm al Qaywan	25.22	55.23	44.0	44.5	17.0	19.2	23.0	23.8	16.0	15.1
				353.0	347.7	39.0	31.8	144.0	153.3	102.0	97.7
17	As Shariqah	25.22	55.23	44.0	44.5	17.0	19.2	23.0	23.8	16.0	15.1
				353.0	347.7	39.0	31.8	144.0	153.3	102.0	97.7
18	Port Rashid	25.15	55.16	44.2	43.19	15.9	19.0	22.0	26.0	16.0	15.5
				353.1	351.47	41.4	35.0	151.0	156.9	97.3	99.5

Table 2: Comparison of Computed (C), observed (O), Key: Above value represent amplitude a (cm) and below value Phase (κ) of the Major Tidal Constituents.

Notations

U, V	components of vertically averaged velocity in the λ and ϕ directions, resp. [ms^{-1}]
ζ	water elevation [m]
A_h	coefficient of horizontal eddy viscosity [m^2s^{-1}]
g	acceleration due to gravity [ms^{-2}]
h	mean water depth [m]
H	$= h + \zeta$(actual depth) [m]
r	bottom friction coefficient
R	radius of the Earth [m]
t	time [s]
λ, ϕ	geographical longitude and latitude
ω	angular velocity of the Earth's rotation [s^{-1}]
∇^2	horizontal Laplacian operator [m^{-2}]

References

1. Lehr, William J., "A brief survey of Oceanographic modelling and oil spill studies in the region". Proc. of the Symp. on Oceanographic modelling of the KAP region, UNEP Report No. 70, 1985, pp. 175-192, 1983.

2. Koske, P., Hydrographische Verhaltnisse im Persischen Golf Grand von beobachtungen von F.S. Meteor in Fruhjahr 1965. Meteor Forsch Ergnbn, Gebruder Borntraeger, Berlin, pp. 58-73, 1972.

3. Hughes P. and Hunter J., A proposal for a physical oceanography program and numerical modelling of the KAP region. Project for KAP 2/2, UNESCO, Paris, 1980.

4. Defence Mapping Agency, USA, Sailing Directions for the Persian Gulf. Hydrographic Center, Washington, D.C., U.S.A., 1975.

5. Lardner, R, Belen, M. and Cekirge, H., Finite Difference Model for tidal Flows in the Arabian Gulf. Comp. and Maths. with Appls., 8, pp 425-444, 1982.

6. Evans-Roberts, D., Tides in the Persian Gulf. Consulting Engineer, June, 1979.

7. Von Trepka, L., Investigations of the Tides in the Persian Gulf by means of a Hydrodynamic-numerical model. Proceedings of Symposium on Mathematical-Hydrodynamical Investigations of the Physical Processes in the Sea. Institute fur Meereskunde der Universitat Hamburg, pp 59-63, 1968.

8. Hunter, J., A Review of the Residual Circulation and Mixing Process in the KAP region, with reference to applicable modelling techniques. Proceeding of the symp/workshop on oceanographic modelling of the Kuwait Action Plan (KAP) region, NUEP Regional Seas Reports and Studies No. 70, UNEP, 1985.

9. Belen, M., Lehr, W. and Cekirge, H., Spreading Dispersion and Evaporation of Oil Slicks in the Arabian Gulf. Proceedings 1981 Oil Slick Conference, Marh 2 - 5, 1981, API, Atlanta, pp 161 - 166, 1981.

10. Le Provost, Models for Tides in the Kuwait Action Plan (KAP) region. Proc. of Symp. on Oceanographic modelling of the Kuwait Action Plan (KAP) region, UNEP Report No. 70, 1985, pp 205 - 230, 1983.

11. Pond S., George L Pickard. Introductory Dynamical Oceanography, 2nd Edition, Pergamon Press, 1983.

12. Taylor, G.I., Tidal Friction in the irish Sea Phil. Trans. Roy. Soc. A 220 (1-3), 1919.

13. Elahi Kh. Z, A numerical model for tides in the Gulf of Oman, AJSE, Vol., 16, No. 24, pp. 173-188, 1991.

14. Co-tidal chart for the Gulf, No. 5091, U.S. Hydrographic Office, Washington, D.C.

15. IHO Tidal Constituent Bank, Tidal harmonic constants, Marine Environmental Data Service, Canada, 1987.

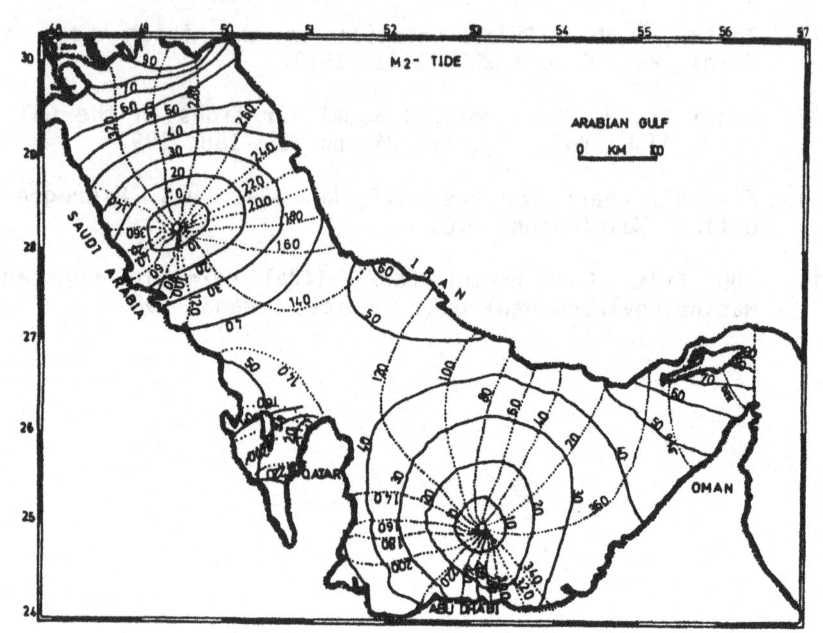

Fig. 1: M_2 - tide in the Arabian Gulf. Co-range lines (——) in cm and co-tidal lines (· · ·) in degree.

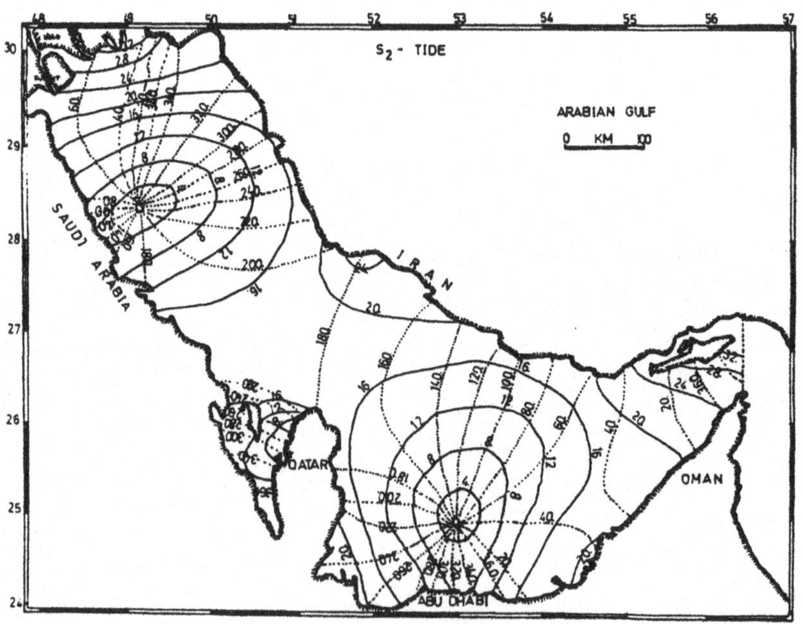

Fig. 2: S_2 - tide in the Arabian Gulf

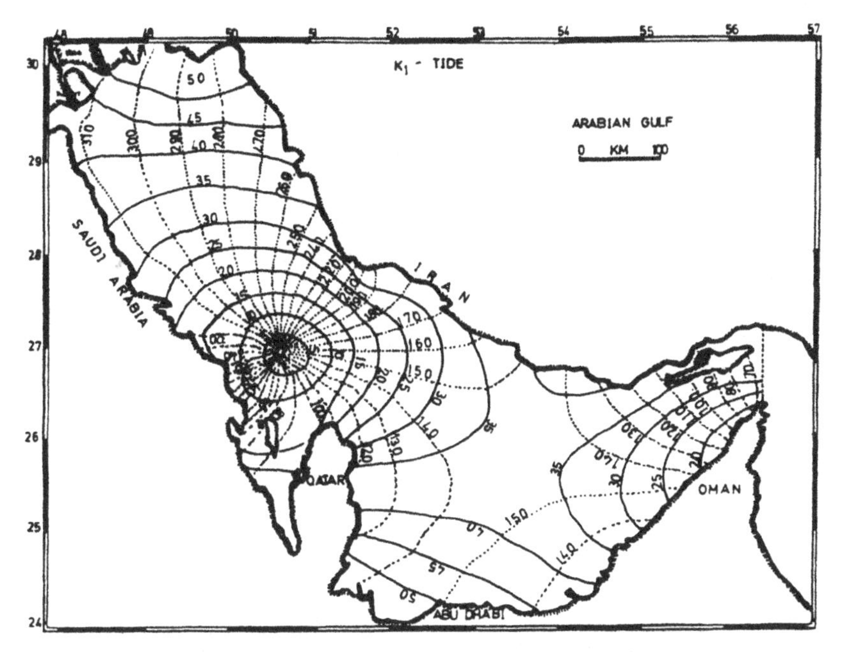

Fig. 3: K_1 - tide in the Arabian Gulf

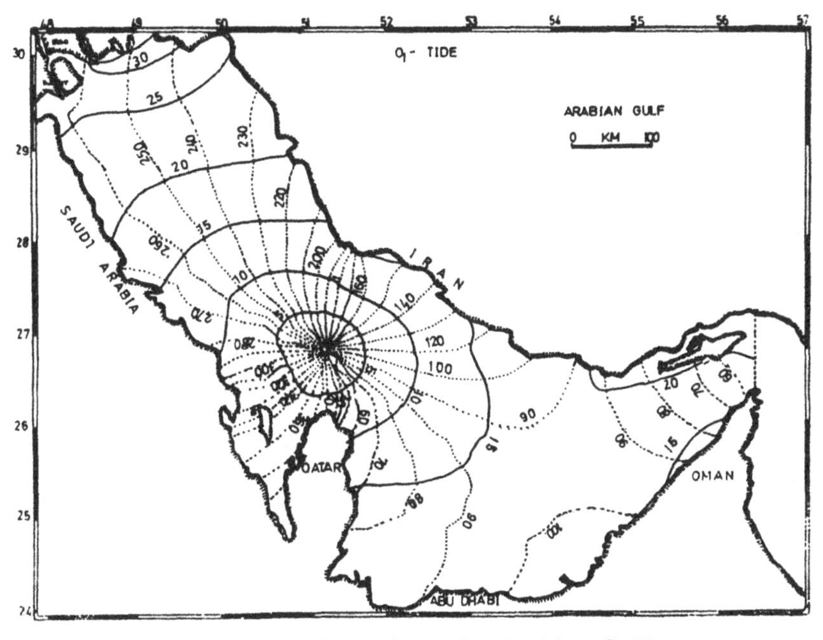

Fig. 4: O_1 - tide in the Arabian Gulf

Numerical Modelling of Tidal Motion in the Southern Waters of Singapore

H.-F. Cheong, N.J. Shankar, C.-T. Chan

Department of Civil Engineering, National University of Singapore, 10 Kent Ridge Crescent, 0511 Singapore

ABSTRACT

Numerical modelling of tidal hydrodynamics in coastal seas has reached the level where it is now possible to obtain very reliable predictions of tidal currents, flows and water levels. The application of a nested computational technique allows for tidal motions in the coastal waters to be solved up to a reasonable degree of resolution. The technique involves an in-depth study of the selected sub domain in the overall problem domain utilizing the boundary data generated from a global/regional model covering a much larger area. This has been made possible on the PC due to the spectacular development in speed, memory capacity and reliability of such computers.

This paper describes in detail the development of a regional spatially two-dimensional numerical tidal hydrodynamics model (1km x 1km grid) involving an alternating explicit and implicit finite difference method suggested by Stelling (1984). The regional model is rigorously calibrated and validated using a 14 day spring-neap cycle of tidal levels and current measurements obtained from several field stations. A finer grid nested model 2DEAST (with mesh size 125m x 125m) is also developed using the boundary data generated by the regional model. The finer grid model is further validated with recent field measurements and the results of the simulations and field experiments are presented.

INTRODUCTION

Singapore lies in a low wave energy environment being protected by the Malaysian peninsula in the north, Sumatra in the west and the Rhiau Archipelago in the south. The maximum wave height recorded in the coastal waters is about 1 m with a period of 2.5 to 3 seconds. (Chew, 1974). However, tidal fluctuations in the Singapore Strait has been found to be of the order of 2.5-3.0 m during springs and 0.7-1.2 m during neaps. These tidal occurrences are semi-diurnal with a strong westerly stream and a strong easterly stream within a day. Velocities of the associated tidal streams vary from about 0.5-1.0 m/s in the open waters of the Singapore Straits to as much as 1.5-2.0 m/s in constricted channels between islands. In view of the large variations in tidal elevations and the associated currents in the coastal waters, careful evaluation of tidal characteristics has become essential for the successful planning, design and implementation of large scale coastal engineering developments such as land reclamation.

It is known that many numerical methods for the approximation of the shallow water equations (SWE) have been developed and well documented in the literature. One of the leading and efficient finite difference method (FDM) for practical problems of SWE is the method proposed by Leendertse (1967). A method which has been reported to be stable, robust and accurate is the finite difference scheme developed by Stelling (1984). The numerical model presented in this paper is based on the numerical scheme introduced by Stelling (1984).

NUMERICAL MODEL

The governing SWEs referred to a rectangular coordinate system with x and y as horizontal axes are given as follows:

$$\frac{\partial u}{\partial t} + u\frac{\partial u}{\partial x} + v\frac{\partial u}{\partial y} - fv + g\frac{\partial \zeta}{\partial x} + gu\,\frac{(u^2+v^2)^{1/2}}{C^2(d+\zeta)} - \nu\nabla^2 u = F^{(x)} \quad (1)$$

$$\frac{\partial v}{\partial t} + u\frac{\partial v}{\partial x} + v\frac{\partial v}{\partial y} + fu + g\frac{\partial \zeta}{\partial y} + gv\,\frac{(u^2+v^2)^{1/2}}{C^2(d+\zeta)} - \nu\nabla^2 v = F^{(y)} \quad (2)$$

$$\frac{\partial \zeta}{\partial t} + \frac{\partial (Hu)}{\partial x} + \frac{\partial (Hv)}{\partial y} = 0 \quad (3)$$

where u,v = depth averaged velocity components in the x, y directions respectively, ζ = water elevation above a plane of reference (mean sea level), d = water depth below a plane of reference (sea bed profile), f = Coriolis force, ν = eddy viscosity, g = gravitational acceleration, C = Chezy coefficient for bed resistance, $F^{(x)}$, $F^{(y)}$ = external forces (eg. wind stress) in the x, y directions respectively.

The solution of Eq. 1 to Eq. 3 is achieved numerically using an alternating direction implicit method. The finite difference approximations are based on a fully staggered grid scheme shown in Fig.1 where the integer indices "m" and "n" run from 1 to the respective limits of the computational domain. All space derivatives are approximated by central differences except for the cross-advective terms $\partial u/\partial y$ and $\partial v/\partial x$, which are approximated by a weighted central difference (explicit) and a second order upwind difference (implicit). The complete forms of the two-stage finite difference method are given below. Verification of the stability conditions for this complete nonlinear set of equations is very complex indeed and can only be ascertained through experience.

In the computational scheme, for a variable A at node (m,n) and time step k, the following nomenclature is used.

$$A_{ox} = [A^k_{m+1/2,n} - A^k_{m-1/2,n}]/\Delta x \quad \text{(simple difference along x)} \tag{4}$$

$$A_{oy} = [A^k_{m,n+1/2} - A^k_{m,n-1/2}]/\Delta y \quad \text{(simple difference along y)} \tag{5}$$

$$\overline{A}^x = [A^k_{m+1/2,n} + A^k_{m-1/2,n}]/\ 2 \quad \text{(simple average along x)} \tag{6}$$

$$\overline{A}^y = [A^k_{m,n+1/2} + A^k_{m,n-1/2}]/\ 2 \quad \text{(simple average along y)} \tag{7}$$

$$\overline{\overline{A}} = [A^k_{m+1/2,n+1/2} + A^k_{m-1/2,n+1/2} + A^k_{m+1/2,n-1/2} + A^k_{m-1/2,n-1/2}]/4 \tag{8}$$

Stage 1:

At (m+1/2,n)

$$u^{[0]} = u^k, \quad v^{[0]} = v^k, \quad \zeta^{[0]} = \zeta^k$$

for p = 1, 2 and q = 1, 2, ... Q;

$$[u^{[q]} - u^k]/(\tau/2) + u^{[q]}\overline{(u^k_{ox})}^x + S_{oy}(\overline{\overline{v}}^{k+1/2}, u^k) - f\overline{\overline{v}}^{k+1/2}$$

$$+ g\zeta^{[q]}_{ox} + gu^{[q]}[(\overline{\overline{v}}^{k+1/2})^2 + (u^k)^2]^{1/2}/(C^2H^k) - \nu[u^k_{oxx} + u^k_{oyy}] = 0 \tag{9}$$

where

$S_{oy}(\bar{\bar{v}}^{k+1/2}, u^k)$ at $(m+1/2, n)$ =

$$\bar{\bar{v}}^{k+1/2}_{m+1/2,n}(u^k_{m+1/2,n+2} + 4u_{m+1/2,n+1} - 4u^k_{m+1/2,n-1} - u^k_{m+1/2,n-2})/(12\Delta y) \quad (10)$$

$$\bar{\bar{v}}^{k+1/2}_{m+1/2,n} = (v^{k+1/2}_{m,n+1/2} + v^{k+1/2}_{m+1,n+1/2} + v^{k+1/2}_{m+1,n-1/2} + v^{k+1/2}_{m,n-1/2})/4 \quad (11)$$

$$u^k_{oxx} = (u^k_{m+1/2,n})_{oxx} = [u^k_{m+3/2,n} - 2\,u^k_{m+1/2,n} + u^k_{m+1/2,n})/(\Delta x)^2 \quad (12)$$

$$u^k_{oyy} = (u^k_{m+1/2,n})_{oyy} = [u^k_{m+1/2,n+1} - 2\,u^k_{m+1/2,n} + u^k_{m+1/2,n-1})/(\Delta x)^2 \quad (13)$$

At $(m, n+1/2)$

$$[v^{[p]} - v^k]/(\tau/2) + v^k(\overline{v^{[p]}}^y)_{\delta y} + S_{+x}[\bar{\bar{u}}^k, v^{[p]}, s(p+p')] + f\,\bar{\bar{u}}^k$$

$$+ g\zeta^k_{oy} + gv^{[p]}[(\bar{\bar{u}}^k)^2 + (v^k)^2]^{1/2}/(C^2H^k) - \nu[v^{[*]}_{oxx} + v^{[p]}_{oyy}] = 0 \quad (14)$$

where

$S_{+x}(\bar{\bar{u}}^k, v^{[p]}, s)$ at $(m, n+1/2)$ =

$$\bar{\bar{u}}^k_{m,n+1/2}(3v^{[p-1+s]}_{m,n+1/2} - 4v^{[p-1+s]}_{m-1,n+1/2} + v^{[p-1+s]}_{m-2,n+1/2}]/(2\Delta x)$$

if $\bar{\bar{u}}^k_{m,n+1/2} > 0$ (15)

$$\bar{\bar{u}}^k_{m,n+1/2}(-3v^{[p-s]}_{m,n+1/2} + 4v^{[p-1+s]}_{m+1,n+1/2} - v^{[p-s]}_{m+2,n+1/2}]/(2\Delta x)$$

if $\bar{\bar{u}}^k_{m,n+1/2} < 0$ (16)

$$s = [1+(-1)^{p+p'}] \tag{17}$$

$$p' = \begin{cases} 0, & \text{if } \sum_{m,n} u^k > 0 \\ 1, & \text{if } \sum_{m,n} u^k \le 0 \end{cases} \tag{18}$$

$$\bar{\bar{u}}^k_{m,n+1/2} = (u^k_{m+1/2,n+1} + u^k_{m+1/2,n} + u^k_{m-1/2,n} + u^k_{m-1/2,n+1})/4 \tag{19}$$

$$v^{[*]}_{oxx} = (v^{[p-1+s]}_{m+1,n+1/2} - 2\, v^{[p]}_{m,n+1/2} + v^{[p-s]}_{m-1,n+1/2})/(\Delta x)^2 \tag{20}$$

$$v^{[p]}_{oyy} = (v^{[p]}_{m,n+3/2} - 2\, v^{[p]}_{m,n+1/2} + v^{[p]}_{m,n-1/2})/(\Delta x)^2 \tag{21}$$

At (m,n)

$$[\zeta^{[q]} - \zeta^k]/(\tau/2) + \zeta^{[q-1]}(u^{[q]}_{ox}) + [\bar{h}^y u^{[q]}]_{ox}$$

$$+ \overline{u^{[q-1]}\zeta^{[q]}_{ox}}^x + [H^k v^k]_{oy} = 0 \tag{22}$$

Stage 2:

At (m+1/2,n)

$u^{[0]} = u^{k+1/2}, \quad v^{[0]} = v^{k+1/2}, \quad \zeta^{[0]} = \zeta^{k+1/2}$

for p = 1, 2 and q = 1, 2, ... Q;

$$[u^{[p]} - u^{k+1/2}]/(\tau/2) + u^{k+1/2}\overline{(u^{[p]})}^x_{ox} + S_{+y}[\bar{\bar{v}}^{k+1/2}, u^{[p]}, s(p+p')]$$

$$- f\bar{\bar{v}}^{k+1/2} + g\zeta^{k+1/2}_{ox} + gu^{[p]}[(\bar{\bar{v}}^{k+1/2})^2 + (u^{k+1/2})^2]^{1/2}/(C^2H^{k+1/2})$$

$$- \nu[u^{[*]}_{oyy} + u^{[p]}_{oxx}] = 0 \tag{23}$$

where

$S_{+y}(\bar{\bar{v}}^{k+1/2}, u^{[p]}, s)$ at $(m+1/2, n)$ =

$$\bar{\bar{v}}^{k+1/2}_{m+1/2,n}(3u^{[p-s]}_{m+1/2,n} - 4u^{[p-s]}_{m+1/2,n-1} + u^{[p-s]}_{m+1/2,n-2}]/(2\Delta x)$$

$$\text{if} \quad \bar{\bar{v}}^{k+1/2}_{m+1/2,n} > 0 \tag{24}$$

$$\bar{\bar{v}}^{k+1/2}_{m+1/2,n}(-3u^{[p-1+s]}_{m+1/2,n} + 4u^{[p-1+s]}_{m+1/2,n+1} - u^{[p-1+s]}_{m+1/2,n+2}]/(2\Delta x)$$

$$\text{if} \quad \bar{\bar{v}}^{k}_{m+1/2,n} < 0 \tag{25}$$

$$s = [1+(-1)^{p+p'}]/2 \tag{26}$$

$$p' = \begin{cases} 0, & \text{if } \sum_{m,n} v^k > 0 \\ 1, & \text{if } \sum_{m,n} v^k \le 0 \end{cases} \tag{27}$$

$$\bar{\bar{v}}^{k+1/2}_{m+1/2,n} = (v^{k+1/2}_{m,n+1/2} + v^{k+1/2}_{m+1,n+1/2} + v^{k+1/2}_{m+1,n-1/2} + v^{k+1/2}_{m,n-1/2})/4 \tag{28}$$

$$u^{[*]}_{oyy} = (u^{[p-1+s]}_{m+1/2,n+1} - 2\, u^{[p]}_{m+1/2,n} + u^{[p-s]}_{m+1/2,n-1})/(\Delta y)^2 \tag{29}$$

$$u^{[p]}_{oxx} = (u^{[p]}_{m+1/2,n+1} - 2\, u^{[p]}_{m+1/2,n} + u^{[p]}_{m+1/2,n-1})/(\Delta x)^2 \tag{30}$$

At $(m, n+1/2)$

$$[v^{[q]} - v^{k+1/2}]/(\tau/2) + v^{[q]}\overline{(v^{k+1/2}_{oy})}^{x} + S_{ox}[\bar{\bar{u}}^{k+1}, v^{k+1/2}]$$

$$+ f\bar{\bar{u}}^{k+1} + g\zeta^{[q]}_{oy} + gv^{[q]}[(\bar{\bar{u}}^{k+1})^2 + (v^{k+1/2})^2]^{1/2}/(C^2 H^{k+1/2})$$

$$- \nu[v^{k+1/2}_{oyy} + v^{k+1/2}_{oxx}] = 0 \tag{31}$$

where

$S_{ox}(\bar{\bar{u}}^{k+1}, v^{k+1/2})$ at $(m,n+1/2)$ =

$$\bar{\bar{u}}^{k+1}_{m,n+1/2}(v^{k+1/2}_{m+2,n+1/2} + 4v^{k+1/2}_{m+1,n+1/2} - 4v^{k+1/2}_{m-1,n+1/2} - v^{k+1/2}_{m-2,n+1/2})/(12\Delta x) \quad (32)$$

$$\bar{\bar{u}}^{k+1}_{m,n+1/2} = (u^{k+1}_{m-1/2,n+1} + u^{k+1}_{m+1/2,n+1} + u^{k+1}_{m+1/2,n} + u^{k+1}_{m-1,n})/4 \quad (33)$$

$$v^{k+1/2}_{oyy} = (v^{k+1/2}_{m,n+3/2} - 2\, v^{k+1/2}_{m,n+1/2} + v^{k+1/2}_{m,n-1/2})/(\Delta x)^2 \quad (34)$$

$$v^{k+1/2}_{oxx} = (v^{k+1/2}_{m+1,n+1/2} - 2\, v^{k+1/2}_{m,n+1/2} + v^{k+1/2}_{m-1,n+1/2})/(\Delta x)^2 \quad (35)$$

$$\zeta^{[q]}_{oy} = [\zeta^{[q]}_{m,n+1} - \zeta^{[q]}_{m,n}]/\Delta y \quad (36)$$

At (m,n)

$$[\zeta^{[q]} - \zeta^{k+1/2}]/(\tau/2) + \zeta^{[q-1]}(v^{[q]}_{oy}) + v^{[q-1]}(\zeta^{[q]}_{oy})$$

$$+ [v^{[q]}\, \bar{h}^{x}]_{oy} + [H^{k+1/2}u^{k+1/2}]_{ox} = 0 \quad (37)$$

where

$$v^{[q]}_{oy} = [v^{[q]}_{m,n+1/2} - v^{[q]}_{m,n-1/2}]/(\Delta y) \quad (38)$$

At stage 1, Eq. 14 is an implicit equation and it is solved column by column (y axis) proceeding along the dominant flow direction of u. If the sign of u is constant, then Eq. 14 is solved in one sweep, otherwise a second iteration is necessary sweeping in the opposite direction. After Eq. 14 is solved, Eq.9 and Eq. 22 are solved. Since these two equations are coupled implicitly, they are solved simultaneously. By substitution of Eq. 9 at nodes $(m-1/2,n)$ and $(m+1/2,n)$ into Eq. 22 at node (m,n), the implicit equations are tri-diagonal.

At stage 2, Eq. 23 is implicit and is solved similarly as Eq. 14. The coupled implicit equations Eq. 31 and Eq. 37 are solved according to Eqs. 9 and 22. By solving Eqs. 9, 14, 22, 23, 31, and 37 in the sequence Eqs.14, (9 & 22), 23, (31 & 37), the computer implementation needs only one array per dependent variable (u,v,ζ) and one working array of the size of the member of "ζ points" of the grid.

Two different types of boundary conditions are encountered in the computational domain. A closed boundary is referred to physical or existing land-water boundary for which the velocity component normal to it is taken to be zero. An open boundary is a water-water boundary judiciously chosen to restrict the extent of the domain. The boundary conditions are mathematically given by

$$u_{\perp} = f^{u}(t) \quad \text{for a velocity boundary} \tag{39}$$

$$\zeta = f^{\zeta}(t) \quad \text{for a water level boundary} \tag{40}$$

$$u_{//} = 0 \tag{41}$$

$$\partial(u_{//})/\partial n = 0 \tag{43}$$

where $u_{\perp}$, $u_{//}$ = velocity normal and parallel to the open boundary respectively, $f^{u}(t)$, $f^{\zeta}(t)$ = velocity and water level at the open boundary.

APPLICATION

Regional Model

The testing of 2DTIDFLO commenced with the establishment of the regional hydrodynamics model. The computational domain of the regional model covers an area of 107 km by 72 km with a grid size of 1km as shown in Fig.2. The model calibration and verification processes were carried out based on 14 days spring-neap tidal cycle involving field measurements in August 1978. The calibration period covered the spring tide conditions between 1200 hrs 5 August 1978 to 0000 hrs 8 August 1978 whilst the verification period covered the neap tide conditions between 1000 hrs 12 August 1978 to 2200 hrs 14 August 1978. The calibration process involved a series of test runs with adjustments of the Chezy resistance coefficient C, the eddy viscosity coefficient and the numerical time step. The model was operated with the water levels prescribed at the four open boundaries with data supplied by the Port of Singapore Authority (PSA). The optimum model parameters were finally calibrated as follows:- Chezy C =65 $m^{1/2}/s$, eddy viscosity ν = 10 m^2/s and Δt = 10 minutes.

Six tidal stations and three current stations with continuous field recordings were used to examine the numerical results. The locations of these stations are shown in Fig.2. The results of the computed water surface elevations are depicted in Fig.3 to Fig.4. It can be observed that the simulated and measured water surface elevations at all the tidal stations showed very good agreement for the calibration and verification runs. The magnitudes and the phase of the water surface elevation were well simulated for the spring and neap tide conditions. Minor deviations of tidal variation can be observed from the comparisons of the results at station Kepala Jernih. These discrepancies may be attributed to the coarse grid approximation error near the Indonesian island group.

The difficult part in the model calibration and verification processes is that of achieving accurate simulation of tide induced currents with regard to both magnitude and direction at the selected stations. As can be observed from Fig.5 and Fig.6, although good correlation is obtained between the computed and measured currents at all the 3 stations, the degree of agreement is inferior to that obtained for tidal elevations. While the current directions are simulated satisfactorily, the computed magnitudes are relatively lower than the measured values. Possible reasons for this deviation could be the coarse grid resolution and the association of stronger current with mid-depth measurements as compared to the depth averaged values obtained numerically.

It was found in the calibration runs that current magnitudes were relatively more sensitive to the Chezy parameter C than the eddy coefficient ν. An increase in the Chezy value would lead to measurable increase in estimates of the current speeds. The calibrated Chezy C of 65 appears to be consistent with Manning sea bed roughness in the clayey silt range. The Chezy C may be related to the Manning coefficient according to

$$C = 1/n \; H^{1/6}$$

where n is the Manning n and H is the water depth. It is obvious that a successful calibration will require a changing Chezy C to reflect the varying bottom roughness due to changing water depths. However, very little is known between Chezy C and the form roughness offered by the sea bed variations.

The eddy coefficient is treated as a semi-empirical factor describing the effects of non-uniform vertical velocity distributions and is kept low to fulfill the requirement of the nearly horizontal flow assumption. Because of its insignificant influence upon the computed water level and current magnitudes, a value of 10 m^2/s was adopted. This value posed no numerical difficulty for the present coarse grid-large time step regional model.

Simulation runs were also carried out on a set of input tidal package and field measurements covering the period 0000 hrs 17 February 1987 to 1200 hrs 23 February 1987. The tidal level and current histories obtained are shown in Fig. 7a, 7b & 7c. It can be observed that the computed water levels compare very well with the measurements taken at Tanjong Pagar Terminal. However, the simulated tidal currents at CUR2 and CUR5 show a consistent under-estimation. Good agreement of current directions are observed. Fig.8 shows the velocity field in the coastal waters of Singapore at 2100 hrs, 5 August 1978.

Nested model-2DEAST

Following the successful simulation with the coarse grid regional model, a nested model 2DEAST is established within the 2DTIDFLO so that a more detailed picture of the tidal streaming pattern can be obtained. 2DEAST encompassed an area 24km x 12 km covering the southeast coast of Singapore coastal water as shown in Fig.2. A grid size of 125m x 125m is used to obtain a better discretization of coastline geometry and sea bed bathymetry. During the execution of the numerical nested model 2DEAST, no further adjustment of numerical coefficients were needed. Although it would be reasonable to reduce the magnitude of eddy coefficient in the finer grid model for similar dispersion representation as in the coarse grid model, the reduction in the value has been found to have very small effects in the results.

Figs. 9a, 9b & 9c depict the time history output of 2DEAST. In Fig. 9a, the comparisons between the computed water levels in 2DTIDFLO and 2DEAST and field measurements show excellent correlation. The results also portray a consistent agreement with the regional model simulated currents for both magnitude and direction. (Figs. 9b & 9c) Circulation pattern around the southern coastal waters for a certain tidal phase is also shown in Fig. 10.

CONCLUSIONS

A numerical model for tidal hydrodynamics has been developed and successfully applied to Singapore coastal waters. The finite difference method proposed by Stelling (1984) has proven to be a suitable choice in view of its state of the art modelling techniques and its excellent performance in simulating the tidal behaviour in the coastal regions of Singapore.

The good results obtained from the regional and nested model simulations undoubtedly indicate that the model is capable of predicting the characteristics of tides and currents in a coastal region with complicated geometry. Considering the required computational labour per timestep and the adaptability of the method for large timestep integration without numerical instability as demonstrated in the regional model simulations with Δt = 10 minutes and Δs = 1 km, the model is considered to be efficient and sufficiently accurate for practical application in coastal hydrodynamics modelling.

ACKNOWLEDGEMENT

This study was undertaken with a research grant ST/85/01 from the Singapore Science Council under the Research and Development Assistance Scheme.

REFERENCES

1. Chew S.Y., "Waves at southeast coast of Singapore", Journal of the Institution of Engineers Singapore, Vol. 14, 1974, pp. 36-40.

2. Leendertse J.J., "Aspects of a computational model for long period wave propagation", RM-5294-PM, The Rand Corporation, Santa Monica, CA., 1967.

3. Stelling G.S., "Improved stability of Dronker's tidal schemes", Journal of the Hydraulic Division, ASCE, No. HY8, 1980.

4. Stelling G.S., "On the construction of computational methods for shallow water flow problems", Rijkswaterstaat Communications No. 35/1984, The Hague, Netherlands, 1984.

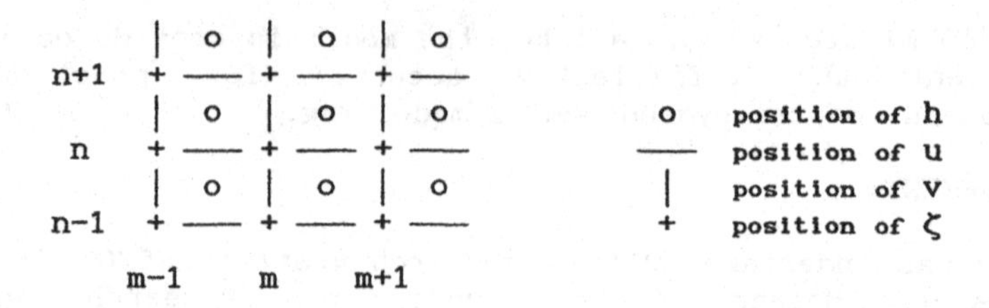

Fig. 1 Staggered Grid Scheme

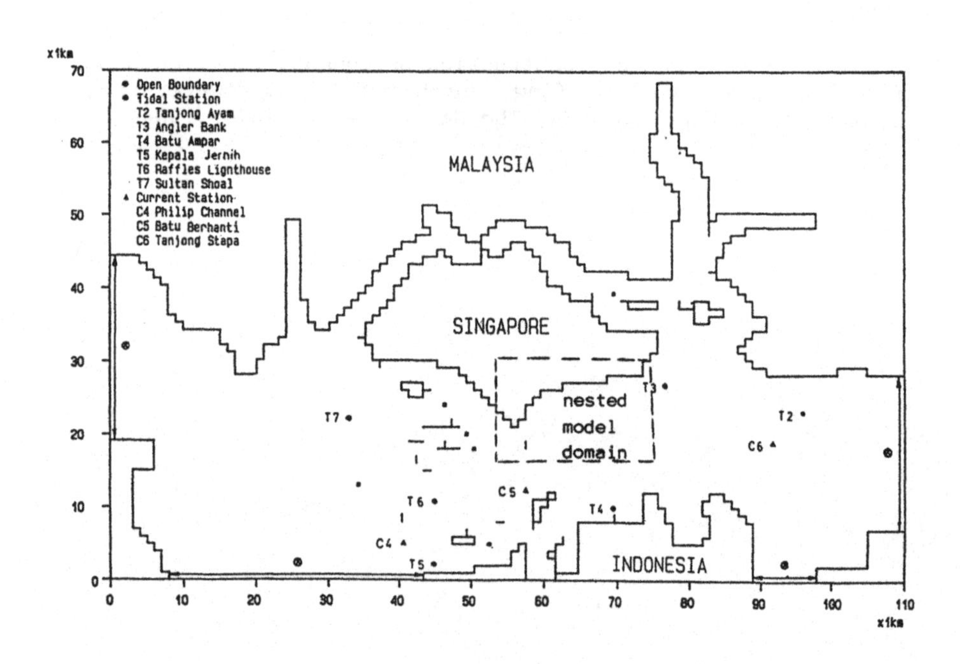

Fig. 2 Computational Domain of Regional Model Showing Locations of Tidal & Current Stations

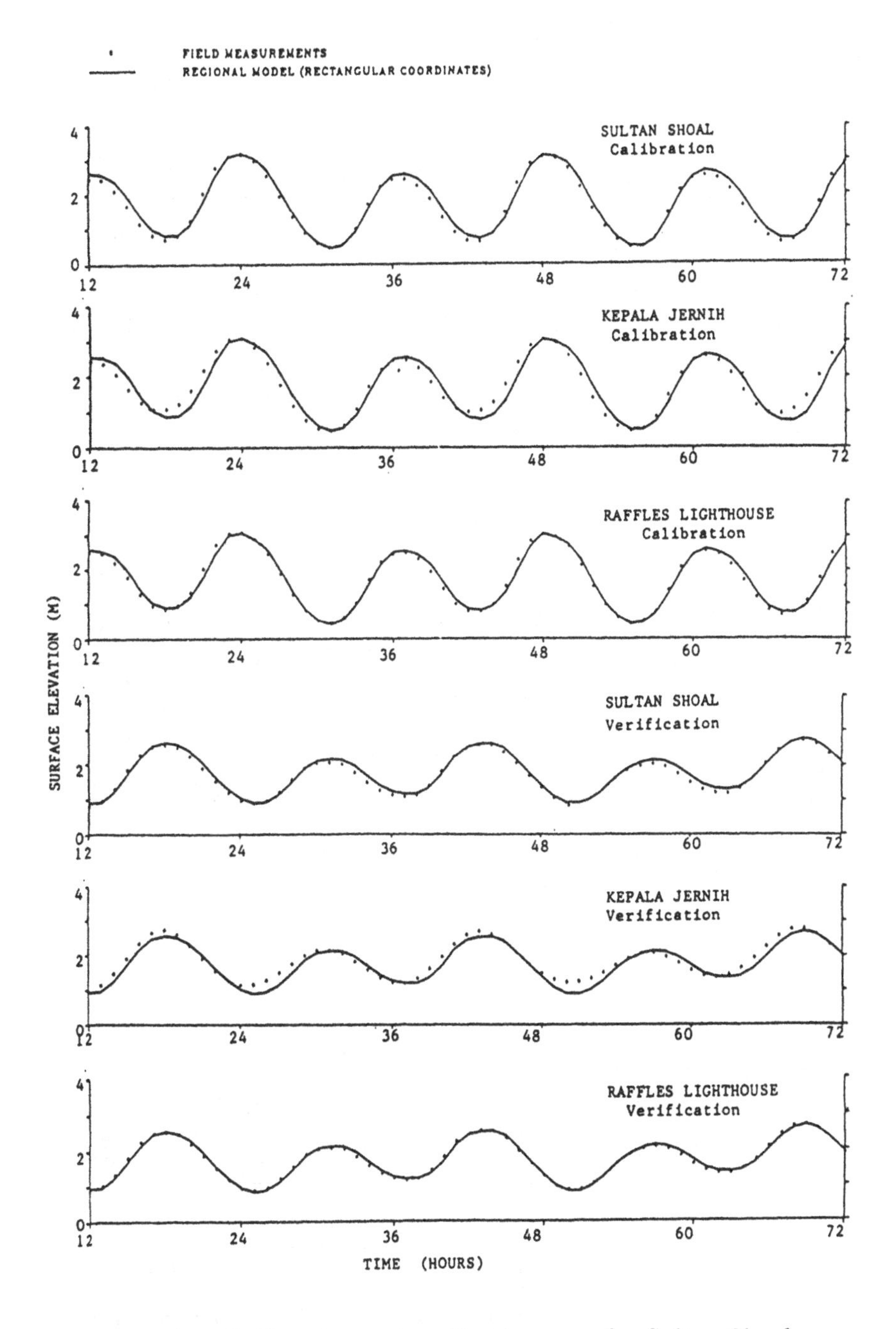

Figs. 3 & 4 Calibration and verification runs for Sultan Shoal, Kepala Jernih and Raffles Lighthouse respectively

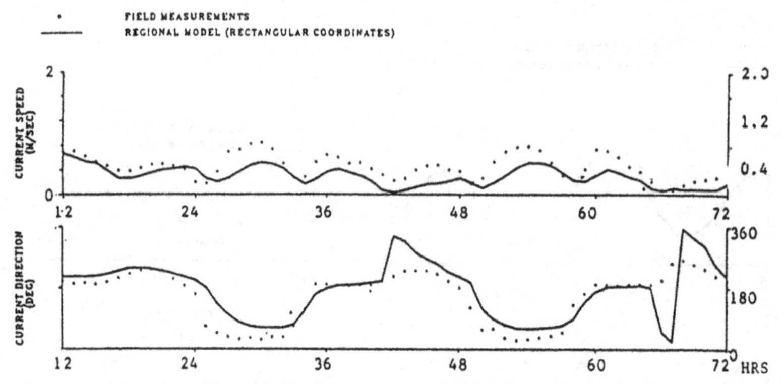

Fig. 5a Measured and Computed Values at PHILIP CHANNEL - Calibration

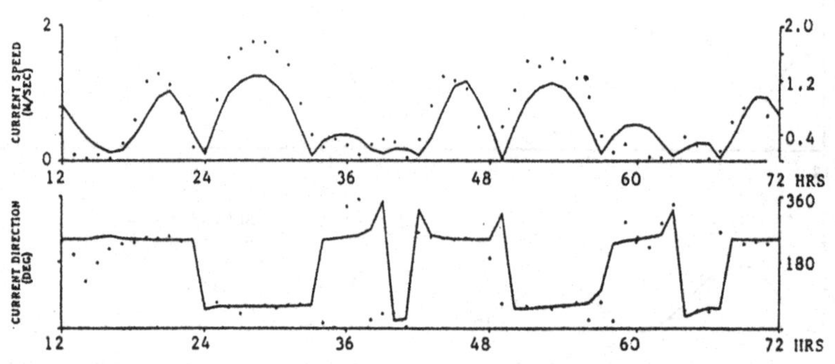

Fig. 5b Measured and Computed Values at BATU BERHANTI - Calibration

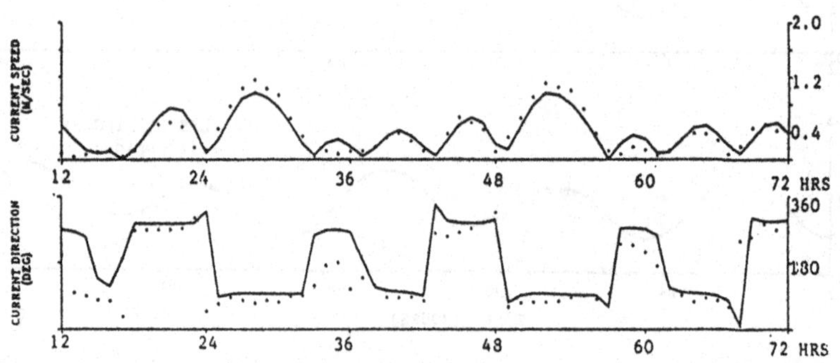

Fig. 5c. Measured and Computed Values at TANJONG STAPA - Calibration

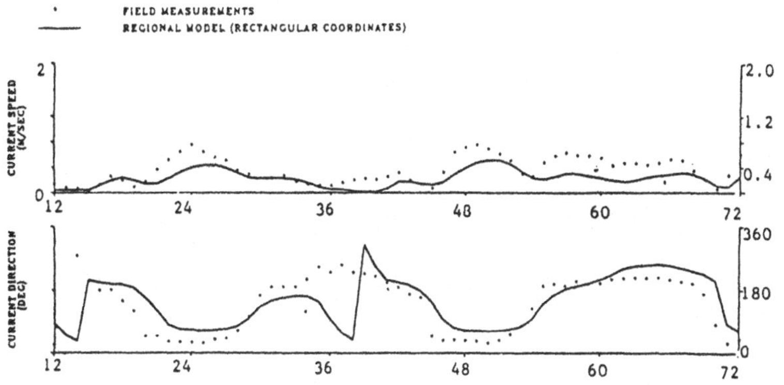

Fig. 6a Measured and Computed Values at PHILIP CHANNEL - Verification

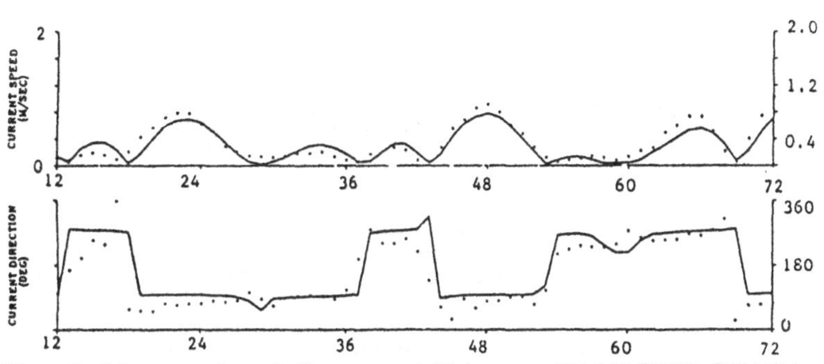

Fig. 6b Measured and Computed Values at TANJONG STAPA - Verification

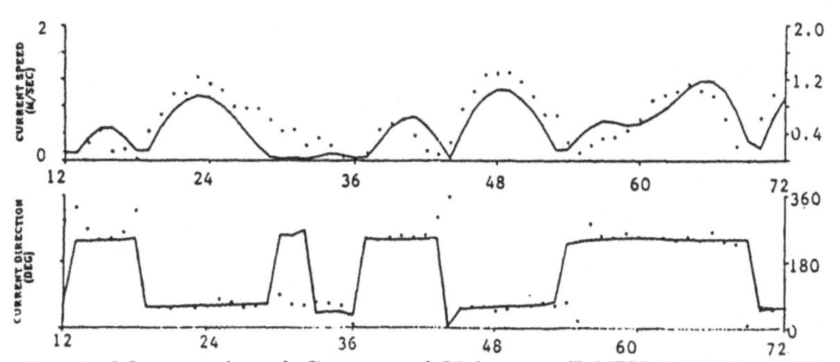

Fig. 6c Measured and Computed Values at BATU BERHANTI - Verification

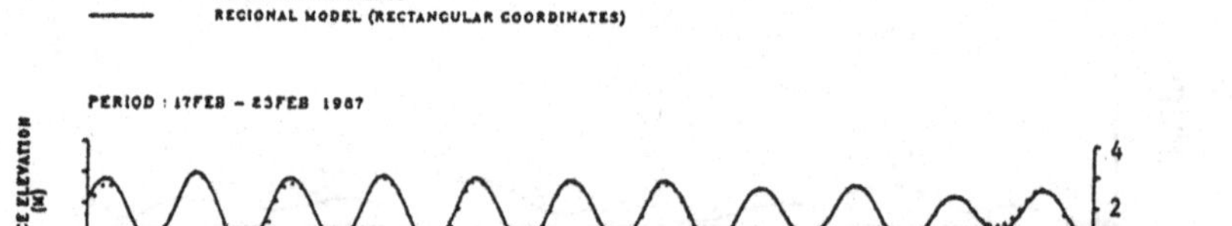

Fig. 7a Measured and Computed Water Level at TANJONG PAGAR TERMINAL

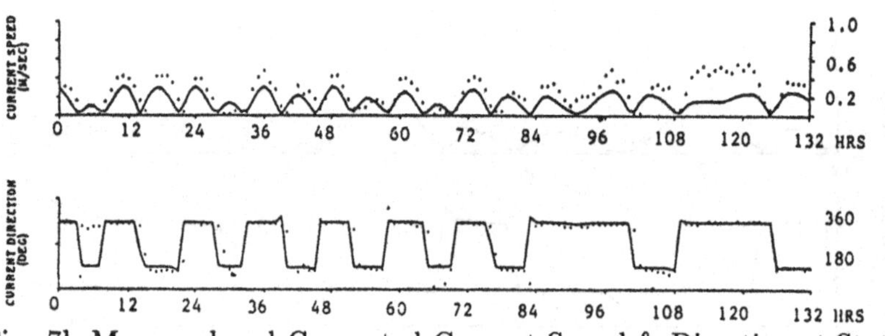

Fig. 7b Measured and Computed Current Speed & Direction at Station CUR 5

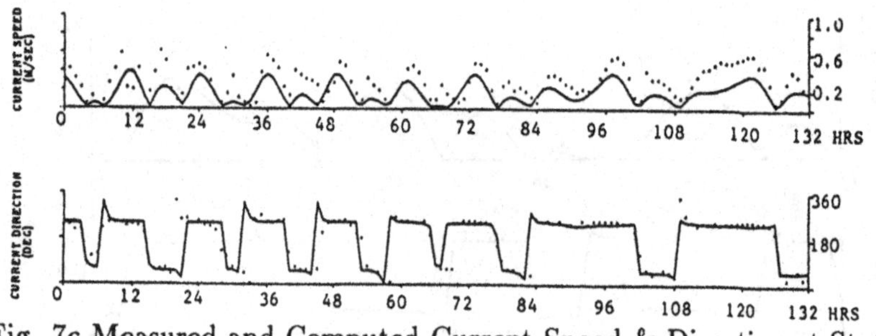

Fig. 7c Measured and Computed Current Speed & Direction at Station CUR 2

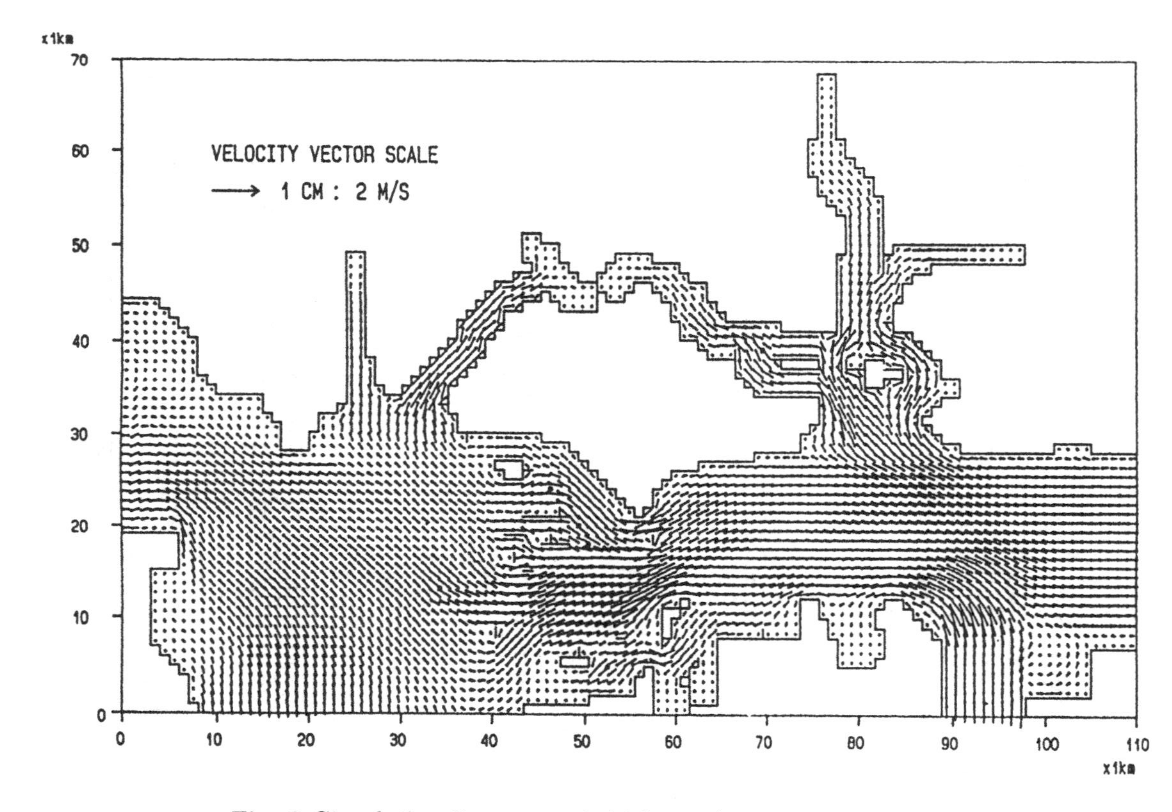

Fig. 8 Circulation Pattern at 2100 hrs 5 Aug 1978 - Regional Model

NUMERICAL OUTPUT FOR REGIONAL MODEL
NUMERICAL OUTPUT FOR NESTED MODEL

• FIELD MEASUREMENTS

PERIOD: 17 FEB - 23 FEB 1987

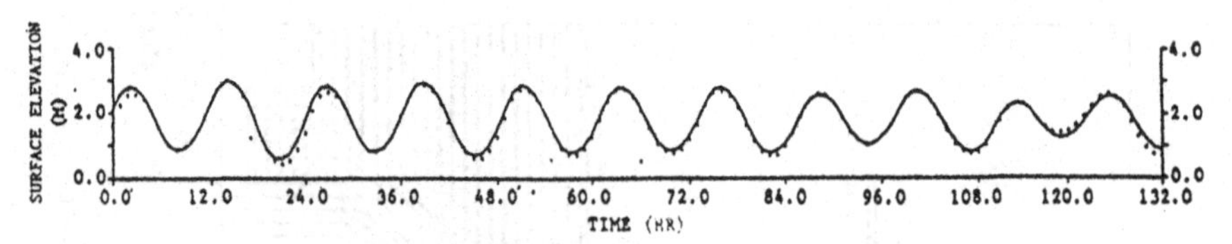

Fig. 9a Comparison of Water Level Among Regional Model, Nested Models and Field Data - TANJONG PAGAR TERMINAL

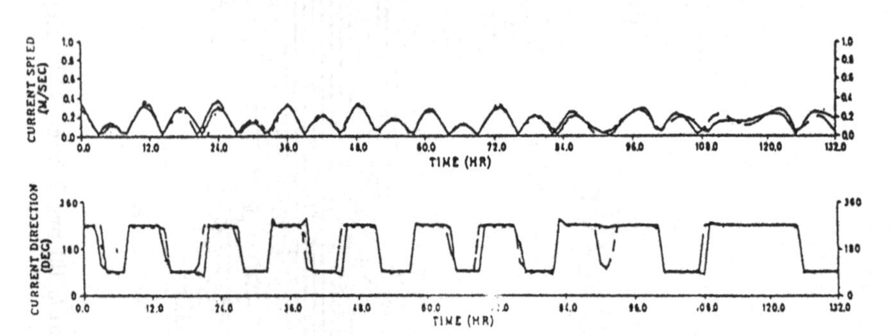

Fig. 9b Comparison of Current Magnitude & Direction Among Regional and Nested Models - Station CUR 2

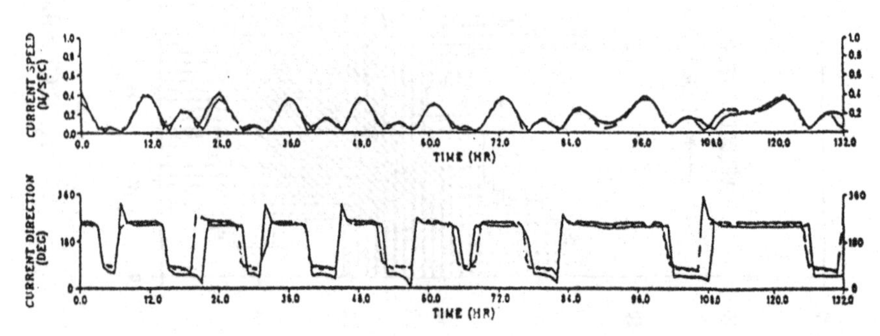

Fig. 9c Comparison of Current Magnitude & Direction Among Regional and Nested Models - Station CUR 5

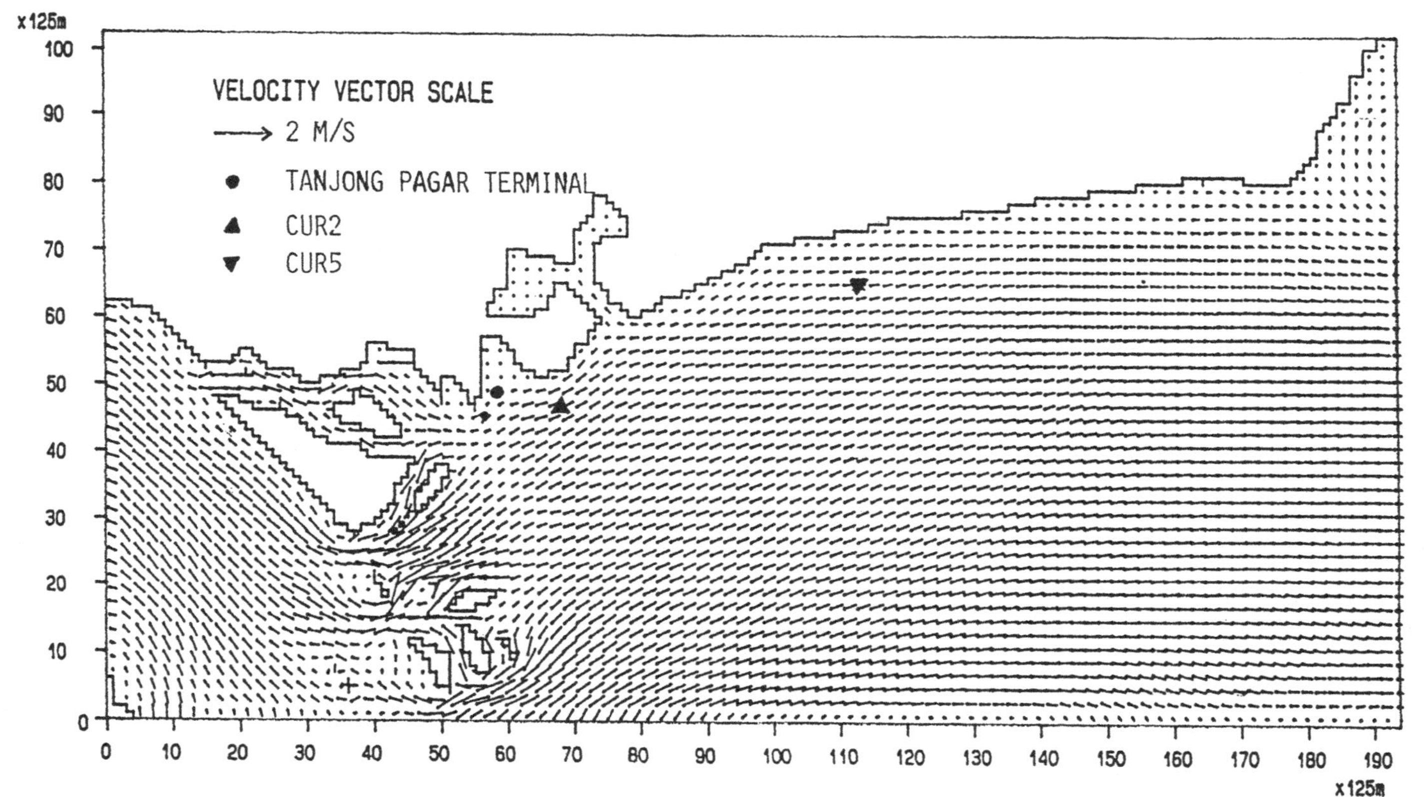

Fig. 10 Circulation Pattern at 2400 hrs 18/8/1978 - Nested Model

SECTION 3: SHALLOW WATER CIRCULATION AND CHANNEL FLOW

Numerical Prediction of Tidal Flow in Shallow Water

D.E. Reeve (*), R.A. Hiley (**)

(*) Sir William Halcrow and Partners Ltd., Burderop Park, Swindon, Wiltshire, SN4 OQD, U.K.

(**) Computational Fluid Dynamics Services, Harwell Laboratory, Oxon., 0X11 ORA, U.K.

ABSTRACT

A numerical model has been developed capable of predicting tidal flows in the southern North Sea. The model solves the depth-integrated equations of motion using an implicit finite-difference scheme on body-fitted coordinates. Viscous dissipation and wetting/drying processes are represented in detail. The calibration and validation procedures are presented and numerical results discussed.

INTRODUCTION

The southern North Sea basin contains a wide variety of mobile bed forms, such as banks, waves, flats and pits. Sediment movement near the East Anglian and Dutch coasts is intimately connected to the prevailing wave conditions and the tidal currents. In particular, the role of residual currents in the evolution of the East Anglian coastline has been highlighted by Robinson [11]. Evidence for sediment circulation patterns in the nearshore region has been reviewed by Carr [3]. A complete picture of the roles of tides and waves on sediment transport is yet to be determined. The present paper describes the development of a model designed not only to provide tidal currents, but also to map the nearshore tidal residuals along the Anglian coast.

MODEL EQUATIONS

Attention is restricted to small domains, so that the tangent plane approximation is valid. Vertical variations in fluid density are neglected, so stratification effects are not represented. At the upper and lower fluid surfaces kinematic boundary conditions are imposed, for zero advective flux. The depth integrated continuity equation is then:

$$(H\rho)_t + \nabla.(H\rho\langle\mathbf{u}\rangle) = 0 \qquad (1),$$

where H is total depth, ρ is density, **u** is fluid velocity, t is time and angle brackets denote a depth mean quantity.

Using the hydrostatic approximation, (see eg. Pedlosky [9]), conservation of momentum implies:

$$(H\rho\langle\mathbf{u}\rangle)_t+\nabla.(H\rho\langle\mathbf{u}\rangle\langle\mathbf{u}\rangle) = -H\rho g\nabla\eta-\tfrac{1}{2}gH^2\nabla\rho-2H\rho\omega\Lambda\langle\mathbf{u}\rangle$$

$$+\nabla.[H(\langle\mathbf{T}\rangle-\rho\langle\mathbf{u'u'}\rangle)]+\tau_s-\tau_b \qquad (2),$$

where **T** is the viscous stress tensor, τ_s and τ_b are boundary stresses, $\mathbf{u'} = \mathbf{u}-\langle\mathbf{u}\rangle$, g is the acceleration due to the Earth's gravity, η is fluid elevation of the free surface above zero datum and ω is the Earth's angular velocity. Atmospheric pressure is assumed uniform over the model domain. Now assume an enhanced viscosity tensor, Γ, may be defined, such that

$$\Gamma.(\nabla\langle\mathbf{u}\rangle+(\nabla\langle\mathbf{u}\rangle)^T) = \langle\mathbf{T}\rangle - \rho\langle\mathbf{u'u'}\rangle \qquad (3),$$

where the superscript T indicates matrix transposition. Equations (1) and (2) are closely analogous to the equations governing flow of a compressible fluid, with variables transformed as

$$(\rho,\ \mathbf{u},\ p,\ \mu) \rightarrow (H\rho,\ \langle\mathbf{u}\rangle,\ \eta,\ H\Gamma) \qquad (4),$$

where p is pressure and μ is viscosity. The principle difference is that the elevation gradient (cf. the pressure gradient) is multiplied by the weight of the fluid column. Also there are extra source terms representing boundary stresses and density gradients.

This analogy is the key to the solution strategy. The base computer code is a well established, general purpose finite-difference Navier-Stokes solver, (see Burns & Wilkes [2]), with the following features: the temporal discretisation uses unconditionally stable backward differences; the equations are solved on a boundary fitted, non-orthogonal, collocated grid, using a coordinate transformation approach; and within each time step the coupled equations are solved iteratively using the SIMPLEC algorithm, modified to account for the factor $Hg\rho$ in the elevation gradient term. The solution procedure then comprises iteration of the following steps until a converged state is reached:

i) solve equation (2) with horizontal components of $\langle \mathbf{u} \rangle$,
ii) solve an elevation correction equation derived from equation (1) assuming a linear relationship (derived from equation (2)) between $\langle \mathbf{u} \rangle$ and η.
iii) update variables $\langle \mathbf{u} \rangle$, η and ρH.

PHYSICAL PROCESSES

Wetting/Drying
The flow in the nearshore regions is of particular interest and thus the wetting and drying of inter-tidal areas must be represented. The following algorithm has been developed:
1) impose a lower bound of 10^{-20}kg/m^2 on areal density $H\rho$;
2) define a wetness function χ to be zero if areal density is not greater than 20kg/m^2, and one if it is not less than 80kg/m^2, with linear variation in between; then impose zero velocity if $\chi < \frac{1}{2}$;
3) if $\chi < 1$ add a non-negative source term to the elevation correction equation, such that the nett mass outflow does not exceed the mass available at the end of the preceeding time step;
4) define cell face wetness χ_f as the wetness of the neighbouring cell with the greater surface elevation, then multiply coefficients of the elevation correction equation by χ_f;
5) use χ_f as a weight factor when calculating the

2δ-centred elevation gradient as a weighted mean of 1δ-centred values;

6) if $\chi_f < 1$ suppress mass flow across cell faces with adverse elevation gradients;

7) during the iterative solution procedure, apply under relaxation to the wetness functions.

Bed Stress

The frictional force at the sea bed is assumed to be related to velocity at the bed. For a quadratic law:

$$\boldsymbol{\tau}_b = C\rho|\mathbf{u}_b|\mathbf{u}_b \qquad (5),$$

where C is the stress coefficient and $\mathbf{u}_b$ is fluid velocity at the bed. The latter is evaluated using a modified version of the spectral technique described by Davies [4]. The vertical variation in velocity is expanded as a cosine series:

$$u'(x,y,z,t) = \Sigma_k \, \mathbf{a}_k(x,y,t)\cos(\pi k(z+h)/H) \qquad (6),$$

where $\mathbf{a}_k$ are the series coefficients, x, y and z are spatial coordinates and h is the depth of the bed below datum. The following equations for the series coefficients are now obtained, by multiplying the three-dimensional momentum equation by the basis functions, then integrating:

$$\tfrac{1}{2}H\rho(\mathbf{a}_k)_t = -H\rho\omega\wedge\mathbf{a}_k - \tfrac{1}{2}\pi^2k^2\mu_V H^{-1}\mathbf{a}_k - (-1)^k\boldsymbol{\tau}_s - \boldsymbol{\tau}_b \quad (7).$$

It has been assumed that advection and horizontal shear are negligible, that effective viscosity for vertical shear, μ_V, is independent of depth and that ρ is uniform. This equation contains no spatial derivatives and may be solved, after discretising the temporal derivative using a backward difference, for each horizontal component of each coefficient, in terms of other flow variables which are either known or become known as the iterative solution procedure progresses. Equation (7) is solved for a finite number of $\mathbf{a}_k$'s, (corresponding to vertical modes), which are then used in (6) to calculate $\mathbf{u}_b$.

BOUNDARY CONDITIONS

The model is driven by tidal elevation and depth mean

normal velocity, $\langle u_n \rangle$, prescribed at open sea boundaries. The primary driving force is the tidal elevation, the normal velocity being used to define a radiation condition, (see eg. Davies & Flather [5]). The boundary conditions are determined from a sum of the five predominant tidal constituents: M_2, S_2, N_2, O_1 and K_1. The Greenwich amplitude and phase of each of these constituents were obtained from the Proudman Oceanographic Laboratory (POL), from their British Isles model, (see eg. Flather [6]). To illustrate the procedure, consider the tidal elevation at a point (x,y), this is given by

$$\eta(x,y,t) = \Sigma_i A_i f_i(t_0)\cos(e_i(t,t_0)+u_i(t_0)-g_i) \quad (8),$$

where A_i and g_i are the Greenwich amplitude and phase of the i'th tidal constituent at (x,y), e_i is its instantaneous phase, f_i and u_i are the amplitude and phase of the nodal modulation and t_0 is the instant of local mean time corresponding to the central time $t = 0$ of the model run.

Two grids were prepared: a fine grid (394 x 104 cells), and a coarse grid (78 x 20 cells) which is shown in Figure 1. In the coarse grid, at the start of each time step, the summation equation (8) is performed for η, and similarly for $\langle u_n \rangle$, for each of the 75 boundary points (73 at the eastern boundary and 2 across The Channel). A second order accurate four point bivariate scheme, Abramovitz & Stegun [1], was used to interpolate the boundary data on to the body fitted mesh.

The bathymetry is defined in terms of the mean bed depth for each cell. Data was obtained from Admiralty charts and nearshore bathymetric surveys. The number of data points (in excess of 170000) was large compared with the number of grid cells. Thyssen's polygon method was used to determine average bathymetry values.

CALIBRATION AND VALIDATION

Calibration and validation are two distinct steps in a modelling study and they serve separate objectives. Calibration is the process of tuning a particular model by altering various parameters in order to obtain the best fit between modelled and observed results for a specific problem. Validation is the demonstration that the calibrated model provides good results when compared against data which is independent from that used in the calibration stage.

Calibration

The calibration procedure included changes to the bed stress calculation and to the boundary conditions, but specifically excluded any modifications to the bathymetry or the use of viscosity. The bed stress can be changed by altering the bed stress law (linear or quadratic), the coefficient C (in equation (5)) and/or the number of vertical modes.

The calibration methodology was as follows: Initially, only M_2 was used to drive the model. Both bed stress and boundary conditions were altered until satisfactory results were obtained. The bed stress formulation was then fixed and calibration of the remaining constituents was performed, adding one constituent at a time, by altering the boundary conditions only.

The above procedure relies upon quasi-linear behaviour, to the extent that it will be most successful when the addition of extra constituents does not substantially alter the amplitudes and phases of existing constituents. To this end, the constituents are added in order of diminishing relative importance; M_2, S_2, N_2, O_1, K_1.

Model elevations and velocities were calibrated against detailed spatial descriptions of each constituent as prepared by POL [7]. Positions of amphidromic points were well captured and constituent amplitudes agreed to within 10% or better throughout the model domain. The phases were in agreement to within 10 to 15 degrees, except in The Wash and the Thames Estuary. These are regions in the coarse grid which are not well resolved in comparison with the rest of the east coast. Tidal currents were calibrated via tidal ellipse parameters, a very stringent test. The maximum (major semi axis length) was within 10% over the model domain for the two dominant semi-diurnal constituents. The direction and phase of the maximum showed good levels of conformity against the observational data. Comparison of the minimum (minor semi axis length) was less good, particularly near the Dutch coast where grid resolution is poorest.

Calibration of the coarse grid was obtained for quadratic bed stress law with two vertical modes and a bed stress coefficient of 0.0025. The time step was 15 minutes, with the transition from wet to dry beginning at a water depth of 8 centimetres and ending at 2 centimetres. Horizontal viscosity, Γ, was zero. The wetting/drying algorithm outlined above proved reliable and exhibited negligible oscillatory behaviour, see Figure 2.

Validation

Tide gauges were deployed at a number of sites along the East Anglian coast for one month. Tidal velocity data were gathered from ship-borne current meters. Results from the measuring stations were compared with those at the nearest model grid cell which remained wet at all times. Figures 3a & b show a comparison, typical of all constituents, of model and observational elevation data for the amplitude and phase of tidal constituent M_2. If agreement were perfect all the points (representing different locations) would lie on a line inclined at 45 degrees to the vertical. The ±0.23m lines are marked on the amplitude plot (corresponding to 10% for the largest amplitudes), and the ±15 degrees lines are drawn on the phase plot. Agreement is good for results predicted by a depth-integrated model. Locations at which the largest discrepancies occur are in sheltered estuaries, which are not well resolved by the coarse grid, and where stratification may be significant.

Published data on experimentally determined tidal residual currents is scarce. However, the residual tidal currents computed over a 32 day period (Figure 4), agree well in both magnitude and direction with previous numerical studies, eg. Nihoul & Ronday [8] and Prandle [10]. These in turn were in good agreement with observed residual flows and thus we may infer that the present model is also in good agreement with observations. Furthermore it provides an improved description of the nearshore current residuals through the body-fitted mesh and inclusion of wetting/drying processes. A 32-day simulation takes 48 hours CPU time on a Hitec10 workstation.

DISCUSSION

A depth-integrated finite difference model has been developed, calibrated and validated for tidal flow in the southern North Sea. It includes detailed representation of viscous dissipation and wetting and drying, important in determining the characteristics of near shore flow. The model has been used to calculate tidally induced residual currents, from which inferences can be made about the long term sediment transport patterns along the East Anglian coast.

ACKNOWLEDGEMENTS

The authors would like to thank Dr. Chris Whitlow for preparing the grid meshes and bathymetry. This work was undertaken as part of the Anglian Sea Defence Management Study on behalf of the National Rivers Authority Anglian Region.

REFERENCES

1. Abramovitz, M. and Stegun, I.A. Handbook of mathematical functions. Dover, 1046p, 1970.
2. Burns, A.D. and Wilkes, N.S. A finite difference method for the computation of fluid flows in complex three-dimensional geometries. UKAEA Harwell Laboratory AERE R 12342, HMSO, 1987.
3. Carr, A.P. Evidence for sediment circulation along the coast of East Anglia. Marine Geology, vol.40, ppM9-M22, 1981.
4. Davies, A.M. On formulating two-dimensional vertically integrated hydrodynamic numerical models with an enhanced representation of bed stress. J. Geophys. Res. 93(C2), pp1241-1263, 1988.
5. Davies, A.M. and Flather, R.A. Computing extreme meteorologically induced currents, with application to the NW European continental

shelf. Cont. Shelf Res., 7(7), pp643-683, 1987.
6. Flather, R.A. A numerical model investigation of the storm surge of 31 January and 1 February 1953 in the North Sea. Q.J.R.Met.Soc, vol.110, pp. 591-612, 1984.
7. HMSO. Atlas of tidal elevations and currents around the British Isles. Proudman Oceanographic Laboratory, OTH 89293, 1990.
8. Nihoul, J.C.J and Ronday, F.C. The influence of the tidal stress on the residual circulation. Application to the Southern Bight of the North Sea. Tellus XXVII(5), pp.484-489, 1975.
9. Pedlosky, J., Geophysical Fluid Dynamics, Springer-Verlag, 1979.
10. Prandle D., Residual flows and elevations in the southern North Sea. Proc.R.Soc.Lond. A359, pp189-228, 1978.
11. Robinson, A.H.W. Erosion and accretion along part of the Suffolk coast of East Anglia. Marine Geology, vol. 37, pp133-146, 1980.

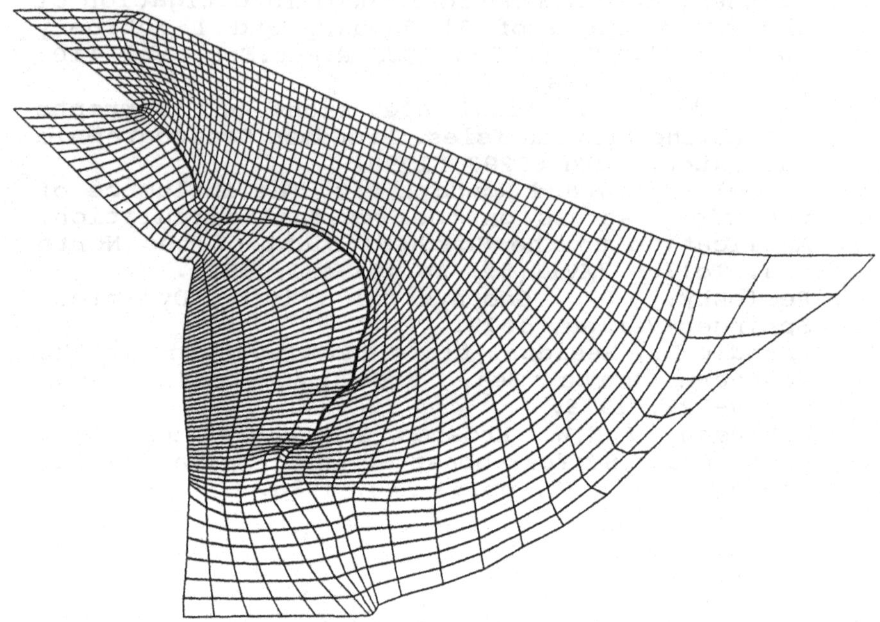

Figure 1: Coarse grid for the Southern North Sea model.

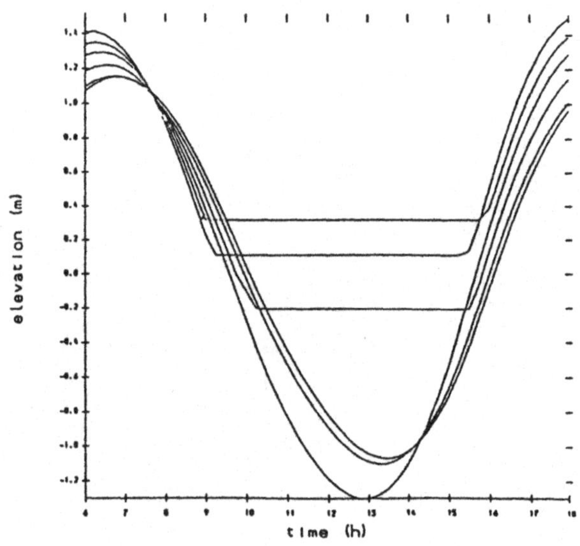

Figure 2: Illustration of the wetting/drying algorithm. Tidal elevations at several neighbouring points near Great Yarmouth.

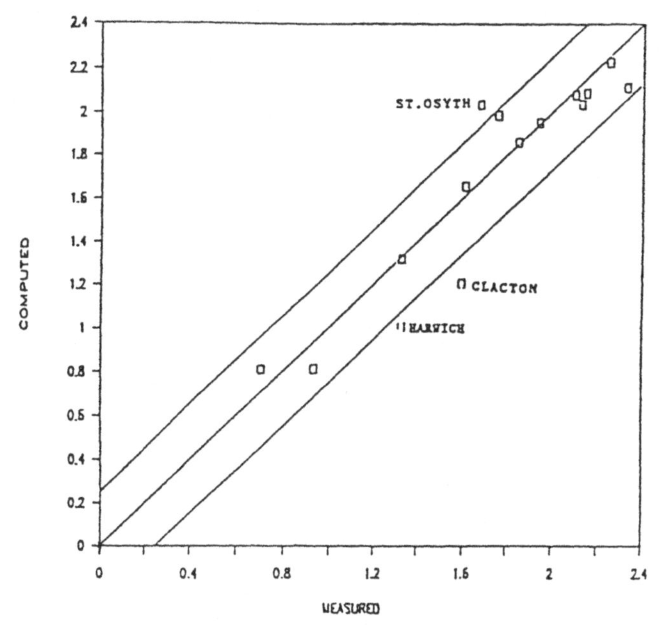

Figure 3a: Comparison of elevation data from the model and observations for the amplitude of tidal constituent M_2 at a number of coastal measuring stations.

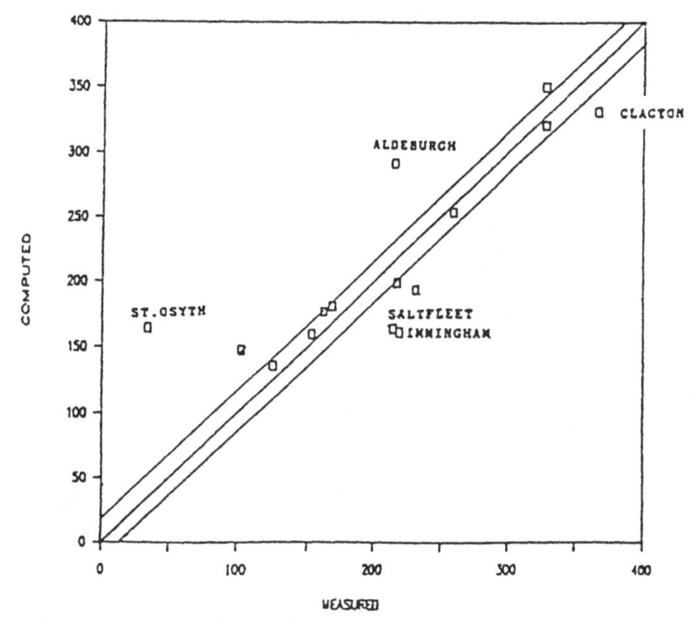

Figure 3b: As in Figure 3a but for the phase of the tidal constituent M_2.

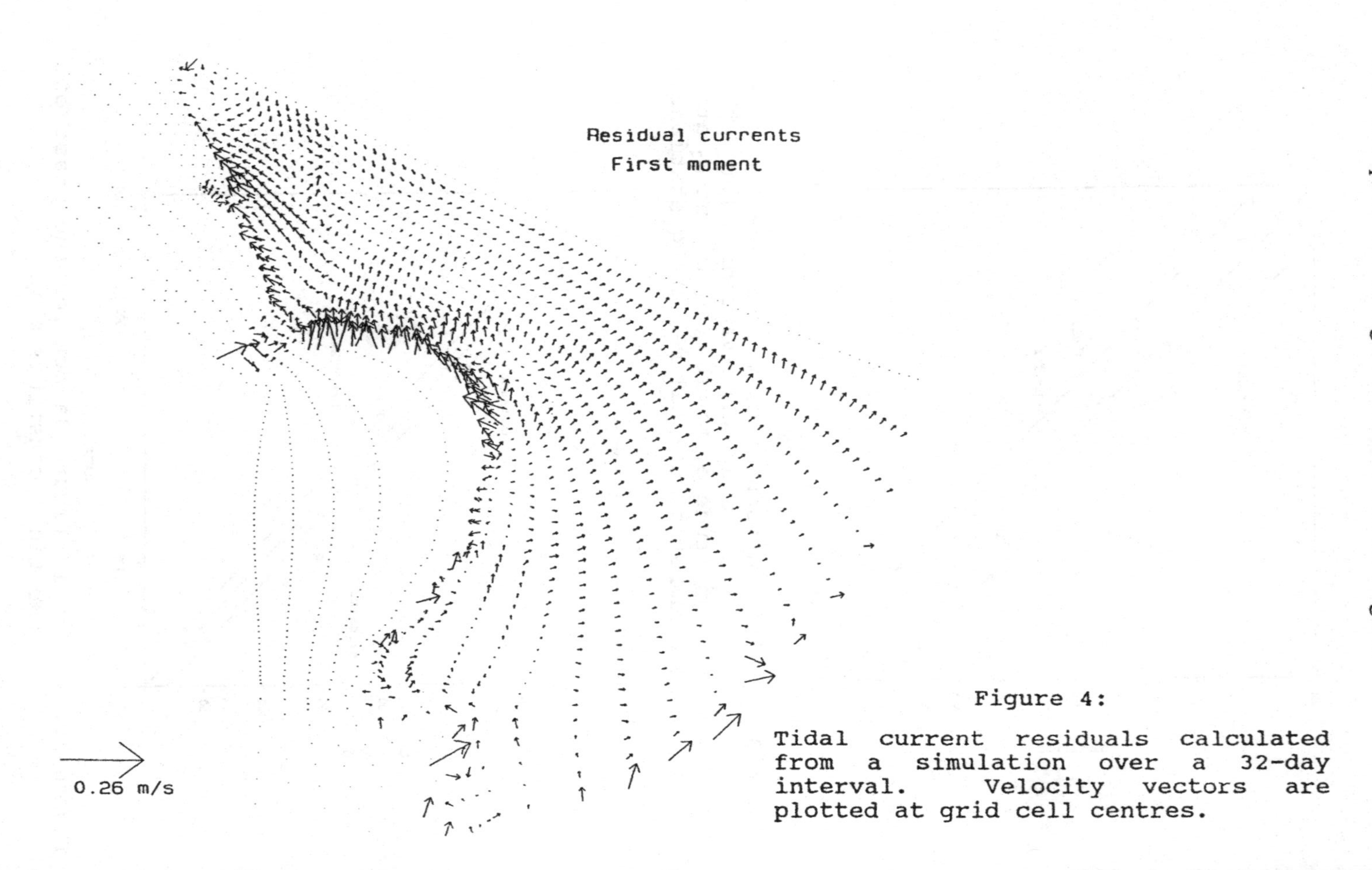

Figure 4:

Tidal current residuals calculated from a simulation over a 32-day interval. Velocity vectors are plotted at grid cell centres.

Modeling Man Made Channels Between Sea and Coastal Lagoons

L. Traversoni

Universidad Autonoma Metropolitana, Iztapalapa, Division de Ciencias Basicas e Ingenieria, Departamento de Ingenieria de Procesos e Hidraulica, Mexico D.F., Mexico

ABSTRACT

Construction of artificial channels between sea and coastal lagoons is a common practice in Mexico in order to improve fisheries or for navigation purposes. However consequences of such channels are still being studied. Our model intends to be a practical contribution to this subject and is devoted specifically to the channel itself and its surroundings.

INTRODUCTION

Sand barriers separating coastal lagoons from the sea are generally formed due to several phenomena the most important of them are near shore sea currents interacting with alluvial deposits coming from rivers.

When a channel is excavated opening a communication between sea and the lagoon, natural equilibrium is broken driving the whole system to an unsteady state that tends to reach a new equilibrium point that can be restored to the previous conditions or move to a new condition. Generally, neither of them is convenient for men use so unsteady conditions must be maintained. Trying to achieve such a purpose several actions are generally used such as building barriers into the sea to protect the mouth of the channel, reinforcing the sand walls of the channel or re-excavating it. Studies like building a physical model are too expensive. Those actions are generally taken empirically.

FOCUSING THE PHYSICAL PROBLEM

We will assume some basic hypothesis emphasizing some phenomena and neglecting others. The assumptions are :

1) Sea waves incident in the shoreline are the main force acting in the model

2) Currents, due to such waves are mainly parallel to the shoreline.

3) Other relevant currents are due to tides or to periodical fluvial floods coming outland.

4) Sand characteristics are homogeneous in the zone.

The main phenomena to be modeled are then wave refraction, reflection and diffraction; currents and sediment deposition and removal.

MATHEMATICAL MODEL

As there are many empirical formulas which describe the phenomena we choose the most commonly used and implemented them all in order the user can elect the one he thinks suits the phenomena best.

Refraction

When the waves arrive to shallow waters refraction deviate them to be parallel to the main bathimetric lines, to calculate such deviation we use Snell law :

$$\frac{\sin \alpha}{\sin \alpha o} = \frac{L}{Lo} = \frac{Cd}{Co} = \text{tgh} \frac{2\pi d}{L}$$

where :

α ....refracted angle
αo ....incidence angle
Co ....celerity of the incident wave
Cd ....celerity of the refracted wave
Lo ....incident wavelength
L ....refracted wavelength
d ....depth

The energy of the wave will be calculated using the relation :

$$\frac{1}{8}\,\gamma\,H_o^2\,b_o\,\frac{L_o}{T} = \frac{1}{8}\,\gamma\,H^2\,b\,\frac{L}{T}$$

where :

Ho .... the height of the wave in deep waters
H ....the height of the wave in shallow waters
bo ....distance between two arbitrary stream lines before refraction
b .... the same distance after refraction

Making some operations we obtain the formulas :

$$\frac{H^2}{H_o^2} = \frac{b_o L_o}{b L} \qquad \frac{H}{H_o} = \sqrt{\frac{b_o}{b}}\ \sqrt{\frac{L_o}{L}}$$

we name :

$$\sqrt{\frac{b_o}{b}} = \sqrt{\frac{\cos\alpha_o}{\cos\alpha}} = K_r \text{ refraction coefficient}$$

$$\alpha = \text{arcsen}(C/C_o \sin\alpha_o) \qquad \sqrt{\frac{L_o}{L}} = K_s \text{ bottom coefficient}$$

then

$$H = H_o\,K_r\,K_s$$

Diffraction

We use the well known formula by M. Larras (A. Frias [1]):

$$H_m = \frac{H_r}{H_i}\ \text{arcotg}\frac{8\alpha}{\pi}\ \left[\ H_i - \frac{H_i}{\pi}\ \text{arccotg}\frac{8\alpha}{\pi}\ \right]\ e^{-4r/L_i}$$

where :

H_m ....height of the diffracted wave
r .... distance from the point being measured to the source of the diffraction

Sediment Transport

We also use other empirical formulas to calculate the sediment transport

As there are many empirical formulas we implemented in our program several of them in order the user can elect one:

CERC Formula

$$S = A\ E_a$$

Where:

S = Littoral transport (m^3 /seg/m)

$E_a = E_o\ K_{rbr}\ \text{sen}\phi_{br}\ \cos\phi_{br}$

here :

$E_o = 1/16\ \rho\ g\ H_o^2\ C_o$

K_{rbr} ....refraction coefficient

ϕ_{br} ....angle between the top of the wave and the shoreline

H_o ....height of the wave in deep waters

C_o ....celerity of the wave in deep waters

A ....proportionality constant approximately 0.02

J. Larras and R Bonefille

$$Q = f(\gamma o, D)\ H/T\ \sin \frac{7}{8}\ \alpha$$

where

$$f(\gamma o, D) = 0.00175 \left(3500 \frac{D}{D^4 + 2}\right) \left(\frac{11 - \gamma o}{10}\right)$$

D ....grain size in mm

γo ....svetlex rate of the wave in percent

Q ....sediment volume m^3

T .... wave period in seconds

α ....angle between the wave front and the shoreline

Bijker (Delft [2])

$$S_b = BD \sqrt{g}\ \frac{V}{C} \exp\left(\frac{-0.27\ \Delta\ D\ \rho\ g}{\mu\ \tau_{cw}}\right)$$

in this case :

S_b ....sediment transport in the bottom (m^3/m/seg)

B ....addimensional coefficient = 5
D ....average diameter of the sediment particles
g ....acceleration due to gravity
v ....average velo:ity of the current
C ....Chezy coefficient
r ....bottom roughness
h ....depth
Δ ....sediment relative density

$$\Delta = \frac{\rho_s - \rho_w}{\rho_w}$$

ρ_w ....water density (kg/m^3)

μ ....buckle coefficient = $(C/C')^{3/2}$
C' ....Chezy coefficient for D_{90} instead of r as in C
τ_{cw} ....shear velocity under the combined effects of waves and current

The idea is that in each particular circumstance one formula could be better than the other for the user so he can elect one or even put one due to himself.

THE NUMERICAL INTERPRETATION

No matter which of the different combinations of empirical formulas we choose there is still the problem of how to use them in a numerical environment, to do this we must make some assumptions:

1) We are dealing with a local phenomena reduced to the mouth of a lagoon (or of the channel communicating it with the sea) and its surroundings.

2) Our main purpose is to simulate what happens with the channel and the sand wall between sea and the lagoon, the other related phenomena are simulated with other program of which this is only a server devoted to calculate a very important bordering condition.

3) Balance of sediment transport is one of the most important problems to deal with because we assume that changes happen very quickly and are very correlated; for example an accumulation of sand means shallower conditions and therefore increments in refraction and even the beginning of diffraction in some places.

Discretization

Always when we use numerical methods our continuous phenomena must be discretized in order to make the calculations, triangular and quadrilateral elements are the most commonly employed when this happens. Neither of them has been chosen by us, we switched to what we think is a "smoother" discretization method because it improves approximation : Covering Circles.

Covering Circles some definitions

If we have a set of points $\mathcal{V}$ in a plane (for example in our case the points where we have measurements of our variables); there exists a set $\mathcal{C}$ of circles we called "Covering Circles" (Traversoni [3]) such that:
1) Every point of $\mathcal{V}$ belongs to the circumference of some circle of $\mathcal{C}$.
2) There is no point of $\mathcal{V}$ inside any circle of $\mathcal{C}$.
3) Every circle of $\mathcal{C}$ can be determined by at least a set of 3 points of $\mathcal{V}$.

This concept is closely related with other well known, we can note that the centers of the Covering Circles are vertices of the Voronoi tiles of the set $\mathcal{V}$ and that every Covering Circle circumscribes a Delaunay triangle.

When triangles or quadrilaterals are used all the zone covered by them has the property that every point of the plane on it belongs to one and only one triangle or cuadrilateral, except in the case they are in the sides or they are vertices. Knowing that when the value in a new point is needed, we know it belongs to for example to one triangle and we interpolate with the vertices of that triangle if our interpolation is linear or with other points, all belonging to the triangle if what we want is cuadratic or higher order interpolation. Circles overlap so a point in the zone may belong to several of them at the same time, the question is then which one has to be used . What at first could be considered a problem can be used as an advantage using Sibson's interpolation (Sibson[4]):

Sibson's Interpolation

If we consider the set $\mathcal{V}$ and its related Voronoi tessellation when a new point is added to $\mathcal{V}$ it forms its own tile, the interception of it with the former tessellation divides it in sections each one belonging to a neighbor tile (see figure 1). When interpolation is implemented the weight of each neighbor is the relative weight of its portion of tile in the new one. This approximation is continuous and quadratic (Farin [5]).

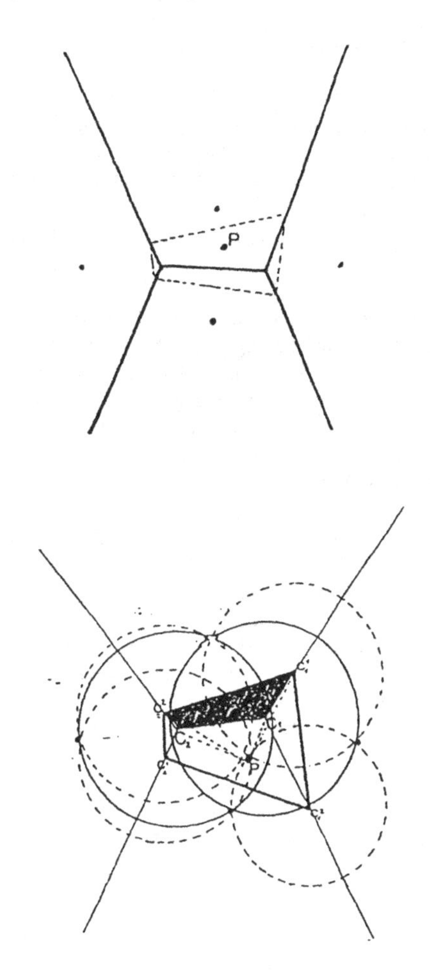

Figure 1 the point P and its tile divided by the ones of its neighbors

It is very important to note the fact that without adding auxiliary points we obtain cuadratic interpolation doing only simple operations as if we were doing linear interpolation.

The relation of the above with the Covering Circles are :

1) Every vertex of the tiles is a center of a Covering Circle.
2) Every circle such that its center is vertex of the former tessellation has P inside.

3) Every circle such that its center is vertex of the new tile has P on its circumference.

It is also very important to note that with the same algorithm we build the Covering Circles we can add new points and, with only local modifications rebuild the set of Covering Circles. That is very useful if we want to refine our net, interpolating at one point and adding it to the set V.

All the above has enormous advantages when we implement computationally the building of the discretization and when we use it for interpolation. However the most important to us are the advantages when we apply it to Finite Element Methods, and when we interpolate to obtain refraction patterns of the waves because it allows us smoother representations.

When Finite Element is used in the equation of a given point P intervenes all its neighbors, understanding that those neighbors are the points that belong to elements that have P as vertex. Consider now that we are using Delaunay triangulation as discretization, in that case the neighbors will be the like the points belonging to the same circumferences to which P belongs. As can be seen, we can forget the triangulation and use the circles for neighbor determination and later on as basis for Sibson interpolation.

However in our case it is also useful for determining the path of a particle floating with the waves under refraction conditions that we will use to build the perpendicular wave fronts.

On figure 2 we can see our graphical method for refraction and diffraction.

The steps are as follow :

1) We begin with a rectilinear wave front in open sea and we locate on it "particles" uniformly distributed (the distance between them can be elected by the user).

2) At a given time we calculate the new position of each particle advancing in a straight line in the direction of the velocity on its former position.

3) In the new position we calculate the refracted or diffracted direction and the new velocity as well as the other parameters of the wave in the point.

4) As we don't know exactly the depth in the new point we interpolate using Covering Circles (previously determined

with the points where we have depths measurements) and Sibson's interpolation.

5) If more detail is needed the new point can be added to the set and their corresponding circles added.

6) To obtain the modified wave front we draw the line passing trough the particles on their new position and we repeat the procedure as many times as needed.

7) The depths are modified using the formulas above to calculate the removal or deposition of sediment using as area the Voronoi tiles of each point.

8) When a new wave front comes, it finds the sea bottom modified by the former one.

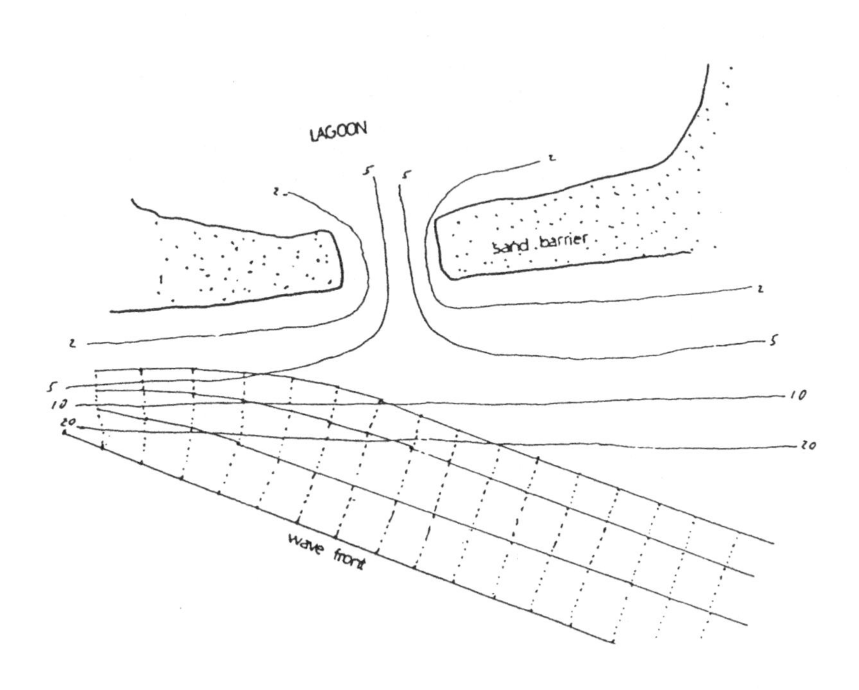

Figure 2 The wave front building

ADVANTAGES

Our method means faster computations and less storage requirements as well as it has good results due to its new interpolation procedure.

Using the Covering Circles we can do several jobs with the same procedure, for instance we use the same procedure to locate a point and to build the circles as well as to calculate areas for interpolation.

As it is a local algorithm it can be used in parallel configurations (although this has not been implemented yet).

REFERENCES

[1] A Frias Ingeniería de Costas Limusa I.P.N. 1988
[2] Delft University of Technology Coastal Engineering 1981
[3] L. Traversoni A Building Method for Hierarchical Covering Spheres of a Given Set of Points. Curves and Surfaces P.J. Laurent, A Le Méhauté, L L Shumaker Eds. Academic Press 1991
[4] Sibson R. A brief description of the natural neighbour interpolant, in V. Barnett, ed. Interpolating Multivariate data, Wiley New York 1981.
[5] G. Farin Surfaces over Dirichler Tessellations. Computer Aided Geometric Design V 7 Nos 1-4 June 1990.

Effect of the Length of the Estuary on its Characteristics

J.S.R. Murthy, T.V. Praveen

Department of Civil Engineering, Indian Institute of Technology, Bombay 400 076, India

ABSTRACT

It is witnessed in the case of Rotterdam Waterway [1], that excessive dredging for constructing the Europort harbor led to a change in the estuary boundary geometry configuration, (viz., L/H ratio) and the estuary characteristics. In the present study an attempt is made to assess the effect of boundary geometry, i.e., ratio of length of the estuary, L to tidally averaged depth, H, on various parameters of the estuary which may affect the characteristics of the regime. The above approach is sought to be validated in the light of the changes in the estuary characteristics after dredging the Rotterdam Waterway.

INTRODUCTION

Man's intense utilization of the marine environment occurs in the estuary. The experience in the past few decades illustrates the growth of the developmental activities in the estuary regions. For whatever may be the purpose of the developmental activities, i.e., dredging for navigation, waste disposal, selective withdrawal of water of specified quality, etc., a knowledge of ambient flow conditions, i.e., space-time distribution of salinity for a given tidal input and fresh water inflow is required. Such studies will help in choosing the place and time for withdrawing fresh water with known and acceptable salinity level.

Many of the world's seaports are situated on estuaries and access to them depends on maintaining navigable channels of sufficient depth. The promotion of trade and industry has led to large-scale alteration of the natural balance within estuaries by alteration of their topography, for providing easy access to large ships and leading to large-scale pollution, with industrialization and population growth. Deforestation of the land leads to increased run-off from the land and flashy floods and increased sediment load in the rivers which may be deposited in the estuary regions leading eventually to the alteration in the boundary geometry.The form of the estuary may constantly alter by erosion and deposition of sediment. It is essential to understand the effect of these changes in the boundary geometry on the characteristics of the flow in the estuary for a planned development.

Pritchard [4] considered estuary depth and width to be important parameters controlling its characteristics in the sequence. If the river flow and tidal range are kept constant and the estuary width is increased, the ratio of tidal volume to river flow is changed, the result of which is similar to a decrease in river flow. This leads to a greater degree of mixing in the estuary. Similarly, increasing the depth by dredging will decrease the ratio of river flow to tidal flow, but the effect of this will be offset by decreasing the effectiveness of vertical tidal mixing leading to a greater degree of stratification in the estuary.

In the present study, an attempt is made to assess the effect of boundary geometry, i.e., ratio of length of the estuary, L to tidally averaged depth, H, on various parameters of the estuary which may affect the characteristics of the regimen. The above approach is sought to be validated in the light of the changes in the estuary characteristics after dredging of the Rotterdam Waterway.

The field data of Rotterdam Waterway was analyzed by Harleman and Abraham[1] on lines similar to those of Ippen and Harleman[2]. Ippen and Harleman[2] conducted a series of salinity intrusion tests at the waterways experiment station (WES), Vicksburg. The studies of Ippen and Harleman[2] dealt with the partially mixed estuaries and developed correlation between parameters which reflect the characteristics of the regimen. The estuary numbers of Rotterdam Waterway field conditions for the year 1908, when the estuary was natural and undredged were within the range of those of WES flume studies. The field data of Rotterdam Waterway for the year 1958 was within the range of the extrapolated values of WES flume studies and the predictions from the analysis were found to be satisfactory. However for the year 1963, the analysis was found to deviate from the trend of the earlier results. The deviation was attributed to change in the mean water depth due to dredging which might have affected the characteristics of estuary regimen, viz., hydrodynamic conditions and salinity transport. The dredging not only effects the boundary geometry, i.e., L/H ratio, but also other dimensionless parameters, viz., fresh water Froude number and amplitude ratio. However, observing that the WES flume studies were conducted for a range of fresh water Froude number and amplitude ratios and for only one L/H ratio (=654), the deviation of the field data can be expected to be dependent on the change in the boundary geometry configuration (L/H) based on the dimensional considerations.

The estuary is assumed to be an 'idealised estuary' as termed by investigators at M.I.T. (USA), which is convenient and suitable in the present study. A time varying one-dimensional numerical model for hydrodynamics and salinity transport in rectangular idealized estuarine reach is used. To be able to run the 1-D numerical model, some basic data consisting of tidal input at the seaward boundary, the fresh water inflow at the upstream and maximum salinity at the downstream end need to be known. First two are known from the quasi-steady-state setting and are independent in nature, but the maximum salinity at the estuary mouth can be known only from laboratory experiments after quasi-steady-state is believed to have been attained in the given setting, as was done by Ippen and Harleman[2]. For a general application of a numerical run the correlation given by Praveen, Murthy and Chandhra[3] is used to evaluate maximum salinity.

To understand the effect of L/H ratio on an estuary several 1-D numerical runs are taken with three L/H ratios, viz., 654, 1308 and 1962 approximately equal to $\lambda/6$, $\lambda/3$ and $\lambda/2$, respectively, where λ is the wave length of the tidal wave. The combination of fresh water Froude numbers (0.005, 0.01 and 0.014) and amplitude ratios (0.2, 0.15 and 0.1) are chosen for the study. The downstream boundary condition during flood flow, viz., the maximum salinity is estimated from the correlations given by Praveen, Murthy and Chandhra[3].

EFFECT OF THE LENGTH OF THE ESTUARY ON THE DEGREE OF STRATIFICATION

The significant non-dimensional parameters representing the degree of stratification, viz., stratification number, G/J, Pritchard number, P_t, Estuary number, $\mathbb{E}$ and densimetric estuary number, $\mathbb{E}_D$ are evaluated from the results of the numerical runs. The variation of these dimensionless parameters with L/H ratio for a given set of fresh water Froude number and amplitude ratio are given in Table 1. From the Table 1 it can be observed that the values of above dimensionless parameters decrease with increasing L/H ratio indicating that the degree of stratification increases with increasing L/H ratio. However, such a trend is not clear in the comparison of stratification number, G/J for varying L/H ratios. The evaluation of stratification number involves the estimation of wave number, k and damping coefficient, μ from the correlations using the numerical results. The possible reason for this aberration is attributed to the approximate estimation of wave number, k and damping coefficient μ from the correlations developed for the evaluation of stratification number.

It is observed that the nodes corresponding to high water level and low water level form only for the lengths 654' ($\lambda/3$) and 981' ($\lambda/2$), following the general trends observed in the co oscillatory tidal flows. However, due to the damping effects in the flume conditions, the nodal formation is of impure nature not forming at mean water level. The deviation from the nature of pure nodal formation is however predominant in the cases of higher amplitude ratios and only marginally affected by the fresh water Froude number and length of the estuary. The temporal variation of sectionally averaged velocities at the estuary mouth are compared and given for the data of Test 14 of fresh water Froude number, 0.005, and amplitude ratio, 0.2, for various L/H ratios in the Fig 1. It is observed that the intra-tidal variations of velocities at the estuary mouth decrease with the increase in the L/H ratio. It is observed that the maximum ebb velocity at the estuary mouth varies only marginally with the length of the estuary from 327' to 654'. However, the maximum flood velocity is observed to rapidly decrease from 327' (λ/6) to 981 (λ/2). The effect of the length of the estuary on the longitudinal variation of velocities at the characteristic conditions of maximum flood velocity (MFV) and maximum ebb velocity (MEV) at the estuary mouth for WES Test14 are shown in Fig 2. Irrespective of the direction of the flow at the estuary mouth, the upstream region is observed to have ebb velocity due to the fresh water inflow. . It is further understood that for the estuaries of large lengths and small fresh water Froude numbers, the upstream reach is in the flood flow conditions, despite the characteristic of MEV prevailing at the estuary mouth. Further, it is observed that in the cases of larger lengths, beyond a certain distance from the estuary mouth, the velocities vary only marginally within the tidal period and more so when the amplitude ratio is small.

The temporal variation of sectionally averaged salinities at 40ft., 80ft. and 120ft. are compared for various L/H ratios, for the data of WES flume Test 14, in the Fig 3. Following a trend similar to that in the convective terms, the intra tidal variation of salinities are observed to decrease with increasing L/H ratios in the down stream regions. The maximum intrusion lengths are observed to increase with the length of the estuary. However, beyond certain length of the estuary, the length of the estuary is inconsequential and the maximum intrusion length remains to be unaffected by it.

MULTIPLE CORRELATION FOR MAXIMUM INTRUSION LENGTH

An attempt is made to express the maximum intrusion length, x_{intru} in terms of readily computable bulk parameters, derived from the dimensional considerations. The functional relationship for the salinity intrusion in an estuary can be written as

$$\frac{x_{intru}}{L} = f\left[\frac{C}{C_{OCN}}, \frac{t}{T}, \frac{a}{H}, \frac{U_f}{\sqrt{gH}}, \frac{L}{H}, n_m\right] \quad ......(1)$$

where

x longitudinal distance from estuary mouth (measured landward)

t time elapsed since beginning of the tidal cycle (taking as beginning of the ebb tide)

L length of the flume representing estuary

H constant depth of water in estuary

a amplitude of tidal variation at the estuary mouth

T tidal period

U_f velocity of fresh water inflow

C_{OCN} the ocean salinity

n_m Manning's roughness coefficient for estuary

Assuming n_m as a constant, the maximum intrusion length, x_{intru} is defined as the distance between the estuary mouth and the point at which the salinity is at least 1% of the ocean salinity during high tide. Hence the eq(1) can be rewritten as

$$\frac{x_{intru}}{L} = f\left[\frac{a}{H}, \frac{U_f}{\sqrt{gH}}, \frac{L}{H}\right] \qquad \ldots\ldots(2)$$

The maximum intrusion length, x_{intru} obtained from numerical runs corresponding to the chosen data are used to develop a multiple correlation between x_{intru}/L and $U_f/\sqrt{gH}$ as a parametric variable of a/H for various L/H ratios as shown in Fig 4. The effect of the length of the estuary on the longitudinal salinity gradients was observed to be significant at lower fresh water Froude number (=.005) than at the higher Froude number (=.014). This effect is reflected in the intrusion lengths, as the intrusion length proportionally increased with the length at lower fresh water Froude numbers. It is observed that as the length of the estuary increases, beyond a certain length of the estuary, the maximum intrusion length essentially depends on fresh water Froude Number and varies only marginally with the amplitude ratio and length of the estuary.

PROTOTYPE VALIDATION

An attempt is made to verify the validity of the multiple correlations developed in the Fig 4 for the maximum intrusion length, x_{intru}, in terms of readily computable bulk parameters. The real estuaries for which the detailed field data are available in literature, viz., Delaware, Hudson and Rotterdam Waterway are chosen for the comparisons. The correlations in the Fig 4 are in terms of bulk parameters,viz., fresh water Froude number, amplitude ratio and L/H ratio. The bulk parameters corresponding to the field data are to be evaluated in dimensionally similar conditions of WES flume type to avoid extrapolations of graphical correlations.

Assuming the bed shear generated near the bottom of the flume is under dimensionally similar conditions as that of the real estuary, the dimensional similitude between model and prototype is shown by Yalin[5] to be governed by Froude number, Strohaul number and a/H ratio. The dimensional parameters can be written as

$$\frac{U_m^2}{gH_m} = \frac{U_p^2}{gH_p} \qquad \ldots\ldots(3)$$

$$\frac{a_m}{H_m} = \frac{a_p}{H_p} \qquad \ldots\ldots(4)$$

$$\frac{U_m T_m}{L_m} = \frac{U_p T_p}{L_p} \qquad \ldots\ldots(5)$$

where subscripts m and p represent model and prototype values respectively.

The irregular width and depth of the real estuary is schematized, considering the mean width over the length, and tidally and longitudinally averaged depth. The evaluation of fresh water Froude number and amplitude ratio of the real estuary is simplified due to the above schematization. The scale distortion in the model (L_r/H_r, where r represents the ratio of variables between model and prototype) can be evaluated from the dynamic similarity of the characteristics of the tidal wave of the real estuary with that of the flume, represented by Strohaul number in the eq(5) as

$$\frac{L_m}{H_m} = \left[\frac{L_m T_m H_p^{1/2}}{L_p T_p H_m^{1/2}}\right] \qquad \ldots\ldots(6)$$

The bulk parameters that are evaluated for the dimensionally similar model of the estuary are superimposed on Fig 4 to estimate the maximum intrusion length ratio, x_{intru}/L. The comparisons between the prototype data and the estimated values from the Fig 4 are given in Table 2.

The estimated values of maximum intrusion length, X_{intru} from the correlation in the Fig 4 are observed to be within 9% error when compared to the field data. However the estimated values of x_{intru} for the field data of Rotterdam Waterway of year 1963 were observed to be in an agreement better than others since the extent of extrapolation needed for the comparison is less than those for other field data. The reason for the disparity can be attributed to the idealization of the natural estuary and the non uniform bed roughness.

APPLICATION OF THE ANALYSIS TO THE FIELD DATA OF ROTTERDAM WATERWAY :

On lines similar to those of Harleman and Abraham[4], a correlation between stratification number and estuary number is developed with L/H ratio as parametric variable as shown in Fig 5. It can be observed from the correlation in Fig 5 that the dimensionless parameters G/J, and $\mathbb{E}$ are uniquely correlated for smaller lengths of the estuary (327'), however as the length of the estuary increases some scatter is observed. This indicates that the correlation between G/J vs $\mathbb{E}$ is possible up to certain length of the estuary beyond which the relationship among the above dimensionless parameters may not be uniquely defined though the trend continues to be the same.

Following the analysis of Rotterdam Waterway data by Harleman and Abraham[1], a correlation between estuary number $\mathbb{E}$ and $\sigma B/U_o$, with L/H ratio as a parametric variable is defined as shown in Fig 6. The effect of the L/H ratio on the correlation between $\mathbb{E}$ vs $\sigma B/U_o$ is observed to be significant.

The correlations developed in the present studies are applied to the field data of Rotterdam Waterway on lines similar to those of Harleman and Abraham[1] considering the effect of L/H ratio. The waterway went through an overall transformation from an undredged natural waterway in 1908 of mean depth, 5.8m to a significantly dredged waterway in 1963 of mean depth, 11m. Further it is observed that the fresh water inflows have increased from 22nd July 1908 to 19th March 1963, with a slight dip on 18th April 1963. The amplitude ratio is also be observed to decrease from 22nd July 1908 to 18th April 1963.

The field data of Rotterdam Waterway is superimposed on Fig 6 to verify the applicability of the correlations which include the affect of L/H ratio as a bulk parameter. The estimated values of $\sigma B/U_o$ from the correlation in the Fig 6 are compared with the field data and the estimated values of Harleman and Abraham[1] as given in Table 3 . A significant improvement in the prediction of the field data in the present study can be observed when compared with those of Harleman and Abraham[1].

CONCLUSIONS

1) It is observed that the increase in the length of the estuary leads to a greater degree of density stratification in the estuary.
2) The impure nodal formation and their deviation from the pure nodes is essentially governed by the amplitude ratio. For larger lengths of the estuary, beyond a certain distance form the estuary mouth the velocity remains unaffected by the conditions at the estuary mouth.
3) An attempt is made to express the maximum salinity intrusion in terms of readily computable bulk parameters. The multiple correlation thus developed is verified with the field data of Delaware, Hudson and Rotterdam Waterway.
4) The validity of the correlations in Fig. 6 is verified with the Field data of Rotterdam Waterway. An improvement in the

prediction of the field data in the present study can be observed when compared with those of Harleman and Abraham[1].

REFERENCES

1) Harleman, D.R.F and Abraham, G., One-dimensional Analysis of Salinity Intrusion in Rotterdam Waterway, Publication No. 44, Delft Hydraulics Laboratory, (Oct., 1966).
2) Ippen, A.T. and Harleman, D.R.F., One-dimensional Analysis of Salinity Intrusion in Estuaries, Technical Bulletin No. 5, Waterways Experiment Station, Vicksburg, (June, 1961).
3) Praveen, T.V, Murthy, J.S.R. and Chandhra, J., Dependence of High Water Slack Salinity on Estuary Length, 2nd Regional Conference on Computer Applications in Civil Engineering, John Bahru, Malaysia, (Feb. 1971).
4) Pritchard, D.W., Estuarine Circulation Patterns, Proc. ASCE, Vol. 81, No. 717, (1955).
5) Yalin, M.S., Theory of Hydrodynamic Models, The McMillan Press, London, (1971).

SYMBOLS

a amplitude of tidal variation at the estuary mouth
B The distance between estuary mouth and a section within the ocean at which the salinity is always equal to the ocean salinity
G Rate of energy dissipation per unit mass of fluid
H constant depth of water in estuary
J Rate of gain of potential energy per unit mass of fluid
L length of the flume representing estuary
T tidal period
t time elapsed since beginning of the tidal cycle (taking as beginning of the ebb tide)
x longitudinal distance from estuary mouth (measured landward)
C_o Maximum salinity that can occur at estuary mouth
C_{OCN} the ocean salinity
F_o Froude number corresponding to maximum velocity ($= U_o/\sqrt{gH}$)
n_m Manning's roughness coefficient for estuary
P_t Tidal prism
Q_f Fresh water inflow
U_f velocity of fresh water inflow
U_o Maximum flood velocity at the estuary mouth
σ $2\pi/T$
$\mathbb{E}$ Estuary number
$\mathbb{E}_D$ Densimetric estuary number
$\Delta\rho$ Density difference between fresh water and sea water
ρ Fresh water density

TABLE 1: VARIATION OF STRATIFICATION NUMBER (G/J), PRITCHARD NUMBER (P_t), ESTUARY NUMBER ($\mathbb{E}$) AND DENSIMETRIC ESTUARY NUMBER ($\mathbb{E}_D$)

$\frac{U_f}{\sqrt{gH}}$	$\frac{a}{H}$	G/J			P_t			$\mathbb{E}$			$\mathbb{E}_D$		
		L(ft)			L(ft)			L(ft)			L(ft)		
		327	654	981	327	654	981	327	654	981	327	654	981
.005	.20	167	157	123	11.2	9.2	7.8	.359	.271	.152	16.2	12.5	7.1
.005	.15	99	107	71	9.0	7.5	5.9	.186	.151	.069	8.5	7.1	3.3
.005	.10	43	51	27	6.4	5.9	4.0	.073	.071	.021	3.3	3.4	1.0
.010	.20	95	90	63	5.4	4.2	3.6	.176	.133	.075	8.6	6.7	3.7
.010	.15	54	50	39	4.3	3.5	2.7	.091	.076	.034	4.5	3.8	1.7
.010	.10	22	25	16	3.0	2.7	1.8	.035	.035	.011	1.8	1.8	0.5
.014	.20	69	73	46	3.7	2.9	2.4	.124	.096	.051	6.1	4.6	2.5
.014	.15	39	40	23	2.9	2.4	1.8	.062	.054	.024	3.1	2.6	1.2
.014	.10	17	20	10	2.0	1.7	1.1	.024	.024	.008	1.2	1.2	0.4

TABLE 2: ESTIMATION OF MAXIMUM INTRUSION LENGTH, x_{intru} FOR THE FIELD DATA

Estuary		$\frac{U_f}{\sqrt{gH}}$	$\frac{a}{H}$	L/H	Intrusion Length Ratio, x_{intru}/L	
					Actual	Estimated
Delaware		.00065	.095	2872	0.5	0.54
Hudson		.00087	.076	2804	0.6	0.52
Rotterdam Waterway	1908	.0159	.130	1364	0.282	0.32
	1956	.0188	.074	1077	0.353	0.28
	1963 March	.023	.067	991	0.315	0.31
	April	.0173	.067	991	0.295	0.31

TABLE 3: ESTIMATION OF $\sigma B/U_o$ FROM THE CORRELATIONS FOR THE FIELD DATA OF ROTTERDAM WATERWAY

Date	Field Data	Harleman and Abraham[1]	Estimated value from Fig. 6
26th July, 1908	1.23	1.35	1.38
22nd June, 1956	1.70	1.70	1.90
18th March, 1963	3.45	1.80	2.25
19th April, 1963	2.90	1.65	2.10

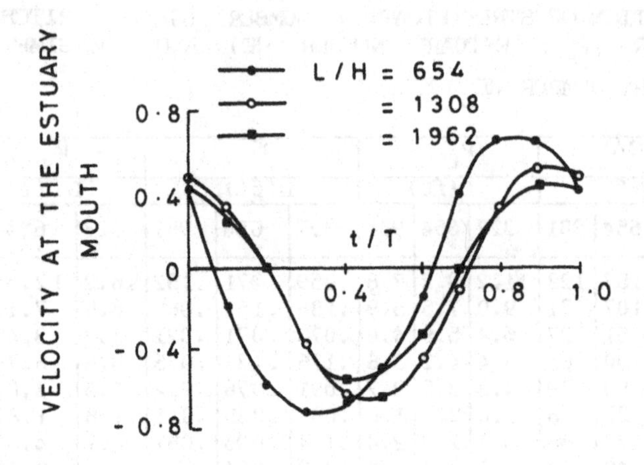

FIG.1. COMPARISON OF SECTIONALLY AVERAGED VELOCITIE AT THE ESTUARY MOUTH WITH VARYING L/H RATIOS (WES FLUME, TEST 14)

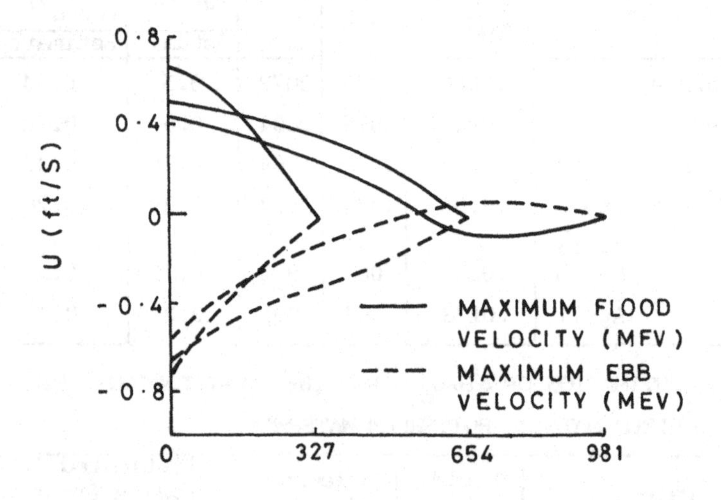

FIG.2. LONGITUDINAL VELOCITY PROFILES AT CHARACTERISTIC TIMES OF MFV AND MEV FOR VARYING LENGTHS OF THE ESTUARY (FRESH WATER PROUDE NO. = 0·005, AMPLITUDE RATIO = 0·2)

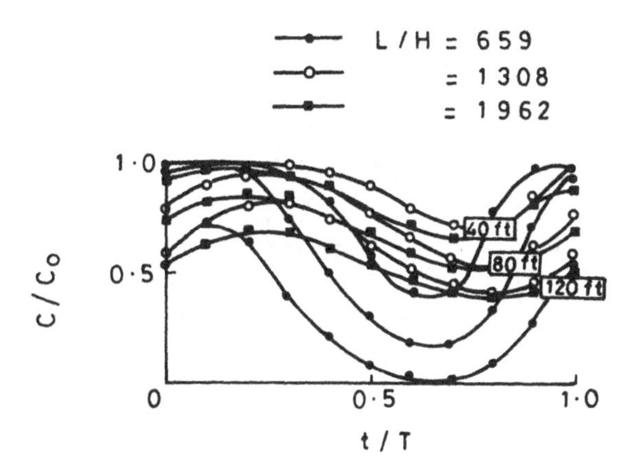

FIG. 3. COMPARISON OF SECTIONALLY AVERAGED SALINITIES WITH VARYING L/H RATIOS (WES FLUME, TEST 14)

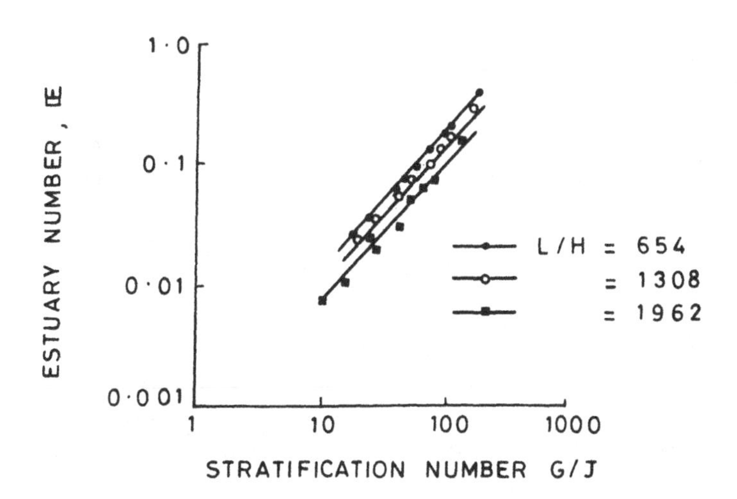

FIG. 4. CORRELATION BETWEEN STRTIFICATION NUMBER, G/J AND ESTURAY NUMBER, $P_t F_o^2 / Q_f T$ WITH L/S AS PARAMETRIC VARIABLE.

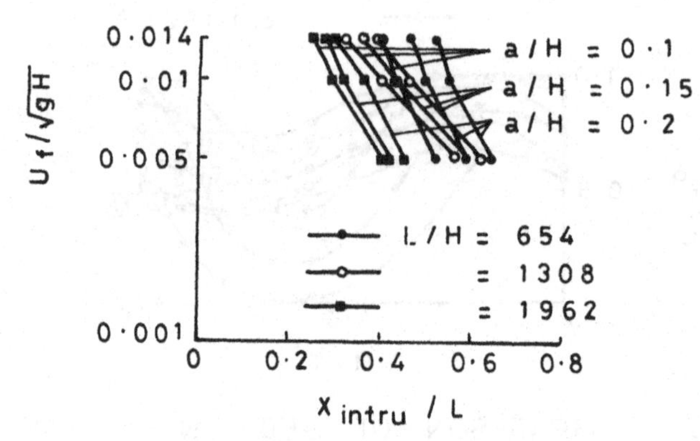

FIG. 5. MULTIPLE CORRELATION BETWEEN X_{intru}/L AND FRN FOR WARYING L/H RATIO WITH a/H AS PARAMETRIC VARIABLE.

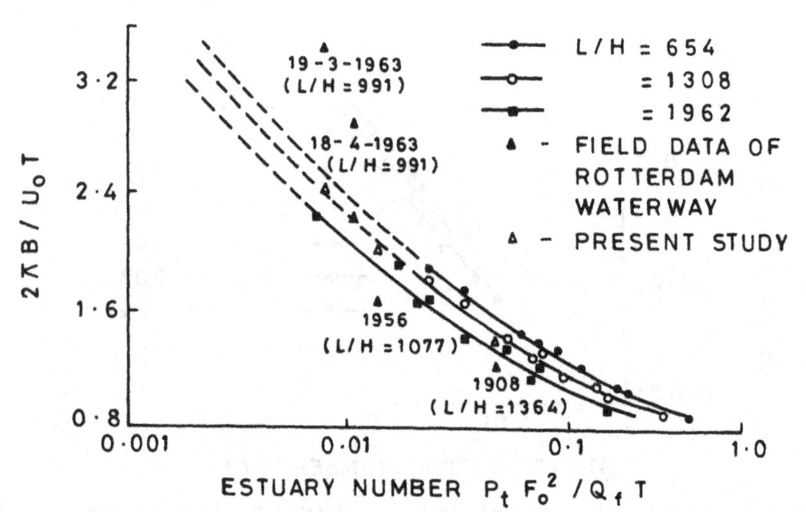

FIG. 6. CORRELATION BETWEEN ESTUARY NUMBER AND $2\pi B/U_o$ WITH L/H AS PARAMETRIC VARIABLE.

A Finite Element Simulation Model for the Study of Wind Driven and Tidal Current in a Shallow Coastal Lagoon

P.R. Cardoso

Departamento de Ingenieria de Procesos e Hidraulica, Division de Ciencias Basicas e Ingenieria, Universidad Autonoma Metropolitana-Iztapalapa, Mexico

ABSTRACT

The equations required for modeling three dimensional hydrodynamics (space two dimension and time) in a shallow coastal lagoon are derived from the three dimensional Navier-Stokes equation and Continuity equation by integrating and taking the average along water depth. The wind stress at the surface, the friction stress at the bottom, the Coriolis parameter, eddy viscosity and shore line geometry are incorporated in the mathematical model. Both the analytical and numerical approaches cannot be used for solving the governing Navier-Stokes equations from the existence of the nonlinear convective terms and complexity of equations and geometry involved, the time dependent shallow water equations are solved using Galerkin's method. A finite element formulation for solving the shallow water equations is presented for the prediction of wind-driven and tidal currents in a coastal lagoon.

INTRODUCTION

Starting from the theory of hydrodynamics, the problem is to develop numerical methods which can be used to reproduce a real word, that means, to reproduce the observed or measured data with the objective to learn something about it. If the model can represent successfully the real world the numerical results of the model must show the results of the actions that have been simulated, it means that is possible to understand what is happening in the situations of a real world without these situations really occurring. When the validity of the model has been proved, this hydrodynamics model may finally be used to forecast the dynamic processes of the coastal lagoon or any other coastal region, finding its practical application.

THE HYDRODYNAMICS DIFFERENTIAL EQUATIONS OF CONTINUITY AND NAVIER-STOKES

The two-dimensional equations needed for shallow water modeling have been derived by Pritchard [1], Proudman [2], Defant [3] and others. A careful development is given by Pinder and Gray [4]. The development presented in this paper follows along the same lines as their work.

The set of equations in a Cartesian coordinate system are derived under the assumption that the vertical acceleration and the shear stress are negligible compared to the gravity and vertical gradient of the pressure of the incompressible fluid. This set of equations take the form (the kind of lagoon that will be considered is shown in cross section in fig. 1.):

$$\frac{\partial u}{\partial x} + \frac{\partial v}{\partial y} + \frac{\partial w}{\partial z} = 0 \qquad (1)$$

$$\frac{\partial u}{\partial t} + u\frac{\partial u}{\partial x} + v\frac{\partial u}{\partial y} + w\frac{\partial u}{\partial z} - f v + g\frac{\partial \zeta}{\partial x} - \frac{1}{\rho}\left(\frac{\partial \tau_{xx}}{\partial x} + \frac{\partial \tau_{xy}}{\partial y} + \frac{\partial \tau_{xz}}{\partial z}\right) = 0 \qquad (2)$$

$$\frac{\partial v}{\partial t} + u\frac{\partial v}{\partial x} + v\frac{\partial v}{\partial y} + w\frac{\partial v}{\partial z} + f u + g\frac{\partial \zeta}{\partial y} - \frac{1}{\rho}\left(\frac{\partial \tau_{yx}}{\partial x} + \frac{\partial \tau_{yy}}{\partial y} + \frac{\partial \tau_{yz}}{\partial z}\right) = 0 \qquad (3)$$

$$\frac{1}{\rho}\frac{\partial p}{\partial z} + g = 0 \qquad (4)$$

Where the notation is as follow:

u, v, w - velocity components in the x, y, z directions, respectively.

t - time.
ρ - water density.
f - Coriolis parameter ($2\omega \sin\phi$).
ω - Earth's angular velocity.
ϕ - latitude.
g - gravity acceleration.
p - pressure.
τ_{xx}, τ_{xy}, ..., τ_{yz} - bottom friction shear stresses.

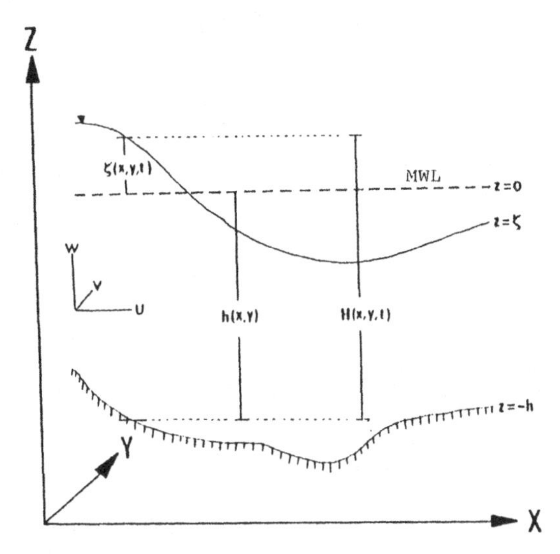

Figure 1. Transversal cross section of a coastal lagoon

THE DEPTH INTEGRATED FLOW MODEL

Integrating an arbitrary depth (z) over the free surface ($z=\zeta$), assuming that the pressure on the free surface is equal to the atmospheric pressure over the surface $p_a(x, y, t)$, i.e.

$$p - p_a = \rho g(\zeta - z) \qquad (5)$$

and taking the average of equations (1)-(4) with the following two boundary conditions: (i) the stress at the surface due to the friction of the wind is represented by a constant function of the wind speed (K), the wind velocity (W) and the direction across the lagoon (φ). (ii) the friction stress at the bottom is represented by the Chezy coefficient. Then the equations take the form:

$$\frac{\partial \zeta}{\partial t} + \frac{\partial (HU)}{\partial x} + \frac{\partial (HV)}{\partial y} = 0 \qquad (6)$$

$$\frac{\partial U}{\partial t} + U\frac{\partial U}{\partial x} + V\frac{\partial U}{\partial y} + g\frac{\partial \zeta}{\partial x} - fV - \frac{KW^2}{H}\cos\varphi$$
$$+ \; - \frac{g(U^2+V^2)^{1/2}}{C^2 H} U - \nu\left(\frac{\partial^2 U}{\partial x^2} + \frac{\partial^2 U}{\partial x^2}\right) = 0 \qquad (7)$$

$$\frac{\partial V}{\partial t} + U\frac{\partial V}{\partial x} + V\frac{\partial V}{\partial y} + g\frac{\partial \zeta}{\partial y} + fU - \frac{KW^2}{H}\sin\varphi$$
$$+ \; - \frac{g(U^2+V^2)^{1/2}}{C^2 H} V - \nu\left(\frac{\partial^2 V}{\partial y^2} + \frac{\partial^2 V}{\partial y^2}\right) = 0 \qquad (8)$$

The notation used in the above equations is:

U, V - depth-integrated components of the velocity in the x, y directions, respectively.
K - constant coefficient function of wind speed.
W - wind velocity.
φ - angle between the wind velocity vector and the x-axis.
C - Chezy coefficient.
ν - cinematic viscosity ($\nu=\varepsilon/\rho$).
ε - eddy viscosity.
H - total distance from the bottom to the surface ($H=h+\zeta$).
h - distance from the mean water level (MWL) to the bottom of the lagoon.
ζ - surface elevation from MWL.

Three unknowns variables are present in this system of equations (6)-(8). Now the problem is to solve these equations when the initial conditions of the variables ζ, U and V and appropriate boundary conditions are given.

A direct method of solving this set of nonlinear partial differential equations is not feasible and the finite difference method faces difficulties by the existence of the complex geometrical configurations and boundary conditions so it is necessary to use nonuniform nets.

In this type of problems the finite element method (FEM) has been used successfully in the last few years.

FINITE ELEMENT FORMULATION

An approximate solution to the equations (6)-(8) can be constructed by Galerkin's method, which is a part of the Weighted residuals method [6]. Galerkin's method uses the approximation of variable functions as functional combinations of weight functions of the residuals and stipulates that the residuals are orthogonal to the weight functions. That means that the integral of the weighted residuals, over the region of interest, must be zero [7].

The finite element solution of the Continuity equation (6) and Navier-Stokes equations (7) and (8) is obtained by choosing a simple trigonometric element and a linear interpolation function as follows,

$$\zeta = \sum_{i=1}^{M} \zeta_i(t)\phi_i(x,y) = \phi_m\zeta_m \qquad (9)$$

$$U = \sum_{i=1}^{M} U_i(t)\phi_i(x,y) = \phi_m U_m \qquad (10)$$

$$V = \sum_{i=1}^{M} V_i(t)\phi_i(x,y) = \phi_m V_m \qquad (11)$$

where M is the number of nodes in the FE domain and

$$\phi_i(x_i,y_i) = \frac{1}{2A}(a_i + b_i x_i + c_i y_i)$$

$$a_i = (x_j y_k - x_k y_j)/2A$$

$$b_i = (y_j - y_k)/2A$$

$$c_i = (x_k - x_j)/2A$$

Analogous equations as above can be obtained for j and k, where i, j, k are the numbers associated to a triangle and A is its area given by,

$$2A = \det\begin{bmatrix} 1 & x_i & y_i \\ 1 & x_j & y_j \\ 1 & x_k & y_k \end{bmatrix} = (x_i y_j + x_j y_k + x_k y_i) - (x_i y_k + x_j y_i + x_k y_j)$$

$$2A = c_j b_i - c_i b_j \qquad (12)$$

Making $(U^2 + V^2)^{1/2} = W$ and applying the Galerkin's condition to equations (6)-(8):

$$Dnm\zeta m = -\ Enm\ Hm\ Um - Fnm\ Hm\ Vm \equiv f_1(\zeta m) \qquad (13)$$

$$DnmUm = -\ Anqm\ Uq\ Um - Bnqm\ Vq\ Um + Dnm\ f m\ Vm - \bar{E}nm\zeta m$$
$$+ \frac{\bar{D}nm\ Km\ Wm^2\ \cos\varphi m}{Hm} - \frac{\bar{D}nm\ Wm}{Hm} Um + Cnm\ Um \equiv f_2(Um) \qquad (14)$$

$$DnmUm = -\ Anqm\ Uq\ Vm - Bnqm\ Vq\ Vm - Dnm\ f m\ Um - \bar{F}nm\zeta m$$
$$+ \frac{\bar{D}nm\ Km\ Wm^2\ \sin\varphi m}{Hm} - \frac{\bar{D}nm\ Wm}{Hm} Vm + Cnm\ Vm \equiv f_3(Vm) \qquad (15)$$

Where the notation is as follows:

$$Anqm = \int_A \phi n\phi q\phi m,x\ dA = \begin{cases} bm/12\ (n = i,j,k;\ n = i) \\ bm/24\ (n = i,j,k;\ n \neq i) \end{cases}$$

$$Bnqm = \int_A \phi n\phi q\phi m,y\ dA = \begin{cases} cm/12\ (n = i,j,k;\ n = i) \\ cm/24\ (n = i,j,k;\ n \neq i) \end{cases}$$

$$Cnm = \int_A (\phi n,x\phi m,x + \phi n,y\phi m,y)\ dA = \frac{\nu}{4A} (bnbm + cncm)$$

$$\bar{D}nm = gDnm/C^2 = g/C^2 \int_A \phi n\phi m\ dA = \begin{cases} gA/6C^2 & (n = m) \\ gA/12C^2 & (n \neq m) \end{cases}$$

$$\bar{E}nm = gEnm = g \int_A \phi n\phi m,x\ dA = gbm/6$$

$$\bar{F}nm = gFnm = g \int_A \phi n\phi m,y\ dA = gcm/6$$

Ensambling system into a matrix form, we obtain,

$$\mathbf{M}_{ij} \frac{d\zeta_j}{dt} = f_1(\zeta_j) \qquad (16)$$

$$\mathbf{M}_{ij} \frac{dU_j}{dt} = f_2(U_j) \qquad (17)$$

$$\mathbf{M}_{ij} \frac{dV_j}{dt} = f_3(V_j) \qquad (18)$$

The coefficient matrix **M** of the resultant algebraic equations are symmetric, non-singular, positive definite and banded. These properties suggest that the Gaussian elimination method may be stable [10], however, Gradient Conjugated method is more practical because reduces the computing time and improves the convergence [12]. Must be taken account that the boundary conditions must be incorporated to the system before it can be solved preserving its symmetry [4,6].

INITIAL AND BOUNDARY CONDITIONS

INITIAL CONDITIONS

Because the water motion is independent of initial conditions after a certain time and becomes influenced only by the specified oscillations of the boundary values, computations can be begun from an initial condition of $\zeta=0$, U=0 and V=0 at all points [4,9].

BOUNDARY CONDITIONS

In coastal lagoons there are two different types of boundaries: the fixed boundary given by the shore line and the open boundary which is given by 'artificial' limits in the contacts with other water bodies, like the mouth of the lagoon (contact lagoon-sea) or the mouth of a river (contact lagoon-river).

On fixed boundary U=0 and V=0.

On open boundaries, either the normal velocity or the value of ζ must be specified.

In the contact lagoon-sea, the water level is specified according to a sinusoidal function: $\zeta=A \sin\omega t$, where A, is the amplitude, $\omega=2\pi/T$, is the angular velocity, T, is the period of the tide and t, is the time.

In the contact lagoon-river the boundary conditions are specified by the hidrogram at time t.

The convective terms play an important role in the description of the dynamic processes in shallow waters therefore they must not be neglected in the numerical model. Unfortunately, convective terms cause numerical disturbs in the model which have to be stabilized by chosen an adequate time interval, Δt.

The convective terms in open boundaries are:

$$\frac{\partial U}{\partial x} = 0, \quad \frac{\partial V}{\partial y} = 0 \text{ normal to the open boundary.}$$

ITERATIVE METHOD TO SOLVE THE FINITE ELEMENT SYSTEM

In the system (16)-(18), the time derivates ζ, U and V must be differentiated over a time interval, Δt to obtain:

$$[M_{ij}]\frac{1}{\Delta t}\left(\{Y_j\}^{\circ}_{t+\Delta t} - \{Y_j\}_t\right) = f(Y_j) \quad (19)$$

from which

$$[M_{ij}]\{Y_j\}^{\circ}_{t+\Delta t} = [M_{ij}]\{Y_j\}_t + f(Y_j)$$

Note that Y represents ζ, U and V for each one of the equations.

From the known solution at t, estimates are made for ζ, U and V at $t+\Delta t$. These estimates are used to make successive approximations with [14],

$$[M_{ij}]\{Y_j\}_{t+\Delta t} = [M_{ij}]\{Y_j\}_t + \frac{1}{2}\Delta t\ f(Y_j)_t + \frac{1}{2}\Delta t\ f(Y_j)^{\circ}_{t+\Delta t} \quad (20)$$

Through each iteration, the prediction of ζ, U and V is refined at $t+\Delta t$. That is, the preceding approximation can be used various times to produce a better approximation of ζ, U and V. It must be understood that this process does not necessarily converge to the right solution but it does to an approximation with a finite truncate error.

A criterion for stopping convergence of the equations is given by [11,14]

$$|\varepsilon_a| = \left| \frac{Y^{j}_{t+\Delta t} - Y^{j-1}_{t+\Delta t}}{Y^{j-1}_{t+\Delta t}} \right|$$

where j-1 and j are the result of the preceding and present iterations of equation (20).

Noteworthy that the error some times become larger as the iterations advance, specially for large Δt. This is the reason for which we must avoid the general conclusion that an additional iteration improves the result. However, for a Δt small enough, the iteration must eventually converge in a single value; but, which is the adequate Δt to use? A stability criterion is the one of Courant-Friederich-Lewy (CFL) [13]: $\Delta t = \min(c_t \Delta x/(gh)^{1/2})$, where c_t, is a dimensionless coefficient from friction water-bottom ($c_t = g/C^2$) and Δx the distance between two nodes.

THE RESULTS

The model above present was applied for the coastal lagoon 'La Mancha', Mexico. The area of the lagoon was subdivided into 78 triangles (figure 2). The mean water depth at each one of the 60 nodal points which compose the net was prescribed. There are three nodes (mouth of lagoon) where the water level has to be specified (open boundary). After a simulation the lines with the same tidal range are shown in figure 3. In figure 4 the depth (0.8m) is shown darker. These are particular results of the simulation, exists general results.

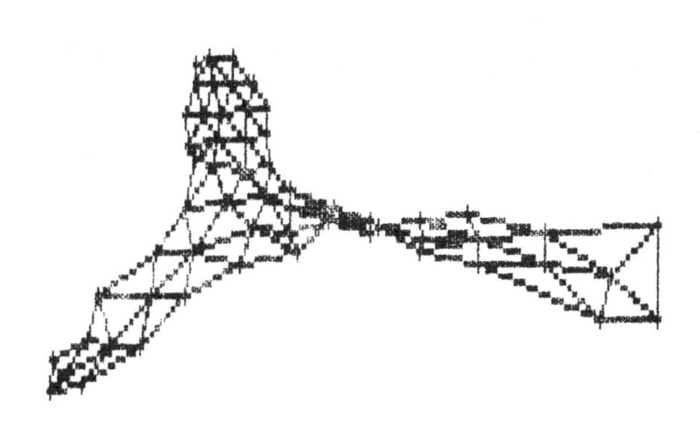

Figure 2. Division of the La Mancha lagoon into 78 triangles with 60 nodal points

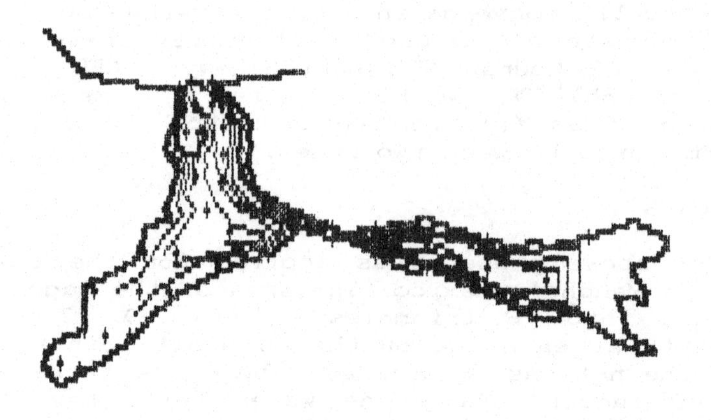

Figure 3. single Lines with the same tidal range

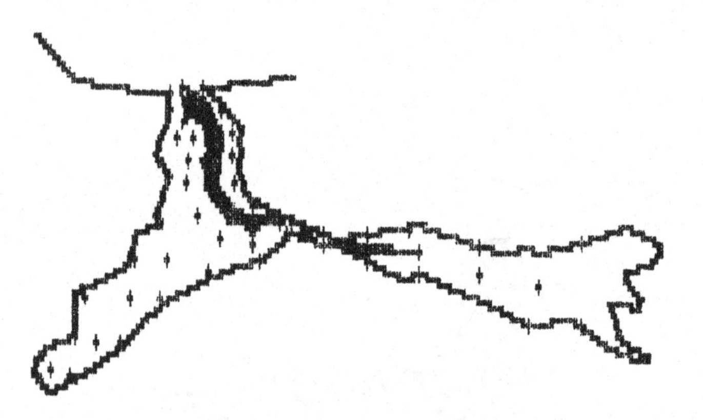

Figure 4. Various lines with the same tidal range

CONCLUSIONS

The type of problem in the mathematical modeling present in this paper can be summarized as follows: given hydrological parameters, inputs, initial and boundary conditions, solve the set of partial differential equations which govern the water flow in a shallow water coastal lagoon.

Some more study is required to analyze the effect of the distinct parameters in the hydrodynamics of the lagoon, being possible to simplify the model making it more efficient.

REFERENCES

1. Pritchard, D.W.,*Estuarine Modeling: An Assessment*, (Ed. Ward, G.H, Jr. and Epsey, W.H, Jr.), Nat. Tech. Inform. Serv. Publ., 1971.

2. Proudman, J. *Dynamical Oceanography* Methuen, London. Willey, New York, 1953.

3. Defant, A. *Physical Oceanography* Pergamon Press, Oxford, Vol.1, 1961.

4. Pinder, G.F. and Gray, W.G. *Finite Element Simulation in Surface and Subsurface Hydrology* Academic Press, New York, 1977.

5. Weare, T.J. 'Finite Element or Finite Difference Methods for the Two-dimensional Shallow Water Equations?' Comput. Methods. Appl. Mech. Eng., Vol.7, pp. 351-357, 1976.

6. Chung, T.J. *Finite Element Analysis in Fluid Dynamics* Mcgraw-Hill, New York

7. Leonhard, J.W. 'Finite Element Analysis of Perturbed compressible Flow' Int. Journ. Num. Methods. Eng. Vol.4, pp. 123-132, 1972.

8. Grotkop, G. 'Finite Element Analysis of Long-Period Water Waves' Comput. Methods. Appl. Mech. Eng., Vol.2, pp. 147-157, 1973.

9. Ramming, H.G. and Kowalik, Z. *Numerical Modeling of Marine Hydrodynamics: Applications to Dynamical Physical Processes* Elsevier, Amsterdam, 1980.

10. Cheng, R. 'Numerical Solution of the Navier-Stokes Equations by the Finite Element Method' The Physics of Fluids, Vol.15, pp. 2098-2105, 1972.

11. Burden, L.R. and Faires J.D., *Numerical* Analysis PWS, Boston, 1985.

12. Traversoni, L., 'Overlapped Conjugated Gradient Method', Internal Report, UAMI, 1990.

13. Yue, J. 'Selective Lumping effects on depth-integrated Finite Element Model of Channel Flow' Adv. Water Resources, Vol.12, pp. 74-78, 1989.

14. Ortega, J. and Rheinboldt, W., *Iterative Solution of Nonlinear Equations in Several Variables*, Academic Press, New York, 1970.

Finite Element Modelling of Moving Boundary Problems in Estuaries and Coastal Waters

T.H. Lan (*), A.G. Hutton (**), J.H. Loveless (*)
(*) Dept. of Civil Engineering, University of Bristol, Bristol, U.K.
(**) Berkeley Nuclear Laboratories, Berkeley, Gloucestershire, U.K.

ABSTRACT

A method and algorithm to deal with moving boundary problems in shallow water is proposed in this paper. A characteristic feature is that the governing equations are solved by means of a transformation to the original computational region (mesh) at $t=t_0$. This is attractive as it uses established numerical methods for fixed-grid problems and has no limitation on the boundary movements. The utility of the approach is demonstrated by application to a one dimensional problems of side to side water sloshing in a canal with parabolic bathymetry.

INTRODUCTION

Many mathematical models have been developed and used to study estuarine hydrodynamics. The majority of these models are based upon finite difference methods. Few finite element based models have yet evolved into practical, widely used tools. However, this is a CFD application area for which finite element methodology offers considerable attractions. Complex shore-line geography can be readily and accurately modelled and the mesh can be arranged to track gradations in bathymetry. Furthermore, since the mesh design is not restricted to regular topologies, the mesh can be focused by non-structured refinement to resolve local details such as sources of pollution. This means that features ranging in scale from metres to several kilometres can be modelled by a single mesh.

A recent programme of research undertaken at Bristol Polytechnic in collabration with Nuclear Electric (NE) was designed to explore this potential in some depth. The outcome was an implementation of estuarine and coastal modelling in NE's general purpose finite element based CFD Suite, FEAT, (Knock, 1990). This has proved very successful both in terms of performance and flexibility on a range of validation problems, including a propagation of a depression in a tank and pollution in the Severn estuary. However, Knock's study assumed a fixed coastline , yet such an assumption becomes invalid in regions with large tidal flats which are submerged only intermittently. Under these circumstances,in order to determine water level fluctuations and currents one needs to know how the coastline changes as the tide advances and recedes.Indeed,in some problems, like the flooding in estuaries, the movement of the coastline is actually the purpose of the study.Various moving boundary models have been developed and they fall into two groups, the fixed grid type and the deforming grid type (Lynch, 1980).

- The fixed grid type is a straight forward application of existing methods. In each time step, elements with at least one dry node are removed from the area of computation. This method was first used by B. Herring (1976). It has the di sadvantage of not following the water boundary closely and also the difficulty of approximating the physics at the moving boundary.

- The deforming grid method has also been used. In this method, the boundary condition (where water depth, h=0) provides the necessary information for the location of the new boundary position at each step of the computation. The finite element nodes on the boundary are moved consistently with the local kinematics and sea floor depth so that the nodes always remain on the edge of the water body. This means the elements with one or more nodes on the boundary keep changing their shape and size with time. Lynch developed the method to account for the effects of grid deformation.However, when this method is used to solve problems involving very large movements of the boundary, the elements can become excessively distorted .

Futhermore, when certain classes of implicit time stepping algorithms are introduced, difficulties arise which are associated with consistent representation of the grid variables at different time levels. This is particularly the case, when implementing the predict-corrector algorithm of Gresho et al (1979), which is adoped in the present work.

These problems can be alleviated by transforming the grid in time back to its initial configuration and solving the transformed equations on this grid. In this way the moving boundary problem becomes for practical purposes a fixed boundary problem. The only variable which needs interpolation or extrapolation is the bathymetry , and this has a very clear physical conception and will not cause any consistency problems. A practical algorithm is developed here based upon the Galerkin Finite Element Method.

TRANSFORMATION OF THE EQUATIONS

The simplified shallow water equations considered here are:

$$\begin{cases} \dfrac{\partial h}{\partial t} + h\dfrac{\partial u_m}{\partial X_m} + u_m\dfrac{\partial h}{\partial X_m} = 0 \\ \dfrac{\partial u_n}{\partial t} + u_m\dfrac{\partial u_n}{\partial X_m} + g\dfrac{\partial \eta}{\partial X_n} = 0 \end{cases}$$

where,

h is the total water depth.

u is the vertically integrated water velocity.

η is the elevation of free surface above a reference datum.

Denote the gridded (i.e.computational) region at $t=t_o$ as Ω_o and that at time $t>t_o$ as Ω_t. Now suppose any point in Ω_o with position vector $\underline{X}^o$(in fixed frame $\overline{OX}$) moves to point $\underline{X}$ at time t. The $\underline{X}$ is related to $\underline{X}^o$ by the analytic function.

$$\underline{X}=\underline{X}(\underline{X}^o,t) \tag{1a}$$

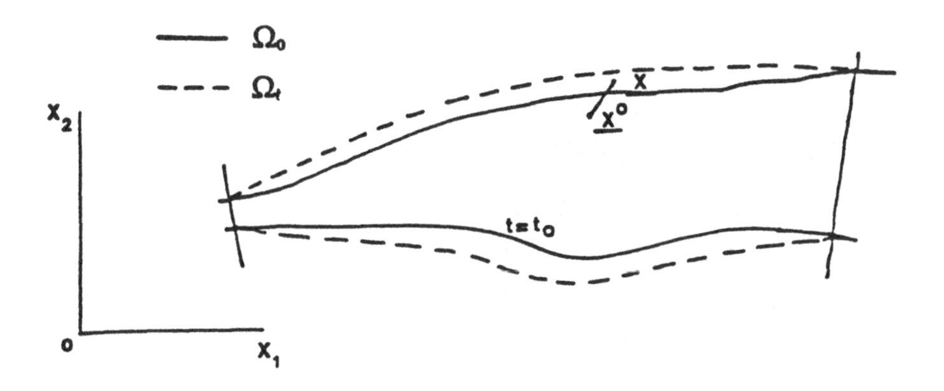

Fig. 1 Definition of the Moving Boundary Problem

It should be explained that the point $\underline{X}$ is not fixed in the fluid but fixed in the gridded region and at inlet and outlet,we have $X(X^o,t)=X^o$. The construction of (1a) is considered later. Now define the grid velocity $\underline{u}^g$ by the relation

$$\underline{u}^g(\underline{X}^o,t) = \left(\frac{\partial \underline{X}(\underline{X}^o,t)}{\partial t}\right)_{\underline{X}^o} \tag{1b}$$

Then

$$\underline{X}(\underline{X}^o,t)=\int_{to}^{t}\underline{u}^g(\underline{X}^o,s)ds+\underline{X}o \tag{2}$$

Now consider a variable $\phi = \phi(\underline{X}^0,t)$. We can write

$$\left(\frac{\partial\phi}{\partial Xn}\right)_t = \left(\frac{\partial\phi}{\partial X_m^0}\right)_t \left(\frac{\partial Xm^0}{\partial Xn}\right)_t \quad (3)$$

From (2)

$$\left(\frac{\partial Xm}{\partial Xn^0}\right)_t = \int_{to}^{t} \frac{\partial um^g(\underline{X}^0,s)}{\partial X_n^0} ds + \delta mn \quad (4)$$

where,

δmn is the kronecker delta.

Eq.(4) defines a 2 × 2 non-symmetric matrix at each point $\underline{X}^0$ at all times.

Set $\Gamma_{mn}(\underline{X}^0,t) = \left(\frac{\partial Xm}{\partial Xn^0}\right)_t$

Then,

$$\left(\frac{\partial\phi}{\partial Xn}\right)_t = \left(\frac{\partial\phi}{\partial Xm^0}\right) \Gamma^{-1}_{mn} (\underline{X}^0,t) \quad (5)$$

We must also evaluate $\left(\frac{\partial\phi}{\partial t}\right)_{\underline{X}}$

$$\left(\frac{\partial\phi}{\partial t}\right)_{\underline{X}} = \left(\frac{\partial\phi}{\partial Xm^0}\right)_t \left(\frac{\partial Xm^0}{\partial t}\right)_{\underline{X}} + \left(\frac{\partial\phi}{\partial t}\right)_{\underline{X}^0} \quad (6)$$

Now, since $\underline{X}=\underline{X}(\underline{X}^0, t)$,

then $$d\underline{X} = \left(\frac{\partial \underline{X}}{\partial Xp^0}\right) dXp^0 + \left(\frac{\partial \underline{X}}{\partial t}\right)_{\underline{X}^0} dt$$

After some manipulation, we can get

$$\left(\frac{\partial Xm^0}{\partial t}\right)_{\underline{X}} = -\Gamma^{-1}_{mq}\, uq^g \quad (7)$$

Thus $$\left(\frac{\partial\phi}{\partial t}\right)_{\underline{X}} = -\left(\frac{\partial\phi}{\partial X_m^0}\right)_t \Gamma^{-1}_{mq} uq^g + \left(\frac{\partial\phi}{\partial t}\right)_{\underline{X}^0} \quad (8)$$

All the relations required by the transformation of the shallow water equations are now available. They can be rewritten as:

$$\frac{\partial h}{\partial t} + \left\{ h \frac{\partial u_n}{\partial X_m^o} + u_n \frac{\partial h}{\partial X_m^o} \right\} \Gamma^{-1}_{mn} - u_q^g \frac{\partial h}{\partial X_m^o} \Gamma^{-1}_{mq} = 0 \tag{9}$$

$$\frac{\partial u_n}{\partial t} + u_q \frac{\partial u_n}{\partial X_m^o} \Gamma^{-1}_{mq} + g \frac{\partial \eta}{\partial X_m^o} \Gamma^{-1}_{mn} - u_q^g \frac{\partial u_n}{\partial X_m^o} \Gamma^{-1}_{mq} = 0 \tag{10}$$

These equations must now be solved on the fixed computational region Ω_o subject to the following boundary conditions.

Riemann:

$\underline{u} \circ \underline{n} + 2\sqrt{gh} = a_1(t)$ on inflow.

$\underline{u} \cdot \underline{n} + 2\sqrt{gh} = a_1(t)$, $\underline{u} \cdot \underline{t} = b_1(t)$ on outflow.

Specified height:

$h = a_2(t)$ on inflow.
$h = a_2(t)$, $\underline{u} \circ \underline{t} = a_2(t)$ on outflow .

On shoreline:

$h = 0$ on inflow.
$h = 0$, $\underline{u} \circ \underline{t} = b_3(t)$ on outflow.

$\underline{n}, \underline{t}$ are normalized vectors in normal and tangential directions respectively.

It remains to construct $\underline{u}^g$, Γ_{mn} then being available from eq.(4).

The grid velocity at any instant in time is chosen as the solution to:

$$\frac{\partial}{\partial X_m^o} \left(\frac{\partial u_m^g}{\partial X_n^o} + \frac{\partial u_n^g}{\partial X_m^o} \right) = 0 \tag{11}$$

$$\forall X^o \in \Omega_o$$

Subject to $\underline{u}^g = \underline{u}$ on the shoreline and $\underline{u}^g \circ \underline{n} = 0$ on all other boundaries (i.e. $\underline{u} \circ \underline{t}$ is fixed by the FE natural boundary condition). Note that if $\underline{u}$ is zero on shoreline, then $\underline{u}^g = 0$ and from eq.(4) $\Gamma_{mn} = \delta_{mn}$. The equations then reduce to the usual shallow water formulation.

FINITE ELEMENT FORMULATION

The first stage is to rewrite the equations as an integral formulation, (Hutton,1974). Two sets of functions are defined in the area of interest Ω_o. One is a set of smooth vector functions, S_u, used with the momentum equations and the second a set of smooth Scalar functions, Sh, used with the continuity equation, (Knock, 1990).

For the sake of simplicity, the superscript zero notation is omitted.

The integer formulation can be written as:

$$\left\{\begin{array}{l} \displaystyle\int_\Omega \frac{\partial h}{\partial t} p d\Omega + \int_\Omega h \frac{\partial u_n}{\partial X_m} \Gamma^{-1}_{mn}\, p d\Omega + \int_\Omega u_n \frac{\partial h}{\partial X_m} \Gamma^{-1}_{mn}\, p d\Omega \\ \displaystyle\quad - \int_\Omega u_q^g \frac{\partial h}{\partial X_m} \Gamma^{-1}_{mq}\, p d\Omega = 0 \qquad (12) \\ \displaystyle\int_\Omega \frac{\partial u_n}{\partial t} V_n d\Omega + \int_\Omega u_q \frac{\partial u_n}{\partial X_m} \Gamma^{-1}_{mq}\, V_n d\Omega + \int_\Omega g \frac{\partial \eta}{\partial X_m} \Gamma^{-1}_{mn}\, V_n d\Omega \\ \displaystyle\quad - \int_\Omega u_q^g \frac{\partial u_n}{\partial X_m} \Gamma^{-1}_{mq}\, V_n d\Omega = 0 \qquad (13) \end{array}\right.$$

$\forall p \in S_h$; $\forall V_n \in S_u$.

p, v are independent weighting functions.

The following discrete FE representation is now introduced.

$$p = \sum_r p_r Y_r , \qquad V_n = \sum_j V_{n,j} W_j, \qquad u_n = \sum_j u_{n,j} W_j,$$

$$u_n^g = \sum_j u_{n,j}^g W_j, \qquad h = \sum_s h_s Y_s, \qquad H = \sum_s H_s Y_s$$

where,

Y_s are linear shape functions and W_j are quadratic shape functions on mid-side noded elements.

Using Green's theorem to rewrite the elevation term,the equations are discretized as:

$$\left\{\begin{array}{l} \displaystyle\frac{\partial h_s}{\partial t} A_{sr} + h_s\, u_{n,j}\, B_{sjr,n} + u_{n,j}\, h_s\, C_{jsr,n} - h_s\, M_{sr}{}^h = 0 \qquad (14) \\ \displaystyle\frac{\partial u_{n,j}}{\partial t} D_{jl} + u_{q,k}\, u_{n,j}\, E_{kjl,q} - g(h_s - H_s) F^1_{sl,n} \\ \displaystyle\quad - g(h_s - H_s) F^2_{sl,n} - u_{n,j}\, M_{jl}{}^u = - \int_{\partial\Omega} g(h_s - H_s)\, Y_s W_l\, \Gamma^{-1}_{mn} \underline{n}_m d\Omega \qquad (15) \end{array}\right.$$

Where: $A_{sr}=\int_{\Omega} Y_s Y_r d\Omega$, $\quad B_{sir,n}=\int_{\Omega} Y_s \frac{\partial W_j}{\partial X_m} \Gamma^{-1}_{mn} Y_r d\Omega$,

$C_{jsr,n}=\int_{\Omega} W_j \frac{\partial Y_s}{\partial X_m} \Gamma^{-1}_{mn} Y_r d\Omega$, $\quad M_{sr}{}^{h}=\int_{\Omega} \frac{\partial Y_s}{\partial X_m} u_q{}^{g} \Gamma^{-1}_{mq} Y_r d\Omega$,

$D_{ji}=\int_{\Omega} W_j W_i d\Omega$, $\quad E_{kji,q}=\int_{\Omega} W_k \frac{\partial W_j}{\partial X_m} \Gamma^{-1}_{mq} W_i \, d\Omega$,

$F^1_{si,n} =\int_{\Omega} Y_s \frac{\partial \Gamma^{-1}_{mn}}{\partial X_m} W_i d\Omega$, $\quad F^2_{si,n}=\int_{\Omega} Y_s \Gamma^{-1}_{mn} \frac{\partial W_i}{\partial X_m} d\Omega$,

$M_{ji}{}^{u} =\int \frac{\partial W_j}{\partial X_m} u_q{}^{g} \Gamma^{-1}_{mq} W_i d\Omega$.

Γ^{-1}_{mn} , $u_q{}^{g}$ are values at Gauss points.

COMPUTATIONAL ALGORITHM

An implicit predictor/corrector time stepping scheme is used here. The predictor is the 2nd order explicit Adams-Bashforth method and the corrector is the second order accurate Crank-Nicholson scheme, (Gresho, 1979).

The computational algorithm is :

1. At time level r, suppose we know

   h^r, $\dot{h}^r$, h^{r-1}, $\dot{h}^{r-1}$

   $\underline{u}^r$, $\dot{\underline{u}}^r$, $\underline{u}^{r-1}$, $\dot{\underline{u}}^{r-1}$

   and Γ^{r}_{mn} , $\underline{u}^{gr}$, Δt_r .

2. Predict h^p and $\underline{u}^p$ with the predictor.
3. Specifying $\underline{u}^g=\underline{u}^p$ on the moving boundary, solve eq.(11) to yield $\underline{u}^{gp}$.
4. Calculate

$$\Gamma^{p}_{mn} = \Gamma^{r}_{mn} + \int_{t_r}^{t_{r+1}} \frac{\partial u_m{}^{g}}{\partial X_n} ds = \Gamma^{r}_{mn} + \left\{ \frac{\partial u_m^{gr}}{\partial X_n} + \frac{\partial u_m^{gp}}{\partial X_n} \right\} \frac{\Delta t_r}{2}$$

5. Calculate

$$\underline{X}^{r+1}(\underline{X}^{o},t) = \underline{X}^{r}(\underline{X}^{o},t) + \int_{t_r}^{t_{r+1}} \underline{u}^{g} dt$$

$$= \underline{X}^r(\underline{X}^O,t) + (\underline{u}^{gr} + \underline{u}^{gp})\frac{\Delta tr}{2}$$

and hence the bathymetry $H^p(\underline{X}^o, t_{r+1}) = H(\underline{X}^{r+1})$

6. Solve eq.(14) (15), using $\underline{u}^{gp}$ $(\Gamma^{p}_{mn})^{-1}$ when r+1 values are required to yield the corrected values $\underline{u}^{r+1}$, h^{r+1}.

7. Return to 3, repeat steps 3 to 6 replacing $\underline{u}^P$ with $\underline{u}^{r+1}$ until certain accuracy is obtained.

NUMERICAL EXAMPLE

A numerical example is shown to verify the applicability of the present method. The example is a 1-D problem of water sloshing from side to side in a canal with parabolic bathymetry (Thacker,1981). The finite element idealization is shown in fig 2. The mesh has 1 element in the X_2 direction and 90 elements in the X_1 direction.

AB=CD=100m, BC=AD=16000m.

The bathymetry is $H = 10 \times \left\{ 1 - \frac{(X_1 - 7680)^2}{8000^2} \right\}$.

The initial elevation is $\eta = 0.0001 \times (X_1 - 7840)$

The boundary conditions specified are:

$h=0$, $u_2=0$ on AB and CD.

$u_2=0$, $\frac{\partial u_1}{\partial X_2}=0$ on BC and AD.

The initial time step size is set to 1.0 s. The computed results of water elevation and velocity (after 554 steps, at t=20316.2 s =5.72 periods), position of moving boundary CD and change of time step size with time are shown in Fig. 3 - Fig. 6.

The time step size inceases very rapidly from 1s, and keeps around 27 - 98s. With an explicit scheme, the time step size should be less than $\Delta t = \frac{\Delta X}{\sqrt{gh}} = 25.1$ s (the averaged water depth 5m is used here).

It can be seen that the computed results are extremely accurate. No instability arises during the course of the computation which extends for more than 5 periods.

CONCLUSIONS

A new method and algorithm to incorporate moving shoreline into a finite element shallow water model has been developed here. A numerical example shows it gives results which are in good agreement with an analytic solution. The method is currently being evaluated against a two-dimensional analytic problem and a practical estuarine problem of tidal dynamics.

ACKNOWLEDGEMENT

This paper is published by permission of Nuclear Electric plc.

REFERENCES

1. Herring, B. Finite Element Model for Estuarines with Inter-Tidal Flats, Proceedings of the conference on coastal Engineering, Chap. 195,1976.

2. Hutton, A.G. A Survey of the theory and Application of Finite Element Mehtod in the Analysis of Viscous Incompressible, Newtonian Flow, C.E.G.B. Report RD/B/N3049, 1974.

3. Knock, C. Finite Element Modelling of Estuarine Hydrodynamics, Ph.D thesis, Bristol Polytechnic, 1990.

4. Lynch, D.R. and Gray, W.G. Finite Element Simulation in Deforming Region, J. of Compu. Phy. 36, 135-153 (1980).

5. Wathen, A.J. Moving Finite Elements, Report of University of Bristol, AM -85 -05.

6. Gresho, P.M. et al, Solution of the Time - Dependent Incompressibl Navier-Stokes and Boussinesq Equations Using the Galerkin Finite Element Method. Preprint for 'IUTAM Symposium on Approximation Methods for Navier-Stokes problems', West Germany Sept. 1979.

7. Thacker, W.C., Some Exact Solutions to the nonlinear Shallow-Water Wave Equations. J. Fluids Mech., Vol 107, 1981.

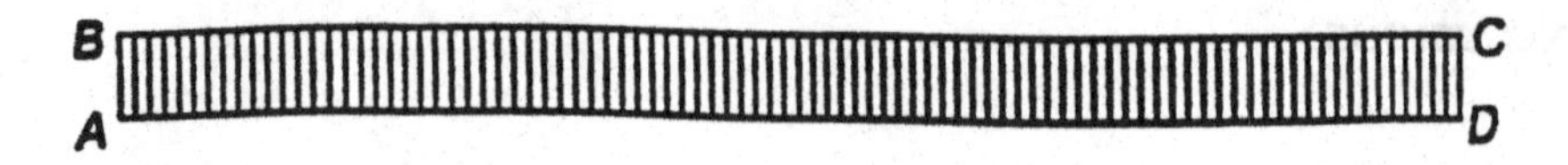

Fig. 2 Finite Element Idealization

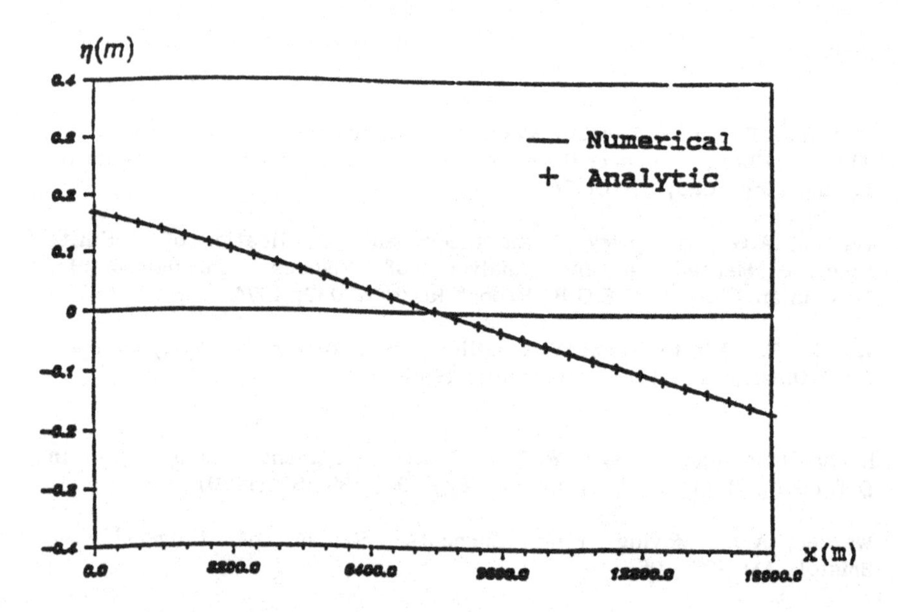

Fig. 3 Solutions of the Water Elevation

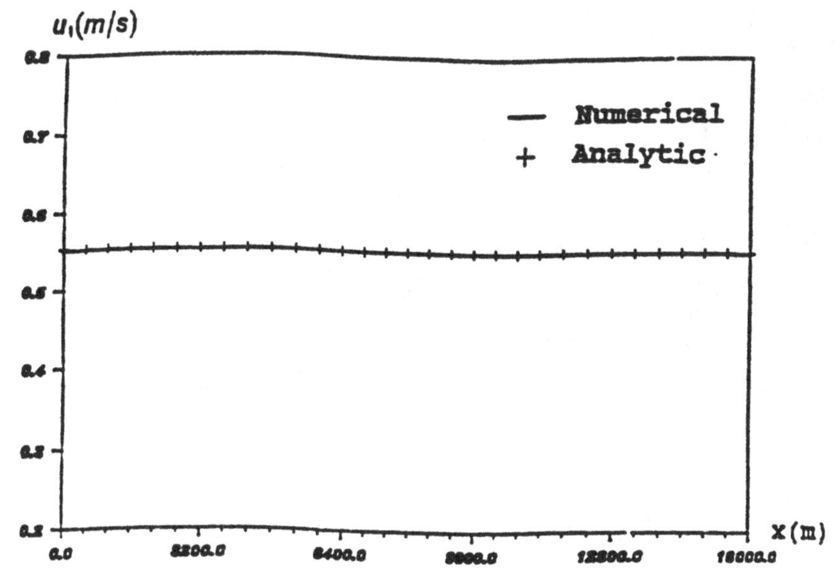

Fig. 4 Solutions of the Velocity

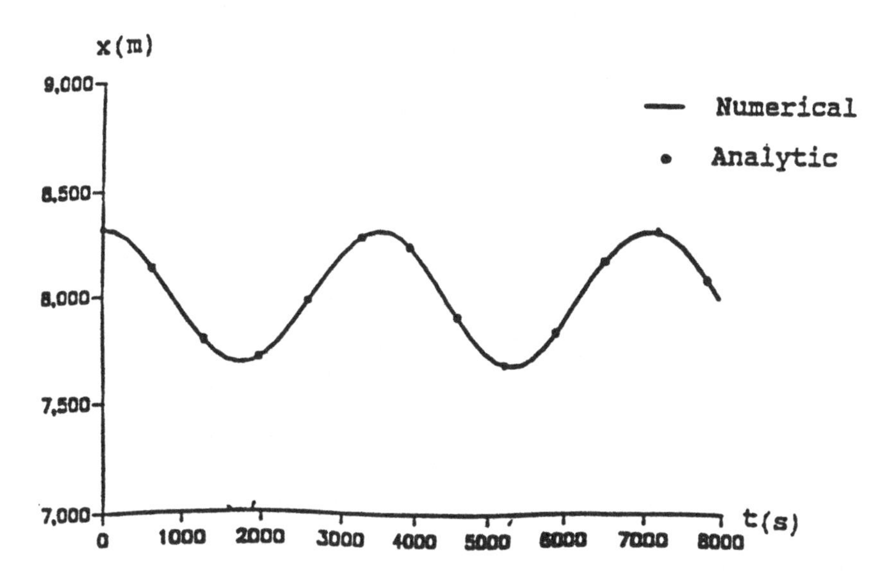

Fig. 5 Position of the Moving Boundary CD

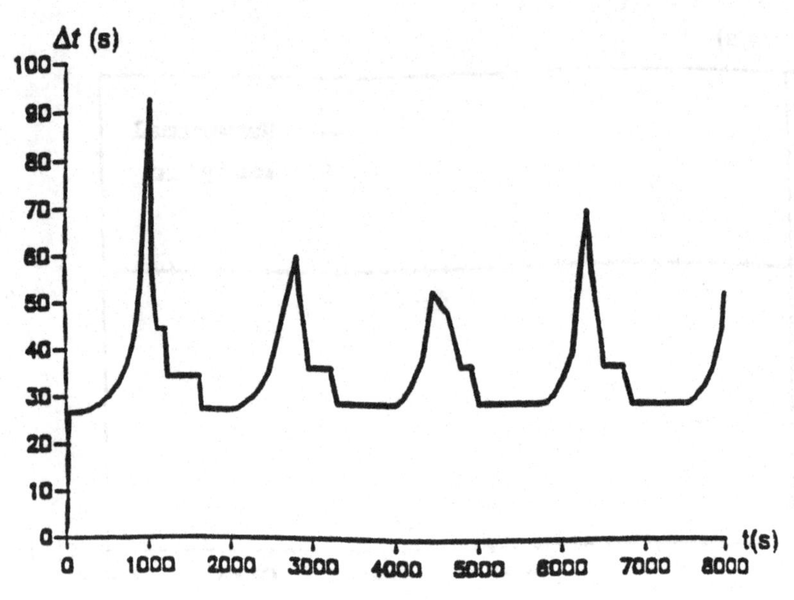

Fig. 6 Change of the Time Step Size with Time

Flux-Vector and Flux-Difference Splitting Methods for the Shallow Water Equations in a Domain with Variable Depth

A. Bermúdez, M.E. Vázquez

Dept. of Applied Mathematics, University of Santiago, 15706 Santiago de Compostela, Spain

ABSTRACT

In this paper we solve the shallow water equations by using flux-vector and flux-difference splitting methods. When the depth of the domain is not constant a source term arises. It is discretized by using explicit upwind schemes in a similar way to the flux. A conservation property is introduced as a requirement for the good behaviour of the discretized problem. Numerical results are presented to compare performances of the different methods.
Subject Classifications: M.R.: 65M99, 76B15.

INTRODUCTION

The shallow water equations are frequently used as a mathematical model for water flow in coastal areas, lakes, estuaries, etc. Thus they are an important tool to simulate a variety of problems related to coastal engineering, environment, ecology, etc. (see Gambolati *et al.* [5])

In the last years many papers have been devoted to the numerical solution of these equations by using finite differences and finite element methods. We mention, for instance, Taylor and Davies [17], Kawahara *et al.* [9], Linch and Gray [12], Zienkiewicz and Heinrich [19], Goussebaile *et al.* [7], Peraire *et al.* [13], Glaister [6], Bermúdez *et al.* [1].

When turbulent and dispersive effects are neglected the shallow water equations become a nonlinear system of first order hyperbolic partial differential equations. The numerical solution of this kind of system has been extensively studied in the recent years mainly because of their applications to some important fields as aeronautical and aerospace engineering.

In particular, finite volume methods combined with approximated Riemann solvers have undergone a quick development in the last decade. The so called flux-vector and flux-difference splitting methods give numerical results with high accuracy and properly capture discontinuities or shocks for Euler equations.

More recently, the great interest of some relevant technological problems like hypersonic reactive flows or combustion have promoted increasing research on numerical solution of hyperbolic systems with source terms (Desideri *et al.* [4], LeVeque and Yee

[11]). When the bottom surface is not flat, the shallow water equations have a source term containing the gradient of the depth from a fixed reference level. Centred discretization of this term leads to nonconservative schemes (see fig. 3 in section 5). In order to avoid this problem we propose, following Bermúdez and Vázquez [2], the use of upwind schemes in a similar way as for flux terms.

Thus the well-known schemes for Euler equations as the Q-scheme of Van Leer, Roe's scheme, the Steger-Warming scheme, etc. are adapted to solve the shallow water equations. The behaviour of the resulting schemes is analyzed in terms of a conservation property. Numerical results for some test examples including the dam break problem and tidal currents simulation are given in order to compare performances of the different methods.

THE SHALLOW WATER EQUATIONS

In this section we recall the shallow water equations which are a useful model to simulate free boundary flows of an incompressible ideal fluid a shallow domain, i.e. a domain with a small depth compared to its other dimensions (see fig. 1).

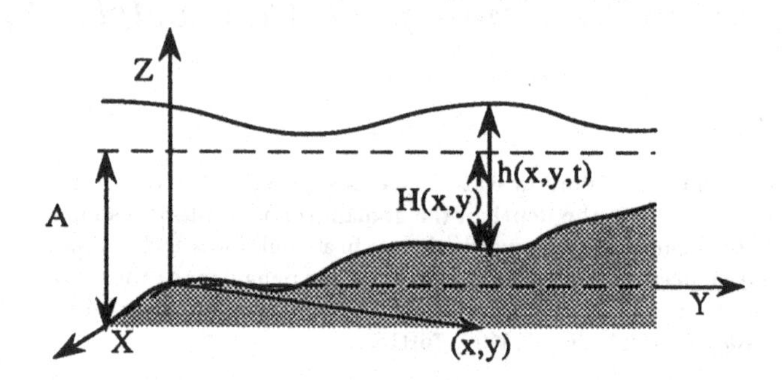

Figure 1.

Let $h(x,y,t)$ be the height of the fluid at point (x,y) at a time t and $H(x,y)$ the depth of the same point but from a fixed reference level.

By integrating the incompressible Euler equations in depth and then taking into account the kinematic and kinetic boundary conditions on both the free and the bottom surfaces one can get (see for instance Stoker [16]) the following generalized conservation law:

$$\frac{\partial w(x,y,t)}{\partial t} + \frac{\partial F_1(w(x,y,t))}{\partial x} + \frac{\partial F_2(w(x,y,t))}{\partial y} = G(x,y,w(x,y,t)), \quad (x,y) \in \Omega, t \in (0,T), \tag{1}$$

where:

$$w = \begin{pmatrix} h \\ hu_1 \\ hu_2 \end{pmatrix} = \begin{pmatrix} h \\ q_1 \\ q_2 \end{pmatrix}, \tag{2}$$

$$F_1(w) = \begin{pmatrix} q_1 \\ \dfrac{q_1^2}{h} + \dfrac{1}{2}gh^2 \\ \dfrac{q_1 q_2}{h} \end{pmatrix} \qquad F_2(w) = \begin{pmatrix} q_2 \\ \dfrac{q_1 q_2}{h} \\ \dfrac{q_2^2}{h} + \dfrac{1}{2}gh^2 \end{pmatrix}, \tag{3}$$

$$G(x,y,w) = \begin{pmatrix} 0 \\ gh\dfrac{\partial H(x,y)}{\partial x} \\ gh\dfrac{\partial H(x,y)}{\partial y} \end{pmatrix}. \tag{4}$$

Vector field (u_1, u_2) denotes the averaged horizontal velocity and Ω is the projection of the domain occupied by the fluid onto the xy plane.

For the sake of simplicity, in this paper we consider the one dimensional version of (1), i.e.

$$\frac{\partial w(x,t)}{\partial t} + \frac{\partial F(w(x,t))}{\partial x} = G(x, w(x,t)), \tag{5}$$

where

$$w = \begin{pmatrix} h \\ hu \end{pmatrix} = \begin{pmatrix} h \\ q \end{pmatrix}, \tag{6}$$

$$F(w) = \begin{pmatrix} q \\ \dfrac{q^2}{h} + \dfrac{1}{2}gh^2 \end{pmatrix}, \quad G(x,w) = \begin{pmatrix} 0 \\ ghH'(x) \end{pmatrix}. \tag{7}$$

However numerical results in section 5 include two-dimensional test cases.

We recall that (5) is a hyperbolic system of nonlinear partial differential equations. Indeed the Jacobian matrix of the flux given by

$$A(w) = \begin{pmatrix} 0 & 1 \\ -\dfrac{q^2}{h^2} + gh & \dfrac{2q}{h} \end{pmatrix} \tag{8}$$

is diagonalizable and its eigenvalues are real numbers. More precisely we have

$$A = T\Lambda T^{-1} \tag{9}$$

with

$$T(w) = \begin{pmatrix} 1 & 1 \\ \dfrac{q}{h} + \sqrt{gh} & \dfrac{q}{h} - \sqrt{gh} \end{pmatrix}, \Lambda(w) = \begin{pmatrix} \lambda_1 & 0 \\ 0 & \lambda_2 \end{pmatrix} \tag{10}$$

and

$$\lambda_1 = \frac{q}{h} + \sqrt{gh} \tag{11}$$

$$\lambda_2 = \frac{q}{h} - \sqrt{gh}. \tag{12}$$

Observe that q/h is the velocity of the fluid and $c = \sqrt{gh}$ is the wave speed.

NUMERICAL SCHEMES

For the numerical solution of (5) we use an explicit finite volume method.
Consider a mesh $\{x_1 \cdots, x_M\}$ of the domain Ω, (see fig. 2).

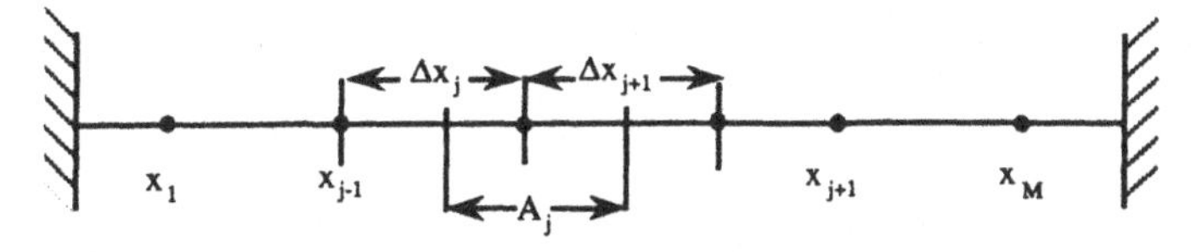

Figure 2.

Denote by $\Delta x_j = x_j - x_{j-1}$ and by A_j the length of the C_j-cell, i.e.

$$A_j = \frac{\Delta x_j + \Delta x_{j-1}}{2}. \tag{13}$$

Let us consider a discretization in which the nodes do not lie on the boundary .

Then, starting from an approximation W^n of the exact solution $w(.,t_n)$ we use the Euler's scheme, to compute W^{n+1}, i.e.

$$\frac{W^{n+1} - W^n}{\Delta t} + \frac{\partial F(W^n)}{\partial x} = G(x, W^n). \tag{14}$$

Spatial discretization is done assuming that W^n is a piecewise constant function given by

$$W^n(x) = W_j^n \text{ if } x \in \left(x_j - \frac{\Delta x_j}{2}, x_j + \frac{\Delta x_{j+1}}{2}\right). \tag{15}$$

Upon integration of (14) over the cell C_j using the trapezoidal rule, the following explicit conservative scheme is deduced:

$$W_j^{n+1} = W_j^n - \frac{\Delta t}{\Delta x}(F_{j+1/2}^n - F_{j-1/2}^n) + \frac{1}{2}\Delta t(G_{j+1/2}^n + G_{j-1/2}^n). \tag{16}$$

Upwind schemes for homogeneous problems are now obtained by replacing $F_{j\pm 1/2}$ by a numerical flux function ϕ :

$$F_{j+1/2}^n = \phi(W_j^n, W_{j+1}^n) \tag{17}$$

$$F_{j-1/2}^n = \phi(W_{j-1}^n, W_j^n). \tag{18}$$

In recent years a great number of papers were devoted to introduce different numerical fluxes. In the homogeneous case, i.e. when $G \equiv 0$, the so called flux-vector and flux-difference splitting methods give numerical results with high accuracy and discontinuities or shocks are properly captured. Moreover theoretical bases are well established. A review of these has been given in Harten, Lax and van Leer [8] and in the book by LeVeque [18].

More recently , research is increasingly devoted to the case of non-strict conservation laws, i.e. hyperbolic systems with source terms (see [4] and [11]). A particular case is the shallow water equations when the depth of the domain is not constant (i.e. the bottom surface is not flat).

Numerical experiments reveal that, in this case, the source term cannot be discretized in a centred way because mass is not conserved and waves are not well propagated. This bad behaviour can be seen in fig. 3 by comparing two different discretizations of the source term: a centred one and an "upwind one". In both cases the van Leer Q-scheme together is used.

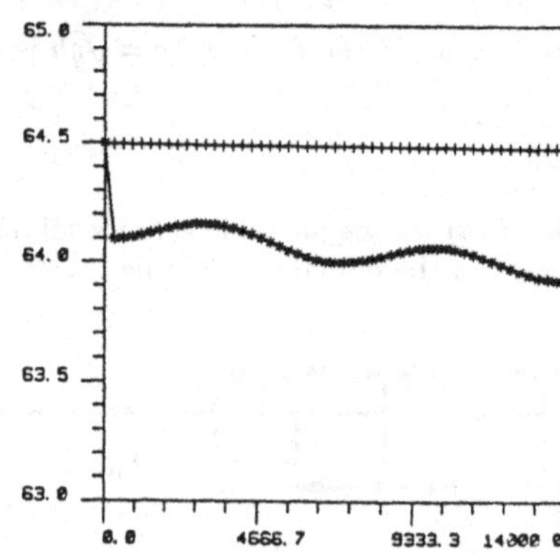

Figure 3. Centred source ****, "upwind" source ++++

Observe that while physical differences of the values of h at the same time must be quite small (depth is about 40m. and then the velocity of propagation of waves is about 20m/s.), the computed values for h exhibit oscillations . This fact leads us to upwind the source term in a similar way to the flux. In what follows this upwinding is done for several flux-vector and flux-difference splitting methods.

In order to do that a technique is proposed in Bermúdez and Vázquez [2] where the source terms are approximated by "numerical source functions", ψ_L and ψ_R. More precisely,

$$G_{j+1/2}^n = \psi_R(x_j, x_{j+1}, W_j^n, W_{j+1}^n) \tag{19}$$

$$G_{j-1/2}^n = \psi_L(x_{j-1}, x_j, W_{j-1}^n, W_j^n). \tag{20}$$

In the above mentioned paper general techniques to obtain these functions in the case of well known flux-vector and flux-difference splitting methods are given. In what follows we consider the application of these techniques to the shallow water equations.

EXTENSION OF FLUX-DIFFERENCE SPLITTING METHODS: Q-SCHEMES

These schemes are a family of upwind schemes corresponding to numerical fluxes of the form:

$$\phi(w,v) = \frac{F(w)+F(v)}{2} - \frac{1}{2}|Q(w,v)|(v-w), \tag{21}$$

where Q is a matrix characteristic of each Q-scheme.

In order to obtain the numerical sources we project the source term evaluated in a reference state $\tilde{w}$, near w and v, onto the eigenvectors of Q and then we define the left and right numerical sources as follows:

$$\psi_L(x,y,w,v) = (I + |Q(\tilde{w})|Q^{-1}(\tilde{w}))G(x,\tilde{w}) \tag{22}$$

$$\psi_R(x,y,w,v) = (I - |Q(\tilde{w})|Q^{-1}(\tilde{w}))G(x,\tilde{w}). \tag{23}$$

This means that the contribution of the source to the solution is given by its projection onto the local eigenvectors of the Q matrix upwinded by using the Q-scheme. In the same way as for the numerical flux, automatically we have the consistency of the numerical source with the source term.

Q-scheme of van Leer

This scheme corresponds to the choice of Q equal to the Jacobian matrix A of the flux and $\tilde{w}$ is the arithmetic mean value of w and v :

$$Q(w,v) = A\left(\frac{w+v}{2}\right). \tag{24}$$

Replacing (24) in (21), (22) and (23) we obtain the expressions of the numerical flux and the numerical source functions, respectively.

Roe's scheme

In [14] Roe defines a matrix Q which satisfies the property:

$$F(v) - F(w) = Q(w,v)(v-w). \tag{25}$$

Q is equal to the flux jacobian matrix A evaluated at some state $\tilde{w} = \tilde{w}(w,v)$ known as "Roe's average of w and v" . In [6] Glaister proposes the following choice of Q for the shallow water equations

$$Q(\tilde{w}) = A(\tilde{w}) = \begin{pmatrix} 0 & 1 \\ \tilde{u}+\tilde{c}^2 & 2\tilde{u} \end{pmatrix} \tag{26}$$

where

$$w_L = \begin{pmatrix} h_L \\ h_L u_L \end{pmatrix}, w_R = \begin{pmatrix} h_R \\ h_R u_R \end{pmatrix}, \tilde{w} = \begin{pmatrix} \tilde{h} \\ \tilde{h}\tilde{u} \end{pmatrix} \tag{27}$$

$$\tilde{u} = \frac{\sqrt{h_L}u_L + \sqrt{h_R}u_R}{\sqrt{h_L} + \sqrt{h_R}} \tag{28}$$

$$\tilde{h} = \sqrt{h_L h_R} \tag{29}$$

$$\tilde{c} = \sqrt{g\left(\frac{h_L + h_R}{2}\right)}. \tag{30}$$

The eigenvalues and eigenvectors of $\tilde{A}$ are

$$\tilde{\lambda_1} = \tilde{u} + \tilde{c} \qquad \tilde{\lambda_2} = \tilde{u} - \tilde{c}$$
$$\tilde{e_1} = \begin{pmatrix} 1 \\ \tilde{\lambda_1} \end{pmatrix}, \quad \tilde{e_2} = \begin{pmatrix} 1 \\ \tilde{\lambda_2} \end{pmatrix}. \tag{31}$$

In order to obtain the numerical sources we take the Q matrix in (22) and (23) as the one defined in (26). On the other hand $G(x, \tilde{w})$ is taken to be

$$G(x, \tilde{w}) = \begin{pmatrix} 0 \\ \tilde{c}^2 H' \end{pmatrix}, \tag{32}$$

with $\tilde{c}$ given by (30).This choice of $G(x, \tilde{w})$ guarantees that the scheme satisfies a conservation property to be defined later on.

EXTENSION OF FLUX-VECTOR SPLITTING METHODS

These schemes are based on a decomposition of the physical flux:

$$F(w) = F^+(w) + F^-(w). \tag{33}$$

Accordingly the numerical flux is given in the form:

$$\phi(w, v) = \phi^+(w, v) + \phi^-(w, v). \tag{34}$$

To extend these schemes to the non homogeneous case we take:

$$\psi_L(x, y, w, v) = \psi_R(x, y, w, v) = \psi(x, y, w, v), \tag{35}$$

and ψ given by

$$\psi(x, y, w, v) = \psi^+(x, y, w, v) + \psi^-(x, y, w, v), \tag{36}$$

where ψ^+ and ψ^- are functions to be defined for each particular method.

Steger and Warming scheme

In [15] Steger and Warming introduced the notion of the flux-splitting for the equations of gas dynamics. They took the advantage of the fact that in gas dynamics F is a homogeneous function of w of degree one, that is

$$F(w) = A(w)w, \tag{37}$$

where A is the jacobian matrix.

Using (37) they defined

$$\phi(w, v) = A^+(w)w + A^-(v)v. \tag{38}$$

The shallow water equations do not have this homogeneity property of the flux. However we can construct a matrix $A^*(w)$ strictly different from the jacobian matrix such that (37) is verified. It is given by

$$A^*(w) = \begin{pmatrix} 0 & 1 \\ \dfrac{-q^2}{h^2} + \dfrac{gh}{2} & \dfrac{2q}{h} \end{pmatrix} \tag{39}$$

Then we define ϕ by

$$\phi(w, v) = A^{*+}(w)w + A^{*-}(v)v. \tag{40}$$

and the functions ψ^+ and ψ^- by

$$\psi^+(x, w, v) = \frac{1}{2}(I + |A^*(w)|A^{*-1}(w))G(x, w) \tag{41}$$

$$\psi^-(y, w, v) = \frac{1}{2}(I - |A^*(v)|A^{*-1}(v))G(y, v). \tag{42}$$

Vijayasundaram's scheme

This author proposes in [18] a flux splitting scheme corresponding to the following choices:

$$\phi(w, v) = A^+\left(\frac{w+v}{2}\right)w + A^-\left(\frac{w+v}{2}\right)v \tag{43}$$

and

$$\psi^+(x, w, v) = \frac{1}{2}\left(I + \left|A\left(\frac{w+v}{2}\right)\right|A^{-1}\left(\frac{w+v}{2}\right)\right)G(x, w) \tag{44}$$

$$\psi^-(y, w, v) = \frac{1}{2}\left(I - \left|A\left(\frac{w+v}{2}\right)\right|A^{-1}\left(\frac{w+v}{2}\right)\right)G(y, v). \tag{45}$$

This scheme also uses the homogeneity of the flux, then again we need to replace A with the matrix A^* in (43), (44) and (45), for the shallow water equations .

A CONSERVATION PROPERTY

As already mentioned the shallow water equations for a variable depth domain are not a system of strict conservation laws. Due to that there is not a reason for the numerical methods we have introduced in the previous section to preserve mass. To guarantee this desirable behaviour we introduce the following definition.

Property C : A scheme satisfies the property C if it is exact when applied to the shallow water equations with the conditions

$$h(x, 0) = H(x) \tag{46}$$

$$q(x, 0) = 0 \tag{47}$$

$$h(0, t) = H(0). \tag{48}$$

It is clear that the corresponding solution is given by

$$h(x, t) = H(x) \tag{49}$$

$$q(x, t) = 0. \tag{50}$$

In Bermúdez and Vázquez [2] it is proved that property C holds for Roe's scheme and the Q-scheme of van Leer.

However the flux-splitting techniques we have extended do not have this property. This is the reason why the corresponding numerical results included in the next section are good for constant depth but present unphysical oscillations for variable depth.

NUMERICAL RESULTS

We describe the three tests we have considered.

Example 1: Dam break problem

Consider a wide channel having a flat bottom surface and a barrier placed across its width. Let h_1 (resp. h_0) the height of the water upstream (resp. downstream) and assume $h_1 > h_0$. At a time $t = 0$ the barried is suddenly removed and the problem is to determine the subsequent motion for all x and t. The solution of this particular problem is known analytically (see Stoker [16]). It represents a flow consisting of a bore travelling downstream and a rarefaction wave travelling upstream.

Figures 5 to 12 show the numerical results obtained with the methods developed in section 3 together with the exact solution to compare them in terms of the numerical dissipation and shock capturing. In all cases $h_1 = 1, M = 50, \Delta = 10^{-3}$ and $t = 1$.

Figures 5 to 8 represent the solutions for the subcritical case $h_1/h_0 = 2$ and those from 9 to 12 correspond to $h_1/h_0 = \infty$, i.e. $h_0 = 0$ which is a supercritical situation.

In the last case the rarefaction wave presents the "dog-leg" phenomenon at a sonic point. Notice thar Vijayasundaram's scheme solves this problem satisfactorily while the others amplify the phenomenon.

Example 2: Tidal current (one-dimensinal)

The second example is the propagation of a tidal wave in a coastal region with variable depth. More precisely we have $\Omega = (0, x_E)$ and :

$$\begin{array}{ll} \text{Initial conditions} & \left\{ \begin{array}{lcl} h(x,0) & = & H(x) \\ q(x,0) & = & 0 \end{array} \right. \\ \text{Boundary conditions} & \left\{ \begin{array}{lcl} h(0,t) & = & H(0) + 4 + 4sen\left(\pi\left(\frac{4t}{86.400} - \frac{1}{2}\right)\right) \\ q(x_E, 0) & = & 0 \end{array} \right. \\ \text{Depth function} & H(x) = 50.5 - \frac{40x}{x_E} + 10sen\left(\pi\left(\frac{4x}{x_E} + \frac{1}{2}\right)\right). \end{array} \qquad (51)$$

Observe in fig. 13 to 16 that the extensions of Q-schemes give qualitative good results in the sense that the tidal wave is well transmitted along the channel, these schemes are in good agreement with those obtained using implicit finite element methods from Bermúdez, Rodríguez and Vilar [1]. On the contrary flux-splitting techniques lead to numerical results with too high differences in the water level. The reason for this bad behaviour as already mentioned is that these last schemes do not satisfy property C.

In figures 13 to 14 ,$x_E = 14.000m$ and we compare the extensions of Roe's scheme with the Vijayasundaram's. We take $M = 50, \Delta t = 0.5s$ and $t = 10800s$ +++ and $t = 21600s$ ***. In order to have a length greater than one wavelength in fig. 15 to 16 we take $x_E = 648.000m$. The results corresponding to $t = 10800s$ are shown; it can be seen that the wave propagates only to 1/4 of the whole domain and the velocity vanishes in the remainder. In this case the results obtained by the extensions of van Leer's Q-scheme and Steger and Warming's scheme are compared. Now we take $M = 50, \Delta t = 360s$ and $t = 10800s$.

Example 2: Tidal current (two-dimensinal)

In this example we consider the two-dimensional equations for the previous example. Now $\Omega = (0, x_E) \times (0, 2y_M)$ (see fig. 4).

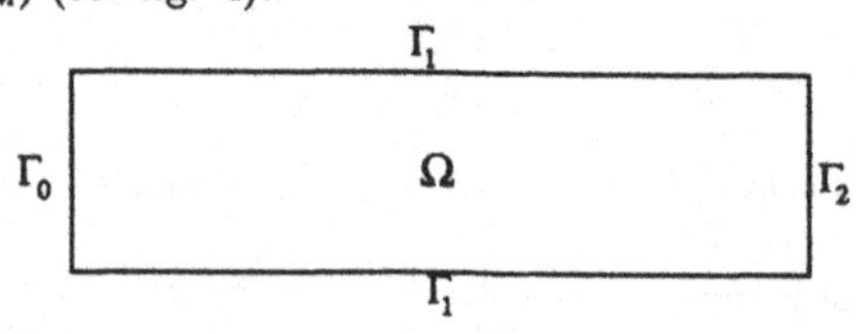

Figure 4.

$$
\begin{array}{ll}
\text{Initial conditions} & \left\{ \begin{array}{lcl} h(x,y,0) & = & H_2(x,y) \\ q(x,y,0) & = & 0 \end{array} \right. \\
\text{Boundary conditions} & \left\{ \begin{array}{lcl} h & \equiv & H_2(0,0) + 4 + 4sen\left(\pi\left(\dfrac{4t}{86.400} - \dfrac{1}{2}\right)\right) \ on \ \Gamma_0 \\ q & \equiv & 0 \ on \ \Gamma_2 \\ q \wedge \nu & \equiv & 0 \ on \ \Gamma_1 \end{array} \right. \\
\text{Depth function} & H_2(x,y) = H(x)\left(1 - \left(\dfrac{y - y_M}{y_M}\right)^2\right).
\end{array}
\tag{52}
$$

where ν is a unit normal vector.

We have used the techniques studied in section 3 for a Delaunay-Voronoi finite volume discretization. Property C can also be proved for the extensions of Q-schemes we have given.

Figures 18 to 19 shows the computed height and velocity field for time $t = 10800$ using the extension of the van Leer Q-scheme. In order to compare these results to the one-dimensional ones fig. 20 and 21 show the h and u_1 values for the section $y = y_M$ at the same time. Observe that as the deth is not constant along the y-direction is greater that obtained for the one dimensional model.

CONCLUSIONS

We have proposed extensions of some upwind schemes to hyperbolic conservation laws with source terms. We have applied these extensions to the shallow water equations: since the flux associated to these equations is not homogeneous we have introduced a matrix different from the jacobian in order to get a factorization of the flux allowing to use some flux-splitting techniques. We have also noticed the importence of a conservation property for the scheme to propagate waves at the right velocity, and proved that the extensions of the Roe's scheme and Q-scheme of van Leer verifies it. Preliminary numerical results and theoretical developements now in progress confirm that this work can be extended to the two-dimensional case.

ACKNOWLEDGMENTS

This work has been supported by DGYCIT(PB89-0566) and "Acción Integrada Hispano-Francesa"(259A).

References

[1] Bermúdez, A., Rodríguez, C., and Vilar, M.A. Solving shallow water equations by a mixed implicit finite element method, *IMA J. of Num. Analysis* **11,** pp. 79-97, 1991.

[2] Bermúdez, A. and Vázquez, M.E. Upwind methods for hyperbolic conservation laws with source terms, preprint, Dep. Applied Math. Univ. Santiago. Spain, 1991.

[3] Brebbia, C.A. (Ed.) *Mathematical Models for Environmental Problems,* Pentech Press, London, 1976.

[4] Desideri, J.A., Glinsky, N. and Hettena, E. Hypersonic reactive flow computations, *Computers and Fluids* **18,** pp. 151-182, 1990.

[5] Gambolati, G., Rinaldo, A., Brebbia, C. A., Gray, W.G. and Pinder, G. F. *Computational Methods in Surface Hydrology,* Computational Mechanics Publ., Southampton, 1990.

[6] Glaister, P. Approximate Riemann solutions of the shallow water equations, *J. Hydraulic Research* **26,** pp. 293-305, 1988.

[7] Goussbaile, J., Hecht, F., Labadie, G. and Reinhart, L. Finite element solution of the shallow water equations by a quasi-direct decomposition procedure, *Int. J. Num. Meth. in Fluids* **4,** pp. 1117-1136, 1984.

[8] Harten, A., Lax, P., van Leer, A. On upstream differencing and Godunov-type schemes for hyperbolic conservation laws, *SIAM Rev.,* **25** ,pp. 35-61, 1983.

[9] Kawahara, M., Takeuchi, N. and Yoshida, T. Two step explicit finite element method for tsunami wave propagation analysis, *Int. J. Num Meth. Eng.* **12,** pp. 331-351, 1978.

[10] LeVeque, R. Numerical Methods for Consercation Laws, Birkhäuser, Basel, Boston, Berlin, 1990.

[11] LeVeque, R. and Yee, H.C. A study of numerical methods for hyperbolic conservation laws with stiff source terms, *J. of Comp. Phys.* **86,** pp. 187-210, 1990.

[12] Lynch, P.R. and Gray, W.C. A wave equation model for finite tidal computations, *Computers and Fluids* **7,** pp. 207-228, 1979.

[13] Peraire, J., Zienkiewicz, O. C. and Morgan, K. Shallow water problems: a general explicit formulation, *Int. J. Num. Meth. Eng.* **22,** pp. 547-574, 1986.

[14] P.L. ROE Approximate Riemann solvers, parameter vectors, and difference schemes,*J. Comput. Phys.,***43,** 1981.

[15] J. STEGER R.F. WARMING Flux vector splitting of the inviscid gasdynamic equations with application to finite-difference methods,*J. Comput. Phys.,***40,** 1981.

[16] Stoker, J.J. *Water Waves.* Interscience, New York, 1957.

[17] Taylor, C. and Davies, J. Tidal and long wave propagation. A finite element approach, *Computers and Fluids* **3,** pp. 125-148, 1975.

[18] Vijayasundaram, G. Transonic flow simulations using an upstream centered scheme of Godunov in finite elements,*J. Comput. Phys.,* **63**,1986.

[19] Zienkiewicz, O. C. and Heinrich, J.C. A unified treatment of the steady-state shallow water and two-dimensional Navier-Stokes equations. Finite element and penalty function approach, *Comp. Meth. Appl. Mech. Eng.* **17/18,** pp. 673-698, 1979.

DAM BREAK PROBLEM (functions h and Q)

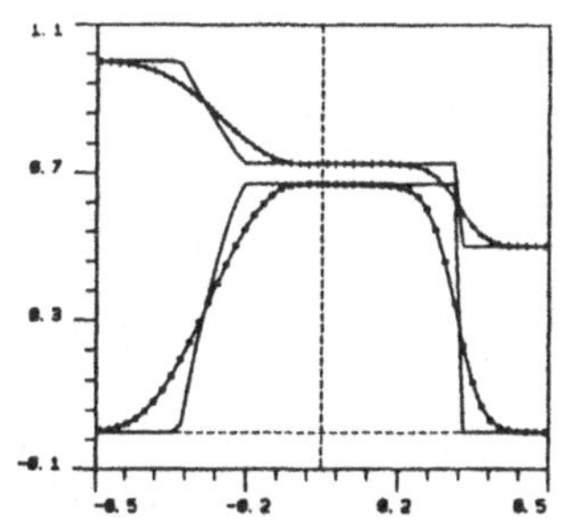

Fig.5 .- Ext. of the Q-scheme of van Leer

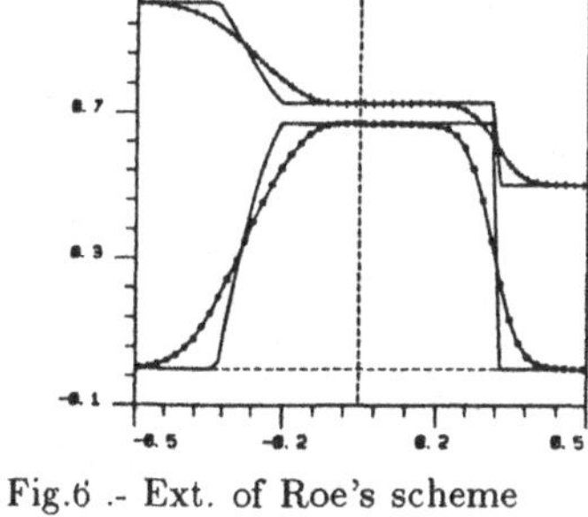

Fig.6 .- Ext. of Roe's scheme

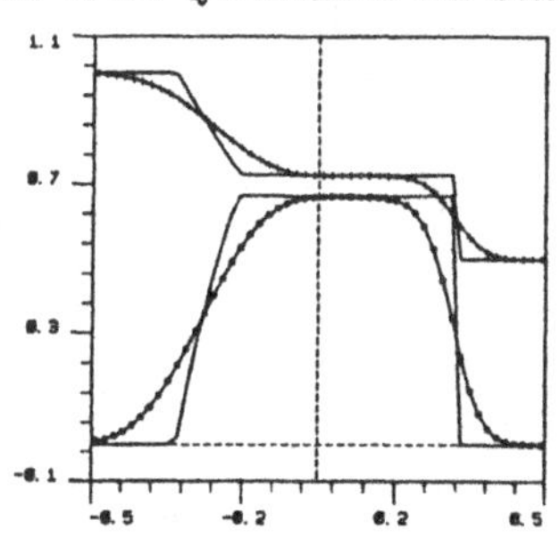

Fig.7 .- Ext. of Steger-Warming's scheme

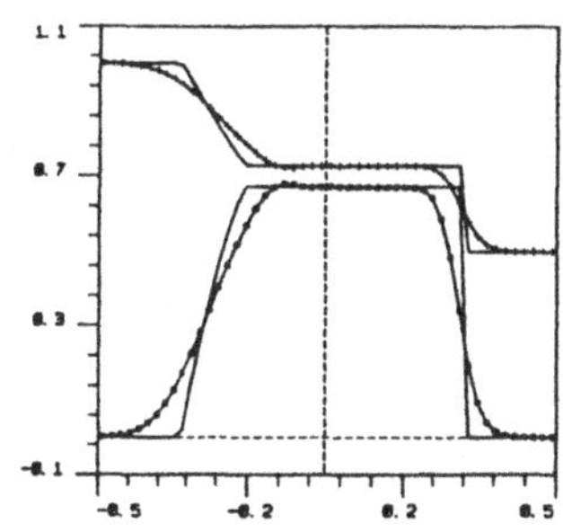

Fig.8 .- Ext. of Vijayasundaram's scheme

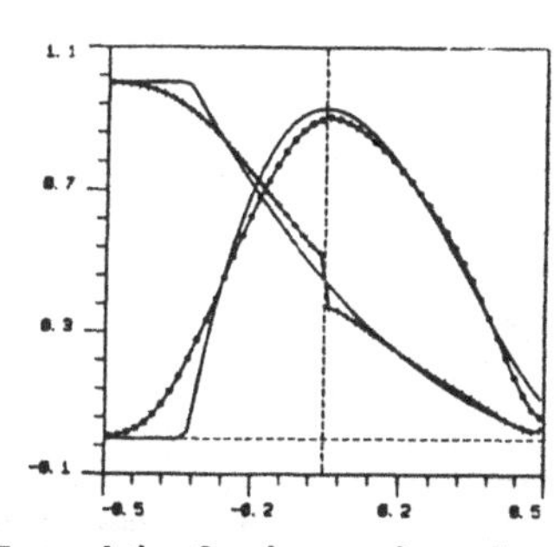

Fig.9 .- Ext. of the Q-scheme of van Leer

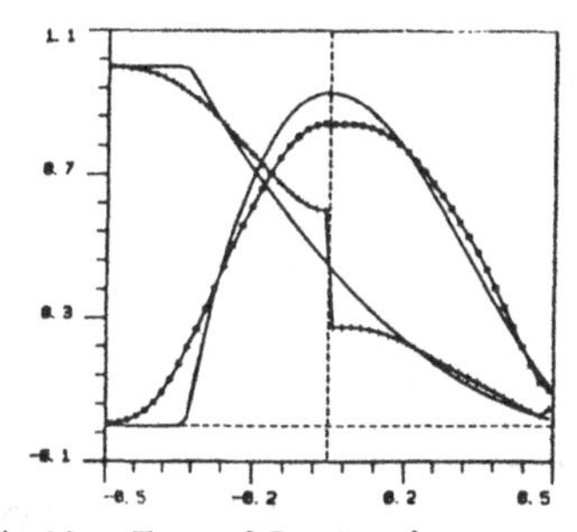

Fig.10 .- Ext. of Roe's scheme

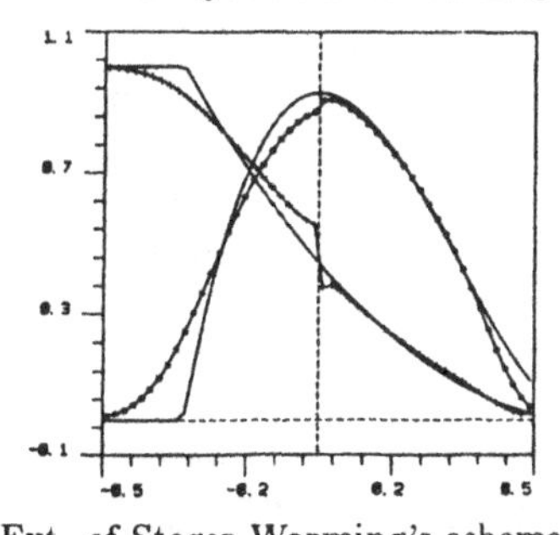

Fig.11 .- Ext. of Steger-Warming's scheme

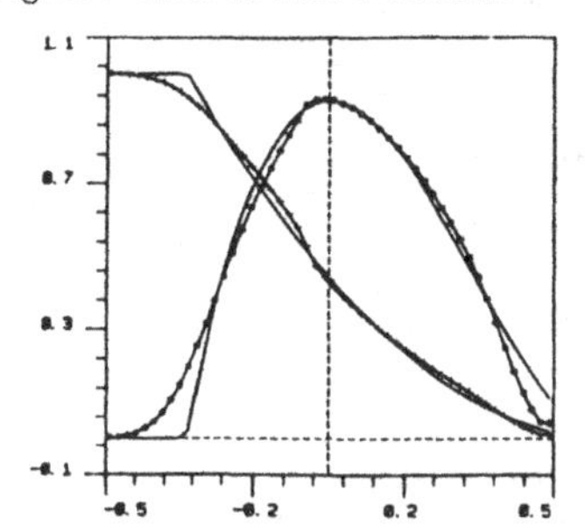

Fig.12 .- Ext. of Vijayasundaram's scheme

TIDAL CURRENT (one-dimensional)

Surface elevation $A - H, A - h$ Velocity u

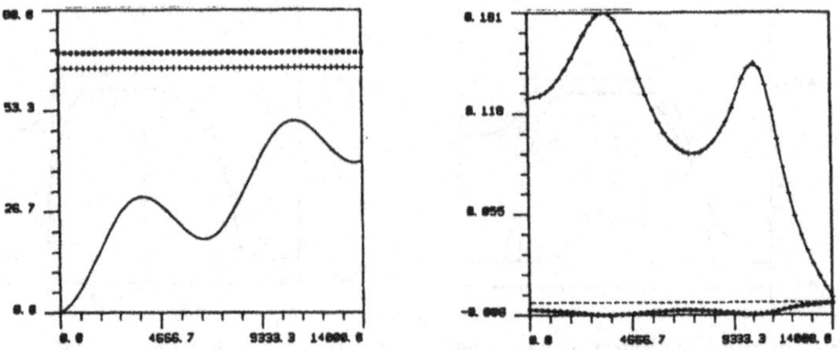

Fig.13 .- Extension of Roe's scheme

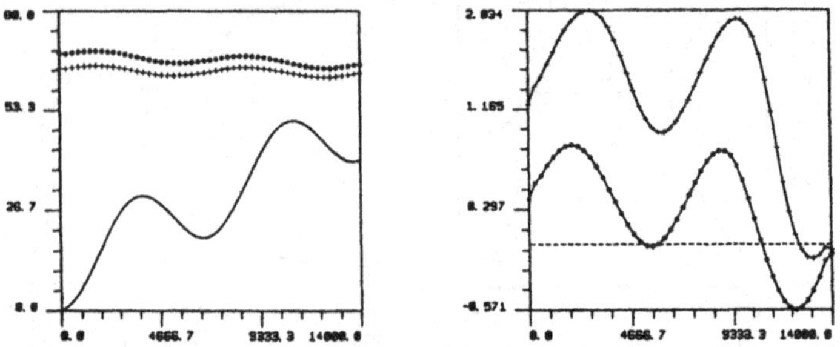

Fig.14 .- Extension of Vijayasundaram's scheme

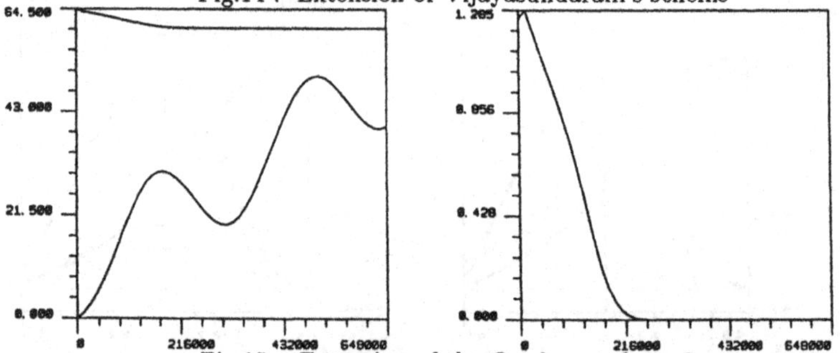

Fig.15 .- Extension of the Q-scheme of van Leer

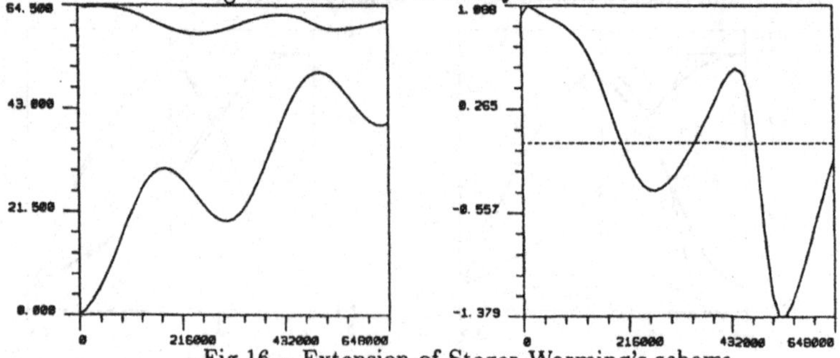

Fig.16 .- Extension of Steger-Warming's scheme

TIDAL CURRENT (two-dimensional)
Extension of the van Leer Q-scheme

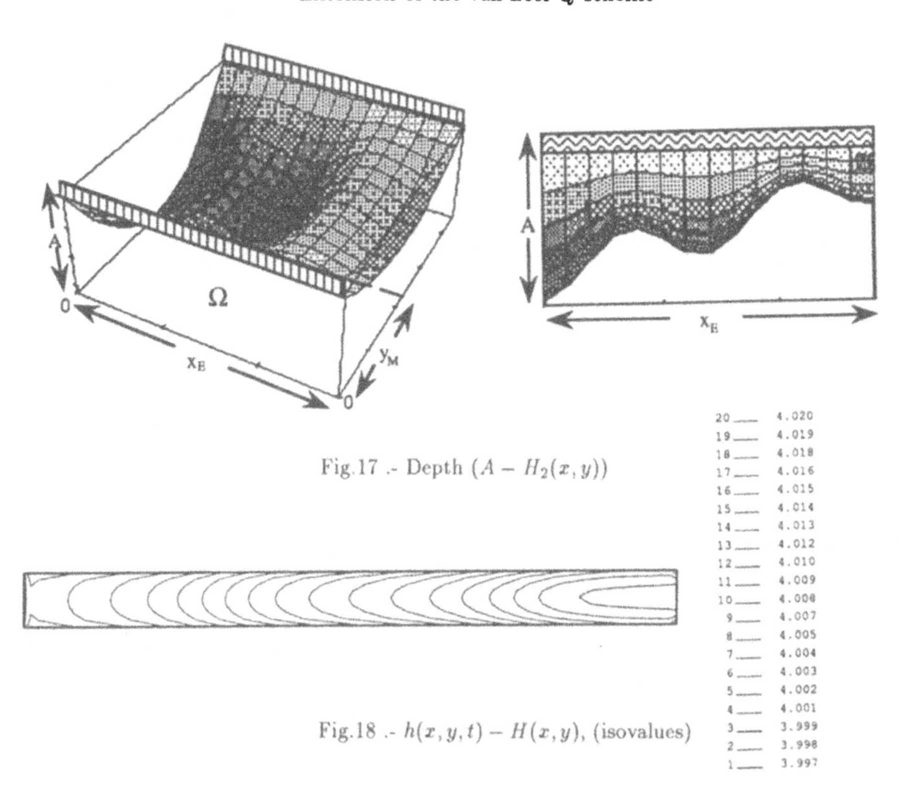

Fig.17 .- Depth $(A - H_2(x, y))$

Fig.18 .- $h(x, y, t) - H(x, y)$, (isovalues)

Fig.19 .- Velocity (u_1, u_2)

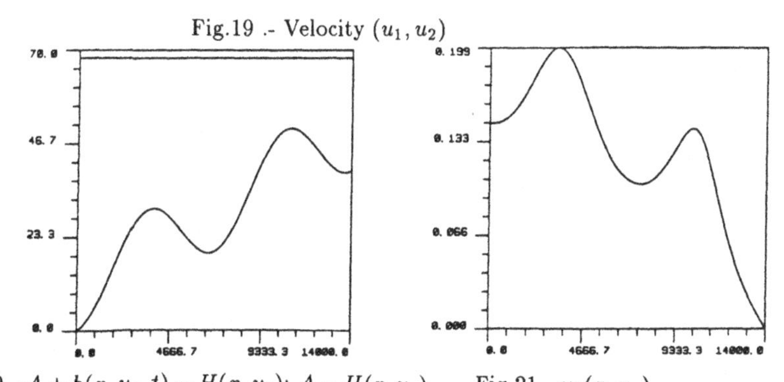

Fig.20 .- $A + h(x, y_E, t) - H(x, y_E)$; $A - H(x, y_E)$.

Fig.21.- $u_1(x, y_E)$.

Three Dimensional Modelling of a Mediterranean Bay

H. Novelli (*), I. Dekeyser (*), P. Fraunié (**)
(*) Centre d'Océanologie de Marseille, Campus de Luminy, Case 901, F 13288 Marseille Cedex 9, France
(**) Institut de Mécanique Statistique de la Turbulence, 12 Avenue Général Leclerc 13003 Marseille, France

ABSTRACT

A three dimensional computation of a Mediterranean Bay is presented. The numerical procedure does not consider the hydrostatic hypothesis and introduces a two equations closure model for turbulence ($\kappa - \varepsilon$ model). Application to the real configuration of the Cannes Bay involves the description of the dynamic and thermal fields, featuring recirculation and upwelling zones, and focuses on the importance of the open sea boundary conditions.

INTRODUCTION

The Mediterranean sea is a deep and closed sea characterized by a broken coastline and large slopes extending from the coast to the shelf. The results presented here concern the complex turbulent flow induced in such coastal areas by offshore currents, water discharge and wind effects.

Taking these factors into account, the local treatment of coastal Mediterranean areas suggests that the hydrostatic assumption is to be discussed and that a non homogeneous turbulent diffusion is to be expected (see Nihoul et al [1]).

So, a complete three dimensional computation of a coastal area has been performed, irrespective of the hydrostatic hypothesis and introducing a two equations closure model for turbulence (a $\kappa - \varepsilon$ model adapted to stratified flows) and including Coriolis effects which are known to be significant in such configurations.

THE NUMERICAL MODEL

Physical assumptions

The fluid is considered as newtonian and incompressible buoyancy effects are introduced through the density variations due to temperature and salinity effects, in keeping with the Boussinesq hypothesis, and a free surface condition is imposed.

The complete three Navier-Stokes momentum equations are considered in the form:

$$\frac{\partial U_i}{\partial t} + U_j \frac{\partial U_i}{\partial x_j} = \frac{\rho - \rho_r}{\rho_r} g \delta_{i3} + g(\delta_{i3} - 1)\frac{\partial \eta}{\partial x_i} - 2\varepsilon_{ijk}\Psi_j U_k - \frac{1}{\rho_r}\frac{\partial P}{\partial x_i} + \frac{\partial}{\partial x_j}\left(\nu \frac{\partial U_i}{\partial x_j} - \overline{u_i' u_j'}\right) \tag{1}$$

associated to the continuity equation:

$$\frac{\partial U_i}{\partial x_i} = 0 \tag{2}$$

The temperature and salinity equations write:

$$\frac{\partial \Theta}{\partial t} + U_j \frac{\partial \Theta}{\partial x_j} = -\frac{\partial \overline{u_i' \theta'}}{\partial x_j} \tag{3}$$

$$\frac{\partial S}{\partial t} + U_j \frac{\partial S}{\partial x_j} = -\frac{\partial \overline{u_i' s'}}{\partial x_j} \tag{4}$$

where U_i are the mean velocity components in a cartesian frame, P the pressure variation from the r reference field corresponding to the hydrostatic equilibrium of the water column, ρ the fluid density, $\vec{\Psi}$ the earth rotation vector, Θ the mean temperature, S the salinity and η is the surface elevation. Note that the molecular diffusivity term is neglected compared to the turbulent one.

The turbulence closure model obeys the Boussinesq assumption introducing the turbulent viscosity concept in the form:

$$-\overline{u_i' u_j'} = \nu_t \left(\frac{\partial U_i}{\partial x_j} + \frac{\partial U_j}{\partial x_i}\right) - \frac{2}{3}\delta_{ij}\kappa \tag{5}$$

where ν_t is the equivalent viscosity coefficient due to turbulence and κ the turbulent kinetic energy $\frac{1}{2}\overline{u_i' u_i'}$.

In the same way, turbulent diffusivity coefficients are introduced for the thermal (K_t) and salinity (D_t) fields:

$$\begin{aligned} -\overline{u_j'\theta} &= K_t\,\partial\Theta/\partial x_j \\ -\overline{u_j's} &= D_t\,\partial S/\partial x_j \end{aligned} \qquad (6)$$

These two coefficients are empirically deduced from ν_t using values of the Prandtl and Schmidt turbulent numbers, $Pr_t = \nu_t/K_t$, $Sc_t = \nu_t/D_t$ equal to unity.

The leading coefficient ν_t is deduced from κ and ε local values using the law (Hanjalic and Launder [2]):

$$\nu_t = C_\mu \frac{\kappa^2}{\varepsilon} \qquad (7)$$

where the C_μ coefficient is related to the production/ dissipation rate as to the flux Richardson number ($R_f = -G/P$) for stratified flows (Rodi [3]). This coefficient takes a constant value in this application.

The transport modelling equations for κ and ε in a stratified flow field write:

$$\frac{\partial\kappa}{\partial t} + U_j\frac{\partial\kappa}{\partial x_j} = P + G + \frac{\partial}{\partial x_j}\left(\frac{\nu_t}{\sigma_k}\frac{\partial\kappa}{\partial x_j}\right) - \varepsilon \qquad (8)$$

and

$$\frac{\partial\varepsilon}{\partial t} + U_j\frac{\partial\varepsilon}{\partial x_j} = C_{\epsilon 1}\left(1 + C_{\epsilon 3}R_f\right)\frac{\varepsilon}{\kappa}(P+G) + \frac{\partial}{\partial x_j}\left(\frac{\nu_t}{\sigma_\epsilon}\frac{\partial\varepsilon}{\partial x_j}\right) - C_{\epsilon 2}\frac{\varepsilon^2}{\kappa} \qquad (9)$$

by introducing:

- the production term

$$P = -\overline{u_i'u_j'}\frac{\partial U_i}{\partial x_j}$$

- the turbulent buoyancy term that writes in a first order approximation:

$$G = -\beta_\Theta\, g_j\, \overline{u_j'\theta} + \beta_S\, g_j\, \overline{u_j's}$$

with $\beta_\Theta = -1/\rho_r\,(\partial\rho/\partial\Theta)_r$ and $\beta_S = 1/\rho_r\,(\partial\rho/\partial S)_r$.

The adopted values for the model coefficients are the classical ones:

$$C_\mu = 0.09 \quad \sigma_k = 1.0 \quad \sigma_\epsilon = 1.3 \quad C_{\epsilon 1} = 1.44 \quad C_{\epsilon 2} = 1.92 \quad C_{\epsilon 3} = 0.7$$

The boundary conditions at the bottom as well as upon the lateral walls introduce a wall correction for the $\kappa - \varepsilon$ model. The free surface boundary condition is obtained in first order by permitting a mesh vertical deformation satisfying the momentum equation (2) and the following equation on the surface ($z = \eta$):

$$\frac{\partial \eta}{\partial t} = -\left(U_S \frac{\partial \eta}{\partial x} + V_S \frac{\partial \eta}{\partial y} \right) - W_S \tag{10}$$

where U_S, V_S, W_S are the mean velocity components on the surface.

The wind effect is deduced from the entrainment law proposed by Sheng (1983):

$$\rho_r \nu_t \left(\frac{\partial U_i}{\partial z} \right) = \rho_a C_{da} \sqrt{U_w^2 + V_w^2} \quad U_i \tag{11}$$

where ρ_a and ρ_r are the density of respectively the air and the reference medium (salt water), U_w and V_w the velocity components of the wind measured ten meters above the air-sea interface, and C_{da} a mean shear coefficient, equal to 0.0015.

The Numerical procedure

In order to solve the complete three dimensional set of equations (1) to (4) and (6) and (7), an iterative algorithm based on the "artificial viscosity" method (Chorin [4]) is performed, using a finite difference discretization in an orthogonal staggered grid, associated to a "MAC" integrating procedure.

The horizontal grid size is 500 meters, when a 10 meters grid was generated on the vertical direction with respect to the real topography.

The numerical code has been validated for different configurations concerning laboratory flows as well as actual on site flows ([5], [6]).

The vectorized code in finite differences was performed using CRAY-2 and YMP computers.

RESULTS

The computational domain (fig. 1) includes from the West to the East: the Cannes Bay, the Lérins Islands and the Juan Gulf, and extends up to 10 kilometers offshore. The circulation is dominated by the entrainment effect of the Ligure current which localisation and intensity are widely variable (Millot [7], Sammari [8]).

The velocity profiles of the modellized current are presented on figure 2. Temperature and salinity profiles as presented on figure 3 correspond to the initial conditions in the whole domain and to the conditions imposed anytime on boundaries. These profiles are provided from measurements realized in the Cannes Bay during the month of May (Dekeyser et al, 1991).

Different configurations have been compared, relative to the Ligure current entrainment effect and the wind occurrence.

At first, two kind of boundary conditions have been introduced to represent the longshore Ligure current, with the vertical profile presented on figure 2 -a.

- The first one corresponds to weak entrainment conditions, i.e. Von Neumann conditions ($\partial U/\partial x = 0$) are imposed at the seaward and outflow boundaries (cases A, B, C).

- The second one corresponds to "moving wall" boundary conditions, i.e. a Dirichlet condition on the seaward velocity component (cases D, E, F, G, H) and a Von Neumann condition at the outflow boundary.

Furthermore, in these two cases, a mass transfer can be imposed through the seaward boundary following the vertical profile of figure 2 -b. Inflow conditions correspond to cases B and E, when outflow conditions are imposed for cases C and F.

In the end, wind effects are accounted in cases G and H.

The surface circulation in the computational domain are presented on figure 4-a to 4-h.

Everytime, the surface circulation is dominated by zones of recirculation located in the Juan Gulf and downstream the Lérins Islands. The main noticeable effect is relative to the shear effect of the offshore current. The comparison between the cases A and D shows clearly the sensitivity of the resulting circulation to the way of modelling this shear effect. While the weak shear effect in case A involves the development of a recirculation region in the Gulf Juan and a skirting of the islands (Fig. 4-a), the strong shear condition in case D accentuates these effects and modifies significantly the circulation inside the Cannes Bay. Indeed, a large vortex is observed in the case D downstream the islands associated to an inverse recirculation effect inside the Cannes Bay. We can notice that the flow penetration observed in the Cannes Bay in both cases is representative of strong three dimensional effects.

By reference to the two cases A and D, only secondary effects are observed when applying mass fluxes across the open sea boundary or a wind entrainment.

An inflow condition (cases B and E) increases the flow penetration effect upstream the Lérins Islands and breaks up the stable vortex observed in the case D (Fig. 4-d and 4-e).

In the same way, an outflow condition across the seaward boundary (cases C and F) modifies the offshore region of the computational domain, acting more significantly in the case F of a strong shear boundary condition, breaking up, one more time, the vortex taking place downstream the Lérins Islands in the case D.

Wind with magnitude of 4 meters per seconds have been imposed from South-West and North-East (cases G and H) associated to a strong shear condition (reference case D).

The surface circulation is then locally modified, in the way of a perturbation when the wind blows against the current (case G) and a regulation when the wind blows the same direction as the current. In both cases, the wind induces the vortex vanishing in surface downstream the islands.

Then, the main forcing effect on the dynamic field arises from the inflow condition acting on the upstream longshore boundary, when the seaward boundary condition only confines the flow. The resulting surface velocity field presents quite different structures, especially in the case D when the sheared boundary conditions are sufficient to induce a structured recirculation region downstream the Lérins Islands.

For every case, a three dimensional flow is observed with vertical components of velocity up to 0.5 cm/s. Vertical velocity profiles are presented on figure 5 for two vertical West - East sections. The figure 5-a corresponds to the first section located south of the Lérins Islands, when the figure 5-b corresponds to a section close to the seaward boundary.

Figure 6 shows the visualization of particles trajectories in the computational domain in the case D. The three-dimensional conical structure of the vortex occuring downstream the islands is clearly observable associated to a significant time of advection over the domain.

Refined mesh

In order to characterize the sensitivity to the mesh size, a $250 \times 250m^2$ horizontal grid has been involved for comparison to the results obtained by using a $500 \times 500m^2$ grid. In the case A simulation, a quite similar surface circulation is then obtained (Figure 7). Detailed comparisons bring out the occurrence of secondary vortices, for example downstream the Lérins Islands. However, the general circulation as well as the local velocity magnitudes are in good agreement between the two simulations.

Hydrostatic hypothesis

A test case has been computed by considering no buoyancy effects in order to verify for coastal flows the hydrostatic hypothesis generally accepted for large and meso-scales circulation models.

Considering equation (1) in the vertical direction, the acceleration terms dW/dt are in balance with the turbulent diffusion terms to generate a vertical pressure gradient.

For the considered bathymetry, the visualization of the pressure field (Figure 8) over the first 100 meters depth confirms that a pressure vertical gradient occur in the strongly shaped regions, in the neighbouring of the islands and in the Gulf Juan. Anyway, the order of magnitude of the different terms is never more than $2\,10^{-4}$ when compared to g.

CONCLUSION

A numerical three-dimensional model, adapted to coastal Oceanography, has been applied to the real configuration of a Mediterranean site. This model considers the complete Navier-Stokes equations associated to two closure equations for turbulence ($\kappa - \varepsilon$ model).

The description of the dynamic field, featuring recirculation and upwelling zones, characterizes the importance of the open sea boundary conditions. Inflow conditions as well as wind effects are analysed, providing secondary effects. Wind effects are shown to modify the surface circulation in a significant way. In the end, the hydrostatic hypothesis is shown to remain acceptable for such configurations.

Acknowledgements

This study was supported by a grant from the PIREN ATP "Modélisation en Milieu littoral et côtier" and a computational support was given by the scientific council of the "Centre de Calcul Vectoriel pour la Recherche".

REFERENCES

1. J. C. J. Nihoul, E. Deleersnijder and S. Djenidi, 1989, "Modelling the general circulation of shelf seas by 3D $K - \varepsilon$ models". Earth Science Reviews, 26, pp 163- 189.

2. Hanjalic K. and Launder B. E., "A Reynolds stress model of turbulence and its application to thin shear flows", J. Fluid Mech. 52, pp 609 - 638.

3. Rodi W., 1980, "Turbulence models and their application in hydraulics: a state of the art review", Book Publication of the Int. assoc. for Hydr. Res., Delft, Netherlands.

4. Chorin A.J., 1967, "A numerical method for solving incompressible viscous flow problems", J. Comp. Phys., vol 2, pp 12-26.

5. Sini J. F. and Dekeyser I., 1987, "Numerical prediction of turbulent plane jets and forced plumes by use of a $\kappa - \varepsilon$ model of turbulence", Int. J. Heat and Mass Transfer, vol 30, 9, pp 1787 - 1801.

6. Marcer R., Fraunié P., Dekeyser I., Andersen V., 1991, "Modélisation numérique d'un couplage physico-biologique en milieu côtier", Oceanologica Acta, vol. SP 11.

7. Millot C., 1987, "The structure of mesoscale phenomena in the Ligurian sea inferred from DYOME experiment", Annales Geophysicae, 5B, pp21 - 30.

8. Sammari C., 1991, "Etude de la dynamique de l'eau d'origine Atlantique dans le bassin Liguro-Provençal à partir de données courantométriques et par télédétection", Thèse de Doctorat de l'Université d'Aix-Marseille II.

9. Dekeyser, I., Fraunié P., Nival P., Sciandra A., Lacroix G., Djenidi S., "Projet de modélisation de la Baie de Cannes", Rapport ATP PIREN CNRS, 1991.

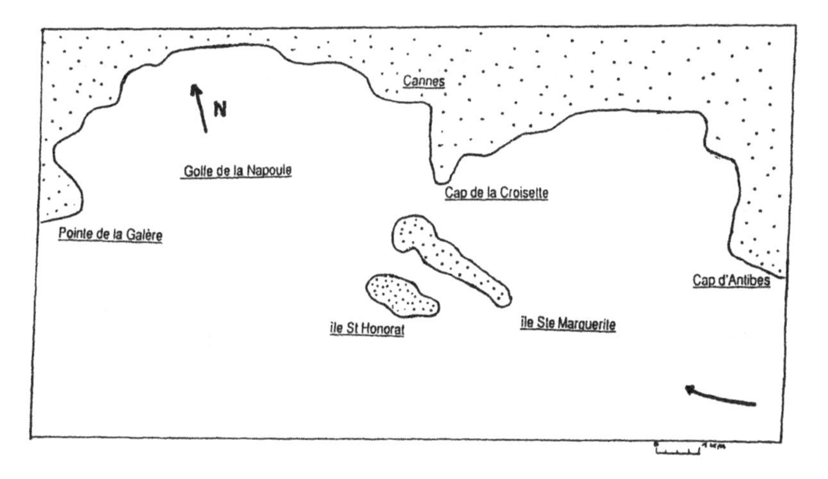

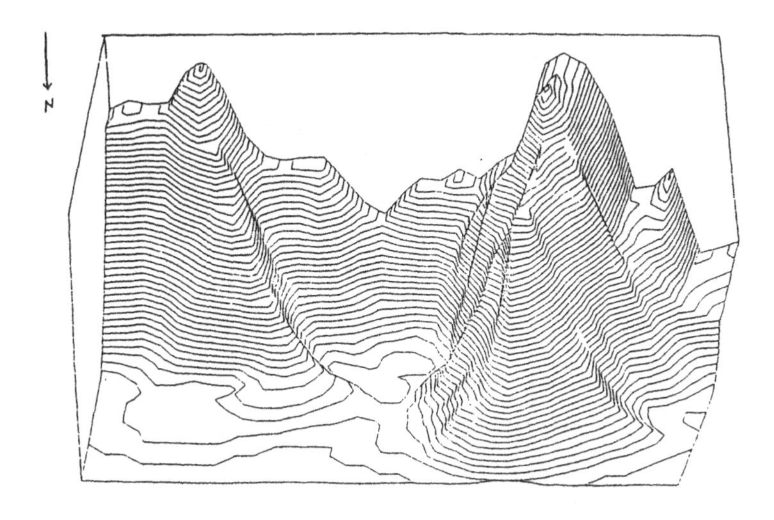

Fig. 1 Computational Domain

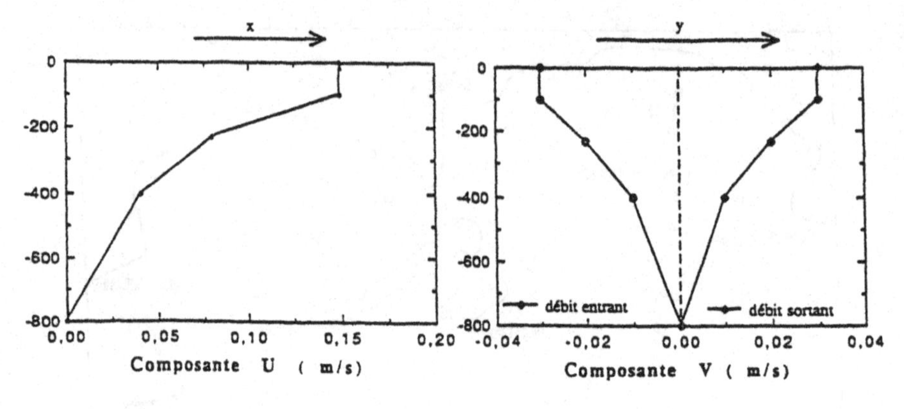

a) streamwise u component

b) crossflow v component

Fig. 2 Vertical velocity profiles of the Ligure current

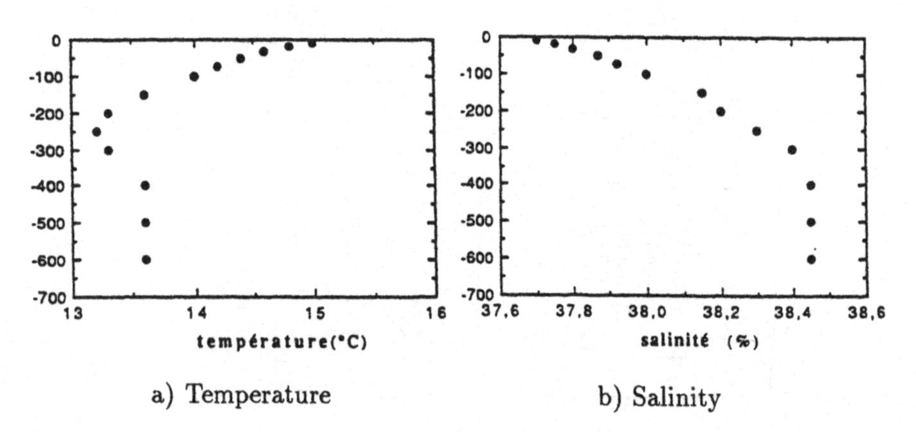

a) Temperature

b) Salinity

Fig. 3 Vertical profiles

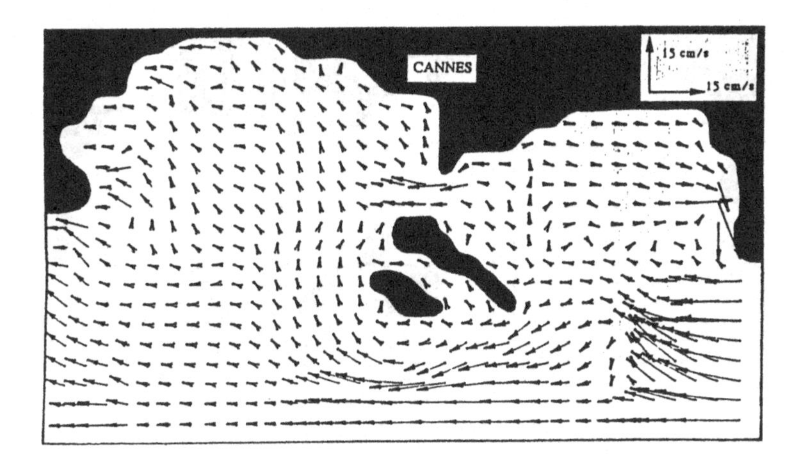

a) Weak shear condition

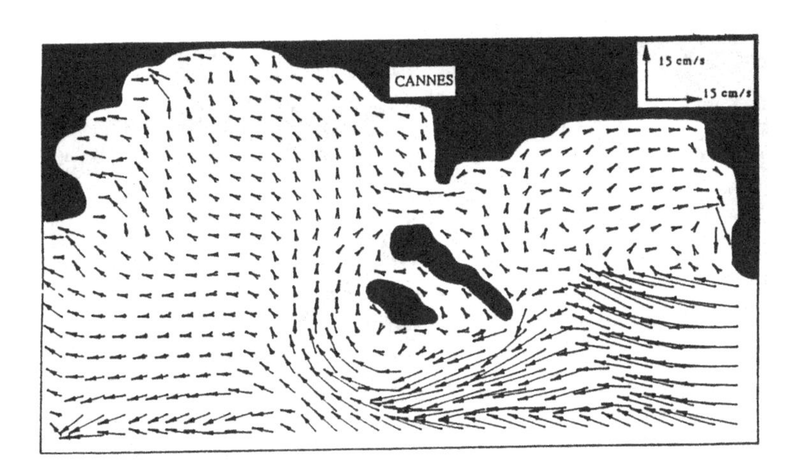

b) Inflow and weak shear conditions

Fig. 4 Surface Circulation for different boundary conditions

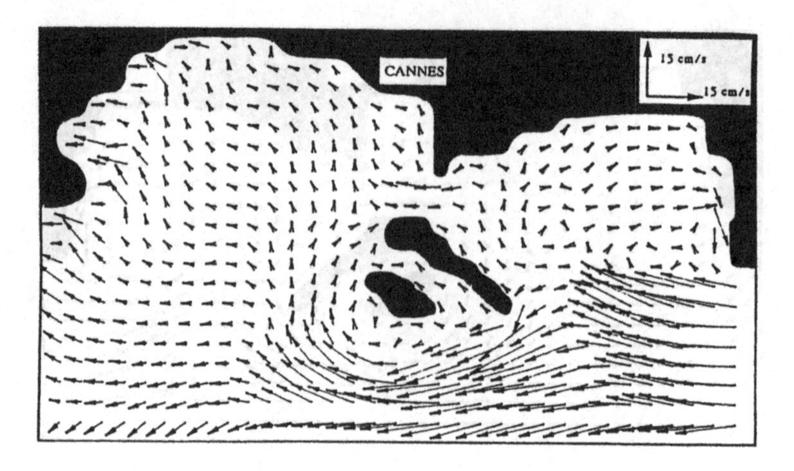

c) Outflow and weak shear conditions

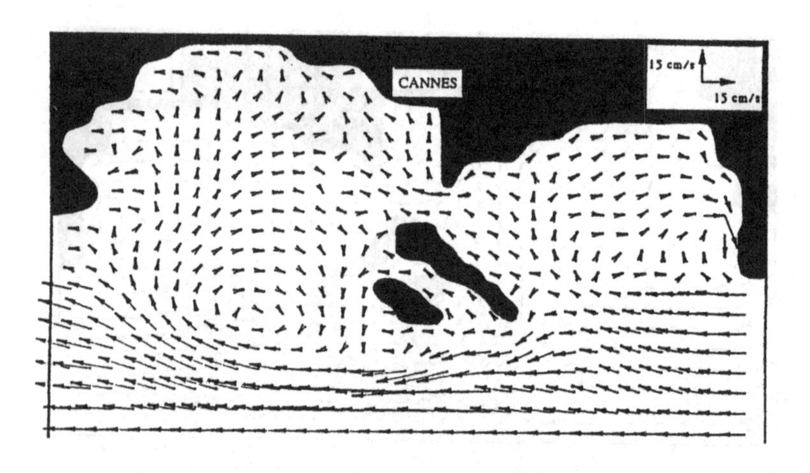

d) Strong shear condition

Fig. 4 Surface Circulation for different boundary conditions

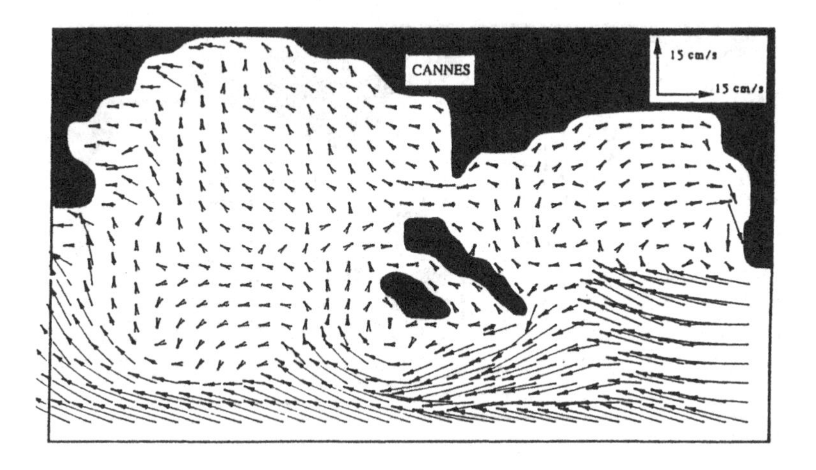

e) Inflow and strong shear conditions

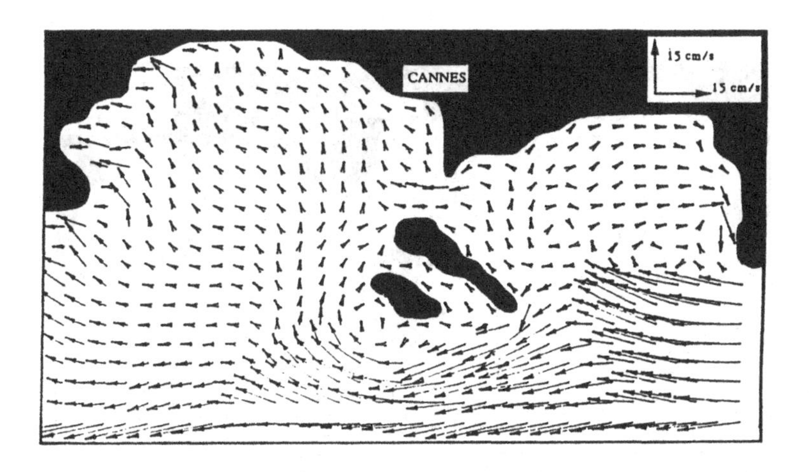

f) Outflow and strong shear conditions

Fig. 4 Surface Circulation for different boundary conditions

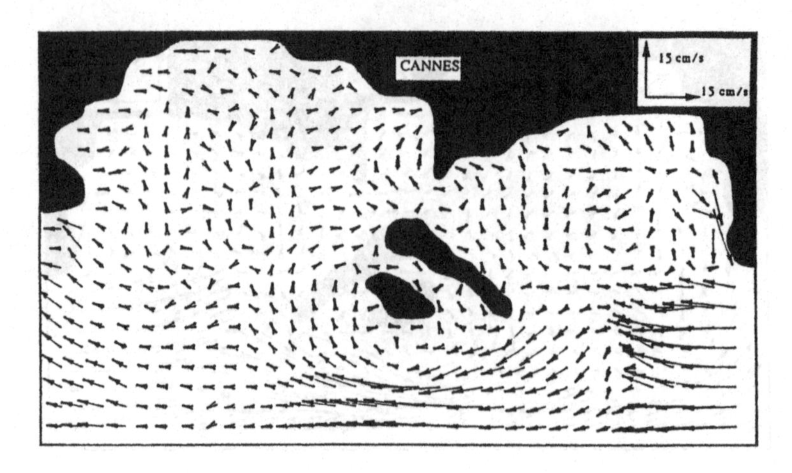

g) Strong shear condition and South - West Wind

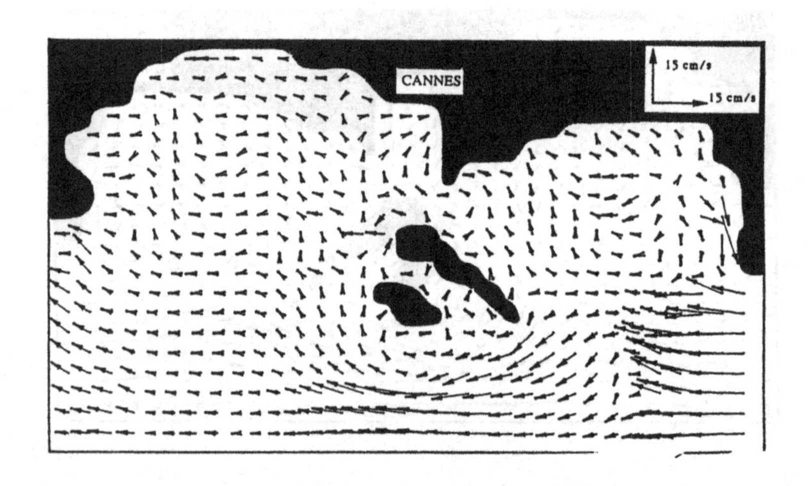

h) Strong shear condition and North - East Wind

Fig. 4 Surface Circulation for different boundary conditions

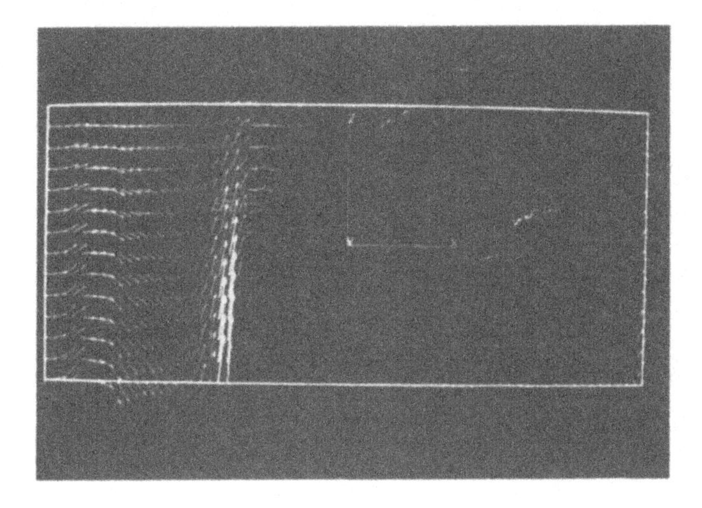

a) South of the Islands

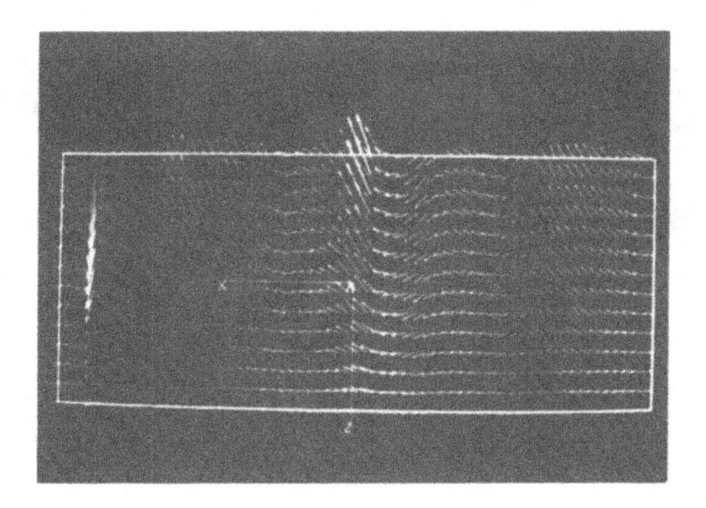

b) Near the seaward boundary

Fig. 5 Velocity fields in West - East vertical sections - Case A

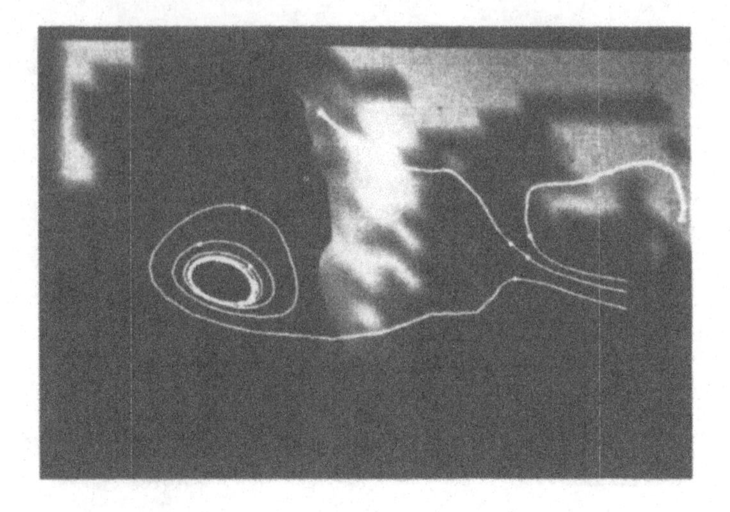

Fig. 6 Particles trajectories in the computational domain - Case D

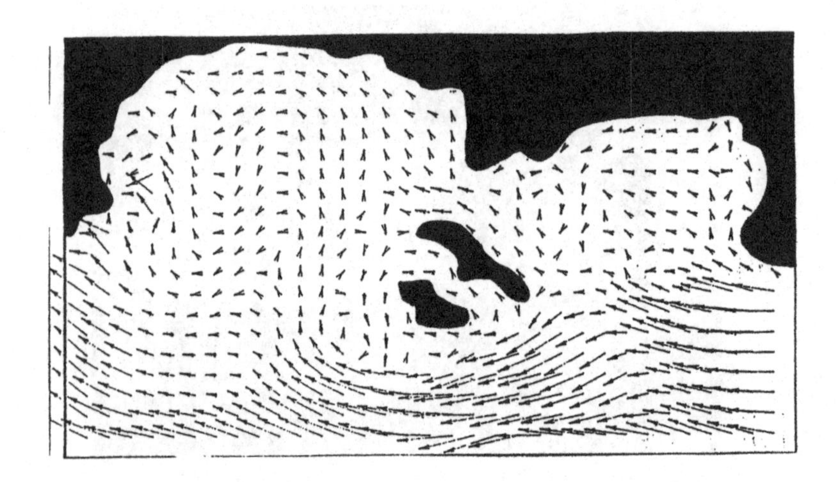

Fig. 7 surface Circulation using a $250 \times 250m^2$ grid - Case A

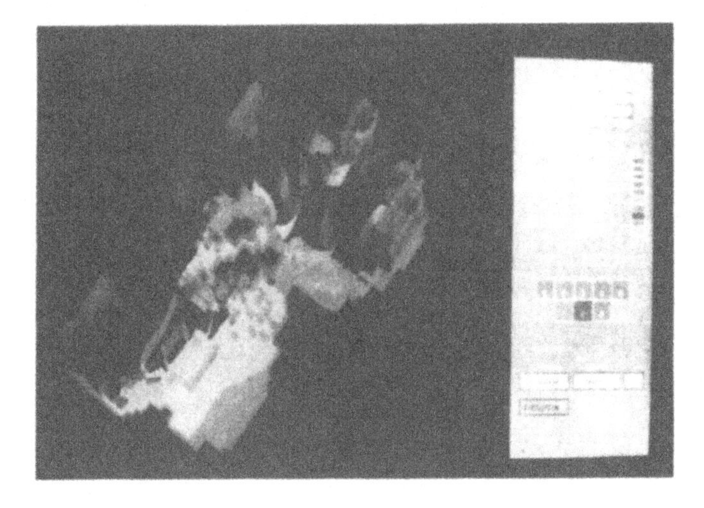

Fig. 8 Pressure field

A Tool for Education and Exploratory Research on Shallow Waters

J.M. Remédio, A.M. Baptista, J.R. Hurst, P.J. Turner

Center for Coastal and Land-Margin Research and Department of Environmental Science and Engineering, Oregon Graduate Institute of Science and Technology, 19600 NW von Neumann Drive, Beaverton, Oregon 97006-1999, U.S.A.

ABSTRACT

We introduce $RITA_1$, a one-dimensional finite element model specifically designed to aid graduate-level teaching and exploratory research on shallow water dynamics and modeling. $RITA_1$ uses a very flexible solution of the momentum and generalized wave equations, and is served by a user-oriented interface with broad scientific visualization capabilities. The application of $RITA_1$ is illustrated through a classroom-oriented study of the generation of shallow water tides and a research-oriented comparison of alternative numerical solutions of the advective terms.

INTRODUCTION

The numerical modeling of shallow water flows has progressed very significantly over the past half-century. Leading depth-averaged models, if properly calibrated and validated, can often claim a satisfactory degree of predictive ability within the requirements of many engineering applications[1]. Computational costs per node and per prototype unit of time have decreased drastically, as a consequence of both an explosive evolution of hardware and a steady evolution of numerical techniques. This reduction of computational costs has in turn fueled fruitful research and applications in fully 3-D modeling, and in long-term, highly resolved, regional modeling.

As in many fast-evolving areas, however, only marginal effort seems to be put into systematic comparative research (The Tidal Flow Forum[1] is a notable exception), advanced education, and practical training. Reflecting this trend, circulation models are typically developed very rigidly around a preferred numerical technique, with emphasis being placed on efficiency rather than on the flexible exploration of alternatives.

$RITA_1$ (River and Tidal Analysis, 1-D version) is a part of an evolving computational structure that is specifically aimed at teaching and education, sensitivity analysis, and exploratory research, rather than at efficiency and problem-solving. Denoted ACE_1 (Analysis of Coasts and Estuaries, 1-D version), this computational structure includes a range of flow and transport-transformation models, and is served by a menu-driven interface that controls grid generation, model selection and execution, definition of input parameters, and scientific visualization (Baptista *et al.*[2]).

The present paper concentrates specifically on $RITA_1$. We will describe first the formulation of the model, with some emphasis on the treatment of non-linear terms. We will then illustrate the use of the model for graduate-level teaching/education and for exploratory research: the generation of shallow water tides and the numerical treatment of advection will be used as sample topics. Elsewhere in the proceedings $RITA_1$ is used in "sensitivity analysis mode" to examine the propagation of locally-generated tsunamis in the Pacific Northwest coast of the United States (Baptista *et al.*[3])

REFERENCE MODEL FORMULATION

$RITA_1$ can be classified as a "generalized wave equation formulation finite element model". This type of model was introduced by Lynch[4] and Kinnmark[5], and is currently very popular among the finite element modeling community, because of the inherent ability to eliminate the spurious spatial oscillations that plague primitive-equation finite element models.

$RITA_1$ solves for the depth-integrated generalized wave equation (GWE) and the momentum equation (ME), neglecting velocities and gradients in the y-direction:

$$\frac{\partial^2\eta}{\partial t^2} - \frac{\partial^2}{\partial x^2}Huu - g\frac{\partial}{\partial x}H\frac{\partial\eta}{\partial x} + G\frac{\partial\eta}{\partial t} + \frac{\partial}{\partial x}(G-\tau)\,Hu = 0 \tag{1}$$

$$\frac{\partial u}{\partial t} + u\frac{\partial u}{\partial x} + g\frac{\partial\eta}{\partial x} + \tau u = 0 \tag{2}$$

where:

η- elevation [m]
g- gravity [ms^{-2}]
τ- friction parameter [s^{-1}]
H- water depth [m]
G-wave equation weighting factor [s^{-1}]

The friction factor, τ, can be either imposed as a constant or defined as a function of the flow, in the form:

$$\tau = \frac{gn^2|u|}{H^{1/3}} \tag{3}$$

where n is the Manning coefficient [$sm^{-1/3}$]. The factor G is imposed as a constant (an option defining G as a direct function of τ will soon be implemented).

The governing equations are solved sequentially: the GWE is solved first for elevations, and the ME is then solved for velocities. This approach prevents an implicit treatment of velocities in the GWE, but otherwise allows a flexible time-discretization in the form:

GWE:

$$\frac{\partial^2\eta}{\partial t^2} \approx \frac{1}{(\Delta t)^2}(\eta^{n+1} - 2\eta^n + \eta^{n-1}) \tag{4}$$

$$G\frac{\partial\eta}{\partial t} \approx \frac{G}{\Delta t}(w_0^1\eta^{n+1} + w_1^1\eta^n + w_2^1\eta^{n-1}) \tag{5}$$

$$g\frac{\partial}{\partial x}h\frac{\partial\eta}{\partial x} \approx g\,(w_0^2\frac{\partial}{\partial x}h\frac{\partial}{\partial x}\eta^{n+1} + w_1^2\frac{\partial}{\partial x}h\frac{\partial}{\partial x}\eta^n + w_2^2\frac{\partial}{\partial x}h\frac{\partial}{\partial x}\eta^{n-1}) \tag{6}$$

$$g\frac{\partial}{\partial x}\eta\frac{\partial\eta}{\partial x} \approx g\,(w_1^3\frac{\partial}{\partial x}\eta^n\frac{\partial}{\partial x}\eta^n + w_2^3\frac{\partial}{\partial x}\eta^{n-1}\frac{\partial}{\partial x}\eta^{n-1}) \tag{7}$$

$$\frac{\partial^2}{\partial x^2}huu \approx w_1^4\frac{\partial^2}{\partial x^2}hu^nu^n + w_2^4\frac{\partial^2}{\partial x^2}hu^{n-1}u^{n-1} \tag{8}$$

$$\frac{\partial^2}{\partial x^2}\eta uu \approx w_1^5\frac{\partial^2}{\partial x^2}\eta^nu^nu^n + w_2^5\frac{\partial^2}{\partial x^2}\eta^{n-1}u^{n-1}u^{n-1} \tag{9}$$

$$\frac{\partial}{\partial x}\tau hu \approx \frac{\partial}{\partial x}w_1^6\tau^nhu^n + \frac{\partial}{\partial x}w_2^6\tau^{n-1}hu^{n-1} \tag{10}$$

$$\frac{\partial}{\partial x}Ghu \approx \frac{\partial}{\partial x}w_1^7Ghu^n + \frac{\partial}{\partial x}w_2^7Ghu^{n-1} \tag{11}$$

$$\frac{\partial}{\partial x}\tau\eta u \approx \frac{\partial}{\partial x}w_1^8\tau^n\eta^nu^n + \frac{\partial}{\partial x}w_2^8\tau^{n-1}\eta^{n-1}u^{n-1} \tag{12}$$

$$\frac{\partial}{\partial x}G\eta u \approx \frac{\partial}{\partial x}w_1^9G\eta^nu^n + \frac{\partial}{\partial x}w_2^9G\eta^{n-1}u^{n-1} \tag{13}$$

ME:

$$\frac{\partial u}{\partial t} \approx \frac{1}{\Delta t}(u^{n+1} - u^n) \tag{14}$$

$$u\frac{\partial u}{\partial x} \approx m_1^1u^n\frac{\partial}{\partial x}u^n \tag{15}$$

$$g\frac{\partial\eta}{\partial x} \approx g\,(m_0^2\frac{\partial}{\partial x}\eta^{n+1} + m_1^2\frac{\partial}{\partial x}\eta^n) \tag{16}$$

$$\tau u \approx m_0^3 \tau u^{n+1} + m_1^3 \tau u^n \qquad (17)$$

where w_i^k and m_i^k are weighting factors, which are constrained only to guarantee consistency; k specifies the term within the equation, and i the time level. We note that three time levels are used for the GWE (because of the second derivative in time of the elevation), while only two time levels are used for the ME. We also note that the time discretization of each term of the governing equations is controlled individually, a feature consistent with the goals of $RITA_1$, but seldom found in "production" codes. Finally, we note that the terms represented by equations (10) through (13) cancel mutually when $G \equiv \tau$.

The spatial problem is handled through a weak-formulation Galerkin finite element model, currently implemented only on linear elements; however, the code is prepared for straightforward extension to quadratic elements. All integrals are evaluated with 3-point Gauss quadrature. The treatment of boundary conditions follows Luettich *et al.*[6].

ALTERNATIVE SCHEMES FOR NON-LINEAR TERMS

Alternatives to several aspects of the reference formulation described above are selectable in $RITA_1$. We will concentrate here on the treatment of non-linear terms (advection, in particular), arguably a leading challenge in circulation modeling.

Element-based definition

In the reference formulation, all non-linear terms vary over each element consistently with the linear variation of velocities and elevations over the same element; in particular, the advective term in the momentum equation, $u\partial u/\partial x$, varies linearly over each element. Luettich *et al.*[6] suggest that better-behaved (smoother and more stable) numerical solutions may be obtained if, instead, the advective term is averaged over the element, in both the momentum and generalized wave equations. This approach, which Luettich *et al.*[6] recommend only for the advective term, leads to:

GWE:

$$\frac{2}{L_e}\int_e \frac{\partial \phi_k}{\partial r} H u \frac{\partial u}{\partial r} dr \approx \frac{2}{L_e}\left\langle H u \frac{\partial u}{\partial r}\right\rangle_e \int_e \frac{\partial \phi_k}{\partial r} dr \qquad k = 1, 2 \qquad (18)$$

ME:

$$\int_e \phi_k u \frac{\partial u}{\partial r} dr \approx \left\langle u \frac{\partial u}{\partial r}\right\rangle_e \int_e \phi_k dr \qquad k = 1, 2 \qquad (19)$$

where $\langle Y \rangle_e$ represents the average of Y over the element.

Time-extrapolation

The reference formulation deals with all non-linear terms explicitly, to allow both the decoupling of the GWE and ME equations, and the time independence of the mass matrices of the associated algebraic systems of equations. However, the same objectives can be achieved by introducing a pseudo-implicit treatment of the non-linear terms. The approach is novel, has inherent potential, and should be applicable to all non-linear terms. It consists in "estimating" the elevations and velocities at time n+1 by linear extrapolation from times n and n+1, i.e.

$$\tilde{\eta}^{n+1} = 2\eta^{n} - \eta^{n-1} \tag{20}$$

$$\tilde{u}^{n+1} = 2u^{n} - u^{n-1} \tag{21}$$

and using the resulting values and their derivatives to estimate the non-linear terms at n+1; those estimates are then appropriately weighted and added to equations (7)-(10), (12), (13), (15) and (17). For instance, Eq. (15) becomes:

$$u\frac{\partial u}{\partial x} \approx m_0^1 \tilde{u}^{n+1}\frac{\partial}{\partial x}\tilde{u}^{n+1} + m_1^1 u^{n}\frac{\partial}{\partial x}u^{n} \tag{22}$$

Upwinding

Upwinding has been often used, although with mixed success, in the solution of both the transport and shallow-water equations. The concept is simple: when dealing with advection, most weight should be given to the information that is upstream (rather than downstream) of the node of interest.

As an alternative to the reference formulation of $RITA_1$, we extend to the shallow water equations the "n+2 upwinding" strategy proposed by Westerink *et al.*[7] for the transport equation. To this end, we introduce special weighting functions when (and only when) dealing with the advective term in the ME. The modified weighting functions are defined as:

$$\begin{aligned} w_i(r) &= \phi_i(r) - \alpha F_2(r) - \beta F_3(r) \\ w_j(r) &= \phi_j(r) + \alpha F_2(r) + \beta F_3(r) \end{aligned} \tag{23}$$

where i, j denote the downstream and upstream nodes of the element, respectively, and $\phi_i(r)$ and $\phi_j(r)$ are shape functions. $F_2(r)$ and $F_3(r)$ are quadratic and cubic correction functions:

$$F_2(r) = 3\phi_i(r)\phi_j(r) \tag{24}$$

$$F_3(r) = \frac{5}{2}r\phi_i(r)\phi_j(r) \tag{25}$$

Coefficients α and β lie in the interval [0,1], and define the desired degree of upwinding. If $\beta = 0$ we revert to a more common "n+1 upwinding", and if $\alpha = \beta = 0$ we revert to the centered Galerkin that constitutes the reference in $RITA_1$.

At an elemental level, the advective term in the momentum equation becomes:

$$\int_e w_k u \frac{\partial u}{\partial x} dx \approx \sum_{i=1}^{2} \sum_{j=1}^{2} u_i u_j \int_{-1}^{1} w_k \phi_i \frac{\partial \phi_j}{\partial r} dr \qquad k = 1, 2 \tag{26}$$

Eulerian-Lagrangian formulation

Eulerian-Lagrangian methods (ELM) have been effectively used to solve both the transport (Baptista[8]) and the shallow water equations (Galland *et al.*[9]). They have never been used, though, in the context of a GWE model. Our implementation of an ELM strategy in RITA1 calls for re-writing and time-discretization of the non-conservative momentum equation as:

$$\frac{Du}{Dt} \approx \frac{u^{n+1} - u^{\xi}}{\Delta t} = -\left(g\frac{\partial \eta}{\partial x} + \tau u\right)\Big|_{\xi} \tag{27}$$

where the superscript ξ denotes the foot (at time n) of a characteristic line that follows the flow. The generalized wave equation remains unaffected, relative to the standard formulation.

To solve the modified ME we use a standard Galerkin method. The location of the foot of the characteristic lines is computed with a second-order Runge-Kutta method; the procedure is iterative, as the position of the characteristic lines is a function of the dependent variable of interest - see Remédio[8] for details.

APPLICATION AND VALIDATION

The interest of $RITA_1$ as a tool for studying both physical processes and numerical methods associated with the shallow water equations is illustrated below.

Generation of shallow-water tides

The propagation of tides in shallow-waters is a problem of significant complexity and practical importance. $RITA_1$ and the associated visualization interface can be used very effectively in a classroom context, to explore and demonstrate several aspects of the problem. We will deal here very narrowly with the role of specific propagation mechanisms on the generation of harmonic and compound tides; a more comprehensive and detailed analysis is presented in Remédio[8] and supporting video sequence.

We consider a longitudinal slice of a closed-end channel, 80 km long and 10m deep. The channel is forced at the open end by a combination of M_2, S_2 and O_1 tides; forcing amplitudes are 1.5, 0.84 and 0.375 m, and all forcing phases are zero. A space-time grid with Δx=500 m and Δt=60 s was used to support several $RITA_1$ simulations, each simulation including only a specific set of physical processes. After several tidal periods of warm-up from analytically-generated initial conditions, 1024 h long, hourly records of elevation were "collected" at two stations: A at the wall, and B at mid-channel. All records were analyzed with a least-squares sinusoidal regression method, to identify amplitudes and phases of the 16 tidal constituents listed in Fig. 1.

The first two simulations are fully linear, differing from each other only on whether linearized friction is included. Frictionless conditions (Fig. 1a) lead to significant, constituent-dependent, tidal amplification at stations A and B; these amplification effects are drastically damped or reversed when linearized friction is considered (Fig. 1b). Both numerical simulations agree very closely with analytical solutions. No energy is detected at any frequency other than the forcing frequencies, an expected but important result that can be effectively used to illustrate the notion that linear propagation mechanisms cannot transfer energy from one frequency to another.

We now add a single non-linear term, either finite amplitude (Fig. 1c) or advection (Fig. 1d), keeping in both cases a linearized representation for friction. It can be easily shown (e.g., Dronkers[9]) that both non-linear terms have the potential to directly transfer energy to frequencies such as:

$$\omega_k = |\omega_i \pm \omega_j| \tag{28}$$

where ω_i, ω_j denote the forcing frequencies and ω_k the shallow-water frequencies. The numerical results are consistent with this expected behavior but provide interesting additional insight. For instance, we note that:

- The M_4 (the first harmonic of the dominant forcing frequency M_2) and the MS_4 (the first order effect of the interaction of the M_2 and S_2 frequencies) are consistently important constituents while the zero-frequency constituent captures significant energy only when advection is accounted; i.e., the finite amplitude contributes only marginally to a residual slope in the channel.

- The second and higher harmonics of the M_2 and S_2 (M_6, S_6, M_8, S_8) are essentially negligible, while the MO_3 and SO_3 (the first order effect of the interaction of the semi-diurnal and diurnal forcing frequency) are more weakly represented than the MS_4, but not negligible.

The individual effect of non-linear friction (Fig. 1e) differs significantly from that of advection or finite amplitude. In particular, we now observe a very significant energy transfer to frequencies of the type

$$\omega_k = |2\omega_i \pm \omega_j| \tag{29}$$

such as $2MS_2$, $2MS_6$, M_6 and $2SM_2$ which are dominant in a fairly full spectrum of shallow-water constituents.

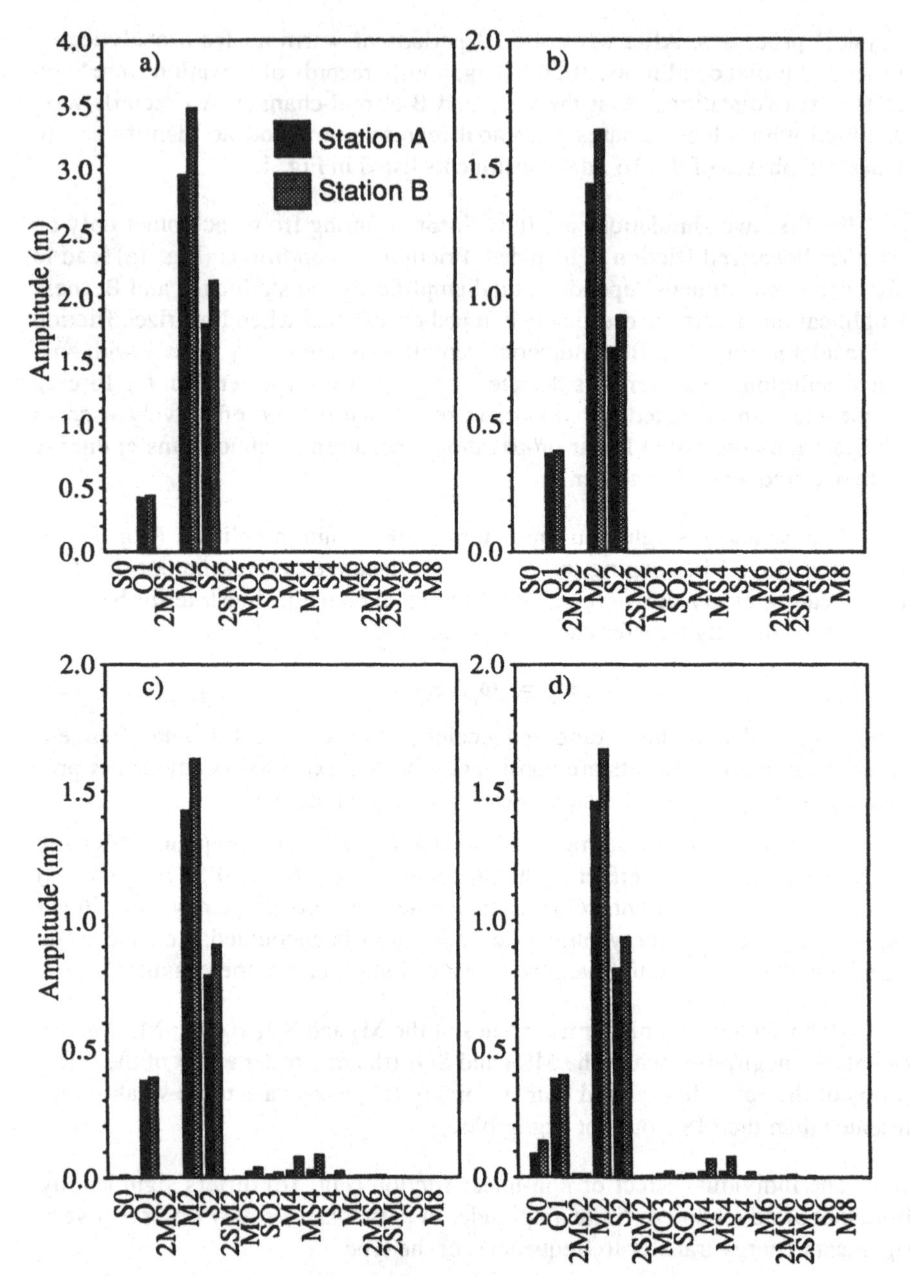

Fig. 1 - Tidal amplitudes: a) Fully linear case (frictionless); b) Linear friction (τ=0.00025s^{-1}); c) Finite amplitude and linear friction (τ=0.00025s^{-1}); d) Advection and linear friction (τ=0.00025s^{-1}).

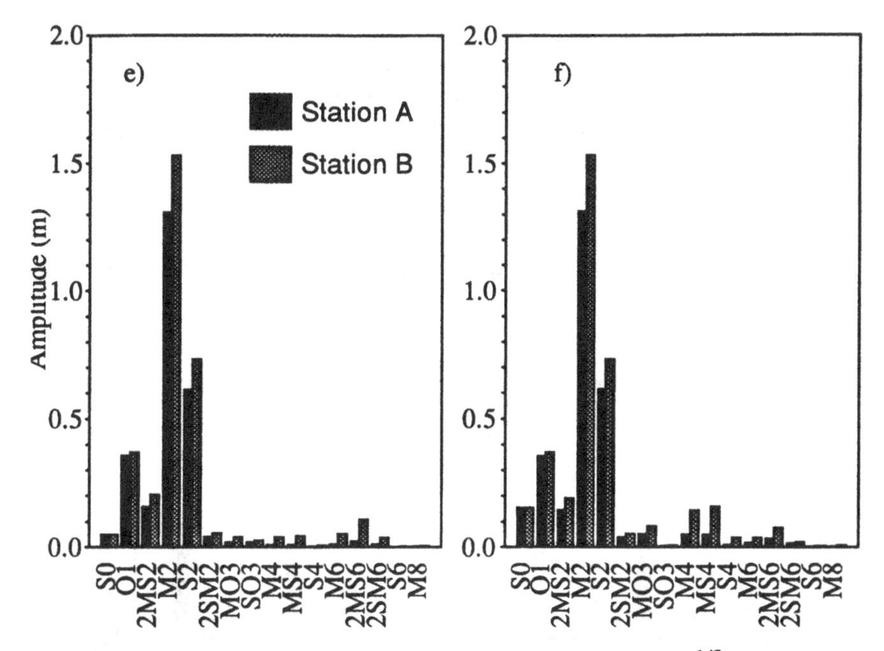

Fig.1 (cont.) - e) Non-linear friction (n=0.025 $sm^{-1/3}$); f) All terms (n=0.025 $sm^{-1/3}$).

The spectrum corresponding to a numerical simulation involving the three non-linear terms of interest (advection, friction, and finite amplitude) is shown in Fig. 1f. Dominant shallow water constituents are the $2MS_2$ (primarily due to friction), the S_0 (primarily due to advection), and the M_4 and MS_4 (contributed by all terms). We note that some shallow water constituents (e.g., S_0) decrease in amplitude when all three (rather than only the dominant) non-linear mechanisms are present, showing an interesting cancellation effect.

Numerical treatment of non-linear terms

The numerical handling of non-linear terms in the shallow water equations remains a very challenging task. This in part because of the complex behavior of some of these terms, and in part because of the scarcity of tools for systematic analysis of their numerical behavior.

We are using $RITA_1$ as a numerical benchmark for testing and comparing alternative solution techniques for non-linear terms. The approach is generally implemented in steps. We first generate a "reference," quasi-exact solution, using the standard technique described earlier together with extremely resolved grids in space and time. The extreme resolution guarantees very high dominant dimensionless wavelengths (e.g. $L_m/\Delta x>1000$) and low Courant numbers (e.g.,

Cu=0.1), hence leading potentially to a quasi-exact numerical solution. We then solve the same problem for an appropriate range of more "practical" values of Δx and Δt, using the specific numerical method(s) of interest. Results are analyzed in terms of traditional error measures (e.g., L_2-norm) and of amplitudes and phases of specific constituents.

Fig. 2 compares several of the alternative treatments of advection, in the context of a specific channel flow simulation. Further results and a consistent analysis of the alternative numeric treatments of the advective term are presented in Remédio[10].

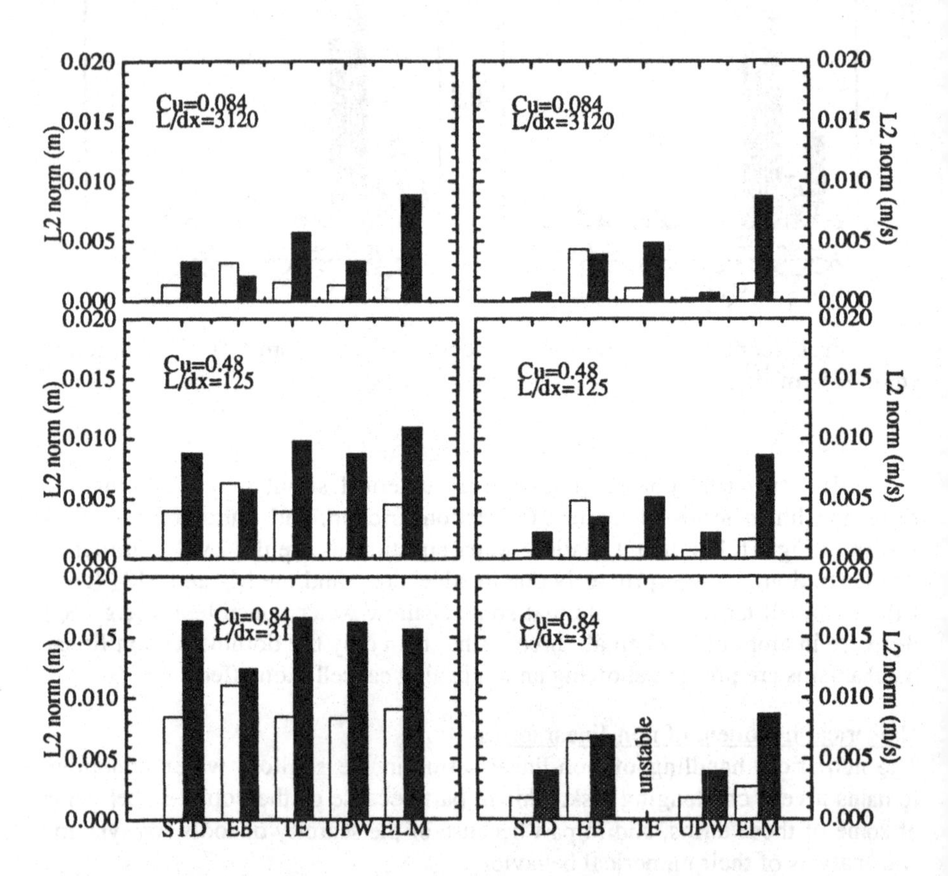

Fig.2 - Comparison of alternative treatments of advection. L_2 error norms for elevations (white bars) and velocities (dark bars) are computed against a reference over-refined simulation (Δx=100 m and Δt=1.2 s). Regular simulations were performed in grids with Δx=10000 m (a, c, e) and Δx=2500 m (b, d, f). (STD - Standard, EB - Element-based, TE - Time-extrapolation, UPW - Upwind, ELM - Eulerian-Lagrangian Method).

FINAL CONSIDERATIONS

$RITA_1$ and the encompassing computational structure ACE_1 were designed specifically to support graduate-level teaching and education, and exploratory research on shallow-waters flow, transport, and transformation. These and similarly oriented tools have a large and mostly unexplored potential for improving our ability to understand and simulate specific processes and their interaction. This should hold true both for leading-edge researchers and for application-minded researchers and engineers.

ACKNOWLEDGEMENTS

The development of ACE_1 has been partially supported by the Software Affiliates Program of the Center for Coastal and Land-Margin Research of the Oregon Graduate Institute of Science and Technology. The activity of the first author was partially supported by Junta Nacional de Investigação Cientifica e Tecnológica, Portugal (grant BM 191-90).

REFERENCES

1 - Advances in Water Resources, Special Issue on the 2nd Tidal Flow Forum, Parts I and II, Vol. 12, pp. 106-222, 1989.

2 - Baptista, A.M., J.R. Hurst, P.J. Turner and J.M. Remédio, ACE_1: An integrated tool for graduate teaching and exploratory research in environmental modeling, CCALMR, SDS#4, in preparation.

3 - Baptista, A.M., C.D. Peterson, J.M. Remédio, J.J. Westerink and P.J. Turner, "Propagation of Locally-Generated Tsunamis on the Pacific Northwest Coast", Proc. Intl. Conf. Computer Modeling os Seas and Coastal Regions, Wessex Institute of Technology, in preparation.

4 - Lynch, D. R., "Finite Element Solution of the Shallow Water Equations", Ph.D. Thesis, Department of Civil Engineering, Princeton University, 1978.

5 - Kinnmark, I.P.E. and W.G. Gray, "An Implicit Wave Equation Model for the Shallow Water Equations", Proc. 5th Intl. Conf. on Finite Element in Water Resources, Springer-Verlag, Berlin, 533-543, 1984.

6 - Luettich, R.A., J.J. Westerink and Scheffner, N.W., ADCIRC: An Advanced Three-Dimensional Circulation Model for Shelves, Coasts and Estuaries. Report 1: Theory and Methodology of ADCIRC-2DDI and ADCIRC-3DL, Coastal Engineering Research Center, Department of the Army, 1991.

7 - Westerink, J.J. and M.E. Cantekin, "Non-Diffusive N+2 Degree Upwinding Methods for the Finite Element Solution of the Time Dependent Transport Equation.", Numerical Methods for Transport and Hydrologic Processes, Computational Methods in Water Resources, 2, Celia et al [eds.], Computational Mechanics Publications, Elsevier, 1988.

8 - Baptista, A. M., "Solution of Advection-Dominated Transport by Eulerian-Lagrangian Methods Using the Backwards Method of Characteristics", Ph.D. Thesis, Massachusetts Institute of Technology, 1987.

9 - Galland, J-C., N. Goutal and J-M. Hervouet, "TELEMAC: A new numerical model for solving shallow water equations", Advances in Water Resources, Vol. 14, pp. 138-148, 1991.

10 - Remédio, J. M., "RITA$_1$ - An Educational Tool for the Analysis of Shallow Waters", MSc Thesis, Department of Environmental Sciences and Engineering, Oregon Graduate Institute of Science and Technology, in preparation.

11 - Dronkers, J.J., Tidal Computations in Rivers and Coastal Waters, North-Holland Publishing Company, Amsterdam, 1964.

Computer Modelling of Coaltar PAH Fate in a Coastal Estuary

R.J. Palczynski (**), J.H. Vandermeulen (*), P.A. Lane (**)

(*) Marine Chemistry Division, Fisheries & Oceans, Bedford Institute of Oceanography, Dartmouth, Nova Scotia B2Y 4A2, Canada

(**) P. Lane and Associates Limited, 5439 Cogswell St., Halifax, Nova Scotia B3J 1R1, Canada

ABSTRACT

The dynamic compartment modelling programs EXAMS II and WASP4 are applied to the aquatic ecosystem of Sydney Harbour, Cape Breton Island, Canada, to describe the input and cycling of PAH through the south arm of the estuary. Input data are based on oceanographic and chemical analytical results obtained over a four year period (1987-1990). Seasonal variations are included. The selected representative PAH for modelling were benzo[a]pyrene and benzo[a]anthracene because of their well-known carcinogenic properties and availability of their property data. The results deal with the fate, exposure, and persistence of the representative PAH in the modelled aquatic ecosystems. Numerical results are provided and EXAMS II and WASP4 modelling capabilities are described.

INTRODUCTION

The operation of the coking plant and associated activities at the Sydney Steel Corporation (Sysco) complex at Sydney, Nova Scotia, Canada, has caused environmental contamination of the Muggah Creek area and of portions of Sydney Harbour [1] (Figure 1). Coking probably began in 1899, together with the start-up of the steel industry, and terminated in 1988. Historically, effluent from the coking plant, together with surface runoff and groundwater from the immediate area, was discharged via Coke Oven Brook into Muggah Creek, a tidal tributary of Sydney estuary. The discharge of effluent from Coke Oven Brook resulted in a buildup of coal and coke fines, and coal tar residues in Muggah Creek, the so-called Tar Pond. From Coke Oven Brook, polynuclear aromatic hydrocarbons (PAH) flowing along Muggah Creek

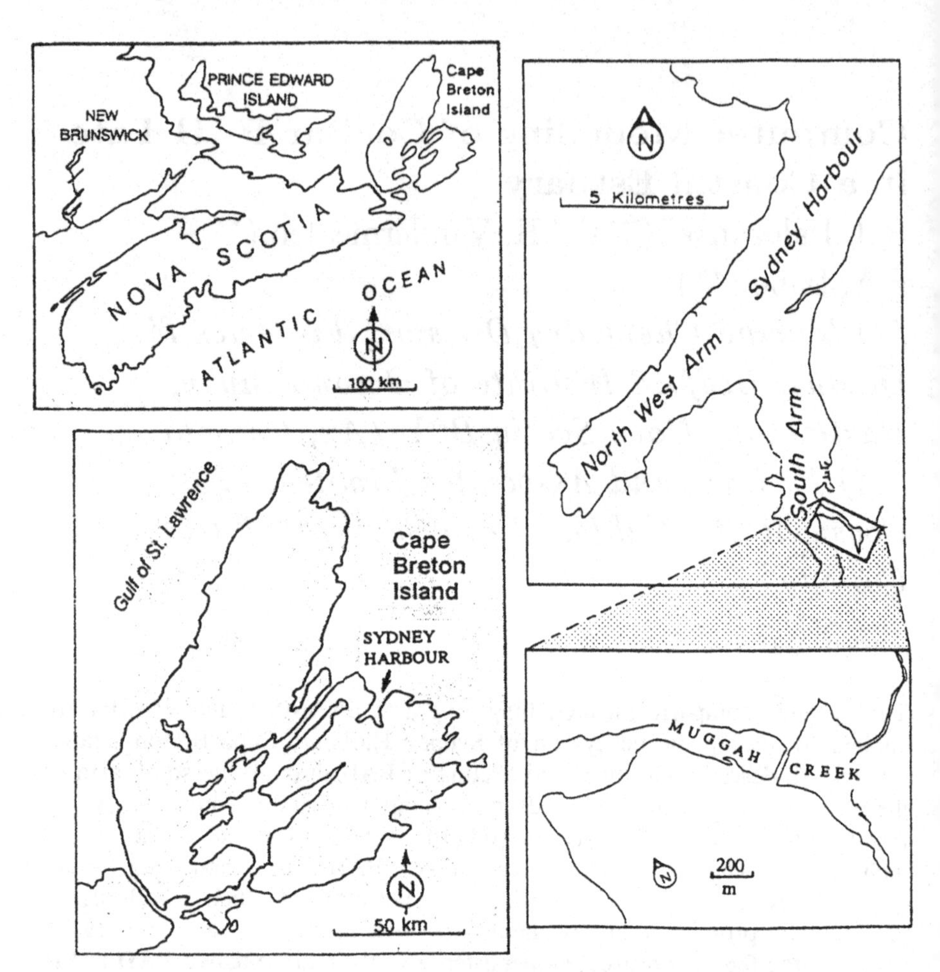

Figure 1 Location of the study area on Cape Breton Island, Nova Scotia, Canada.

together with the PAH contribution from the Tar Pond deposits were being discharged into the South Arm of Sydney Harbour. The coal tar material is contaminated with various organics including polynuclear aromatic hydrocarbons (PAH) and nitrogenous PAH (N-PAH), as well as heavy metals.

Polycyclic aromatic hydrocarbons are one of the more significant classes of pollutant chemicals which give rise to concern over their harmful effects to man and other living organisms when released into the environment as a consequence of man's activities. This concern arises primarily from the fact that even small concentrations of the PAH have been shown to be carcinogenic to a wide range of animals [2-5].

It is generally agreed that the majority of PAH in the environment are derived during incomplete combustion of organic matter at high temperatures. Several mechanisms have been proposed for the formation of PAH by pyrolysis and pyrosynthesis [6,7]. In pyrolysis, complex organic molecules are partially cracked to lower molecular weight free radicals. Pyrosynthesis of PAH then proceeds by the rapid combination of free radicals containing one, two, or many carbon atoms.

The production of coke at Sysco complex involved exposing hard coal to high temperatures in a reducing atmosphere, provided ideal conditions for pyrosynthesis of PAH. Coal tars were also produced by the high temperature treatment of coal. Thus, coal tar containing PAH was produced that could be derived either from PAH indigenous to the coal or from pyrolysis of coal hydrocarbons. The gaseous emissions associated with coal tar production might also contain significant quantities of particulate PAH, which probably contributed to contamination of the Sydney Harbour environment.

The toxic effects of PAH can be evaluated if field data and computer modelling are coupled in a framework of realistic exposure evaluation and risk analysis. In establishing the rate of release of chemicals in an aquatic system, the tools of environmental chemistry, toxicology and ecosystem modelling and simulation can be integrated to estimate exposure, fate, and persistence of the contaminants. We applied two modelling programs, EXAMS II (Exposure Analysis Modelling System, Version II) and WASP4 (Hydrodynamic and Water Quality Model, Version 4) [8], to PAH data obtained from water and sediments from Sydney Harbour.

MODELLING OF AQUATIC ECOSYSTEMS

EXAMS and WASP are public domain models available from the EPA Environmental Research Laboratory in Athens, Georgia, USA. EXAMS was used to assess the fate, exposure, and persistence in the estuary of two representative PAH, benzo[a]pyrene (BaP) and benz[a]anthracene (BaA).

Study Area
The region hydrodynamically modelled with WASP4 was the South Arm of Sydney Harbour including Muggah Creek. Less accurate geometry was established with the EXAMS model, although the area modelled was close to the actual areas of segments arbitrary chosen. The modelled site is of sufficient volume to provide dilution of PAH to below trace-level values.

The entire Sydney estuary (harbour, north and south arms) is primarily saline and tidal. Freshwater input is primarily from the Sydney River which enters at the head of the South Arm. Muggah Creek receives some small freshwater input from lesser tributaries (e.g. Coke Oven Brook), and is tidal (including the Tar Pond).

Overview of Available Data
A concern regarding contamination of Sydney Harbour lobsters by hydrocarbons and heavy metals, resulted in a number of studies designed to reveal a variety of physical, chemical, toxicological and ecological parameters of the ecosystem components. As a result, data sets were available on the PAH and metal pollution of Sydney Harbour [1] and these data were used as input to a modelling program to assess long-term impact of coal-tar on the aquatic environment of the Sydney estuary.

Additional physical data were collected during oceanographic cruises conducted in Sydney Harbour during 1987/1988. Measurements taken during these cruises included temperature, salinity, and dissolved oxygen profiles, current measurements, results of drogue studies, analyses of water column and sediment samples for PAH, as well as vertical plankton tows, chlorophyll measurements, and benthic samples for sorting and speciation. In other cruises measurements included temperature and dissolved oxygen profiles, salinity, pH, trace metals, solids concentration, carbonates, PAH, vertical plankton tows, chlorophyll measurements, bacterial abundances, and benthic samples for sorting and speciation. Sediments were analyzed for PAH, total organic carbon and trace metals.

Modelling Procedure
EXAMS and WASP both have a steady-state option which requires that residual currents be specified. In their treatments of toxic concentrations in water and sediments, WASP and EXAMS are conceptually similar, but EXAMS is simpler in its hydrodynamic compartment. Exposure and fate of PAH which involves kinetics of advective and dispersive transport, volatilization, photolysis, hydrolysis, oxidation, and microbial transformations, however, all are well incorporated in this model. A helpful feature of the EXAMS model is the existence of a 'library' of environmental datasets, which includes the archived chemical data for selected PAH. WASP includes more options involving the bed volume (constant or variable), and

the characteristics of the suspended sediments.

The modelling option initially chosen was to provide an overview of the hydrodynamics of Sydney Harbour and the exposure and fate of PAH as simply as possible. The spatial extent of the model was the South Arm, and the time scale was large and steady-state. With such a spatial extent, constraints on segment length arise from considerations of advection and dispersion. Steady-state implies that the source strength is constant, the currents are steady, and the distribution of PAH is equilibrated. Tidally averaged values were employed for dimensions of segments. In successive refinements more field data were incorporated and utilized, the space scale of resolution altered, and the time scale reduced to one year (with EXAMS) or one tidal cycle (with WASP4).

The sorption of PAH onto the suspended sediment was modelled with the assumption that the organic carbon of the suspended sediment behaves as if it were octanol. The sorbed PAH was taken as the product of the concentration of suspended particulates, of the fraction of this concentration comprised by organic carbon, and of the octanol:water partition coefficient for the PAH.

The WASP4 model was used to simulate the time series, in upper and lower layers over a tidal cycle, of a water quality parameter (for example, salinity). Therefore, the model had to be dynamic (as opposed to steady-state EXAMS) to account for the variability over the tidal cycle, and be two-dimensional (resolved axially and vertically) rather than one-dimensional to account for differences between upper and lower layers. These simulated time series can be compared with, fitted to, and substituted for observed time series in the hydrodynamics calculation.

WASP4 was also the model chosen for simulating Muggah Creek fluxes. WASP4 consists of hydrodynamic and water quality components, DYNHYD4 and TOXI4, respectively. WASP4 is time-variable and success has been achieved in operating it two-dimensionally even though DYNHYD is a one-dimensional model. DYNHYD has been used for simulating tidal flows which were barotropic (that is, uniform over the water column). In order to reflect the stratified, two-layer opposing flow nature of the mouth of Muggah Creek estuary, estuarine residual flows and observed stream flows were superimposed upon the tidal flows from DYNHYD4 in TOXI4.

The task of providing appropriate input information was organized into modules on geometry, hydrologic and environmental data, advection, dispersion, suspended sediment distribution and properties, choice of chemical, chemistry of the pollutant and its loading rate.

Configuration and Bathymetric Inputs The EXAMS model contained three sequential columns of water. Each column was divided into three segments: the top one was termed upper layer type, the middle was lower layer and the bottom one, which models sediments, was termed benthic layer. The geometry is shown in Figure 2.

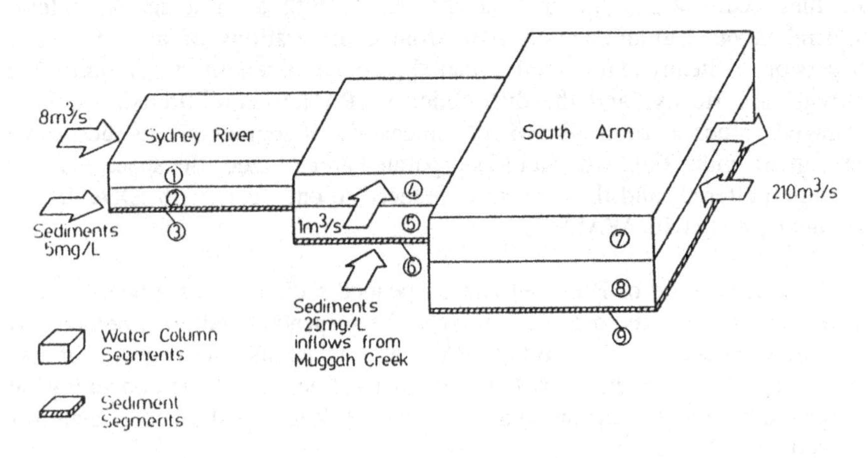

Figure 2 Schematic of EXAMS II model geometry.

The configuration of currents being modelled follows from the results of the oceanographic surveys and moored current meters. The advective stream in the lower layer flows landward while being entrained upward. The upper layer flows seaward with an average speed of 1 to 2 cm/s.

The tides in the Sydney Harbour are mainly semi-diurnal, both lunar and solar, with a maximum 1.4 m tidal range. There are seiches of amplitude as large as 0.6 m, excited by storms [9]. The fundamental period is 120-130 minutes. The harbour is subject to freezing in winter, and in spring is frequently choked with ice which may persist until late April.

Effluxes were sought at the mouth of Muggah Creek estuary, in the interior of the model. In terms of geographical configuration, DYNHYD4 was set up with eight segments, each extending through the whole water column and varying in volume over the tidal cycle, to model tidal currents in the South Arm of Sydney Harbour and Muggah Creek. For TOXI4 the eight hydrodynamic segments were subdivided into upper and lower layer water quality segments. In addition, the hydrodynamic segment at the mouth of Muggah Creek estuary was subdivided into outer (segments 11 & 12) and inner (segments 13 & 14) units. Finally, bed segments were specified to underlie the water column segments within the estuary of Muggah Creek for a total of 22 segments (Figure 3).

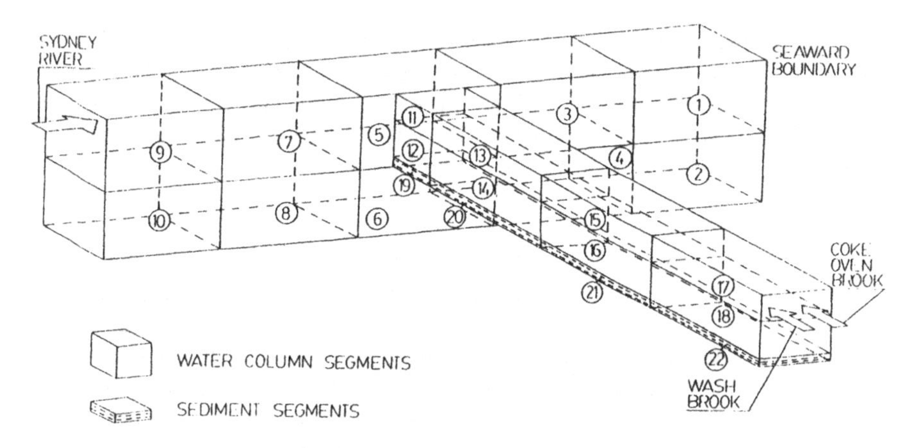

Figure 3 Schematic of DYNHYD4 model geometry.

In Muggah Creek, data collected from a bathymetric survey were used for the modelling effort. For the South Arm, bathymetry was taken from hydrographic charts. From the Muggah Creek bathymetry, a table was calculated relating water surface area and water volume to water level. This table was used to transform water elevation differences to tidal flows.

Flow Inputs In order to run the hydrodynamic DYNHYD4 model, flow information was required for tidal flows, estuarine residual flow and stream flows. Daily average stream flow, prorated from MacAskill's Brook, was used for Sydney River and Muggah Creek runoff. Wash Brook and Coke Oven Brook were grouped together as one freshwater discharge into segment 17.

Tidal flow information for TOXI4 was derived from observations at the mouth of Muggah Creek, either of currents (preferred) or water levels. In effect, the tidal flows for the whole South Arm, the model domain, were keyed to observations at the mouth of Muggah Creek. The tidal flows were prorated to represent tidal flows at the seaward boundary of the modelled region for modelling purposes. Thus, TOXI4 was forced using the flows (measured/modelled) at the mouth of Muggah Creek. The role of DYNHYD in this application was to produce the route coefficients for TOXI4 which relate tidal flows at all segment boundaries of the model to those at the mouth of Muggah Creek. Specifically, DYNHYD4 was used to compute the flows resulting from tidal forcing only.

The magnitudes of estuarine flow and stream flow were measured to be 0.51 m^3/s and 0.34 m^3/s, respectively.

EXAMS flow inputs are represented geometrically in Figure 2. An important 'stream' input is given for segment 8, lower layer. This 'stream' of seawater flowing into the harbour, initiates, in the lower layer, the two-layer circulation at a rate of approximately 1 cm/s as observed from the moored current meter records, and as implied from the estuary classification schemes.

Source Strength Inputs Matheson et al. [10] have identified the Sydney Steel Corporation as the only significant source of PAH into Sydney Harbour (via Muggah Creek). Acres International Limited [11] has estimated source strength for PAH from Muggah Creek into Sydney Harbour, at a time when the coke ovens were not operating, to be 145$\pm$ 80 kg/yr. P. Lane and Associates Limited [12] measured the flux of suspended particulates from Muggah Creek over one tidal cycle under conditions of extreme runoff and found 50 tonnes of suspended particulates being exported via the runoff mode, the (two-layer) shear effect, and the tidal pumping mode, in order of decreasing importance. Although PAH were not measured in the latter, the underlying sediments in the entrance to Muggah Creek are known to exhibit PAH concentrations of as high as 2800 ppm [10]. Therefore, if an average PAH concentration of 1000 to 2000 ppm in the Muggah Creek sediments is considered to be a valid estimate and if the equivalent of ten to thirty tide cycles with extreme runoff occur annually, then the source strength into the harbour can be reasonably estimated at 500 to 3000 kg/yr.

Initially it appeared that the only source of PAH considered active was the influx via contaminated streamflow from Coke Oven Brook into the Tar Ponds. Later it became apparent from chemical analyses (GC/MS profiles) that PAH components were quite different in Coke Oven Brook from those in Muggah Creek and that profiles varied with conditions of, for example, runoff. More processes and "sources" had to be postulated so leaching was included in TOXI4 as diffusive exchange of interstitial water. It was more convenient to include scour/resuspension of contaminated Tar Pond sediment in the model as a source rather than a vertical flux of bed sediments; either choice anticipated the resuspension of particles which might be associated with the excavation of the Tar Ponds.

Stream source strengths for PAH were set to correspond to observed or simulated influx values for a station at Coke Oven Brook. The diffusion coefficient for leaching from Tar Pond sediments was set such that about 2 mg/s PAH entered the water column. The source strength of scour/resuspension of Tar Pond sediments was difficult to establish but varied with runoff and was adjusted for each date simulated. Later, when excavation of the Tar Pond begins as part of planned cleanup work, that activity can be represented as a strengthened scour/resuspension process.

Choice of Representative PAH In EXAMS modelling, the chosen PAH were benzo[a]pyrene (BaP)and benz[a]anthracene (BaA). The former was chosen as the representative PAH because of its well-known carcinogenic properties. These properties have made it one of the few PAH for which a reasonable number of the chemical and physical properties required by EXAMS are actually known. Furthermore, its concentration in any water body suspected of being polluted by PAH has generally been determined [6]. Its chemical nature is sufficiently like that of most PAH so that it can serve as a reasonable indicator of the fate of the whole suite of these pollutants. BaA was chosen because of its well known photodegradation properties and because sufficient data exists on its occurrence in the Sydney Harbour environment.

Constants The EXAMS model compares dispersion with advection by the Peclet number, Pe = uL/K, where u is the speed, L is the scale length of the segment and K is the dispersion coefficient. For Sydney Harbour, K has been measured to be in the range of 3 to 30 m^2/s [10]. Segments have lengths of approximately 3000 m. The value for Pe falls in the range 1 to 20, meaning that advection tends to predominate in the transport of dissolved conservative substances [13]. For numerical stability, it was chosen to set numerical dispersion equal to measured dispersion.

Vertical bulk dispersion coefficients are estimated to range from 1 cm^2/s in a stratified condition to 10 cm^2/s in an unstratified condition. Interstitial sediment pore water diffusion coefficients are estimated as 2 x 10^{-5} cm^2/s [13]. Other constants, such as solubility and octanol-water partition coefficient values for selected PAH, were taken from the literature [14] or used as listed in the user database of the EXAMS model.

Calibration, Sensitivity, and Uncertainties TOXI4 was calibrated for vertical and horizontal dispersion coefficients by reference to salinity data, for settling speed by reference to suspended particulate matter (SPM) observations, and for dissolved/particulate partitioning of PAH by reference to PAH dissolved and particulate (fluorescence) observations. Numerical horizontal dispersion appeared adequate to match salinity observations; the bulk dispersion coefficient was therefore set to a low value, 1.0 m^2/s. Muggah Creek showed stratification under most conditions; therefore the vertical dispersion coefficient was set to a low value, 1.0^{-6} m^2 /s; however, the model's individual compartments and layers were each completely mixed. One could consider increasing vertical dispersion during periods when stratification was small; this was not done at this stage. Choice of settling speed affects concentrations of solids and associated organic chemicals at the mouth, and populations of particles and their settling speeds may vary. In this modelling, settling speed of 0.3 m/tide cycle was used for all runs.

RESULTS

Hydrodynamics

The hydrodynamic processes modelled for the South Arm were fluxes due to the residual flow of water, due to tidal pumping, and due vertical shear processes. Theory and limited observations suggest that in the South Arm of Sydney Harbour shear currents are common and that the salinity distribution varies between low and high water. The residual current estimates imply Hansen- Rattray classification class 3, with 50% to 99% of the up-estuary salt flux being attributed to gravitational circulation producing a shear effect. The variation in salinity distribution between low water and high water suggests that tidal pumping accounts for the remainder of the up-estuary salt-flux, its magnitude depending upon tide, wind, and runoff conditions. The advective pathway consistently emerges as dominant in the computer runs. The pathway from Muggah Creek leads both upstream in the lower layer, and downstream in the upper layer.

Suspended sediment transport, which is the sum of residual sediments, tidal pumping, and vertical shear processes, was also modelled. It was found in the partially-mixed Tamar Estuary [15], that tidal pumping can be directed either up-estuary or down-estuary in association with a turbidity maximum. In that case, tidal pumping was dominant for transport of both salt and suspended sediment. If it should also prove to be the case in Sydney Harbour that the dominant longitudinal dispersive mechanisms are the same for salt and suspended sediments then one would anticipate that shear effects dominate intermittently with tidal pumping also operative, albeit at a reduced level. From the measured distribution of PAH in sediments there is the suggestion that landward bottom currents have indeed transported suspended particulates upstream.

EXAMS and WASP4 both have the capability to predict purging time scale and space scales of PAH contamination, given the required inputs. There is, however, a distinction to be made in that TOXI4 operates with one-dimensional hydrodynamics (DYNHYD4) while EXAMS accepts two-dimensional hydrodynamics. Residual gravitational circulation, enhanced or attenuated by wind stress, was lumped together with tidal pumping and expressed as the product of a diffusion coefficient and a longitudinal gradient.

EXAMS accepts fluxes due to tidal pumping and fluxes due to vertical shear processes inputs as separate diffusive and advective components according to the Hansen-Rattray estuary classification schemes. The advective components were estimated from the Prandle theory and checked from current meter results.

Fate

The fate of PAH as modelled in the base case is that the majority of the load (99%) is exported, water-borne (suspended sediments including), from Muggah Creek into the South Arm. Of the remainder, roughly equal portions are accumulated in the creek bottom sediments and some are broken down by photolysis. The prediction of total PAH accumulated in segments of the water column and sediments of the modelled environment is possible from EXAMS-computed distributions of the two representative PAH, BaP and BaA. The model confirmed that total (water and sediments combined) steady-state accumulation of each PAH depends on mass influx and on mechanisms of degradation. Resident mass in water and sediments, total steady-state accumulation, and steady-state concentrations in the water column and in sediment segments are shown for BaP in Table 1 and for BaA in Table 2. Computer runs were completed for loads of each chemical from 50 to 300 kg/year with 50 kg intervals. It can be seen that the ratio of the resident mass in the water column to that in the benthic sediments is independent of the load values. PAH half-lives also are independent of loadings.

Values of volatile solids, or the organic fraction of the sediments, were found to be highest in the South Arm. This was consistent with the particle size data which showed that these sediments had higher percentages of silt than sediments from other locations within the Harbour. Areas with lower organic fractions had higher percentages of sand in the sediment samples collected.

Sediment-water concentration data for the Muggah Creek area over the last ten years suggests that sorption of PAH is a first-order kinetic process with an equilibrium partition coefficient K_p which increases linearly with the organic content of the sorbed material in the body of water. The partition coefficient also correlates with the octanol/water partition coefficient (K_{ow}) of the PAH.

Exposure

One of the most important characteristics of PAH relative to their incidence in water is solubility. Solubilities of PAH in water are generally quite low, reflecting their nonpolar hydrophobic nature. Thus, as might be expected, solubility tends to decrease as the number of aromatic rings or molecular weight increases. Benzo[a]pyrene (M.Wt. = 252) has a solubility of about 4.0 ppb and benz[a]anthracene (M.Wt. = 228) about 10 ppb. There are other factors affecting solubility. The limited data available indicate that PAH are slightly less soluble in seawater than in freshwater due to salting out. These differences are not large however, and temperature has a much greater effect on the aqueous solubility of PAH. It has also been shown that, in some cases, the presence of one or two PAH in solution affected the solubility of an additional PAH in that solution [16].

Table 1 Summary of benzo(a)pyrene loadings and mass balance in the South Arm of Sydney Harbour as predicted by the EXAMS II model.

Load of B(a)P kg/y	Resident Mass kg		Steady-State Concentration					
	In Water Column	In Bottom Sediments	Mean In Water & Sediment Column ng/L	Mean Dissolved in Water ng/L	In Bottom Sediments, mg/kg			
					Segment No.:			Mean
					3	6	9	
50	1.6	885.7	14.11	0.45	1.56	1.88	0.63	1.36
100	3.2	1771.8	28.21	0.90	3.11	3.76	1.26	2.71
150	4.9	2655.1	42.30	1.35	4.66	5.64	1.89	4.07
200	6.5	3541.5	56.40	1.79	6.22	7.52	2.52	5.42
250	8.1	4426.9	70.51	2.25	7.78	9.40	3.15	6.78
300	9.7	5312.3	84.64	2.69	9.33	11.3	3.78	8.13

ng/L = nanogram per litre (1 ng = 10^{-9} g; 1 ng/L = 10^{-3} ppb);
kg/y = kilogram per year;
mg/kg = milligram per kilogram;
g/h = gram per hour.

Table 2 Summary of benz(a)anthracene loadings and mass balance in the South Arm of Sydney Harbour as predicted by the EXAMS II model.

Load of B(a)A kg/y	Resident Mass kg		Steady-State Concentration					
	In Water Column	In Bottom Sedi-ments	Mean In Water & Sediment Column ng/L	Mean Dis-solved in Water ng/L	In Bottom Sediments, mg/kg			
					Segment No.:			Mean
					3	6	9	
50	1.7	952.8	14.64	0.018	1.66	2.01	0.68	1.45
100	3.4	1905.6	29.28	0.036	3.32	4.02	1.37	2.90
150	5.1	2856.9	44.12	0.055	4.97	6.03	2.06	4.35
200	6.8	3809.2	58.40	0.073	6.63	8.04	2.74	5.85
250	8.5	4762.5	72.87	0.091	8.29	10.0	3.43	7.26
300	10.0	5715	87.80	0.110	9.95	12.1	4.12	8.77

The series of tabular and graphical results for BaP and BaA, generated by EXAMS, suggests that equilibrium concentrations of PAH in the South Arm water are very low (2×10^{-5} to 6×10^{-5} mg/L) while those in sediments are approximately 2 to 6 mg/kg, representing an approximate total PAH value of 40 to 120 ppm (average). These sediment concentrations are comparable to and consisted with, both in magnitude and spatial pattern, observed concentration levels in the South Arm (excluding Muggah Creek) as measured in grab samples analyzed by Matheson et al. [10].

Persistence

Persistence of a chemical in the environment may be estimated on the principle of conservation of mass which can be stated as follows

[Rate of change of mass in control volume] =
[Rate of change of mass in control volume due to advection] +
[Rate of change of mass in control volume due to diffusion] -
[Transformation reaction rate (Degradation)]

Persistence, which is here taken as the rate of change of mass of PAH in an element of the environment, is subject to several transformation processes. Persistence may also be further modified by such physical processes as burial of contaminated sediments which may result in longer half-life times; this option was not considered by EXAMS nor WASP models. Several variables may be influencing each process, leading to multi-term and often non-linear lumped transformation rate.

The modelling results indicate that PAH in the South Arm may persist for years rather than months. The computed half-life for BaP is approximately 13 years. Kinetics of transformation rate, which is a good indicator of persistency, are shown in Figure 4. BaA is much less persistent, with an average half-life of approximately 23 months. Kinetics of transformation for this chemical are shown in Figure 5. Both figures are shown as produced by the model.

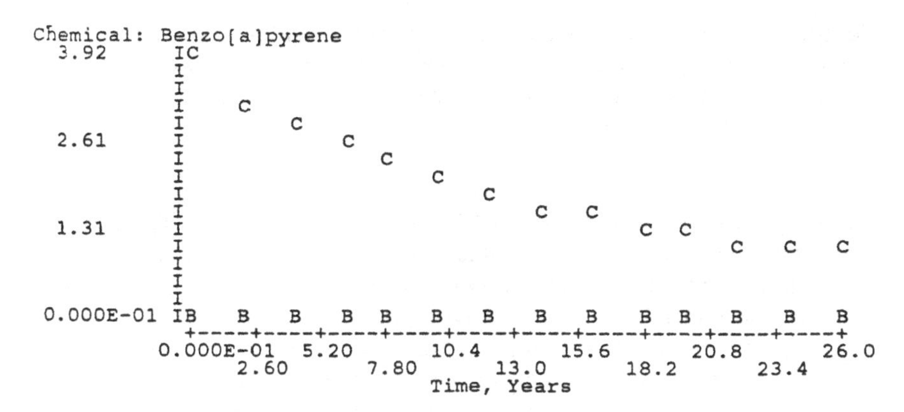

Figure 4 Average concentration (mg/kg) of sorbed BaP in water (B) and benthic sediments (C) versus time.

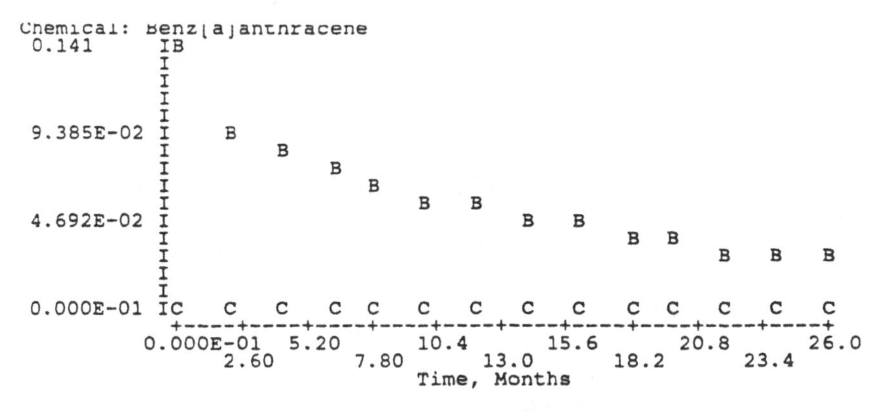

Figure 5 Total concentration (mg/kg) of sorbed BaA in water (C) and benthic sediments (B) versus time.

DISCUSSION

EXAMS appears to be adequate for an assessment of the exposure and fate of PAH in Sydney Harbour, taking into account certain qualifications. The first is that the chemical parameters presently available for benzo[a]pyrene are for freshwater conditions. Unfortunately, almost all of these parameters are unknown for PAH in seawater. In an attempt to compensate for this lack of information, the values for solubility and the octanol:water partition coefficient were adjusted to more accurately reflect seawater conditions.

Sensitivity to these adjustments was low, however, and potential errors arising from using inappropriate coefficient values for PAH in seawater must be considered. Further, the use of BaP and BaA as surrogates for total PAH provides another source of uncertainty.

To complete the initial model, calibration and validation were carried out and a tuned-up model was used for final computation of exposure, fate, and persistence of benzo[a]pyrene and benz[a]anthracene. The results suggest that both modelled chemicals are accumulated in sediments while only small fractions (less than 1%) are dissolved in water. This may be due to small input of chemicals from Muggah Creek into the large body of Sydney Harbour water, the low solubility of BaP and BaA in water and the salting out effect of the sea water. Steady-state concentration of BaP and BaA in sediments are modelled to be in the range of parts per million (ppm) and are highest in sediments of segment no. 3. This seems reasonable as this segment receives PAH directly from Muggah Creek. On the other hand, sediments of segment no. 9 are subjected to PAH-free tidal water, and therefore accumulated PAH would be expected to become washed out, giving lowest concentrations of BaP and BaA in these outer segments of the South Arm of Sydney Harbour.

Table 3 shows results of concentration values in sediments of modelled segments as compared with published field data for BaP and BaA. It can be seen that results obtained with EXAMS model and measured concentrations are similar in spite of the fact that field results vary greatly from place to place (even within the same sector), which would indicate a large concentration spectrum of PAH.

Table 3 Concentrations of BaP and BaA obtained with EXAMS model and from field sampling.

PAH	Seg-ment No.	EXAMS (Range) mg/kg	Reference [10]		
			Range, mg/kg	Mean, mg/kg	% of PAH
BaP	3	1.56 - 9.33	0.27 - 23	7.35	11.2
	6	1.88 - 11.3	0.30 - 28	13.1	10.8
	9	0.63 - 3.78	0.40 - 8.3	3.3	7.3
	Mean	1.36 - 8.13	---	8.77	10.6
BaA	3	1.66 - 9.95	0.25 - 20	5.0	8.0
	6	2.01 - 12.1	0.24 - 21	7.5	6.2
	9	0.69 - 4.12	0.30 - 11	3.8	8.4
	Mean	1.45 - 8.77	---	5.7	6.9

Earlier, we estimated annual source strength input from Muggah Creek into the harbour's South Arm between 500 to 3000 kg (viz. Source Strength Inputs above). More recent estimates by P. Lane and Associates of PAH efflux from Muggah Creek is of the order of 770 ± 360 kg/year, which is on the low end of this range. The model results in fact are more consistent with the scenario that the average rate of release of PAH has been 3000 kg/year rather than releases of 500 kg/year. For example, a comparison of concentration ranges and means for the different segments, as estimated from the model, with those reported by Matheson et al. [10] suggests that the higher range values are more representative. Also, a later review of PAH data [1] lists higher PAH concentrations in the South Arm, argueing that input values may ever be higher.

In this study we restricted our modelling to the South Arm only because our best data set, including trans-segment flows were only available for the South Arm. For the other outer parts of the estuary and its North Arm, PAH concentrations in the water were below trace amounts and could not be adequately treated in the EXAMS model. However, the bottom sediments in the other parts of the estuary do contain substantial amounts of PAH and in a subsequent study we will have to attempt to use these data. Another area that requires further attention is the water surface micro-layer which, because of its surfactant character, may contain uniquely higher PAH concentrations than other parts of the water column.

These results do suggest that the EXAMS model can be used with PAH data in an estuarine context. For more detailed modelling of the dispersion of PAH within a tidal cycle, or the interaction of PAH with suspended solids, the hydrodynamic and water quality model WASP4 was more appropriate than the EXAMS II model. Advection of water through the modelled system was satisfactorily computed in DYNHYD4 (within the WASP4 model) from user-defined data entry for junctions and channels. The EXAMS II is indicated as a convenient modelling tool used to assess the fate, exposure, and persistence of synthetic organic chemicals in aquatic ecosystems in which the chemical loadings can be time-averaged and chemical residuals are at trace levels.

REFERENCES

1. Vandermeulen, J.H. PAH and heavy metal pollution of the Sydney estuary: summary and review of studies to 1987. Canadian Technical Report of Hydrography and Ocean Sciences, No. 10, 48 pp, 1989.

2. Korotkova, G.P., and Tokin, B.P. Simulation of the process of somatic embryogenesis in some porifera and coelentearata. I. Effect of carcinogenic agents on some porifera. *Acta Biol. Hungary* 19, 465-474, 1968.

3. Wynder, E.L. Nutrition and cancer. *Fed. Proc.* 35, 1309-1315, 1976.

4. Sanders, C.L. Toxicological Aspects of Energy Production. Battelle Press, Columbus, Ohio,1986.

5. Uthe, J.F. Polycyclic aromatic hydrocarbons in the environment. *Can. Chem. News* 43(7), 25-27, 1991.

6. Neff, J.M. Polycyclic Aromatic Hydrocarbons in the Aquatic Environment-Sources, Fates and Biological Effects. Allied Science Publishers, London, 1979.

7. Suess, M.J. The Environmental Load and Cycle of Polycycling Aromatic Hydrocarbons. *Sci. Total Environ.* 6, 239-250, 1976.

8. Burns, L.A., and Cline, D.M. Exposure Analysis Modelling System: Reference Manual for EXAMS II. EPA/600/3-85/038, U.S. Environmental Protection Agency, Athens, Georgia. 83 pp., 1985.

9. Easton, A.K. Seiches of Sydney Harbour, N.S. *Can. J. Earth Sci.* 9(7), 857-862, 1972.

10. Matheson, R.A.F., Trider, G.L., Ernst, W.R., Hamilton, K.G. and Hennigar, P.A. Investigation of Polynuclear Aromatic Hydrocarbon Contamination of Sydney Harbour, Nova Scotia. Surveillance Rept. EPS-5-AR-83-6. Environment Canada 86 pp., 1983.

11. Sydney Tar Pond Study, Final Report. Report prepared for the Nova Scotia Department of the Environment. Acres International Limited, 1985.

12. Flux Measurement of Muggah Creek, Sydney Harbour. Report prepared for Acres International Limited. P. Lane and Associates Limited, 1988.

13. Processes, Coefficients, and Models for Simulating Toxic Organics and Heavy Metals in Surface Waters. U.S. Environmental Protection Agency, Environmental Research Laboratory, Athens, GA., EPA/600/3-87/015, 303 pp., 1987.

14. Schnoor, J.L., Sato, C., McKechnie, D. and Sahoo, D. Processes, Coefficients and Models for Stimulating Toxic Organics and Heavy Metals in Surface Waters. EPA/600/3-87/015 Environmental Research Laboratory, U.S. E.P.A., Athens, GA, 1987.

15. Uncles, R.J., Elliott, R.C.A., Weston, S.A., Pilgrim, D.A., Ackryod, D.R., and M.M. Lynn. Synoptic observations of salinity, suspended sediment and vertical current structure in a partly-mixed estuary. *In*: Van de Kreeke, J. (ed.), *Physics of Shallow Estuaries and Bays*. Springer-Verlag, pp. 58-70, 1986.

16. Eganhouse, R.P., and Calder, J.A. The solubility of medium molecular weight aromatic hydrocarbons and the effects of hydrocarbon co-solutes and salinity. *Geochim. Cosmochim. Acta* 40, 555-561, 1976.

SECTION 4: SILTATION AND SEDIMENTATION

Prediction of Seabed Sand Waves

B.A. O'Connor

Civil Engineering Department, The University of Liverpool, Brownlow Street, Liverpool, L69 3BX, U.K.

ABSTRACT

The present paper describes the details of a simple engineering computer model for the prediction of the seabed features to be expected at a particular site using only limited field information.

The model is based upon an extension and simplification of earlier work of O'Connor and Duckett, which now includes the effects of wave action. The model output is compared with field observations at three coastal sites, the German Bight, Skerries Bank and Middelkerke Bank. The model, in combination with statistical information from the sites, is shown to give useful predictions for engineering use.

INTRODUCTION

The development of modern surveying equipment, including side-scan sonar, has revealed the presence of a variety of bed features (bedforms) both within estuaries and in coastal seas. Such features vary in height from a few centimetres (ripples) to many metres (mega-ripples or tidal dunes), and may exceed tens of metres (sandwaves) for the largest features, which often have wavelengths in excess of a hundred metres. The larger bedforms may also be associated with the flanks of coastal sandbanks, which themselves are distinct features with heights of some tens of metres, widths of some 500-1000m and lengths of some tens of kilometres. All these large-scale features are a considerable navigation hazard and can cause problems with the installation and maintenance of seabed pipelines.

There have been many previous studies of bedforms, (see for example, Yalin [1], Allen [2], Langhorne [3], and O'Connor and Duckett [4]. A useful summary has recently been provided by van Rijn [5]. Despite much research effort, (see Langhorne [3]), it has proved difficult in the past to establish precise relationships between local sediment transport rates and bedform size and movement. It is generally recognized that bedforms are asymmetric in shape in the direction of dominant sediment movement and that ripples rapidly adjust their height and wavelength to changing flow conditions. Mega-ripples are found to react more slowly, changing size and shape within a tidal cycle, but showing much less change from one cycle to the next. By contrast, sandwaves move extremely slowly, perhaps at rates of a hundred metres per year in the direction of net sediment transport, while sandbanks show almost no movement at all, although recent research as part of the EC MAST programme, de Moor and Lanckneus [7], shows some small movements during severe storms, see Fig. 4.

Some attempts have been made in the past to predict bedforms, for example Straub and Bijker [5] and van Rijn [6]. O'Connor and Duckett [4] have shown that the sizes of ripples and mega-ripples can be related to local flow conditions and sediment properties. The present paper extends this lattermost work to show that the prediction of sandwave dimensions is also related to local conditions and that waves also play a moderating role in limiting bedform heights. Confirmation of the present ideas is provided by a study of conditions at a number of field sites. Unfortunately, insufficient data exists to fully test the present ideas, although future work under the EC MAST programme at Middelkerke Bank, Belgium, should help to develop and refine them.

The O'Connor and Duckett [4] (OCD) model does not attempt to describe bed form properties from a detailed knowledge of sediment transport processes in complex flows but seeks to link together flow and bed form properties in a comparatively simple manner using evidence from laboratory and field studies. Details of the model have been described elsewhere. However, an outline is presented herein to assist in describing its extension to sandwaves. The basic concepts of the model are summarized below.

(a) The total drag force produced by the bed forms is assumed to be in local equilibrium with the fluid-produced drag, represented by the total fluid shear stress (τ).

(b) The drag force from bedforms is composed of a component from the sediment grains (τ'') and a form drag component from the bed forms themselves, (represented by a form-induced shear stress, τ', whereby $\tau = \tau' + \tau''$).

(c) Since ripples are rather small in size compared with mega-ripples, the ripple form drag is assumed to produce an equivalent grain drag and its effect is included in τ''. A similar approach is used for wave-induced bed ripples.

(d) The steepness (height/wavelength) of both ripples and mega-ripples are assumed to be directly related to the fluid-induced shear stress (τ) and the critical shear stress for sediment movement (τ_c) as shown by Yalin [1]. In addition, the maximum steepness of ripples is related to sediment size and water temperature. Mega-ripple steepness is not allowed to exceed 6.4%.

(e) The wavelength of ripples is assumed to be fixed by the median sediment grain-size (D_{50}). The wavelength of mega-ripples is controlled by the tidal-mean water depth (h_*), the maximum depth-mean tidal velocity (u_m); the time available to move sediment during the flood/ebb phase of the tidal cycle; the sediment grain size (D_{50}) and its relative density(s) and the water temperature (θ°C). In addition, it is assumed that the maximum attainable mean (50%) wavelength (L_o) is depth-limited ($5-7.3h_*$) and that the mega-ripple wavelength is unchanged during a tidal cycle, as also is its height: the flood or ebb tide merely re-arranges bedform shape.

(f) Both ripples and mega-ripples can co-exist for certain flow conditions, but at higher flow rates, only mega-ripples exist.

(g) During a Spring/Neap cycle of tides it is assumed that the maximum Spring tide produces a dominant mega-ripple wavelength, which persists for the full cycle. In addition, it is assumed that the height of mega-ripple that exists on any local tide within the Spring/Neap cycle is the larger value of the height produced by the local tide or by earlier tides: all mega-ripple heights are allowed to decay in size with time.

Testing of the model for fine-grained sediment situations (D_{50} = 144μm and 230μm) at Sizewell/Dunwich, East Anglia (U.K.) and in the Ribble Estuary (U.K.) showed that only ripples existed at Sizewell while both ripples and mega-ripples existed in the Ribble. However, at maximum flows in the Ribble, only mega-ripples existed.

MODEL MODIFICATION

The existing model has demonstrated the importance of sediment grain size. However, Langhorne [3] has shown the importance of waves. A simple modification to the present model can be done by allowing for the interaction effects of currents and waves and, in particular, the increased drag (τ_{wc}) produced by combined wave and current action. Many studies have been made of such interaction effects, see for example, O'Connor and Yoo [8]; van Rijn [6] and Soulsby [9]. In the lattermost case, Soulsby gives a simple expression for τ_{wc}, based on field data at the Isle of Wight, that is,

$$\tau_{wc} = \tau(1 + 1.5v_o/u) \tag{1}$$

where v_o is the nearbed orbital wave velocity, and u is the depth-mean current velocity.

A more general expression can be obtained by considering the combined shear stress in waves and currents. Bijker [10] produced a simple expression of the form:

$$\tau_{wc} = \tau + \tau_w/2; \quad \tau_w = \rho f_w v_o^2/2; \quad \tau = \rho f u^2/8 \tag{2a}$$

or

$$f_{wc} = f + 2f_w(v_o/u)^2 \tag{2b}$$

where f, f_w, and f_{wc} are the current, wave and combined wave/current friction factors, respectively, O'Connor and Yoo [8].

Alternative, more complex, expressions can be used in place of Eqs. (1) and (2). However, since Eq. (2) is likely to over-estimate effects, (see O'Connor and Yoo [8]), and, therefore, provides an upper limit of the effect of waves, only Eqs. (1) and (2) are considered herein.

The effect of wave action is included, therefore, in the model by replacing the total friction factor by its wave/current value (f_{wc}).

The influence of the various model parameters on the size of mega-ripples is demonstrated in Table 1. Conditions for the Ribble are given for comparison purposes. Only coarse sediment is seen to produce fully-developed (σ = 100%) wavelengths, as in rivers.

It is also clear from Table 1 that grain size and water depth have a major effect on the size of the mega-ripples. In fact, it would appear that in deep water, the mega-ripples have the dimensions of sand waves, suggesting that sand waves are also related to local conditions. However, because of their size they can only change shape very slowly. In fact, it is suggested that the rate of change is too slow to accommodate the variability in tidal flow produced by the Spring/Neap cycle. Thus, sandwave size is controlled by conditions

Table 1. Model bed form dimensions

D_{50} (μm)	h_* (m)	$\bar{\Delta}$ (m)	$\bar{L}$ (m)	$\bar{\sigma}$ (%)	SWS (%)	$\bar{\Delta}/h_*$ (%)
230*	6.85	0.46	7.0	20	6.6	6.7
230	6.85	0.30	7.2	21	4.2	4.0
230+	6.85	0.42	7.1	20	5.8	6.3
150	8.3	0.07	1.5	3.7	4.6	0.9
500	8.3	1.5	41	98	3.7	8.3
600	8.3	1.63	42	100	3.9	18
600	30	5.5	134	98	4.1	18
600**	27.5	3.9	125	91	3.1	15

u_m = 1.5m/s; θ = 10°C; θ^+ = 17°C; *Ribble values
**2m waves/7.2s period, equations 2b; SWS = $\bar{\Delta}/\bar{L}$*100

existing on the largest Spring tides; any slow net movement being controlled by long-term tidal residual currents.

In order to test the ideas further, the model is applied to three field situations, as explained below.

MODEL APPLICATION

(i) German Bight (Lister Deep)

Pasenau and Ulrich [11] have collected together various navigational echo sounder runs, in the German Bight to show that many hundreds of square kilometres of the inshore areas are affected by large-scale bed forms. They also indicate some relationship between grain size and depth and ripple size. For example, they found ripples with heights < 2m in areas with medium to fine sands but 2-10m in areas with coarse to medium sand and depths of 6-20m.

Pasenau and Ulrich also made a detailed study of the Lister Deep tidal channel between the Islands of Romo and Sylt. In this region they found large bed forms with heights up to 11m and wavelengths of more than 300m overlain by mega-ripples with heights of 2m. Water depths varied in the range 10-40m and tidal currents had maximum values of the order of 1m/s. A typical echo-sounding trace is shown in Fig. 1 for water depths of some 25-30m. Sediment sizes were found to vary between 200-1000μm. Crest bed samples had D_{50} values of some 850μm and 570μm while trough bed samples had values of 640μm and 290μm.

Operation of the model for the four observed grain sizes and a range of depths shows very similar results to the observed values, see Table 2. The different scales of the fine and coarse sediment bedforms is very clear, suggesting that mega-ripples are associated with the fine sediment moving over the top of the underlying coarser material, which is responsible for the main large-scale bedforms (sandwaves).

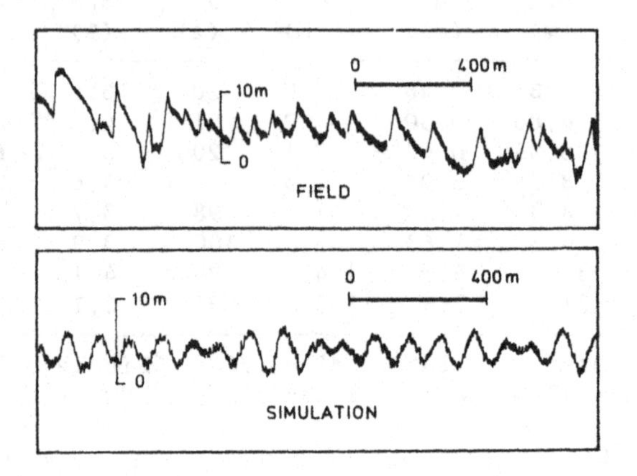

Fig. 1 Sandwaves, Lister Deep

Table 2. Predicted mean bed form sizes, Lister Deep

h_* (m)	D_{50} (μm)	$\bar{L}$ (m)	$\bar{\Delta}$ (m)	σ (%)	SWS (%)	$\bar{\Delta}/h_*$ (%)
20	290	11.5	0.73	11.5	6.3	3.6
	570	81.2	4.39	81.2	5.4	22.0
	640	92.7	5.78	92.7	5.5	25.7
	850	100	5.78	100	5.8	28.9
26.5	290	11.7	0.74	8.3	6.3	2.8
	550	91.8	5.14	69.3	5.6	19.4
	640	114	6.44	86.2	5.6	24.3
	850	132	7.65	99.7	5.8	28.9
40	290	11.9	0.75	5.9	6.3	1.9
	570	113	6.68	56.7	5.9	16.7
	640	146	8.57	73.0	5.9	21.4
	850	196	11.4	98.0	5.8	28.5

u_m = 1m/s; θ = 10°c; L_o = $5h_*$

If the study area is now imagined to be covered by deposits with varying D_{50} sizes, then different sized bedforms will be generated at each location. Once one bedform is generated, it will quickly trigger off a series of other bed forms as with water waves. All such bedforms will then interact so that the final observed spatial distribution of bedforms will have varying heights and wavelengths. Analysis of the limited data presented by Pasenau and Ulrich [11] shows the variable nature of the bedforms, see Table 3, while Tables 4 and 5 show that both the height and wavelength

Table 3. Sandwave characteristics, Lister Deep

Δ (m)	$\Delta/\bar{\Delta}$	L (m)	$L/\bar{L}$	SWS (%)	RS	Δ (m)	$\Delta/\bar{\Delta}$ (m)	L (m)	$L/\bar{L}$	SWS (%)	RS
6.8	1.34	111	1.10	6.0	1.20	4.7	0.95	111	1.1	4.2	0.84
7.4	1.46	65	0.64	11.4	2.28	4.1	0.81	75	0.73	5.5	1.10
9.1	1.6	75	0.73	10.9	2.18	4.1	0.81	148	1.46	2.8	0.56
2.0	0.40	65	0.64	3.1	0.62	6.1	1.20	139	1.37	4.4	0.88
3.4	0.67	130	1.28	2.6	0.52	5.1	1.0	167	1.65	3.1	1.61
4.1	0.81	65	0.64	6.3	1.26	5.1	1.0	176	1.74	2.9	0.58
4.1	0.81	75	0.73	5.5	1.10	6.1	1.2	56	0.55	10.9	2.18
3.4	0.67	93	0.92	3.1	0.62	2.7	0.53	83	0.82	3.3	0.66
						2.8	1.74	83	0.82	10.6	2.12

$\bar{\Delta}$ = 5.07; Δ_s = 7.22m; $\bar{L}$ = 101m; L_s = 145m;
SWS (Local sandwave steepness); RS = SW/($\bar{\Delta}/\bar{L}$)

Table 4. Sandwave height variability, Lister Deep

Range, Δ (m)	Values (m)	No. (n)	No. (>) (Obs.)	No. (>) (Ray.)	n (Ray.)
0-1	0	0	17	17	1
1-2	0	0	17	16.4	2
2-3	2, 2.7	2	17	14.6	3
3-4	3.4, 3.4	2	15	12.0	3
4-5	4.1, 4.1, 4.7 4.1, 4.1	5	10	9.2	3
5-6	5.1, 5.1	2	8	6.5	2
6-7	6.1, 6.1, 6.8	3	5	4.3	2
7-8	7.4	1	4	2.6	1
8-9	8.1, 8.8	2	2	1.5	2

Δ_s = 7.22m; $\bar{\Delta}$ = 5.07m; N = 17

Table 5. Sandwave length variability, Lister Deep

Range, Δ (m)	Values (m)	No. (n)	No. (>) (Obs.)	No. (>) (Ray.)	n (Ray.)
0-25	0	0	17	17	1
25-50	0	0	17	16	2
50-75	56, 65, 65, 65	4	17	13.4	3
75-100	75, 75, 75, 83, 83, 93	6	13	10	3
100-125	111, 111	2	7	6.6	3
125-150	130, 139, 148	3	5	3.9	3
150-175	167	1	2	2	1
175-200	176	1	1	0.9	1

L_s = 145m; $\bar{L}$ = 101m; N = 17

more or less follow a Rayleigh-type statistical distribution, so that

$$n = N.\exp\left[-2(x/x_s)^2 \right] \tag{3}$$

where x is the height or wavelength; x_s represents a "significant" value, that is, the average value of the third highest values in the record; N is the total number of waves in the record; and n is the number of waves of larger size than x.

Equation (3) also allows the prediction of the maximum size in any record, that is:-

$$x(max) = x_s \surd\{\ell n(N)/2\} \tag{4}$$

For N = 17, Eq. (4) indicates values of L(max) = 173m and Δ(max) = 8.6 compared with observed values of 176m and 8.8m, see Tables 4 and 5.

The mean sand wave dimensions can be obtained from the use of the bedform model by using a representative grain size and water depth. For example, using the observed mean-depth of 26.5m and an assumed grain-size of 550μm gives values of mean height ($\bar{\Delta}$) and wavelength ($\bar{L}$) of 5.14m and 92m, respectively, and compares very well with observations of 5.1m and 101m, see Table 2.

Clearly, the Rayleigh distribution allows other sandwave sizes to be determined and provides a good prediction of maximum values, once the significant values are known. However, Table 2 shows that observed mean and significant ratios are somewhat lower than indicated by the Rayleigh distribution ($\Delta_s/\bar{\Delta} = L_s/\bar{L} = 1.43$, cf. 1.70 from Rayleigh).

Unfortunately, the Rayleigh distribution predicts minimum Δ, L values of zero. Consequently, the observed values must be used for guidance. Table 2 suggests ($\Delta(min)/\bar{\Delta} = 0.40$; $L(min)/\bar{L} = 0.55$).

The local steepness of individual waves also shows a statistical variation, see Table 3. To determine the local wavelength of the maximum wave requires a further correlation. Table 3 suggests that the wavelength of the highest waves (82-100%) may be constant. ($\Delta(max)/L_m = 0.11$; $[\Delta(max)/L_m]/[\bar{\Delta}/\bar{L}] = 2.19$, see Table 3). Use of these field expressions with the mean model values predicts L_m of 78m and 70m respectively, compared with observed values of 83m.

Ideally, it would be useful to have information on the occurrence frequency of individual sandwaves. Table 2 suggests that the larger waves may occur in groups, although it appears that 'groupiness" is not so pronounced as with wind-produced water waves. Simulation of groupiness can be done, if details are available of the spectral density distribution of surface elevations as a function of wave-number, in a similar manner to that done for water waves, see for example, O'Connor and Ellis [12]. In the absence of such data, it is possible to illustrate such ideas by combining together just four wave numbers corresponding to the grain sizes observed in the field. The simulation of bed surface elevations is thus given by the equation:-

$$\eta = \sum_{i=1}^{4} a_i \cos(k_i x + \delta_i) \qquad (5)$$

where a_i is a representative amplitude of the ith wave component; k_i is the wave number ($= 2\pi/L_i$) and δ_i is a relative phase shift between components.

For water waves, the a_i values are held constant and related to Δ_s while the k_i values are determined from the spectral-density function. δ_i are usually taken as randomly occurring over the range $0\text{-}2\pi$. For present purposes, it is proposed to phase lock the mega-ripples (produced by the finest grains, $D_{50} = 290\mu m$), with the largest sandwave frequency ($D_{50} = 850\mu m$; $\delta_1 = \delta_4 = 0$), and to allow a random choice of wave number for the intermediate frequencies ($D_{50} = 550\mu m$, $\delta_2 = 4$, $D_{50} = 640\mu m$, $\delta_3 = 1$). Amplitudes have also been varied to emphasize the mean sand wave height.

Figure 1 shows the result of the crude simulation. The groupy nature of the record is clearly seen with quite good reproduction of wavelengths. Improvements in this technique can clearly be made once better field data is available. At present, it appears that the use of the bedform model to predict average conditions in combination with the tidal/Rayleigh statistical information allows a good representation of likely site conditions.

(ii) Start Bay, Skerries Bank

The German Bight data contained no information on the effect of waves. Langhorne [3] has presented data from an extensive, long-term study of bed level changes for an individual sand wave crest located on the north-west flank of the Skerries Bank off Devon (U.K.). The Bank was found to have a series of large, south-facing sandwaves (4m high, 260m wavelength) and mega-ripples (1m high, 8-10m wavelength), in water depths of some 10-15m, see Fig. 2. Tidal flow velocities were measured during the study period at 1m above the bed at crestline A, Fig. 2, as also was the wave climate near the bank (at 2.3km). Limited bed grain size information was reported. The D_{50} size at crest A was found to be 320μm.

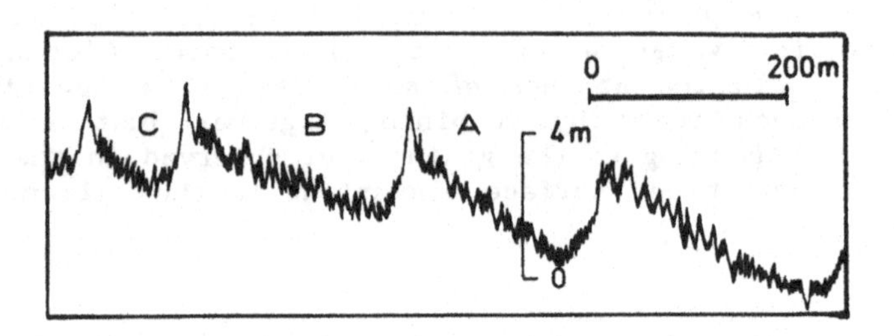

Fig. 2 Mega-ripples and sandwaves, Skerries Bank

The study showed that sandwave asymmetry was controlled by the residual tidal flow and, hence, sediment transport: ebb flow dominated on Spring tides and flood flow on Neap tides (< 3.3m range). Wind-induced flows were found to enhance or reduce the residual movement, while surface waves enhanced bed transport rates. However, detailed surveying of crestline movement using a series of vertical, closely-spaced (500mm) steel rods running normal to the crestline, showed rather small-scale movement (some 4m in length, 2% wavelength, and some 500mm in elevation, 14% sandwave height), with sediment movement being confined to a surface layer, some 500mm thick.

These field observations confirm the present view that the maximum Spring tides are responsible for the main persistent size of the sandwaves and that small changes occur on a tidal basis controlled by tidal flow asymmetry enhanced by wind-induced flows and wave-action. Figure 2 also shows that the tidal mega-ripples are phase-locked to the sandwave wavelength, with maximum elevations coinciding at the sandwave crest. The effect of wave action on the height of the mega-ripples at the crestline can be tested by the use of the bedform model. In the absence of surface waves, the model provides information on the mean height and wavelength. For sandwave B, Fig. 2, the model predicts values of $\bar{\Delta}$ = 0.75m, $\bar{L}$ = 11.8m using h_* = 12m; u_m = 0.91m/s as suggested by the field data and an assumed D_{50} = 305μm, since average values along the sandwave are likely to be less than crest values. The predictions are seen from Table 6 to agree well with the field values for height (within 7%) and wavelength (10%).

Using the German Bight and Rayleigh statistics also enables predication of the minimum, significant and maximum values (Δ(min) = 0.30 (compared with 0.6)m, L(min) =

Table 6. Mega-ripple characteristics, Skerries Bank (B)

Δ (m)	$\Delta/\bar{\Delta}$	L (m)	$L/\bar{L}$	SWS (%)	RS	Δ (m)	$\Delta/\bar{\Delta}$	L (m)	$L/\bar{L}$	SWS (%)	RS
1.4	1.73	16.8	1.59	8.3	1.10	0.70	0.86	8.4	0.79	8.3	1.10
1.0	1.24	14	1.32	7.1	0.94	0.60	0.74	11.2	1.05	5.4	0.70
0.9	1.11	17.3	1.62	5.2	0.68	0.80	0.99	11.2	1.05	7.1	0.94
1.0	1.24	11.8	1.11	8.5	1.11	0.75	0.93	5.6	0.53	13.4	1.76
0.8	0.99	11.2	1.05	7.1	0.94	0.75	0.93	11.2	1.05	6.7	0.88
1.2	1.48	11.2	1.05	10.7	1.41	0.70	0.86	8.4	0.79	8.3	1.10
0.6	0.74	5	0.47	12.0	1.58	0.70	0.86	6.2	0.58	11.3	1.48
0.75	0.93	8.4	0.79	8.9	1.17	0.75	0.93	8.4	0.79	8.9	1.17
0.80	0.99	9.0	0.85	8.9	1.17	0.70	0.86	8.4	0.79	8.3	1.10
0.50	0.62	9.0	0.85	5.6	0.73	0.75	0.93	8.4	0.79	8.9	1.17
0.70	0.86	5.6	0.53	12.5	1.64	1.0	1.24	11.2	1.05	8.9	1.17

$\bar{\Delta}$ = 0.81m; $\bar{L}$ = 10.7m; Δ_s = 1.07m; L_s = 13.4m; SWS = Δ/L^*100; RS = SWS/($\bar{\Delta}/\bar{L}$)

6.5 (5.0)m; Δ_s = 1.07 (1.07)m, L_s = 16.9 (13.4)m; Δ(max) = 1.33 (1.4)m, L(max) = 21(17.3)m; as well as the wavelength of the highest waves, L_m = 12.1(16.8)m. It is clear that very good predictions are obtained for wave height, while wavelengths are somewhat over-predicted, apart from the wavelengths of the longest waves, which are under-predicted. Table 6 suggests that these smaller mega-ripples are associated with a smaller degree of randomness than sandwaves, since the larger elevations tend to be more associated with the larger wavelengths ($\Delta(max)/L_m$ = 0.073).

In order to illustrate the effect of waves, data on mean tide crestline elevations for the period 5-11th September 1979 have been used. Tidal conditions changed slowly over this period. Consequently, the major effects can be ascribed to wave action. Table 7 shows model values, and observed crest levels (hc) as

Table 7. Effect of wave action, Skerries Bank (A)

Date	Tide Range (m)	H_s (m)	T (s)	hc (mm)	$\bar{\Delta}^1$ (mm)	MHC^1 (mm)	MHC^2 (mm)
(9/79)							
5th	4.3	0.5	4.8	735	682	779	777
6th	4.9	0.6	5.5	718	614	711	733
7th	5.2	0.6	6	710	574	671	712
8th	5.5	0.5	6	680	614	711	731
9th	5.3	0.5	5.3	720	654	751	756
10th	5.0	0.45	5	755	680	777	775
11th	4.3	0.3	4.5	820	716	813	808
(9/77)							
19/20th	3.3	1.3	5.0	505	447	544	675
(11/77)							
6th	2.0	1.25	5.5	315	330	427	636
		1.25*	6.0*	315	214	311	596

[1]Eq. (2b); [2]Eq. (2a); *assumed value

well as the average wave-condition on, and immediately prior, to the day's observations. The model crest height values are assumed to be given by:-

$$MHC(mm) = hc + \bar{\Delta} \quad (6)$$

where hc is a datum height, taken as 97mm, from crest level observations on 7th September 1977.

It appears that the model provides a realistic description of the reduction in mega-ripple height due to wave action with Eq. (2b) being better than Eq. (1). The field data also shows that the mega-ripples recover quite quickly from the effects of wave action.

Table 7 also indicates the effect of larger waves for conditions on Neap tides in 1977. It appears that Eq. (2b) again provides a better estimate of observed conditions.

It appears that the wave-modified model is capable of providing quite accurate predictions of mega-ripple dimensions. Unfortunately, it appears less good at predicting the wavelengths of the sandwaves at the Skerries site. No data was recorded by Langhorne on the size of the coarse material making up the Bank. If the same spatial-mean grain size is used as at the Lister Deep site (D_{50} = 570μm), the model shows $\bar{\Delta}$ = 3.23m and $\bar{L}$ = 56.3m. Allowing for mega-ripple crests and sandwave crests to coincide, as suggested by the field data, indicates a total crest bedform height of some 4m (3.23 + 0.75), which compares very well with the field values of 3.5-4.4m, see Fig. 2. However, the wavelength is under-predicted since sandwaves A, B have wavelengths of 194m and 230m, respectively, although sandwave C is much shorter at 103m. Allowing for statistical variations, the maximum expected wavelength over a 1km stretch would be 97m, using Eq. (4) with N = 18. It appears, on the basis of the very limited data set available, that the observed wavelengths have a similar value on the crest of the Bank but are twice as long at deeper depths. Further information is clearly required on grain sizes and other transects in order to explain these wavelength discrepancies.

(iii) <u>Belgian Coastline, Middelkerke Bank</u>

In order to explore the wave length question further, the model was also applied to conditions on Middelkerke Bank which has similar general characteristics to the Skerries Bank, and where preliminary data was available from the EC's MAST research programme RESECUSED, de Moor and Lanckneus [7]. Side scan sonar traces show an extensive pattern of sandwaves on the flanks and crest of the Bank, Figs. 3 and 4. It also has a similar-sized superficial sediment size (D_{50} = 290-300μm), similar maximum tidal velocities (u_m ≈ 1m/s) and somewhat similar tidal-mean depths (10-20m). Maximum bedform heights vary between 0.45-3.85m. Bedform wavelengths along the section-line shown in Fig. 3 are detailed in Table 8, along with associated bedform heights.

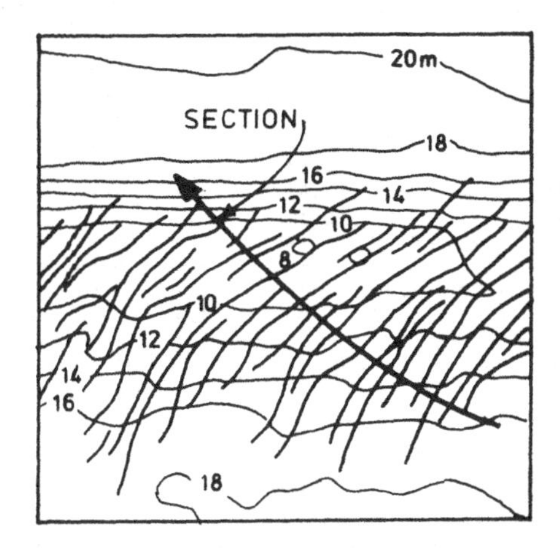

Fig. 3 Bedforms, Middelkerke Bank

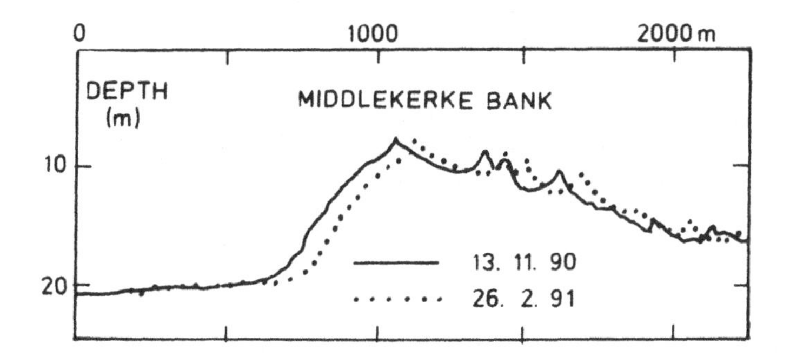

Fig. 4 Movement of the Middelkerke Bank

Using the bedform model with D_{50} = 295μm for mega-ripples and assuming D_{50} = 570μm for the sandwaves, gives mean bedform dimensions of $\bar{\Delta}$ = 0.71m, $\bar{L}$ = 11.9m; and $\bar{\Delta}$ = 3.23m; $\bar{L}$ = 61.8m, respectively. Again Table 8 suggests that the model gives realistic estimates of bedform height but under-estimates wavelengths. However, unlike the Skerries Bank example, the model can readily be adjusted to provide more reasonable answers for wavelength. All previous model examples have used a fully-developed (equivalent to uni-directional river conditions), maximum wavelength of $L_o = 5h_*$. However, other authors, see van Rijn [6], have used a value of $7.3h_*$ for

Table 8. Bedform wavelengths, Middelkerke Bank

h_* (m)	Δ (m)	$\Delta/\bar{\Delta}$	L (m)	$L/\bar{L}$	SWS (%)	RS	h_*	Δ (m)	$\Delta/\bar{\Delta}$ (m)	L	$L/\bar{L}$ (m)	SWS	RS (%)
10	3.5	2.03	112	1.13	2.90	1.79	14	1.6	1.0	87	0.88	1.84	1.14
	1.0	0.63	71	0.72	1.41	0.87		1.47	0.92	168	1.70	0.88	0.54
	2.0	1.25	76	0.77	2.63	1.62		1.9	1.19	97	0.98	1.96	1.21
10	0.85	0.53	132	1.34	0.64	0.39		2.1	1.31	97	0.98	2.16	1.33
	2.5	0.64	102	1.03	2.45	1.51	16	1.5	0.94	86	0.87	1.74	1.07
	1.12	0.70	102	1.03	1.1	0.68		1.05	0.66	112	1.13	0.94	0.58
12	1.42	0.89	117	1.19	1.21	0.75		1.7	1.06	87	0.88	1.95	1.20
	0.70	0.44	51	0.52	1.37	0.85	17	1.4	0.88	82	0.83	1.71	1.05

$\bar{\Delta} = 1.6m$; $\bar{L} = 98.7m$; $SWS = \Delta/L*100$; $RS = SWS/(\bar{\Delta}/\bar{L})$; $\Delta_s = 2.31m$; $L_s = 126.6m$

rivers. In the present case this would improve the sandwave predictions to $\bar{L} = 90.2m$, which is within 9% of observed values. Allowing for statistical variations (N = 16), this would produce significant and maximum wavelengths of some 130m and 152m, respectively, which is within 2-9% of observed values, see Table 8. However, use of these same arguments still does not explain the extra-long wavelengths on the Skerries Bank, where predicted maximum wavelengths are still too low (142m compared with 194-230m).

Examination of the statistics of the Middelkerke bedforms shows similar results to the other sites, see Table 9, although it appears to have lower relative bedform steepness values (F) and local steepness values (E) than the other sites. Further data is clearly needed to resolve these details.

Table 9. Bedform statistics at various sites

Site	Type	A	B	C	D	E	F
Lister	SW	1.43	1.43	0.40	0.55	0.11	2.19
Skerries	MR	1.32	1.25	0.62	0.47	0.073	1.58
Middelkerke	SW/MR?	1.44	1.28	0.44	0.52	0.029	1.23

$A = \Delta_s/\bar{\Delta}$; $B = L_s/\bar{L}$; $C = \Delta(min)/\bar{\Delta}$; $D = L(min)/\bar{L}$;
$E = \Delta(max)/L_m$; $F = E/(\bar{\Delta}/\bar{L})$

SEDIMENT TRANSPORT

Observations by Langhorne [3] on the Skerries Bank sandwave (A) show that only the superficial sediment is moved by the local tidal currents. Langhorne's [3] field observations indicated that most sediment moved as bed load and he found a good correlation with the following Bagnold-type equation, which can readily be included in the OCD model:-

$$q_b = 1000(u_*'' - u_{*c})^3 \ \text{kg/m/s} \qquad (7)$$

where u_{*c} is the critical shear velocity (m/s) for sediment movement and is given adequately by Shields' relationship (1); u_*'' is the local bed shear velocity ($u_*'' = \surd(\tau''/\rho)$).

Equation (7) can be recast in terms of the local depth-mean velocity ($\bar{u}$) by means of the equation

$$u_*'' = \bar{u}\surd(f''/8) \qquad (8)$$

where f" should include the effect of any local current or wave-induced ripples.

f" may also be related to the local effective Nikuradse roughness height (k_s) of the bed. Thus:-

$$f'' = 0.242/[\log(0.364h/z_o)]^2 \qquad (9a)$$

$$z_o = k_s/33 \qquad (9b)$$

Using Langhorne's [3] suggested z_o and u_{*c} values (5.5 x 10^{-3}m and 1.69 x 10^{-2}m/s) for the Skerries Bank indicates a maximum local bed transport rate of some 0.04kg/m/s for u_m = 0.86m/s and h = 12.7m. The OCD model gives a value of 0.036 kg/m/s.

Equation (7) indicates that tidal average transport rates are dominated by the transport from the maximum tidal currents, when conditions are more like those in large rivers. Consequently, it is likely that van Rijn's [6] equations will also apply. Thus:-

$$q_b = 8\ Q_+\ u_+^{2.4}\ D_+^{1.2} \ \text{kg/m/s} \qquad (10)$$

where $Q_+ = \bar{u}h$, the local flow discharge rate (m^3/s); $u_+ = (\bar{u} - u_c)/\surd[(s - 1)gD_{50}]$; $D_+ = D_{50}/h$; u_c is the local depth-mean critical velocity for sediment to start moving along the bed. Van Rijn [6] suggests:-

$$u_c = 0.19(D_{50})^{0.1} \ \log(12h/(3D_{90})) \qquad (11)$$

where D_{90} is the 90% finer fraction of the bed sediment.

Equations (10) and (11) indicate a maximum transport rate of some 0.024kg/m/s, assuming $D_{90} = 2D_{50}$, and compares favourably with Langhorne's results, bearing in mind that Eq. (7) would be expected to over-predict bed load since its constants have been determined from changes in crestline volume.

Suspended load (q_s) can also be estimated by van Rijn's equations [6]. Thus:-

$$q_s = q_b [2.4 . D_+^{-0.20} . D_*^{-0.6}] \text{ kg/m/s} \tag{12a}$$

where

$$D_* = D_{50} [(s - 1)g/\nu^2]^{0.333} \tag{12b}$$

s is the relative density of the sediment grains; and ν is the kinematic viscosity of the water.

For the maximum flow conditions at the Skerries Bank, Eq. (12) indicates a value of 0.156 kg/m/s, which suggests that the suspended load may be more important in providing nett movements of sandwaves and, indeed, of nearshore banks, see Fig. 4, rather than the smaller mega-ripples whose movement is dominated by the bed load. Such nett movements are, of course, controlled by the dominance of the flood and ebb flows and requires that equations (10) and (12) be integrated over a tidal cycle in order to obtain them.

CONCLUSIONS

Application of the O'Connor and Duckett (OCD) engineering bed form model at three coastal sites for maximum spring tide conditions indicates that local bedforms are in equilibrium with applied drag forces. The model suggests that the difference between mega-ripples and sandwaves is only a question of scale, which is controlled largely by water depth and bed sediment grain size. Field data from each site shows that both mega-ripples and sandwaves have statistical distributions about mean values which conform quite well to a Rayleigh distribution, as suggested by laboratory data [13]. Use of the OCD model, together with statistical data from the field enables the mean and maximum bedform characteristics to be determined and should be useful in studying the effects of local dredging works, etc. on navigable depths. Some under-prediction of the sandwave wavelengths was found at the Skerries site and further work is required to clarify this discrepancy. Net movement of bedforms is controlled by the tidally-averaged transport of sediment by both bedload and suspended load, which are related to local conditions. Implementation of the general drag relationships from the OCD model into a large-area depth-average tidal current model would enable local depth-mean velocities to be estimated and hence sediment transport rates using appropriate equations.

Modification of the OCD model to include the extra drag effects from wave action and its application to the Skerries site suggests that waves can cause a reduction in bedform height but that such effects are rather transitory with recovery within a few tides.

ACKNOWLEDGEMENTS

The author is indebted to Frances Zimmermann and Barbara Cotgreave for assistance with the preparation of the present paper. The author is also grateful for financial assistance from the Commission of the European Directorate General for Science, Research and Development under Contract Nos. MAST-0035(C) and 0036(C).

REFERENCES

1. Yalin, M.S. Mechanics of Sediment Transport, Pergamon Press, Oxford, 1972.

2. Allen, J.R. Sandwaves: A Model of Original and Internal Structure. Sedim. Geol., 26, pp. 281-328, 1980.

3. Langhorne, D.N. A Study of the Dynamics of a Marine Sandwave, Sedimentology, 29, pp. 571-594, 1982.

4. O'Connor, B.A. and Duckett, F.J.L. Bed Friction in Tidal Flows, in 'Advances in Water Modelling and Measurement', (Ed. Palmer, M.H.) pp 231-244, BHRA, Cranfield, 1989.

5. Staub, C. and Bijker, R. Dynamic Numerical Model for Sandwaves and Pipeline-burial, in Proc. 22nd Coastal Eng. Conf. (Ed. Edge, W.L.) pp. 2508-2521, American Society of Civil Engineers, New York, 1990.

6. Rijn, van, L. Handbook, Sediment Transport by Currents and Waves. Delft Hydraulics, Report H461, 1989.

7. Moor, de, G. and Lanckneus, J. Relationship Between Sea Floor Currents and Sediment Mobility in the Southern North Sea, Annual Report, MAST Project 25, State University of Ghent, 1991.

8. O'Connor, B.A. and Yoo, D. Mean Bed Friction of Combined Wave-current Flow, Coastal Engineering, 12, pp. 1-21, 1988.

9. Soulsby, R. Sediment Transport by Strong Wave Plus Current Flows. Paper No. 48, Hydraulic Research, Wallingford, Oxfordshire, 1991.

10. Bijker, E.W. Some considerations About Scales for Coastal Models With Movable Beds. Delft Hyd. Lab., Rep. 50, 1967.

11. Pasenau, H. and Ulrich, J. Giant and Megaripples in the German Bight and Studies of their Migration in a Testing Area (Lister Tief), in Proc. 14th Coastal Eng. Conf., Copenhagen, June 1974, 2, pp. 1025-1035, American Society of Civil Engineers, New York, 1974.

12. Ellis, G., O'Connor, B.A. and McDowell, D.M. Generation of lab. waves using a micro-computer, in Engineering Software II (Ed. Adey, R.A.), pp 899-913 Proceedings of the 2nd Int. Conf. on Engineering software, Imperial College, London, 1981, CML Publications 1981.

13. Ashida, K. and Tanaka, Y. A statistical study of sand waves, in Proc. 12th Cong. IAHR, 2, Colorado State University, Fort Collins, Colorado, USA, 1967.

Application of N Line Model at Songkhla Port, Thailand

S. Weesakul, S. Charulakhana

Department of Civil Engineering, Chulalongkorn University, Bangkok, Thailand

ABSTRACT

The upcoast topographic changes at Songkhla port is simulated using the numerical N line model. The merit of the model is to predict shoreline change in 3 dimensions. The local constants for both longshore and cross-shore transport are determined in the present study during calibration test. Verification and prediction of bathymetric changes are conducted to access model capability with the synthetic wave input data from selected wave hindcast model providing the least error from measured values. The yearly average longshore and cross-shore transport rate are computed as well.

INTRODUCTION

Songkhla port is located at the southern part of Thailand as illustrated in Fig.1. The port is at the front of Songkhla lake as shown in Fig.2. It serves as both domestic and international port especially for inbound petroleum and agricultural products and outbound for rubber as raw material. At the lake inlet, the jetty was constructed in 1969 in order to maintain the navigation channel of the inlet. The jetty was made of rubble mound with the length of approximate 700 m and extended to a length of 1000 m in 1987 when the Songkhla port and offshore breakwater with the length of 500 m were completed. The upcoast of the jetty began to be deposited since construction was completed but depletion at the downcoast (north of inlet) was not found because sediment supply from lake is continued. The objective of the study is to simulate bathymetric changes in front of the jetty using numerical model. The study area as illustrated in Fig.2 is restricted at the upcoast of jetty . One boundary condition is the jetty itself providing condition to stop sediment transport along shore and another boundary is at a distance far from the jetty approximately 3.5 km. which provides sediment supply from wave.

The numerical model used herein is so called N line model first developed by Perlin et al [3]. The model can simulate topographic changes with time resulted from sediment transport for both longshore (parallel to shore) and cross-shore (normal to shore) direction which generated by wave only. The type of sediment is limited to sand, not clay or gravel and sediment transport by tidal current is not considered in the model.

Since the wave is the predominant factor to produce sediment transport, hence wave climate and wave data will be analysed in more detail.

WAVE CLIMATE

It is unfortunate that at present there is no wave data which has record long enough to provide valuable statistical analysis of wave climate. Even the wava data for jetty construction was obtained from wave hindcasting. Wave data ,i.e wave height,period are therefore hindcasted using wind data with the appropriate method. However, the method proposed by U.S. Army Corps of Engineers in Shore Protection manual [7] ,so called SPM(1984), seemed to be good enough to be applied in the southern part of Thailand compared with other methods as suggested by Weesakul et al [8] but two more theories,i.e, Sverndrup Munk and Bretschneider (SMB) and Peirson and Moskowitz (PM) are also applied to access degree of accuracy. The former is widely used as a standard method as proposed in the Shore Protection Manual of 3rd edition [5] while the latter is modified for practical use by Silvester et al [3],[4]. Charulukhana[1] did the comparison test with the significant measured wave height (Hs) and period (Ts) during 5-22 November 1989. The results are shown in Figs.3 and 4. All theories show underestimated values for both Hs and Ts. The Root Mean Square Error (RMSE) of Hs for SMB,PM and SPM(1984) are 0.61,0.56 and 0.46m respectively while that of Ts are 1.25,1.33 and 1.29s respectively. SPM(1984) provide the least RMSE for Hs but for Ts, SMB show a little bit better result. Hence,SPM(1984) is selected as the wave hindcast method to be used further.

SPM(1984) is applied with recorded wind data for 7 years from 1981 to 1987. The recorded interval is 3 hours. Wave direction is considered as wind direction. Figure 5 illustrates wave roses for each season and all year. It is found that wave generated by wind occurred about 48% of time in a year,the rest is calm or wind blown from land to sea. Direction of wave propagation mostly come from East 18% and North-East (NE) 12%. Time of occurrence in North-East monsoon ,(Nov-Feb), is 85%, Transition period ,(Mar-Apr,Sep-Oct),is 44%,and South-West monsoon ,(May-Aug), is only 14%. Maximum Hs is 3-3.5 m in 1982 and Ts is 9-10 seconds. Figure 6 illustrates graphs for design Hs and Ts at various return periods or percentage exceeding. For example, Hs and Ts for 50 year return periods (0.000228 %

exceeding) are 3.5 m and 10.4 s.

The 7 years hindcasted wave data can be digested to provide a yearly representative set of wave data as follows:

The summary of wave height,wave period and direction in each seasons will be determined for their percentage of occurrences. The wave data in each season will be obtained by generating random numbers which will subsequently fall into a range of accumulative percentage of occurrences for a set of Hs,Ts and direction of propagation. This set of wave data will be assumed to be constant for 6 hours then the same procedure will be repeated until wave data fulfil the time in that season. With the above procedure, the yearly average wave data will be obtained and ready to be used for verification and prediction test in the numerical model.

NUMERICAL MODEL

The objective of numerical N line model is to solve the bathymetric changes. The model is divided into two parts as shown in Fig.8. The first part mainly deal with wave transformation from deep water to shallow water. Computation of wave shoaling, refraction and wave diffraction due to coastal structures are included in computer program. The second part is to compute sediment transport produced by wave and result in shoreline change. The simulated beach profile and beach plan are shown in Fig.7. Longshore sediment transport is denoted by Qx while cross-shore sediment transport is denoted as Qy in that figure. Due to the variation in Qx and Qy ,the beach will be eroded or deposited ,the unknown to be solved is the length of a specific depth or contour line,y, whether it will be decreased or increased at that section. More detail is elaborated in Perlin et al [3].

The study area is formulated into x,y and h axis as shown in Fig.7. The x axis is divided into 18 sections,each has a distance longshore of 200 m and each section has values of y (distance normal to shore) 7 values corresponding to a depth 1 to 7 meters respectively. There is an island called Ko Nua located about 2.5 km from shore. The influence of this obstruction island will diffract the wave propagated to the study area causing low wave height and alter wave direction as well. It is therefore necessary to make this obstructed island into a model by simulating it as a spur jetty as depicted in Figs.9. The spur jetty in model will interrupt the wave in a northerly direction which can pass through the gap between the actuall jetty and Ko Nua but since the wave in this direction is small as shown in Fig.5, this effect is considered to be negligible.

RESULTS

In this section,the model will first be calibrated and the verification and prediction will be made consequently.

Model calibration is essential for the present problem since parameters for sediment transport formulae should be quantified at the specific study area to provide acceptable result. There are two parameters to be calibrated, first in the longshore transport formula using energy flux concept, the dimensionless parameters, K which related the amount of sediment and energy flux at breaking point, second,the Coff so called activity factor in the cross-shore transport. Unit of Coff is in m/year (eroded or deposited rate). When beach profile is deviated from equilibrium beach profile (proposed by Dean [2]),the amount of cross-shore transport can be computed by product of Coff and the deviation distance from equilibrium beach profile whether it is deposition or erosion. Value of Coff may vary from small up to hundred meter/year depend on beach slope, sediment size, etc. At this present stage,there are no research papers concerned with relating Coff to any parameters,so it will be treated as a calibration constant here. For value of constant K,it had been studied and proposed in SPM [7] to be 0.77 but it will not be used as that number here and will be treated as a calibration constant at a local area because computational scheme in this model include the cross-shore transport into considered ,bathymetric changes occurred by interactive of K and Coff parameters and another reason is that wave data used is the synthetic data , not the measured values. These effect will be absorbed in calibration constant,K.

Calibration is made with the available topographic survey map during April 1988 and July 1989. Figure 10 depicted wave propagation influenced by diffraction and Fig.11 summaries result of calibration. Parameter K is varied from 0.1 to 0.77 and Coff change between 1 to 10 m/year. It is found that K equals 0.25 and Coff equals 3m/year provide the least average RMSE of y values,i.e 66 m. Figure 12 illustrates comparisons of computed and measured beach profiles which is in fair agreement.

Verifation of numerical model using bathymetric survey map during September 1971 and November 1972 is done. The recorded wind data of three hours is lacked, simulated wave data as described in WAVE CLIMATE section will be used herein. The average RMSE of y is 75 m which greater than that in calibration test, one reason is because of synthetic wave data.

Prdictions of shoreline evolution is done for a further 5 years from 1990 to 1995. The deposition rate of 1 m contour depth at jetty is approximately 6.6 m. The 2 m contour depth advances the same rate at jetty but greater rate about 50-100 m/year is found far from jetty. The same result is found for 3 m contour depth while Y at 6 and 7 m does not change much.

The average net longshore sediment transport as shown in Fig.13 is 175,460 cu.m/year which is 73% of the gross transport (175,460 cu.m/year). Net transport direction is from south to north which mostly occur in the North-East monsoon and corresponded to wave climate as shown in Fig.5. The gross cross-shore transport is approximately 346,626 cu.m/year and 12% of this value (16,352 cu.m/year) is the average net onshore transport.

CONCLUSIONS

1. The wave hindcast methods so called SPM(1984) could provide the lowest root mean square error (RMSE) of significant wave height compared with SMB and PM methods. Wave climate at the study area therefore were analysed using this appropriate method with 7 years of recorded wind data. It was found that each year waves generated by wind propagate into the study area about 48% of the time and usually come in NE monsoon, that is around 85% of time in that season.
2. The topographic change at the upcoast of Songkhla jetty could be simulated with the numerical N line model. The calibration of constants K and Coff for longshore and cross-shore transport showed values of 0.25 (dimensionless) and 3 m/year respectively. The former is lower than the theoretical value of 0.77 but both calibrated parameters were appropriate for studying in this area such a way that they were combined and interacted for the amount of transport in both direction to provide the average least RMSE of contour lines from 1 to 7 meters and they included the effect of synthetic wave data as well.
3. The net longshore and cross-shore transport were obtained with the values of 128,080 cu.m/year and 16,352 cu.m/year with the average direction moving from south to north and onshore transport direction respectively.
4. The computed depth at 1 meter adjacent to jetty showed gradual deposition rate of approximately 6.6 m/year while the greater rate were found for 2 and 3 meter depth at some distance from jetty.

REFERENCES

1. Charulukhana,S. Characteristics of Wave and Shoreline Change at Songkhla,M.Eng Thesis,Dept of Civil Engineering, Chulalongkorn University,Bangkok,Thailand,304p.,1991.(in Thai)
2. Dean,R.G. Equilibrium Beach Profiles: U.S. Atlantic and Gulf Coasts,Ocean Engineering Report No.12,University of Delaware Press, Newark, Del.,1977.
3. Perlin,M.and Dean,R.G. A Numerical Model to Simulate Sediment Transport in the Vicinity of Coastal Structures,U.S. Army Corps of Engineers, CERC,MR 83-10, 119p.,1983.
4. Silvester,R.and Vongvisessomjai,S. Computation of Storm Waves

and Swell, Proc. Inst. Civ. Engrs.,Vol.48,p.259-283,1971.
5.Silvester,R. Coastal Engineering,Vol.I,Elsevier,457p,1979.
6.U.S. Army Corps of Engineers, Shore Protection Manual,CERC,3rd edition,3 Vols, 1977.
7.U.S. Army Corps of Engineers, Shore Protection Manual,CERC,4th edition 4 Vols, 1984.
8.Weesakul,S and Charulukhana,S. Comparisons of Wave Hindcast Methods for Lower Gulf of Thailand,22nd ICCE,Delft,1990.

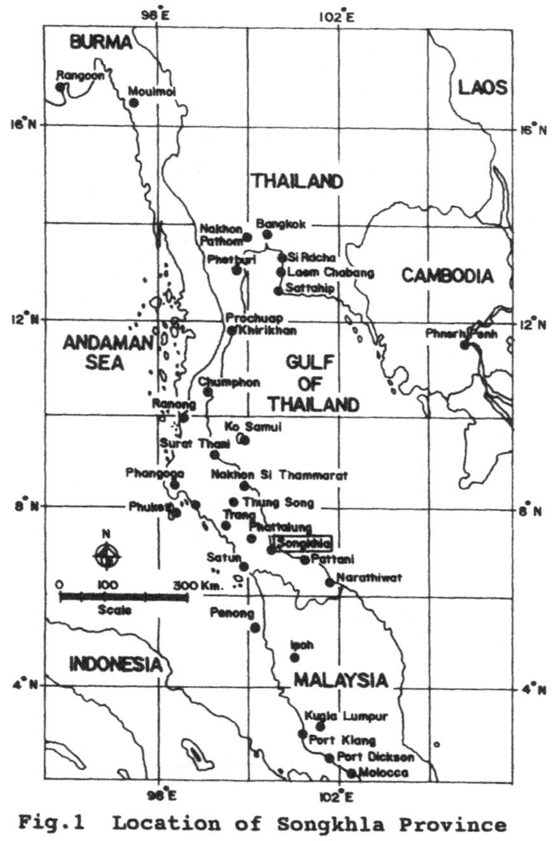

Fig.1 Location of Songkhla Province

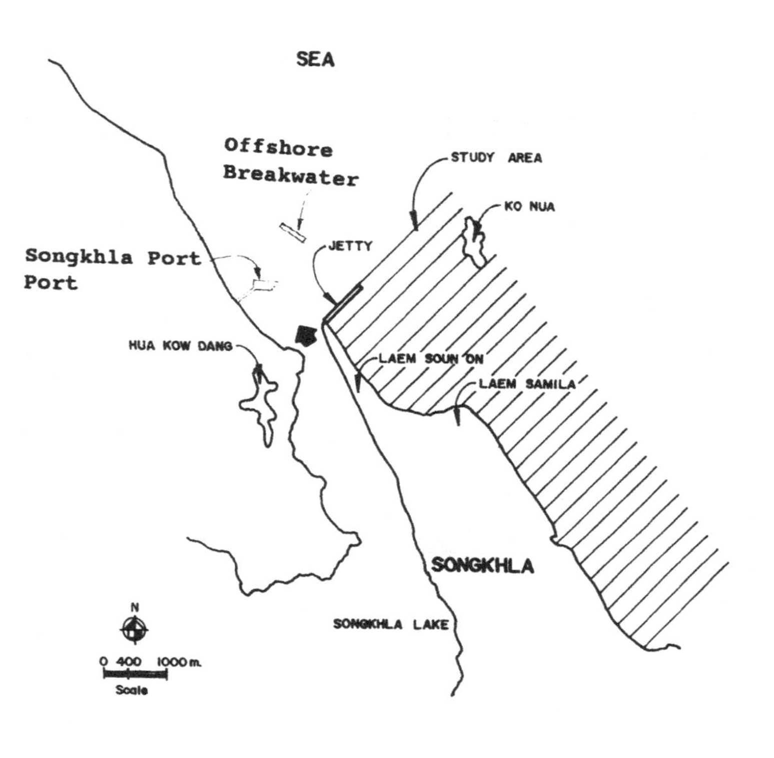

Fig.2 Location of Study Area

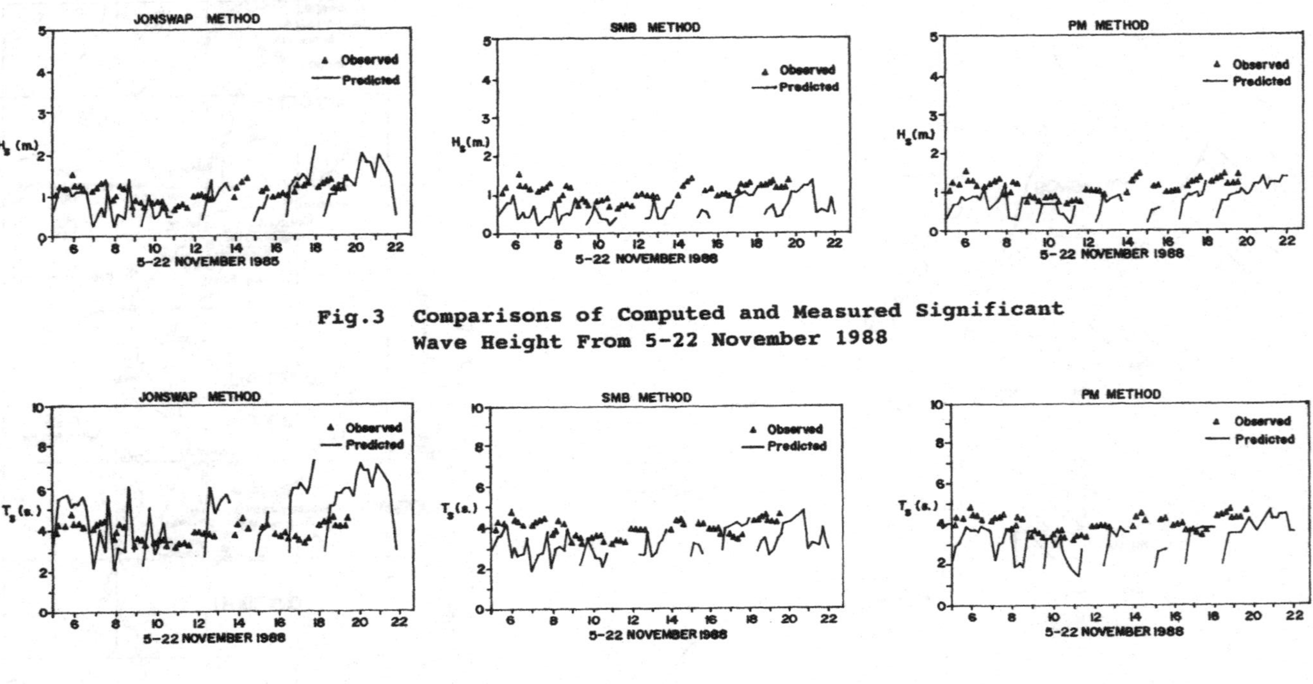

Fig.3 Comparisons of Computed and Measured Significant Wave Height From 5-22 November 1988

Fig.4 Comparisons of Computed and Measured Significant Wave Period From 5 to 22 November 1988

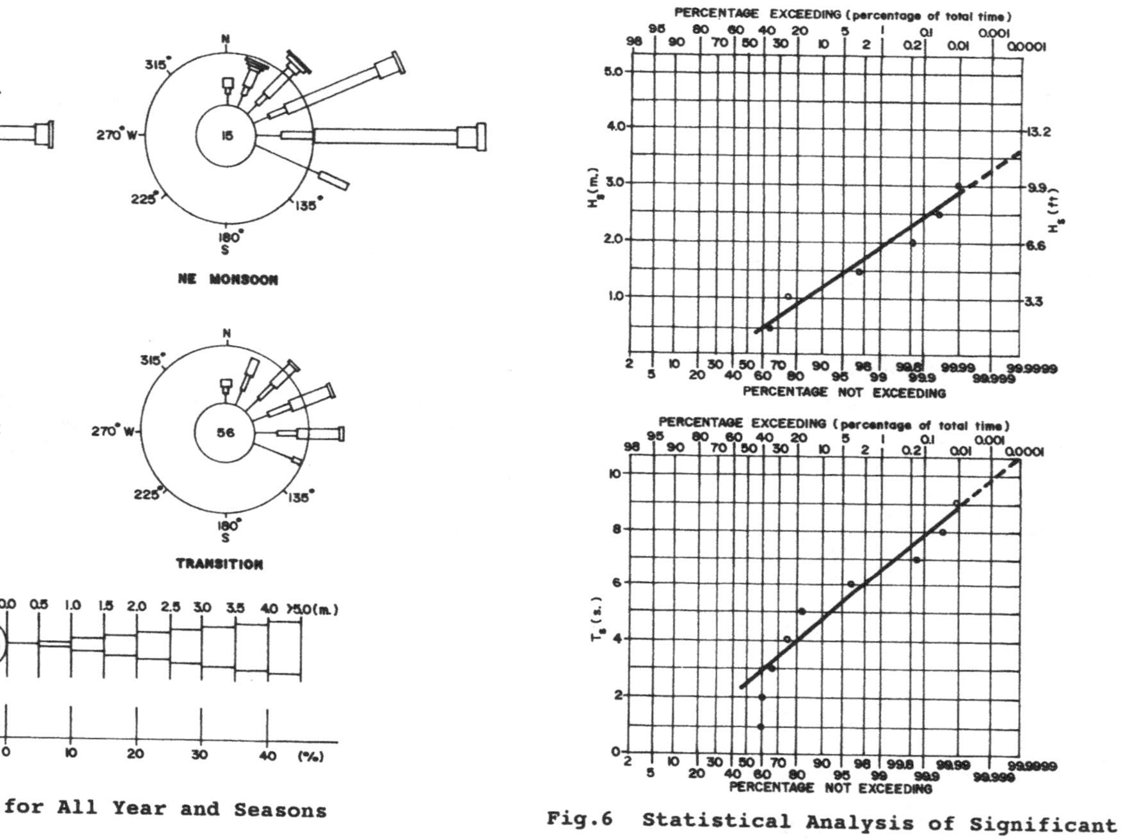

Fig.5 Wave Rose for All Year and Seasons

Fig.6 Statistical Analysis of Significant Wave Height and Period

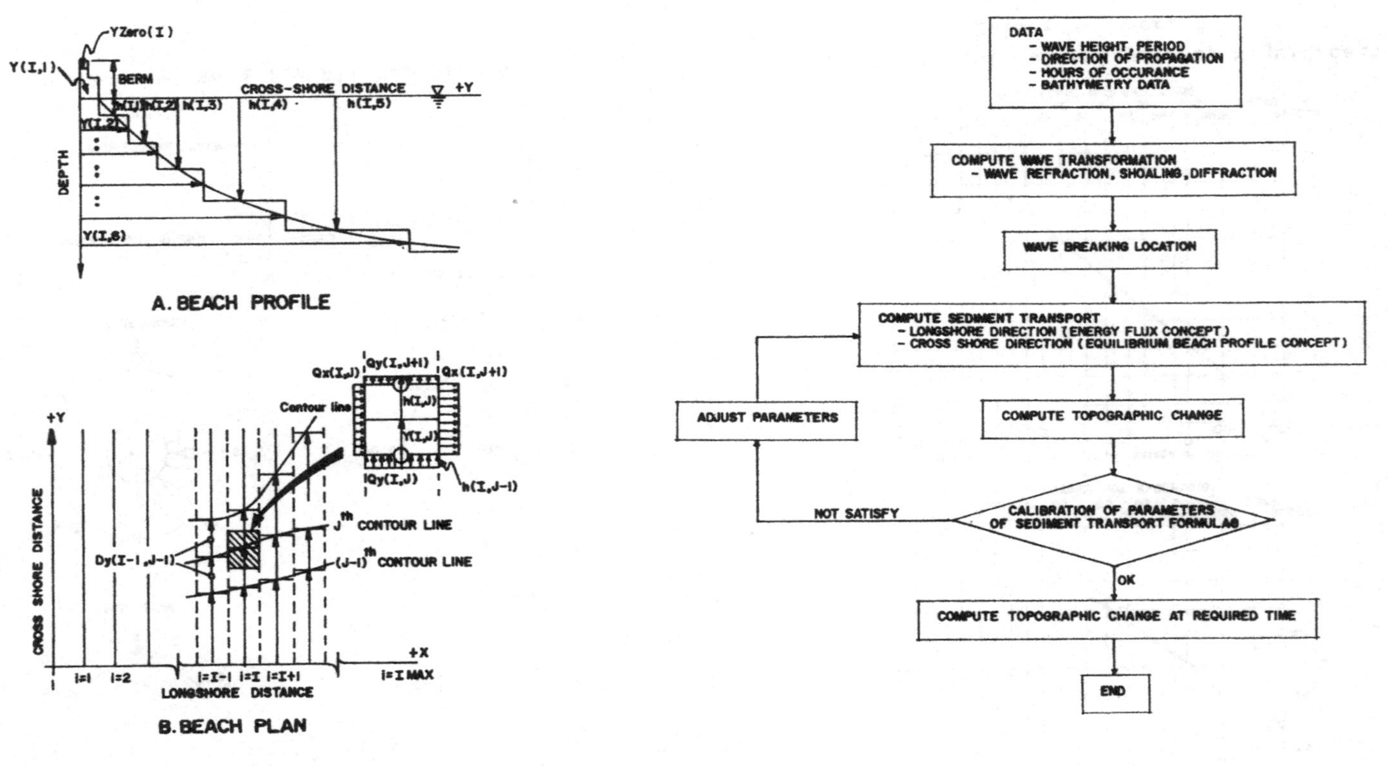

Fig.7 Bean Profile for Sediment Transport Study (After Perlin (1984))

Fig.8 Flow Chart for N line Model

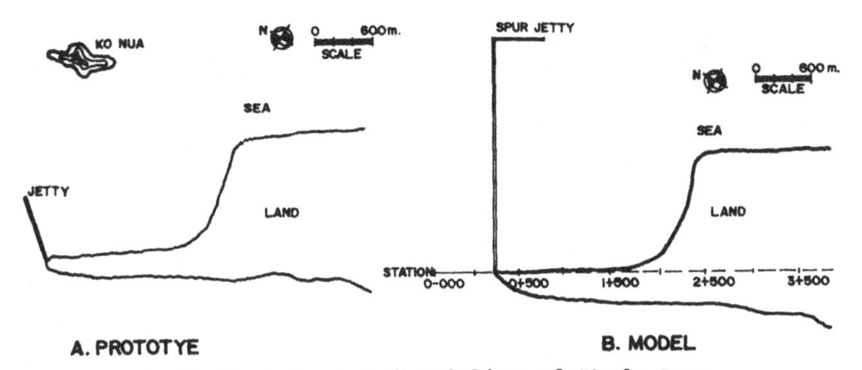

Fig.9 Prototype and Modeling of Study Area

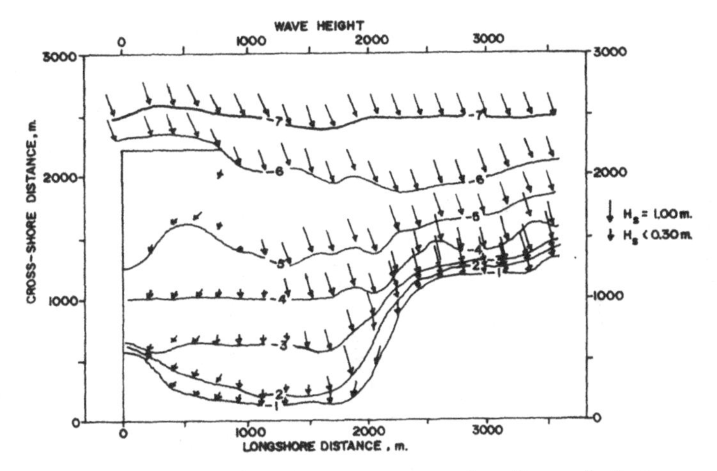

Fig.10 Result of Wave Propagation for H_s = 1.5 m T_s = 5 sec, Direction ENE

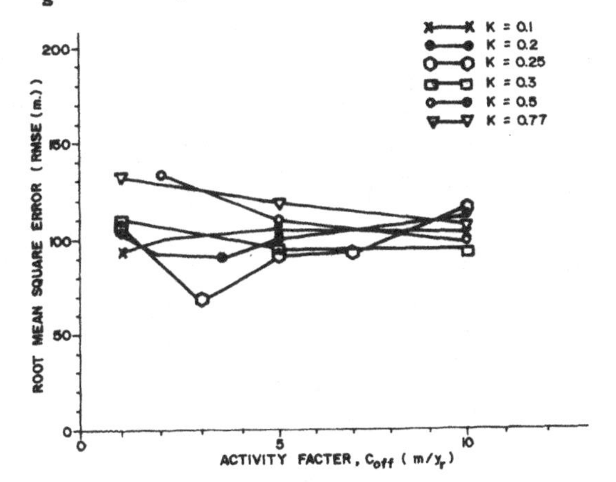

Fig.11 RMSE for Different K and Coff

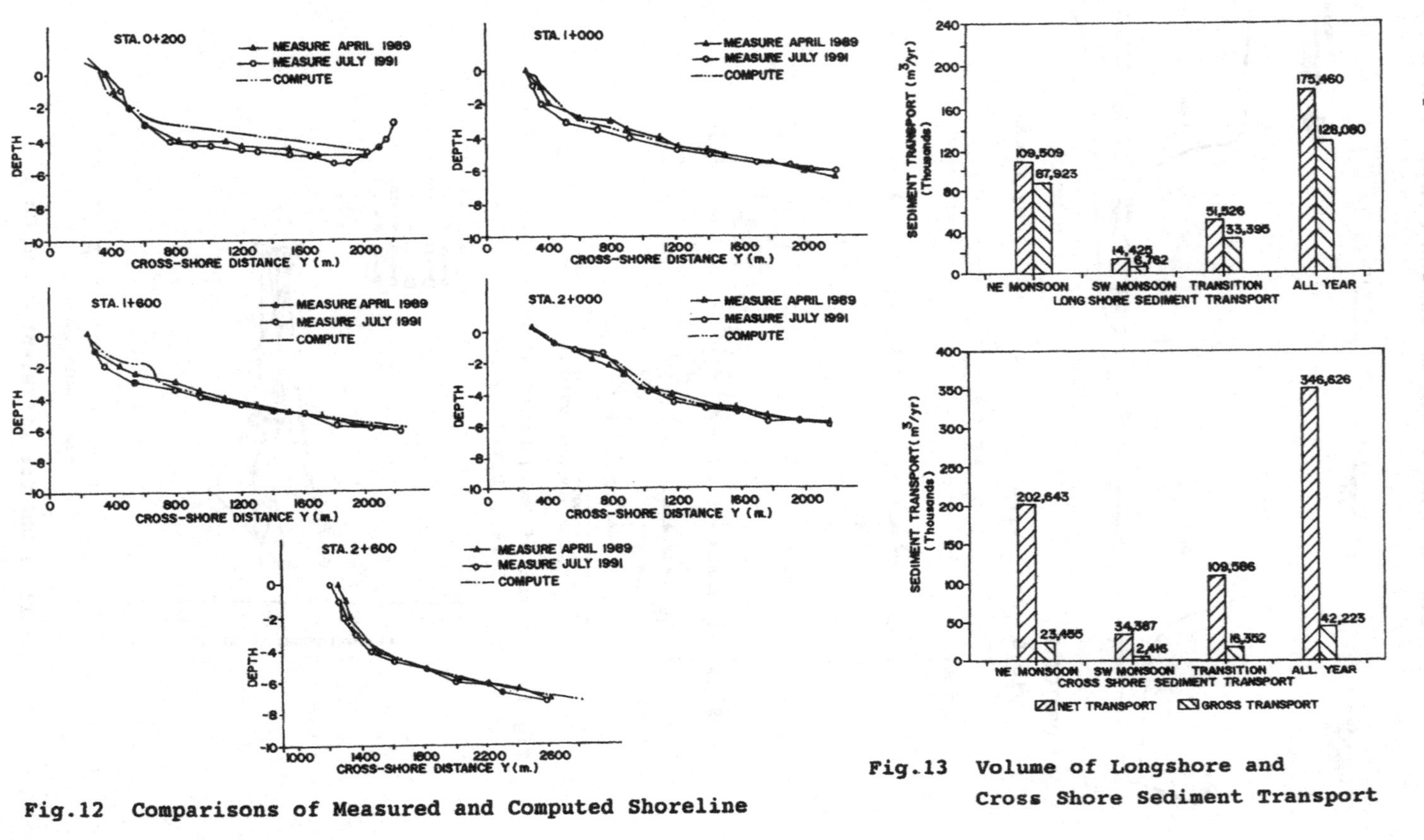

Fig.12 Comparisons of Measured and Computed Shoreline

Fig.13 Volume of Longshore and Cross Shore Sediment Transport

A Two Dimensional Finite Element System of Sediment Transport and Morphological Evolution

B. Latteux, E. Peltier

Electricité de France, Laboratoire National d'Hydraulique, Chatou, France

ABSTRACT

We introduce here TELEMAC-2DST, a two dimensional finite element system of sediment transport and morphological evolution. This system has the same finite element structure as the 2D code TELEMAC solving the shallow water equations. TELEMAC-2DST is made of two models : one which deals with bed-load transport and the other one with suspended sediment transport. Results of these models are described in the present paper.

INTRODUCTION

In most of the models, hydrodynamic and sedimentological equations are uncoupled. This is made possible by the difference between hydrodynamic and morphological evolution time scales.

But an accurate current field is required at a suitable computation cost. Therefore the LNH has worked out a two-dimensional system of sediment transport, TELEMAC-2DST, with the finite element structure of TELEMAC, model developed by the LNH to calculate the hydrodynamic characteristics of open channel flows and having achieved many studies covering a large range of applications in fluvial or maritime hydraulics ([1] and [2]).

This paper presents this system including two models which allow to compute bed evolution induced respectively by bed-load transport and by suspended sediment transport.

For each model, after a presentation of the treatment of the sedimentological equations, results of simulations are displayed.

BED-LOAD TRANSPORT

Presentation of the problem

Once the hydrodynamic pattern is known, sediment transport capacity can be estimated as a function of flow rate per unit width (q) and of water depth (h) from empirical or semi-empirical formulas of bed-load transport (Meyer-Peter, Engelund-Hansen and Einstein-Brown are used in TELEMAC-2 DST).

The bed-load transport induced by currents is computed using one of these transport formulas. Continuity equation applied to the bed-load transport gives the bed evolution :

$$\frac{\partial Z_b}{\partial t} + \operatorname{div}\left(\overrightarrow{Ts}\right) = 0 \qquad (1)$$

with t time,

$\overrightarrow{Ts}$ bed-load transport rate,

Z_b bottom level.

Two types of boundary conditions are taken into account :

- no flux on solid boundaries,
- free output or no evolution on open boundaries.

It is assumed that, as long as the computed bed evolution is small compared to water depth, flow direction and water level remain the same at the same time of the tide. In that case, flow rate is not modified and at any moment velocity can be derived from the initial flow rate and the present water depth :

$$u = \frac{q_x}{h} \qquad (2)$$

$$v = \frac{q_y}{h} \qquad (3)$$

with u, v components of the mean velocity over the depth,

q_x, q_y components of the flow rate per unit width.

Advection treatment

The aim of the following calculations is to transform equation (1) into an equation of advection.

For solving it we use the characteristic method which is naturally upwinded in space and unconditionally stable.

Let :

- q_x, q_y be the two components of the flow rate per unit width : $q = \sqrt{q_x^2+q_y^2}$
- Ts_x, Ts_y be the two components of the bed-load transport,
- Z be the surface elevation,

The bed continuity equation can be transformed into :

$$\frac{\partial Zb}{\partial t} - \left(\frac{q_x}{q}\frac{\partial Zb}{\partial x} + \frac{q_y}{q}\frac{\partial Zb}{\partial y}\right)\frac{\partial Ts}{\partial h} = FF \tag{4}$$

with

$$FF = -\frac{\partial Ts_x}{\partial q_x}\frac{\partial q_x}{\partial x} - \frac{\partial Ts_x}{\partial q_y}\frac{\partial q_y}{\partial x} - \frac{\partial Ts_x}{\partial h}\frac{\partial Z}{\partial x}$$
$$- \frac{\partial Ts_y}{\partial q_x}\frac{\partial q_x}{\partial y} - \frac{\partial Ts_y}{\partial q_y}\frac{\partial q_y}{\partial y} - \frac{\partial Ts_y}{\partial h}\frac{\partial Z}{\partial y} \tag{5}$$

This equation is split into two steps :

a) advection without source term (calculated with the characteristic method which has the advantage of being unconditionnally stable) :

$$\frac{\partial Zb}{\partial t} - \left(\frac{q_x}{q}\frac{\partial Zb}{\partial x} + \frac{q_y}{q}\frac{\partial Zb}{\partial y}\right)\frac{\partial Ts}{\partial h} = 0 \tag{6}$$

b) treatment of the source term :

$$Zb^{n+1} = Zb^{tild} + FF^n \,.\, \Delta t \tag{7}$$

with Zb^{tild} result of the advection step,

Zb^{n+1} bottom level at time t_{n+1},

FF^n source term at time t_n,

Δt time step.

We undertook several numerical simulations with this scheme but there were two main problems : numerical diffusivity and lack of conservativity.

Predictor-Corrector scheme

To solve the problems encountered before, we have proposed a predictor-corrector scheme [3].

In a first phase we use the previous scheme to get a first estimate of the bottom level $(\widetilde{Zb^{n+1}})$ at time t_{n+1} ($t_{n+1} = (n+1)Dt$).

In a second phase, we calculate an estimation of the bed-load transport at an intermediate time $t_{n+\theta}$, located between t_n and t_{n+1} :

$$\vec{Ts}^{n+\theta} = \theta \, . \, \tilde{\vec{Ts}}^{n+1} + (1-\theta) \, . \, \vec{Ts}^{n} \tag{8}$$

$$\text{with} \quad \tilde{\vec{Ts}}^{n+1} = \vec{Ts}\,(\,\widetilde{Zb}^{n+1}\,) \tag{9}$$

$\vec{Ts}^{n}$ calculated explicitly at time t_n.

In a third phase, we solve the equation more or less centered in time :

$$\frac{\partial Zb}{\partial t} = -\operatorname{div} \vec{Ts}^{n+\theta} \tag{10}$$

This step done following a finite element space discretization allows this scheme to be conservative. In the end, we set the following linear system :

$$M\,Zb^{n+1} = M\,Zb^{n} - \Delta t \int_{\Omega} \left(\frac{\partial Ts_x}{\partial x} + \frac{\partial Ts_y}{\partial y} \right) \psi_i \, d\Omega \qquad (11)$$

with M the mass matrix,

ψj basis function for Ts on point i,

Ω domain of computation.

In a last phase, we use the formulation of the slope effect proposed by Koch and Flokstra [4] ; the magnitude of the transport is then multiplied by the factor :

$$\left(1 - \beta \frac{\partial Zb}{\partial s} \right) \qquad (12)$$

The bottom slope effect on the direction of the transport can be approximated by :

$$\tan \alpha = \tan \delta - \beta \frac{\partial Zb}{\partial n} \qquad (13)$$

with s stream-oriented coordinate,

α transport direction with respect to the depth-averaged flow direction,

δ shear stress direction,

n coordinate perpendicular to the flow direction,

β constant value.

Numerical example

Within the frame of the European project MAST (MArine Science and Technology), a schematic estuary opening in a coastal region dominated by strong currents has been simulated in order to test various methods of input filtering, i. e. of choosing a limited set of tides which can represent the whole tide variability[5]. This is a severe case because the non-linearities of hydrodynamics in such a configuration lead current patterns to depend significantly on tidal conditions.

The long-term (19 years) cycle of the tide has been discretised in about 20 classes of tide range with their associated occurence frequency. For each class, hydrodynamic (figure 1) and morphodynamic computations over 1 tide have been performed ; bed changes for each class have been combined, with weighting coefficients corresponding to the occurence frequency of the class, to determine the 1-year reference bed evolution (figure 2).

This reference evolution has then be compared to the morphologic evolutions computed from the various filtering methods.

SUSPENDED SEDIMENT TRANSPORT

Numerical model

Providing that transport quantity is well mixed throughout the water depth, it is therefore sufficient in suspended sediment modelling to consider depth-averaged equations. This approximation is particularly appropriate for a wash-load which consists of fine cohesive particles with very low settling velocities, and when sediment concentration remains moderate.The bidimensional depth-averaged equation of suspended sediment is then the following :

$$\frac{\partial C}{\partial t} + u\frac{\partial C}{\partial x} + v\frac{\partial C}{\partial y} = \frac{\partial}{\partial x}\left(K_x\frac{\partial C}{\partial x}\right) + \frac{\partial}{\partial y}\left(K_y\frac{\partial C}{\partial y}\right) + \frac{S}{h} \qquad (14)$$

with C depth-averaged concentration of suspended sediment,
u, v depth-averaged components of velocity,
K_x, K_y dispersion coefficients,
h water depth,
S source-sink term accounting for erosion or deposition.

This last term is in fact the difference between the erosion rate given by Partheniades's formulation [6] and the deposition rate evaluated by Krone's one [7].

In sediment equation (14), hydrodynamic conditions (u,v,h) are known, being previously computed by the model TELEMAC.

Treatment of the equation

Equation (14) is splitted in two steps :

a) advection without source term (calculated with the characteristic method) :

$$\frac{\partial C}{\partial t} + \vec{V} . \overrightarrow{\text{grad } C} = 0 \tag{15}$$

b) treatment of the dispersion and of the source term :

$$\frac{\partial C}{\partial t} = \frac{\partial}{\partial x}\left(K_x \frac{\partial C}{\partial x}\right) + \frac{\partial}{\partial y}\left(K_y \frac{\partial C}{\partial y}\right) + \frac{S}{h} \tag{16}$$

This equation is solved implicitly by :

$$\frac{C^{n+1} - C^{tild}}{\Delta t} = \theta \text{ div } (\overline{\overline{K}} . \overrightarrow{\text{grad } C^{n+1}}) + (1-\theta) \text{ div } (\overline{\overline{K}} . \overrightarrow{\text{grad } C^{n}}) + \frac{S}{h} \tag{17}$$

Consolidation of the bed is simulated thanks to a new model developed within the frame of MAST G6 MORPHODYNAMICS [8]. It allows to know the consolidation out of equations of the soil mechanics.

Numerical example

A settling basin in St Chamas, France, has been studied by means of this model. The calculations have allowed to design this basin and to evaluate its efficiency.

At first, the currents were computed with TELEMAC, and in a second step the suspended load model was run till to obtain an equilibrium state for the concentrations, as it is shown on figure 3.

REFERENCES

1. Daubert O., Hervouet J.M. and Jami A. Description of some numerical tools for solving incompressible turbulent and free surface flows, International Journal for Numerical Methods in Engineering, Vol. 27, 3-20, 1989.

2. Hervouet J.M. Comparison of experimental data and laser measurements with the computational results of TELEMAC code (shallow water equations), Computational Modelling and Experimental Methods in Hydraulics, Hydrocomp'89, Dubrovnik, 1989.

3. Peltier E., Duplex J., Latteux B., Pechon P. and Chausson P. Finite element model for bed-load transport and morphological evolution, Computational Modelling in Ocean Engineering, Barcelona, 1991.

4. Koch F.G. and Flokstra C. Bed level computations for curved alluvial channels, XIXth Congress of the International Association for Hydraulic Research, New Delhi, India, 1981.

5. Latteux B. and Villaret C. Long-term morphodynamics : techniques in use and under development at LNH, MAST-G6M Workshop, Edinburg, 1991.

6. Partheniades E. Erosion of cohesive soils, Journal of the Hydraulics Division, ASCE, Vol. 91 n°HYI, 1965.

7. Krone R.B. Flume studies of the transport of sediment in estuarine shoaling process, Techn. Rep. Hydraulics Engineering Laboratory, University of California, Berkeley, 1962.

8. Alexis A., Bassoullet P., Le Hir P. and Teisson C. Consolidation of soft marine soils : unifying theories, numerical modelling and in situ experiments, Paper submitted to ICCE'92, 1992.

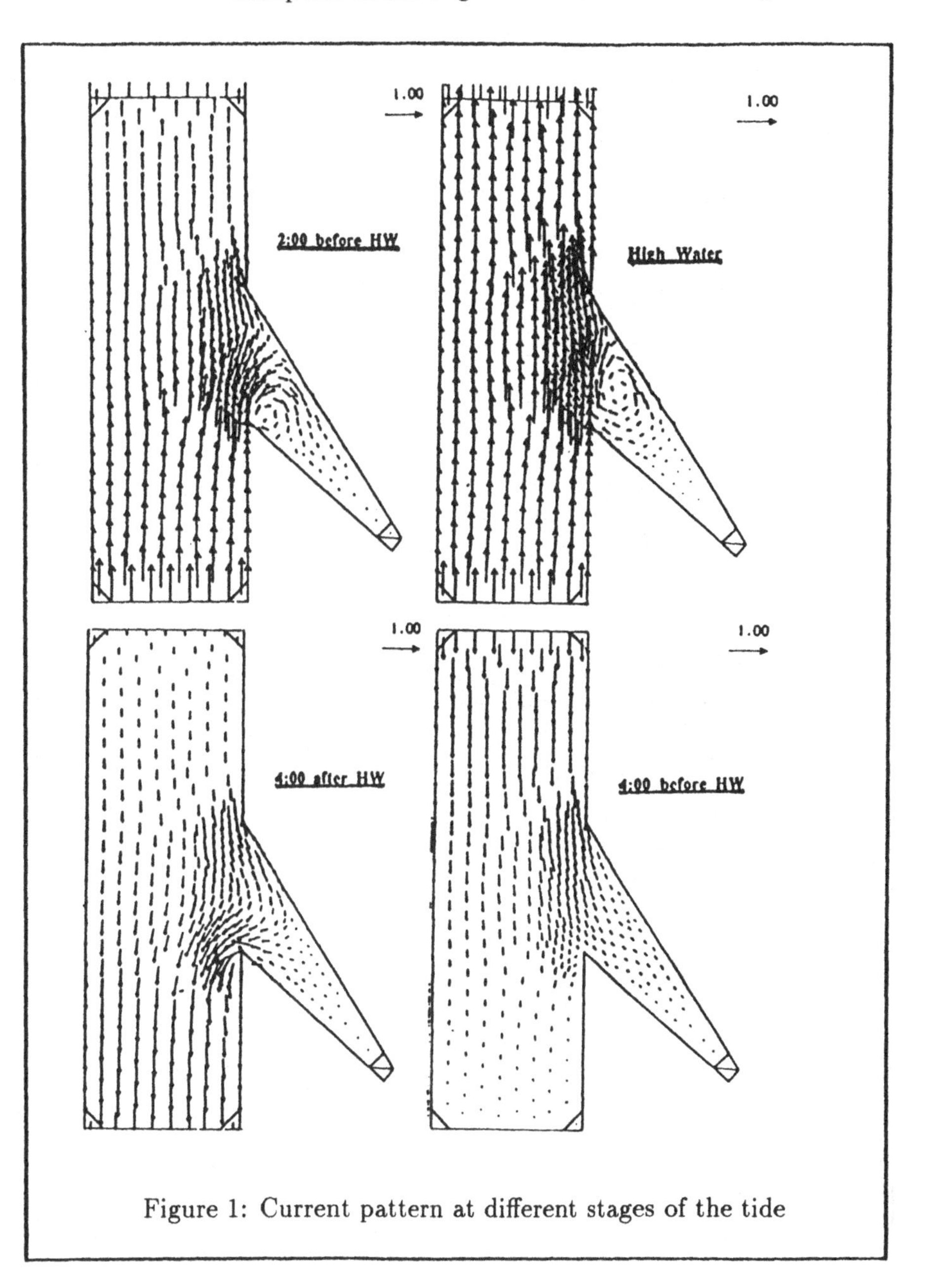

Figure 1: Current pattern at different stages of the tide

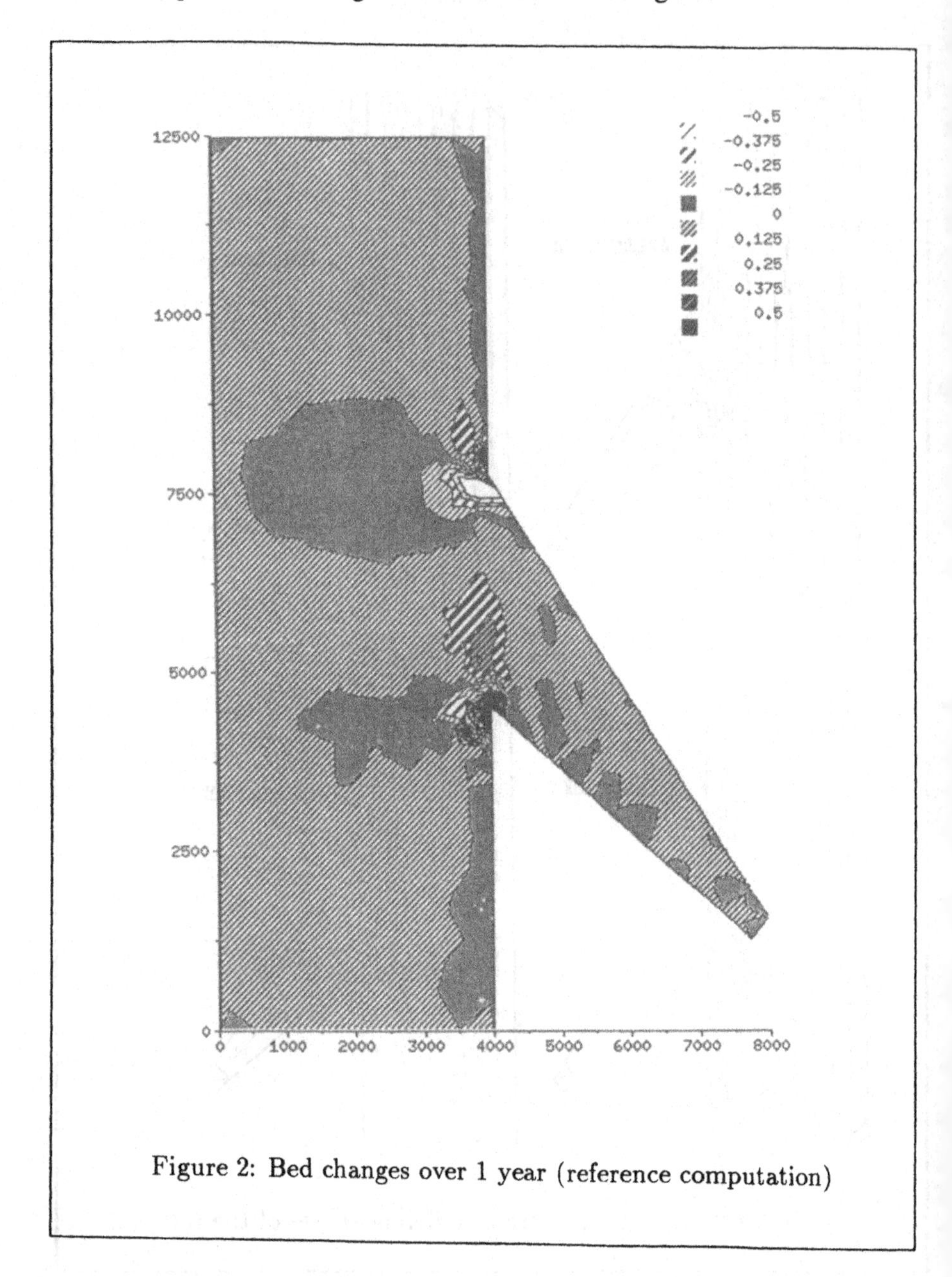

Figure 2: Bed changes over 1 year (reference computation)

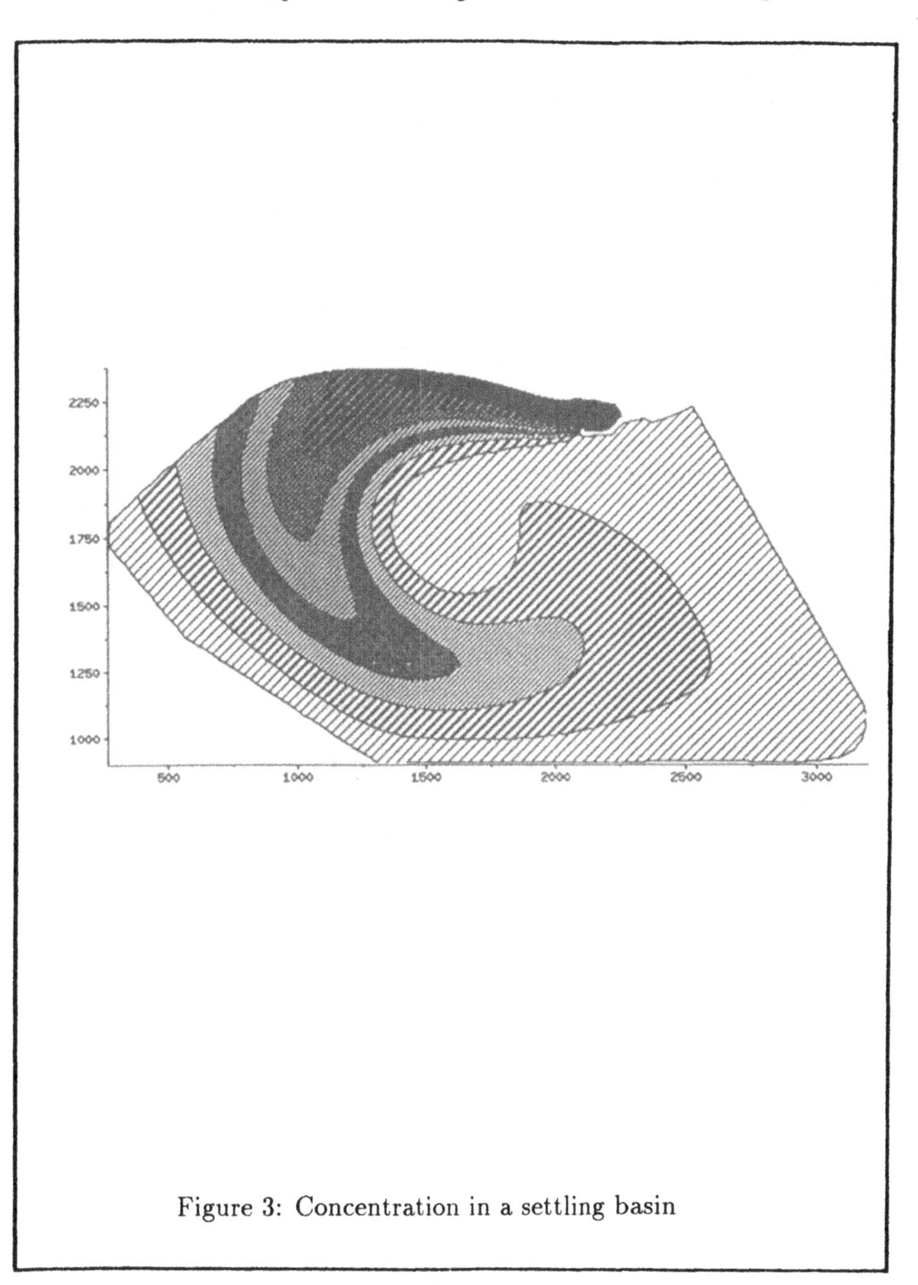

Figure 3: Concentration in a settling basin

Long-Term Simulation of Cohesive Sediment Bed Erosion and Deposition by Tidal Currents

C. Villaret, B. Latteux

Electricité de France, Laboratoire National d'Hydraulique, Chatou, France

ABSTRACT

Numerical experiments of the action of fortnightly tidal currents on erodible sediment beds are presented using a depth-averaged numerical model. Effects of consolidation properties on long-term behaviour are observed in the case of the Loire estuary. After transition time, the evolution of suspended sediment concentration becomes periodic and is determined by the vertical structure of the sediment bed. Based on these observations, new filtering methods of the hydrodynamic forcing are proposed in order to simplify the calculations.

INTRODUCTION

In numerical models of sediment transport, the time step is imposed in order to describe the time variation of the relevant hydrodynamic forcing. In the case of semi-diurnal tidal forcing (T_m=12 hours), the time step must be less than about 30 minutes. For long-term morphodynamics, this often leads to prohibitive computational cost. Fortunately, the difference of time scales between hydrodynamic and sedimentologic processes allows for some simplification of the calculations. Different techniques for input filtering of the tidal forcing are under development at L.N.H. within the frame of the European project MAST G6M. In the case of non cohesive sediment transport, it was found possible to capture the most important features of the response of the system, by representing the whole fortnightly tidal cycle by a unique "representative" tide whose time period is equal to the period of the cycle, which allows to lengthen the time step (x28) of the simulation (e.g. Latteux [1,2]).

In the case of estuarine cohesive sediment beds, the problem is further complicated by the time-dependent properties of the sediment bed which undergoes consolidation. In this paper, we present long term numerical simulations for the erosion-deposition cycle of a cohesive sediment bed by fortnightly tidal currents. The principal objectives of this study are:

- to observe the response of the system, in terms of suspended sediment concentration and bed structure evolutions,

- to perform a sensitivity analysis in order to determine the influence of the input parameters on the solution,

- to develop and compare new filtering methods of the hydrodynamic forcing.

Cohesive sediment bed properties (erosion and deposition laws, bed consolidation) are included in a 0-D depth-averaged numerical model which will first be presented. Input parameters are based on measurements of the Loire estuary. Effects of consolidation on the long-term behaviour of a cohesive sediment bed will then be observed based on numerical simulations. Tentative conclusions on filtering methods for long-term simulations will be drawn from a sensitivity analysis

DEPTH-AVERAGED SEDIMENT TRANSPORT MODEL

A depth-averaged sediment transport model for the erosion-deposition of cohesive sediment beds by tidal currents is developed, based on the following simplifying assumptions:

(1) Uniform flow and concentration : advection terms in the sediment transport equation are neglected.

(2) Constant depth : depth variations due to tidal motion and bed level changes are assumed to be small.

(3) Constant friction factor : effects of sediment induced stratification on bottom turbulence are assumed to be small.

(4) Constant settling velocity : flocculation and hindered settling effects are assumed to be negligible. The flux of sediment D deposited at the bed surface is then proportional to the product of a constant settling velocity w and depth-averaged concentration C.

For the mean flow variation, we impose a semi diurnal sinusoidal variation (T_m=12.4 hours) with a modulated amplitude at the fortnightly period (Tc=14 days).

$$U(t)=U_0(1+B\cos(\omega_c t))\sin(\omega_m t) \qquad (1)$$

The factor B is proportional to the amplitude variation of the maximum tidal velocity during the spring-neap cycle. The turbulent bottom friction velocity u_*, which governs sediment dynamics, is assumed to be proportional to the imposed depth-averaged flow velocity U(t), using a constant friction velocity Fr (u_*=Fr U). In this equation, the erosion and deposition fluxes E and D at the sediment bed surface are empirical functions of the hydrodynamic bed friction velocity u_*, and of sediment and bed properties. (e.g., Mehta and Partheniades [3])

$$E = M\left(\frac{u_*(t)}{u_{*e}} - 1\right) \quad \text{if} \quad u_*(t) > u_{*e}$$

$$E = 0 \qquad \text{if} \quad u_*(t) < u_{*e}$$

$$D = w\,C\,(1 - \frac{u_*(t)}{u_{*d}}) \quad \text{if } u_*(t) < u_{*d}$$

$$D = 0 \quad \text{if } u_*(t) > u_{*d}$$

In these relations, the critical erosion velocity of freshly deposited material u_{*e} is constrained to be greater than the critical deposition velocity u_{*d}, so that erosion and deposition do not occur simultaneously.

The continuity equation is discretised using a finite differences semi-implicit scheme. The deposition flux is discretised by introducing a weighting coefficient β:

$$D \approx w\,(\beta\, C^{i+1} + (1-\beta)\, C^{i})$$

For $\beta > 0.5$, the scheme is unconditionally stable. In order to represent accurately tidal variations, we use a time step of about 20 mn.

The hydrodynamic and sediment transport models are coupled to a consolidation algorithm developped at L.N.H. (Teisson and Fritsch [5]). The bed is discretised in layers of increasing concentrations C_s to which different characteristic residence times are assigned. When the residence time of sediment particles in layer n is over, it empties into the underneath layer n+1 whose thickness E(n+1) is increased in order to conserve the total mass of sediment:

$$\Delta E(n+1) = \frac{C_s(n)\,E(n)}{C_s(n+1)}$$

Considering the difference in time scales between hydrodynamic forcing and consolidation process, the model allows the use of a larger time step for the consolidation algorithm. In most applications, we use a time step of 2 hours which corresponds to the residence time in the surface layer.

NUMERICAL SOLUTIONS FOR LONG TERM EVOLUTION

Input parameters

Effects of bed consolidation are observed in the reference case of the Loire estuary, by comparing two numerical simulations. In the first simulation, the sediment bed deposits as a uniform layer of constant density and erosion velocity while, in the second experiment, the sediment bed undergoes consolidation. For the bed properties, we use characteristic values of a freshly deposited bed (t=1 h): C_s=110 kg/m^3, u_{*e} = 0.008 m/s , and M=7.34 10-4 ((kg/m^3)x(m/s)). In the second simulation, the bed is made of a series of 10 layers of increasing density, critical erosion velocity and erosion parameter which are empirical functions of the cumulated residence time t (in hours) in the overlaying layers (e.g.Migniot [4]):

$$\text{Log } C_s = 136.2 \log_{10}(t + 5.4) \quad \text{if } t \leq 24 \text{ h}$$

$$\text{Log } C_s = 200 + 70 \log_{10}(\frac{t}{24}) \quad \text{if} \quad t > 24\ \text{h}$$

$$u_{*e} = 3.2\ 10^{-5}\ C_s^{1.175}$$

$$M = 0.55\ 10^{-9}\ C_s^3$$

The same numerical values of the different input parameters, and the same initial conditions are used in both simulations. The mean tidal amplitude U_0 is set equal to 0.5 m/s, and parameter B in equation (1) is 0.4. With a friction coefficient Fr=0.036, the corresponding maximum friction velocity varies between 0.025 m/s at spring tide and 0.011 m/s at neap tide. Characteristic values of the settling velocity w and critical deposition velocity u_{*d} are for the Loire estuary: w=0.25 10^{-3} m/s, u_{*d}=0.007 m/s.

At t=0, surface layers of the sediment bed are all empty, so that the bed is formed during transition regime from the suspended material. Initial concentration C_0 is equal to 100 g/l for a non consolidating bed and 10 g/l when consolidation effects are included. Calculations are started at maximum spring tidal amplitude and zero mean current velocity (t=0, in equation (1)). Hydrodynamic forcing U(t) is plotted on figure 1 for the first fortnightly cycle.

Non consolidating sediment bed

The cycle-averaged suspended sediment concentration decreases rapidly during the three first fortnightly cycles from C_0=100 g/l to a constant value C_{eq}=36 g/l. When equilibrium is reached, concentration variations become periodical at the period of the fortnightly tidal cycle, as shown on figure 2.a for the fourth fortnightly cycle. Maximum amplitude variation of about 56 g/l is observed at the frequency of the fortnightly cycle. The signal is modulated by smaller amplitude variations at the frequency of half the tidal period ($T_m/2$). Amplitudes of modulations are proportional to the concentrations. After spring tide, both concentration and concentration modulation reach their maximum values C_{max}=64 g/l with a variation of 5g/l at half the tidal frequency. After neap tide, both concentration and concentration amplitude have minimum values of respectively 8 g/l and about 0.5 g/l.

On figure 3.a, the energy spectrum of the steady state concentration C(t) shows a maximum energy peak at the frequency of the fortnightly tidal cycle ω_c; the smaller energy peak at 2 ω_c indicates that the response is dissymetric between spring and neap tides. The next high frequency peak corresponds to twice the tidal frequency $2\omega_m$.

For each tide of the fortnightly cycle, erosion and deposition alternate at the frequency of half the tidal period. Even at neap tide, the maximum friction velocity is greater than the critical erosion velocity of the surface layer (u_{*max}= 0.0085 m/s). During each half tidal period ($T_m/2$), the concentration evolution C(t) can be decomposed into three different phases:

* Erosion only, for $u_* > u_{*e}$: concentration increases.
* E=0 and D=0, for $u_{*d} < u_* < u_{*e}$: concentration remains constant.
* Deposition only, for $u_* < u_{*d}$: concentration decreases.

During the first half of the cycle, when the tidal current amplitude increases, tidal-averaged erosion fluxes are greater than tidal-averaged deposition fluxes, so that tidal-averaged concentration increases up to a maximum value. As seen on figure 2.a, the maximum concentration is reached when the tidal current amplitude starts to decrease. This time lag of about 2 days between maximum current intensity and maximum concentration can be interpreted as the inertia of the system due to the settling process.

Effects of bed consolidation

Starting from an initial concentration C_0=10 g/l, the suspended sediment concentration keeps decreasing at a very slow rate until the 10th fortnighly tidal cycle. The duration of the transition regime is then longer than in the preceding case.

When steady state is reached, the system response, as regards concentration, becomes periodic at the period of the fortnightly tidal cycle. As seen on figure 2.b, equilibrium concentrations are an order of magnitude smaller than for a non consolidating bed. After spring tide, maximum concentrations stay constant at 0.45 g/l during the first half of the cycle while amplitudes of modulation at half the tidal period increase up to 0.05 g/l near neap tides. After neap tides, concentrations are smaller, with maximum values of 0.35 g/l during the second half of the cycle.

A spectral analysis of the concentration on figure 3.b shows a maximum energy peak at the fortnightly tidal period with a second slightly lower peak at half the tidal period. The signal is more noisy than in the case of a non consolidating bed. This can be explained by the fact that the concentration response looks like a series of step functions instead of sinusoidal variation. Maximum concentration is reached almost instantaneously and limited by the depth of the active surface layer.

The vertical structure of the sediment bed can be schematized in a two layer structure:
- a thin active surface layer, which is entirely eroded when suspended sediment concentrations are maximum. Its critical erosion velocity remains smaller than the maximum friction velocity.
-A bottom consolidated layer, which slowly consolidates at much larger time scale.

The concentration response at steady state is entirely determined by the properties of the surface active layer which is formed during transition regime. Its density and thickness determine the maximum concentration. The observed dissymetry in the concentration response is due to consolidation of the active surface layer during neap tides.

SENSITIVITY ANALYSIS

In the case of a consolidating sediment bed, the sensitivity analysis is made difficult by the number of input parameters. In the following, we only vary the initial concentration C_0, the amplitude of the maximum tidal velocity (parameter B), and the settling velocity w, while the sediment bed parameters, critical deposition velocity and hydrodynamic parameters are kept equal to their reference values.

Influence of initial concentration
As shown in figure 4, the initial concentration has little effect on the duration of the transition regime and no effect on the steady state response of the concentration if C_0 is greater than the maximum value of the equilibrium concentration (C_{max}=0.45 g/l). If C_0 is smaller than C_{max}, equilibrium is reached very rapidly and maximum concentrations remain equal to the initial concentration ($C_{max}=C_0$). This can be explained by the fact that the active layer is formed by deposition of the suspension. The maximum concentration is then constrained to remain smaller or equal to the initial concentration of suspended material.

Influence of tidal amplitude variation
The parameter B has a major influence on the equilibrium concentration response as shown in figure 5. For B<0.4, maximum concentration remains constant during the whole cycle so that the active layer is entirely eroded during each tide and does not have time to consolidate. Amplitude variations of the concentration remain small. For B=0.4, a dissymetry in the concentration reponse starts to be observed which indicates that consolidation of the active layer is occurring at neap tides. For B>0.4, all suspension deposits during neap tide and consolidates until currents intensities become larger than the critical erosion velocity of the surface layer. Concentration varies then between 0 and a maximum value which is equal to the initial concentration C_0=10 g/l.

The parameter B has two opposing effects on the cycle-averaged concentration and its amplitude of variation at steady state. As B increases, the spring tidal amplitude increases while the neap tidal amplitude decreases. Erosion (intensity) fluxes are then larger during spring tides while the duration of the deposition phase is also increased, so that consolidation of the active surface layer becomes important. These two effects appear clearly on figure 6 where both maximum and minimum concentrations are plotted as a function of B. The first effect (increase of deposition duration) explains the diminution of the cycle-averaged concentration, while the amplitude of concentration variation remains small, for B less than 0.4. For B=0.4, both concentration and concentration amplitude are minimum. Above 0.4, maximum concentration increases rapidly with increasing spring tidal amplitude up to the initial concentration (C_0=10 g/l). Amplitude of variation is also increased, because all suspended material deposits during neap tide (C_{min}=0).

Influence of settling velocity
As seen on figure 7, the settling velocity w influences both the duration of the transition regime and the values of the equilibrium concentration. The duration of transition regime is approximately proportional to w, which is natural since the deposition process is enhanced by larger settling velocities. At equilibrium, cycle-averaged concentrations are approximately inversely proportional to w while the concentration amplitude variations remain approximately equal to 0.1 g/l. The active layer structure is then little influenced by the value of settling velocity. Erosion fluxes, which are functions only of bed properties and hydrodynamic, are then independant of the value of settling velocity. At equilibrium, the cycle-averaged deposition flux, which is proportional in first approximation to the product of settling velocity and cycle-averaged equilibrium

concentration, must compensate the cycle-averaged erosion flux. The product wxC is then independent of the settling velocity which explains the fact that the equilibrium concentration is approximately inversely proportional to the settling velocity.

FILTERING METHODS

The choice of a filtering method depends on the parameters of the system response that have to be reproduced. In the following, we choose to represent the concentration time variation.

Non consolidating sediment bed

We first try to represent the whole tidal cycle by a unique mean tide of constant amplitude ($B=0$, $U_0=0.5$ m/s in equation (1)). We find that the equilibrium concentration, which is reached after about 5 tides, is 25% smaller than C_{eq} = 36 g/l of the reference case. The concentration amplitude is now less than 3g/l, which is an order of magnitude (/20) smaller than what was obtained for a fortnightly tidal cycle. We find that the most advantageous method consists in representing the whole tidal cycle by a unique tide whose period is twice the fortnightly cycle period. With a constant amplitude of 0.37 m/s , we can represent satisfactorily with less than 1% difference both maximum and mean concentrations. The frequency peak of the energy distribution at the fortnightly cycle frequency is also respected. This method is very economical, since it allows to multiply the time step by a factor 2x28.

The ratio of the lengthened tidal amplitude to the mean fortnightly tidal amplitude is equal to 0.74 in the reference case. It is believed to vary with the tidal amplitude variation over the neap-spring cycle (parameter B), as well as the other input parameters of the simulations. A sensitivity analysis would be necessary in order to get a predictive tool which could be applied for long term simulations in a 2-D case. On the other hand, the lengthening factor only depends on the ratio between the fortnightly cycle period and the tidal period, and can then be predicted at once.

Consolidating sediment bed

The concentration evolution is entirely determined at steady state by the bed structure which is formed during the transition regime. So far, there is no simple method which allows to simplify the calculation over the transition regime. A detailed calculation over the transition regime is required in order to predict accurately the bed structure.

Once equilibrium is reached, simplifications of the calculations can be made. The concentration and surface layer responses, which become periodic, can be extrapolated from a detailed calculation of the first equilibrium neap-spring cycle. Properties of the bottom consolidated layer can be simulated using very large (economical) time steps for long term simulations.

Further simplification of the calculation may be necessary for long-term simulations, particularly if hydrodynamic computation over the fortnightly cycle turns out to be too expensive. Based on the observation that the equilibrium concentration amplitude variations remain small between neap and spring tides, the hydrodynamic forcing can be simplified by a mean tide of constant mean amplitude (U_0=0.5 m/s and B=0 in equation (1)). The active layer is made of a non consolidated surface layer (C_s=110 g/l) whose thickness is equal to hxC_{max}/C_s= 4.5 cm.

CONCLUSIONS

Consolidation effects have a major influence on the response of the sediment bed and suspended sediment concentration to a fortnigtly tidal forcing. The bed structure, which is formed during transition regime, can be schematised in a two layer structure, with a thin active surface, and a bottom consolidated layer. Once equilibrium is reached, suspended sediment concentration and active layer structure variations become periodic at the period of the spring-neap cycle, with amplitude modulations at half the tidal period. The thickness and density of the active layer determine the amplitude variation of the suspended sediment concentration. Consolidation effects of the active layer during neap tides explain the hysteresis of the bed response to symetrical hydrodynamic forcing. The bottom consolidated layer undergoes consolidation at much larger time scale.

Detailed calculations over the transition regime are required by using a small time step, in order to determine accurately the bed structure. Once periodic steady state is reached, the system response calculations for long term simulation can be simplified in order to minimize computational costs. The active surface layer and sediment concentration time variation can be derived from a detailed calculation over the first neap-spring equilibrium cycle. The bottom layer properties can be determined by using much larger time scales.

Calculations can be further simplified, by schematising the hydrodynamic forcing by a "representative" mean tide of constant amplitude in order to reproduce the main features of the concentration response. The proposed schematization is valid for the reference case. The amplitude of maximum tidal current variation B is 0.4, so that concentration amplitude is small. We expect coefficient B to have a major influence on the choice of a filtrering method. For B>0.4, amplitude variations of concentration become important. A different filtering method, using for example a lengthened tide at twice the period of the fortnightly tidal cycle, will have to be tested in order to represent both the averaged concentration and the concentration amplitude variation.

In order to be able to generalize our results, the influence of the input parameters on the proposed filtering methods has to be further examined. More studies are also needed in order to predict the formation of the bed structure and to simplify calculations of the transition regime.

ACKNOWLEDGEMENTS

The present research is undertaken as part of the MAST G6 Coastal Morphodynamics research program. It is funded partly by the French Sea State Secretary and by the Commission of the European Communities, Directorate

General for Science, Research and Development under contract MAST-0035-C.

REFERENCES

1. Latteux,B., Synthesis of work carried out at L.N.H. on input filtering for morphodynamic computation under tidal condition. Preliminary internal report MAST-G6M, Project 5, 1990.

2. Latteux,B., Synthesis of work carried out at L.N.H. on tidal averaging of sediment transport. Preliminary internal report MAST-G6M, Project 5, 1990.

3. Mehta, A.J., Partheniades, E., An investigation of the deposition properties of flocculated fine sediments, Journal of Hydraulic Research, Vol.12, N°4, pp. 361-381, 1975.

4. Migniot, C., Etude de la dynamique sedimentaire marine, fluviale et estuarienne, Ph.D.Thesis, E.N.S.M., Universite de Nantes, 1982.

5. Teisson, C., Fritsch, D., Numerical modelling of suspended sediment transport in the Loire estuary, 21th ICCE, Malaga, 1988.

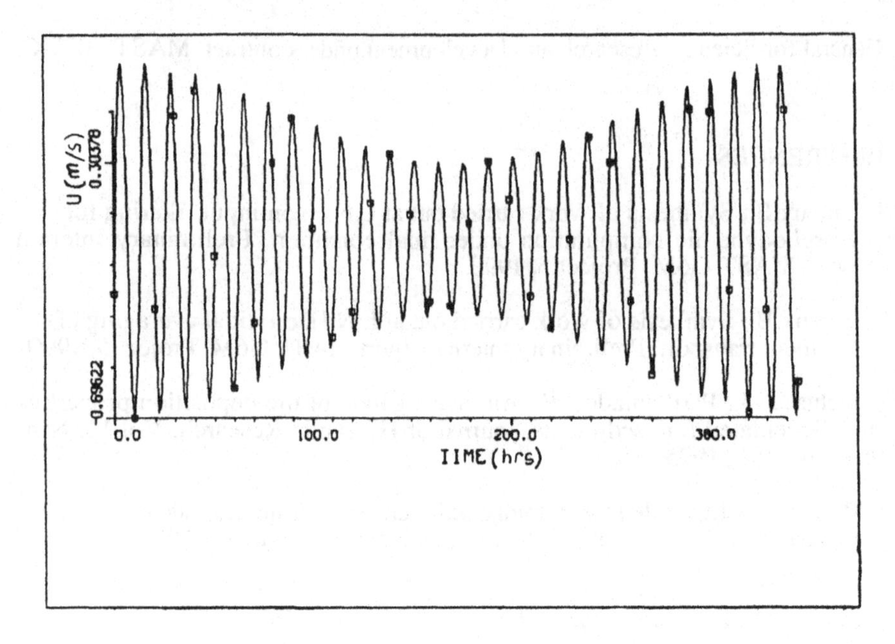

Figure 1: Hydrodynamic forcing U (t) for the Loire estuary. (B=0.4 and U0=0.5 m/s in equation (1)).

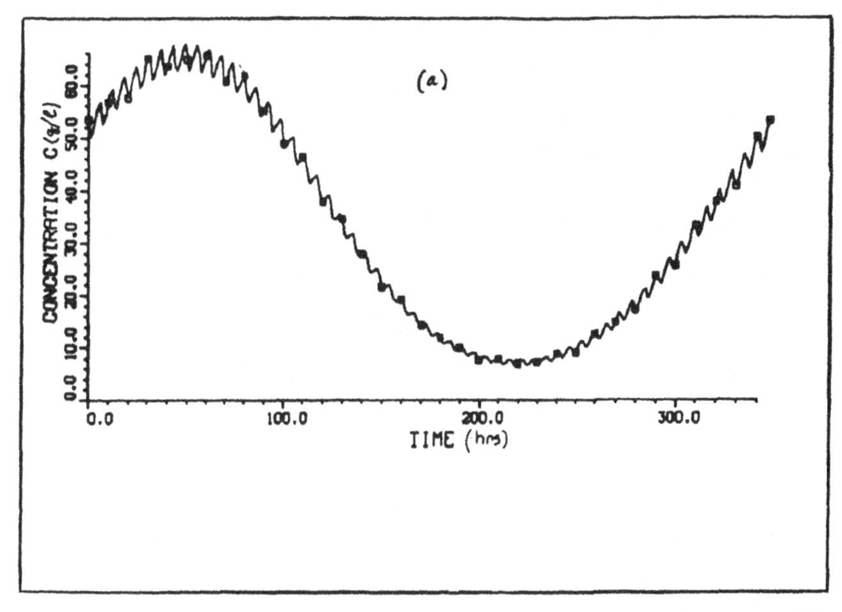

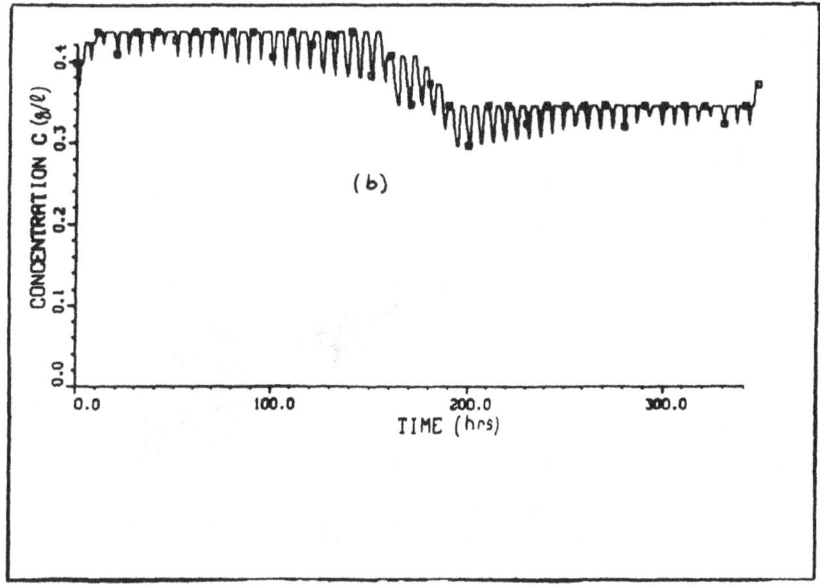

Figure 2: Equilibrium suspended sediment concentration evolution
a) Non consolidating bed b) Consolidating sediment bed.

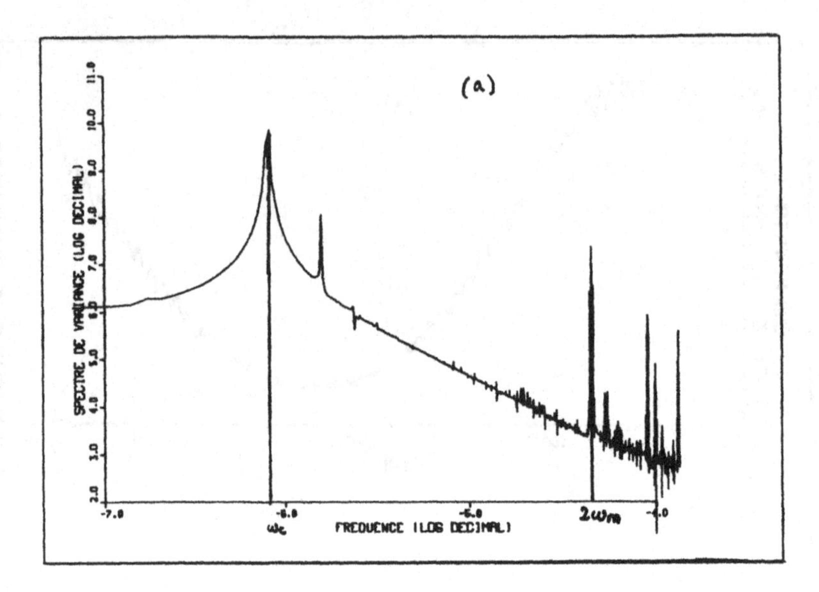

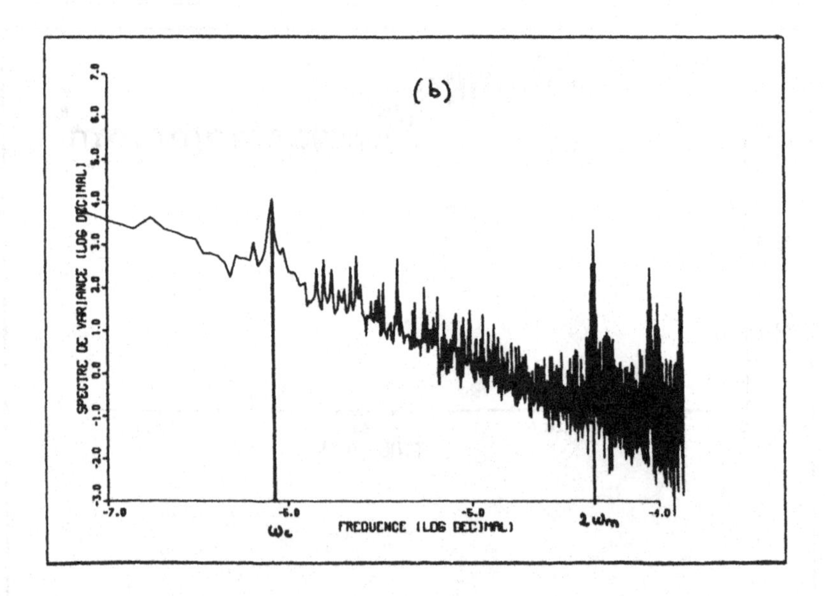

Figure 3: Spectral analysis of the equilibrium suspended sediment concentration a) Non consolidating bed b) Consolidating sediment bed.

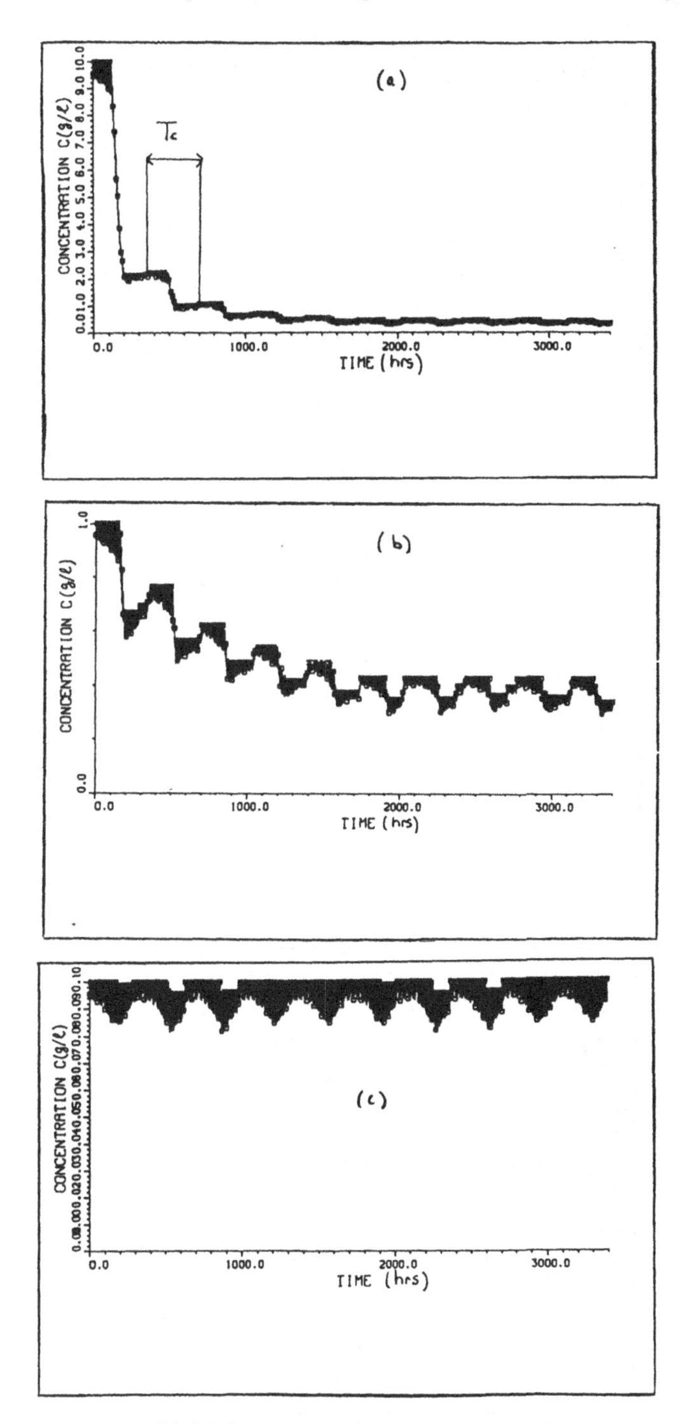

Figure 4: Influence of initial concentration on suspended sediment concentration a) C0=10 g/l , b) C0=1 g/l, c) C0=0.1 g/l.

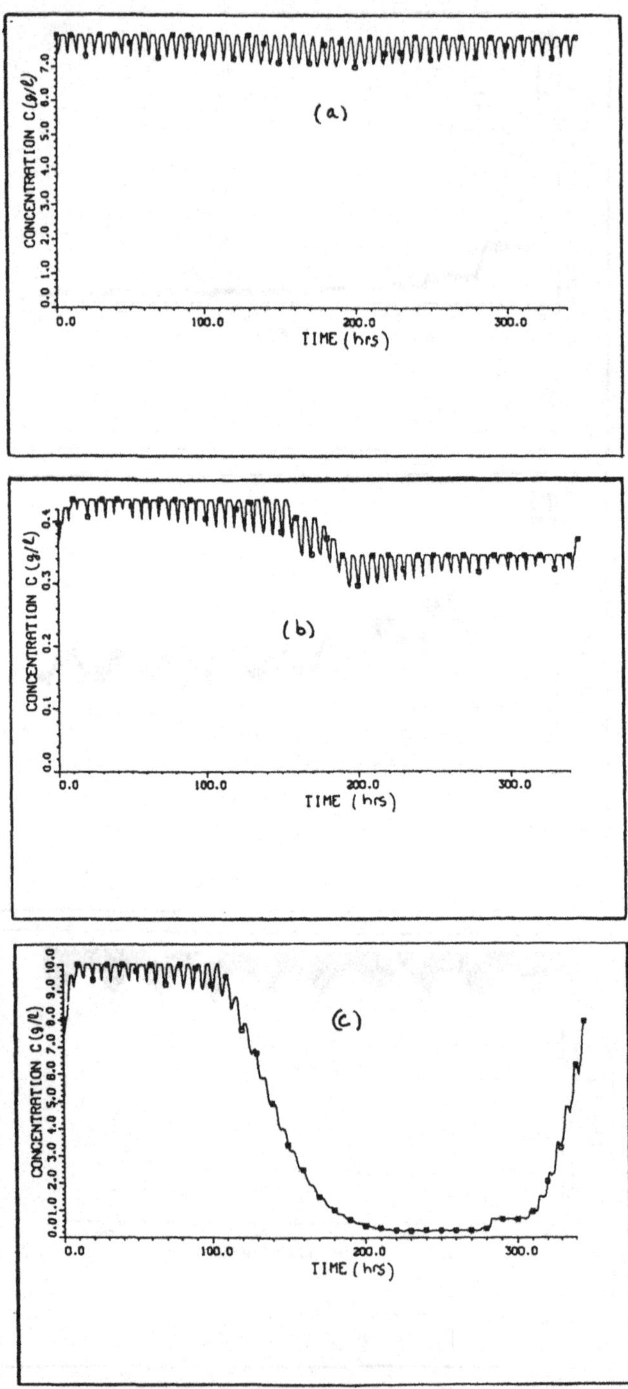

Figure 5: Influence of tidal amplitude variation over the fortnightly tidal cycle on suspended sediment concentration: a) B=0.2, b) B= 0.4, c) B=0.6.

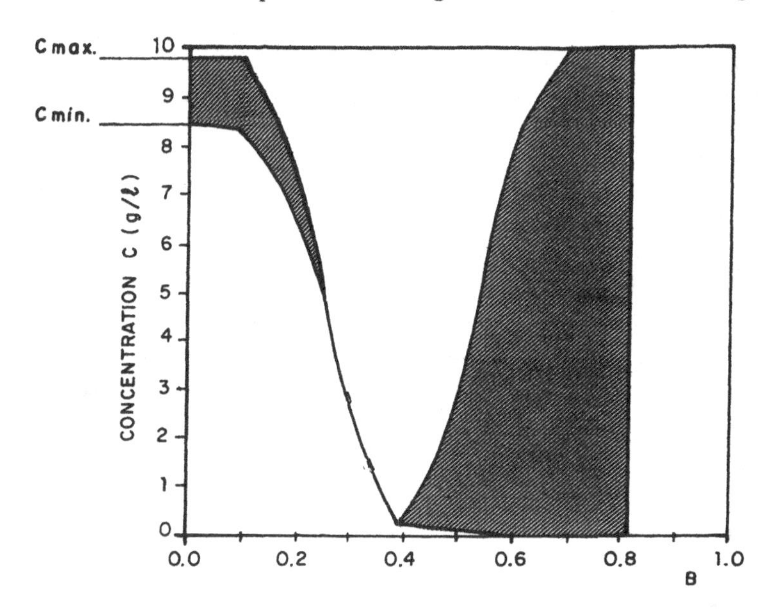

Figure 6: Influence of tidal amplitude variation over the fortnightly tidal cycle on maximum and minimum equilibrium concentrations.

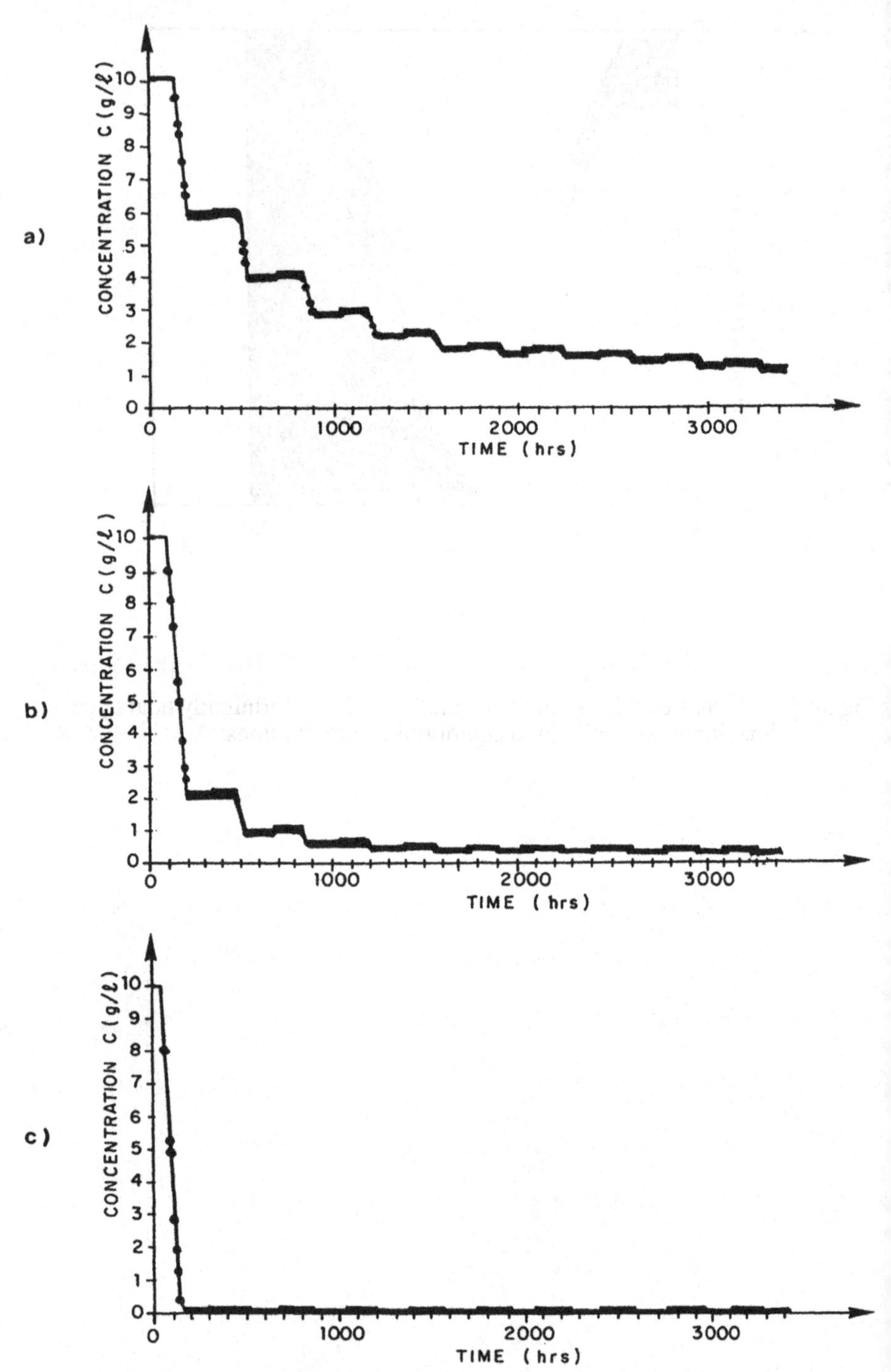

Figure 7: Influence of settling velocity on suspended sediment concentration
a) $w=0.1\ 10^{-3}$ m/s ; b) $w=0.25\ 10^{-3}$ m/s; c) $w=1.10^{-3}$ m/s.

Shoreline Change Modeling

F.M. Abdel-Aal
Dept. of Hydraulic Engineering, Cairo University, Giza, Egypt

ABSTRACT

This paper investigates physical, analytical, and numerical models used for shoreline change modeling. The equilibrium beach profile, longshore sediment transport, and the effect of structures on shoreline are also discussed.

1. INTRODUCTION

Shoreline changes are determined by the interaction of sea waves and the sediment forming the beach. In general, beaches are in a state of dynamic equilibrium. Their deformation due to change in wave climate is rather seasonal causing erosion or accretion. Prediction of the shoreline change is essential in order to take protective measures. This can be achieved by the use of the physical, analytical, and numerical models.

The prediction of shoreline change due to longshore wave induced sediment transport has traditionally been acquired using physical modeling techniques. These models provided special advantages when detailed problems were to be solved.

However, physical models have scaling problems, and are generally too small to reproduce sufficiently long reaches of the shore.

With the recent development of mathematical models, many problems dealing with shoreline changes are now solved numerically. However, these models still depend on quantification involving field measurements and physical model studies.

2. EQUILIBRIUM BEACH PROFILE

Beach profiles in nature are continuously evolving under the varying action of waves, currents, tides, and sediment transport. When these effects are maintained constant, the profile will stabilize into a so - called equilibrium beach profile.

Studies, encompassing several thousand beach profiles from nature and the laboratory have demonstrated that most beach profiles can be represented well by the monotonic form (7) ,

$$h(x) = A x^{2/3} \qquad \text{............} \qquad (1)$$

in which h(x) is the water depth at a distance, x , offshore and A is a so - called "scale parameter". It is noted that the parameter A has dimensions of length to the one - third power ($m^{1/3}$) , Fig. (1). It is clear from this figure that beaches composed of larger diameter sediments are steeper with larger A values, whereas finer sediment beaches are flatter with smaller A values. Figure (2) gives A as a function of the sediment fall velocity, w, the breaking wave height, H_b, and the wave period, T. It is obvious from this figure that steep slopes are associated with large fall velocities, small

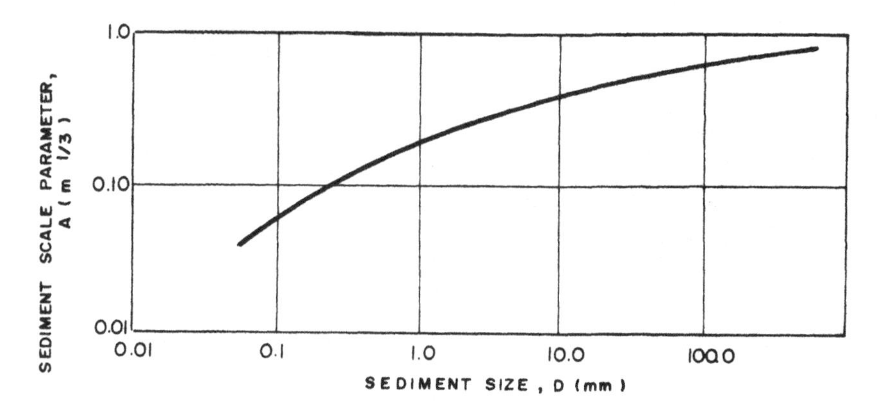

Figure(1) BEACH PROFILE PARAMETER , A , [7].

wave heights, and long wave periods. whileas, mild slopes are associated with small fall velocities, large wave heights, and short wave periods.

The equilibrium beach profile resulting from an initially plan beach could be schematized by two different cases (12) as shown in Fig. (3). In this figure m is the initial plane beach slope, R is the shoreline advancement with negative R indicating recessing, h_c is the depth of closure at which no measurable change in bottom elevation occurs, W_c is the offshore distance to h_c, a is a dimensionless berm height, and positive x is directed offshore with its origin at the intersection of water level and initial plane beach. A model for shoreline advancement or recession was developed by using conservation of sand argument and extending the equilibrium profile concept in the surf zone to the depth of closure (12).

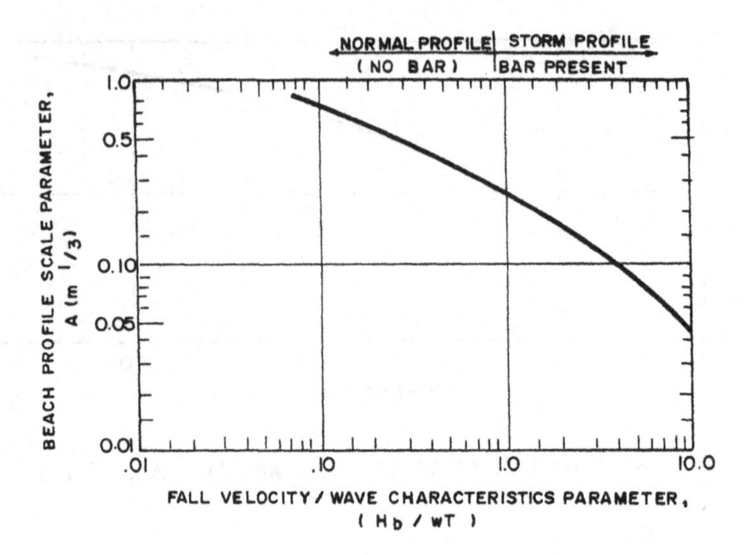

Figure (2) BEACH PROFILE SCALE PARAMETER, A, [7].

Referring to Fig. (3), conservation of sand requires the following volumetric balance per unit beach width:

$$V_1 + V_2 + V_c = V_3 + V_c \qquad \text{............} \qquad (2)$$

Using the expression for the equilibrium beach profile,

$$h(x) \;=\; A\,(x - R\,)^{2/3} \qquad \text{.............} \qquad (3)$$

Where h is water depth, and A is an empirical constant depending on sediment characteristics and wave steepness, the conservation equation can be written as,

$$\frac{m_i R^2}{2} + \int_0^{W_c} A\, u^{2/3} du = \frac{m_i (W_c + R)^2}{2} \qquad \text{.........} \qquad (4)$$

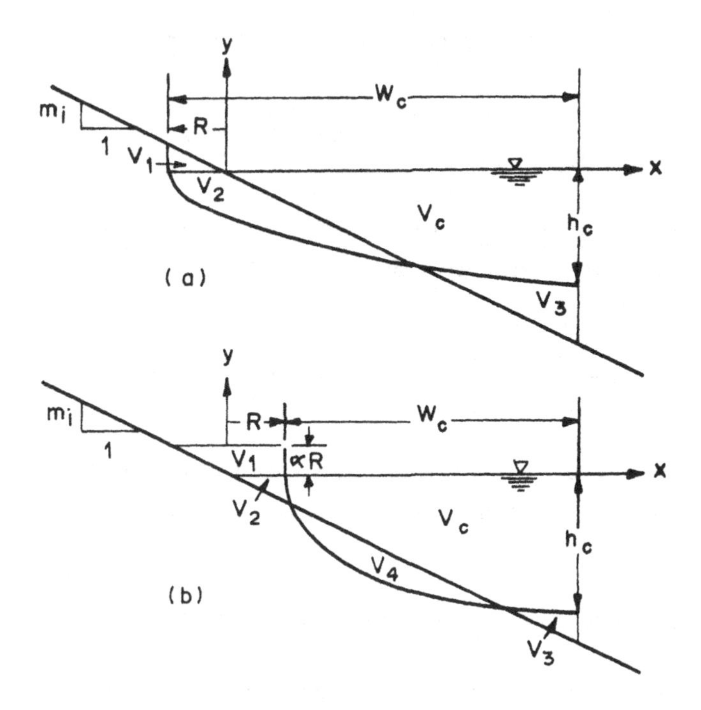

Figure (3) EQUILIBRIUM PROFILE a- EROSION; b- ACCRETION [12]

in which u is a dummy variable of integration. A little algebra and integration then gives,

$$\frac{R}{W_c} = \frac{3}{5}\,\frac{h_c}{m_i\;W_c} - \frac{1}{2} \qquad \ldots\ldots\ldots\ldots\ldots \quad (5)$$

If we define the overall beach slope in the equilibrium profile, m_e , as ,

$$m_e = \frac{h_c}{W_c} \qquad \ldots\ldots\ldots\ldots\ldots \quad (6)$$

Equation (5) can be written in terms of m_i and m_e as,

$$\frac{R}{W_c} = \frac{1}{2}\left(\frac{6\;m_e}{5\;m_i} - 1\right) \qquad \ldots\ldots\ldots\ldots\ldots \quad (7)$$

For an erosional beach, therefore, we have,

$$\frac{m_i}{m_e} \geq \frac{6}{5} \quad (\text{For} \quad R \leq 0) \quad \ldots\ldots\ldots\ldots \quad (8)$$

For an accretional beach, the same procedure is followed and we have,

$$\frac{m_i}{m_e} \leq \frac{6}{5} \quad (\text{For} \quad R \geq 0) \quad \ldots\ldots\ldots\ldots \quad (9)$$

Equation (7) therefore, can be used for shoreline advancement as well as recession.

3. LONGSHORE SEDIMENT TRANSPORT

The relationship governing the transport of sediment along a straight shoreline as expressed by Komar and Inman(9) is,

$$I = K \, P_{\ell s} \quad \ldots\ldots\ldots\ldots \quad (10)$$

in which I is the immersed weight transport rate, K is a dimensionless constant, and $P_{\ell s}$ is the longshore component of wave energy flux at the breakeline, given by,

$$P_{\ell s} = \rho g \frac{H_o^2}{8} C_{gb} \sin \alpha_b \cos \alpha_b \quad \ldots\ldots \quad (11)$$

in which ρ is the density of water, g is the acceleration of gravity, H_b is the breaking wave height, C_{gb} is the group celerity at breaking, and α_b is the wave crest angle relative to shoreline at breaking.

The in - place volumetric flux, Q , is related to I as,

$$Q = \frac{I}{\rho g(S_s-1)(1-p)} P_{\ell s} \quad \ldots\ldots\ldots\ldots \quad (12)$$

Where S_s is the sediment specific gravity, and P is the inplace sediment prosity.

Combining Eqs. (10) and (12) gives,

$$Q = \frac{K}{\rho g(S_s-1)(1-p)} P_{\ell s} \quad \ldots\ldots\ldots\ldots\ldots \quad (13)$$

It is well known that the model results for longshore sediment transport yield values of K that are approximateiy one - third of the prototype values (4) . Dean (6) presented Fig. (4) for the variation of the parameter , K, interpreted form

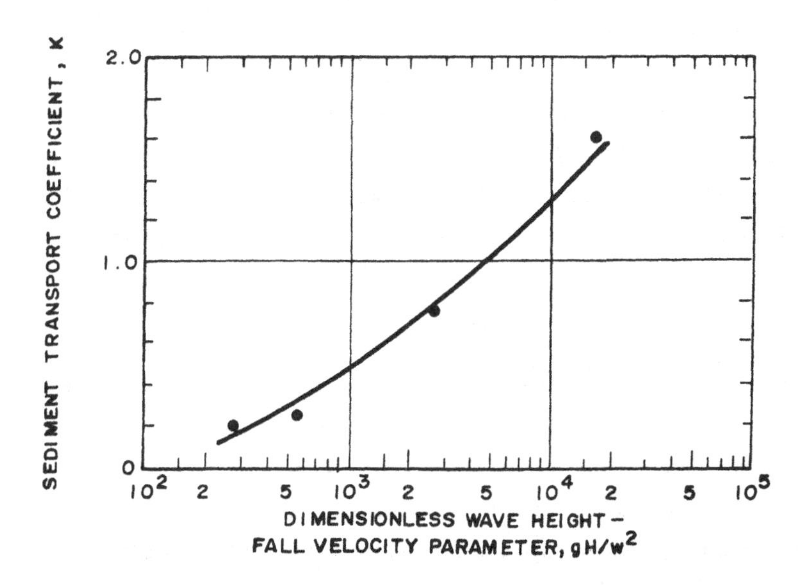

Figure (4) LONGSHORE TRANSPORT COEFFICIENT , K, [6]

the limited available field data. It is seen that the parameter provides a reasonable explanation of the difference in the laboratory and prototype data.

4. EFFECT OF STRUCTURES ON SHORELINE

Coastal structures are commonly constructed for beach erosion control. It is important to know the impact of such structures on shoreline fluctuations. Theoretical solutions are available for idealized cases of simple littoral barriers, unidirectional waves, and linearized transport (18). Most actual conditions are considerably more complex due to reversals of longshore transport, a time - varying wave height and possibly more than one structure present.

There is little quantitative information available on the behavior of shorelines protected by coastal structures. It has been known that under certain wave conditions, a vertical seawall will accelerate erosion of the beach in front of it. The end result is that the beach in front of the seawall can no longer maintain the natural equilibrium profile and the beach slope becomes steeper. An intensive field monitoring effort is needed as a first step towards achieving quantitative understanding of the influence of coastal structures on the shoreline.

Perlin (17) investigated geometric parameters determining the influence of a breakwater on the shoreline by using numerical simulation model. Kraus (10) obtained good agreement in a comparison of breaking wave height and direction and shoreline change behind a detached breakwater calculated by a numerical model and measured in a physical model.

The most obvious shore protection property of detached breakwaters is the wave sheltering afforded to the shoreline. The wave height and longshore current velocity are reduced behind these structures, and sediment carried by the longshore current is deposited in the shadow zone resulting in seaward progression of the shoreline.

5.PHYSICAL MODELS

Physical models provide two advantages when studying coastal processes. First, is that nature is used to integrate the appropriate equations which govern the phenomena. Second,is that the size of the model is much smaller than the prototype, permitting easier aquisition of relevant data.

However, there are drawbacks such as the introduction of scale effects, which are due to changes in the relative importance of various forces as the model is smaller than the prototype. The model is also more simplistic than the prototype when monochromatic waves are used to study real sea states.

When a model involves sediment transport, additional scale effects result from dissimilarities in the modeling of the sediment phase specially with respect to the particle size. Prototype large diameter particles can be readily scaled down, but scaling small diameter particles create a totally different sediment transport regime.

Many investigations have been carried out to define the requirements for physical modeling of coastal processes. Kamphuis (8) employed dimensional analysis to identify four parameters for dynamic similitude:

$$\text{Particle Reynolds Number} = \frac{u_* . D}{\nu} \quad \ldots. \quad (14)$$

$$\text{Shear Stress} = \frac{\rho . u_*^2}{(\rho_s - \rho) g D} \quad \ldots. \quad (15)$$

$$\text{Relative Density} = \frac{\rho_s}{\rho} \quad \ldots. \quad (16)$$

$$\text{Particle Geometric Similitude} = \frac{\lambda}{D} \quad \ldots. \quad (17)$$

Where u_* is the shear velocity, D is particle diameter, ν is the fluid kinematic viscosity, ρ and ρ_s are the water and sediment densities respectively, g is the acceleration of gravity, and λ is a typical length.

Kamphuis noted that it was impossible to satisfy the above four requirements except at full scale. He identified the "best" model as one in which the particle Reynolds Number requirement was relaxed, while the others were satisfied.

In the Froude modeling with an undistorted length ratio, L_r, the time ratio, T_r, and velocity ratio, V_r, are scaled according:

$$V_r \equiv T_r \equiv \sqrt{L_r} \quad \ldots\ldots\ldots\ldots. \quad (18)$$

The scale relationships for any other physical parameter of interest can be determined directly. The fall velocity, w, seems to be a relevant parameter and it must follow the Froude scaling,

$$w \equiv \sqrt{L_r} \quad \ldots\ldots\ldots\ldots. \quad (19)$$

Dean (5) developed a functional relationship for the role of particle size in the formation of beaches with summer and winter profiles. He showed that the dimensionless Fall Time parameter, F_o, should be the same for model and prototype,

$$F_o = \left(\frac{H_o}{w.T} \right)_{model} = \left(\frac{H_o}{w.T} \right)_{prototype} \quad .... \quad (20)$$

where H_o is the wave height in deep water, and T is the wave period.

6.ANALYTICAL MODELS

These models are closed-form mathematical solutions of simplified differential equations for shoreline change derived under assumptions of steady wave conditions, idealized initial shoreline and structure positions, and simplified boundary conditions. Longshore sediment transport is represented, whileas cross-shore transport is omitted, yielding a one-dimensional model. Analytical models serve as a means to examine trends in shoreline change and to investigate basic dependencies of the change on waves and initial and boundary conditions. Larson, Hanson, and Kraus (13) gave a survey of more than 25 analytical solutions of the shoreline change equations .

The theory of the shoreline model originated with Pelnard-Considere (16) . He assumed that the beach bottom, not necessarily of planar slope, always remains in equilibrium and, as a consequence, moves in parallel to itself down to a certain depth, herein called the depth of closure. Therefore, one-contour, or one-line is sufficient to describe changes in beach planform. This line is conviently taken as the shoreline. Pelnard - Considere gave closed - form mathematical solutions for certain idealized cases and verified the results through laboratory experiments.

The governing equation for the shoreline position in the shoreline model is obtained from the continuity equation of sediment, for which a predictive equation for sediment transport rate is necessary. Sediment transport and the resultant shoreline position depend on the local wave conditions, beach planform, boundary conditions and possible constraints such as the ones produced by coastal structures.

The governing equation for Fig. (5) is,

$$\frac{\partial Q}{\partial x} + h \frac{\partial y}{\partial t} = 0 \qquad \text{............} \qquad (21)$$

Where Q is the longshore sediment transport rate, X is the longshore coordinate, h is the depth at which sediment is first moved by wave action and is known as the depth of closure, y is the shoreline position measuring scour or accretion of the beach, and t is the time. Equation (21) may be solved analytically by finite differences for simple problems, such as the accretion updrift a groin or behind a detached breakwater.

Bakker, Bretler and Ross(2) extended the theory of Pelnard Considere by representing the beach profile by two contours, say y_1 and y_2. In this case there are two governing equations which incorporate the effect of the onshore or offshore

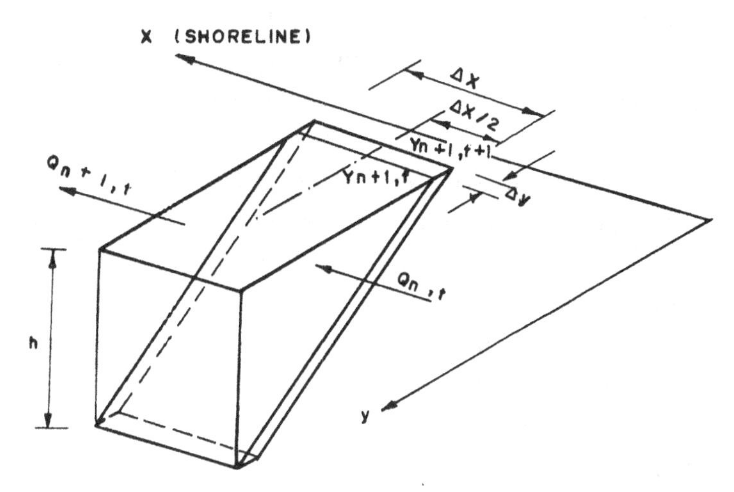

Figure (5) SHORELINE MODEL OF BEACH PLANFORM [16]

motion between the two contours due to nonequilibrium beach slope. These equations were solved for a number of cases including those of a single and multiple groins along a shoreline.

7. NUMERICAL MODELS

To avoide the assumptions and limitations of analytical models, several investigations (1,2,18,19) have been carried out to appiy numerical models to shoreline change problems. The advantages of these models are the capability to include changes in wave climate, nonlinearity of equations, and conditions of coastal structures.

Shoreline models, often referred to as one - line models, may not be used to predict local change of bottom topography, for which three - dimensional models are required. Numerical models now exist which account for cross - shore transport in a schematic way . The two - line model of Bakker (1), Bakker et al.(2), and the n - line model of Perlin and Dean (18,19) are examples.

Borah and Balloffet (3) formulated a numerical model according to the procedure developed by Le Mehaute and Soldate (14,15). Figure (6) shows a schematic representation of the beach evolution model. The continuity equation of beach sediment, equation (21) can be written as,

$$\frac{\partial y}{\partial t} = - \frac{1}{h} \frac{\partial Q}{\partial x} \qquad \text{.............} \qquad (22)$$

in which Y is shoreline distance from the x - axis, t is the time, h is the depth of closure, Q is the longshore transport rate as given by equation (13), and x is the longshore ordinate.

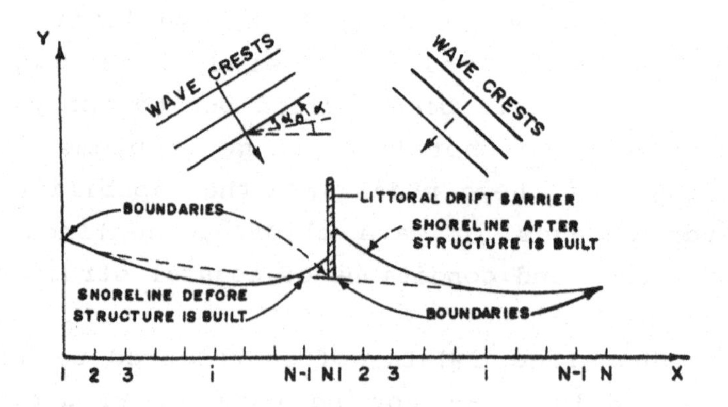

Figure (6) REPRESENTATION OF SHORELINE CHANGE [3].

The shoreline orientation with respect to the x - axis may be expressed in terms of the wave crest angle with respect to the x - axis, α , and the angle between the crest and shoreline, α_o,

$$\frac{\partial y}{\partial x} = \tan(\alpha - \alpha_o) \qquad \ldots\ldots\ldots\ldots\ldots \quad (23)$$

Combining equation (22) and equation (23), a diffusion - type sediment flux equation is obtained (14,15),

$$\frac{\partial Q}{\partial t} = \frac{\dfrac{dQ}{d\alpha_o}}{h[1+(\dfrac{\partial y}{\partial x})^2]} \frac{\partial^2 Q}{\partial x^2} \qquad \ldots\ldots\ldots\ldots\ldots \quad (24)$$

Equation (22) and equation (24) are numerically solved with given initial and boundary conditions to compute the shoreline position and the sediment flux distribution at discrete shoreline increments, and time intervals.

Numerical models, however, have their limitations for engineering use. They require special expertise, powerful computers, extensive field data collection, and verification (11).

8. CONCLUSIONS

This paper investigates shoreline change modeling. The following conclusions are reached:

1. There are still many problems in coastal hydrodynamics which need to be solved. A better understanding of surf zone hydrodynamics is important for modeling of sedimentary processes. An improved understanding of the dist ribution of longshore sediment transport

across the surf zone, mechanics of onshore - offshore sediment transport, and sediment transport under the action of both waves and currents, is needed.

2. Physical models provide advantages when detailed coastal problems are to be solved. However, because of scaling problems, an undistorted scale has to be used, and the model should be of sufficient size to minimize the effect of surface tension.
3. Analytical models are used for simplified conditions. They serve as a means to examine trends in shoreline change.
4. Numerical models are likely to evolve slowly, depending on field measurements and physical model studies to improve their accuracy and verification.

9. LIST OF REFERENCES

1. Bakker, W. T., " The Dynamics of a Coast with a Groin System" Proceeding of the Eleventh Conference on Coastal Engineering, ASCE, pp. 492 - 517, 1968.
2. Bakker, W. T. ,E.W.J. Klein Bretler and A. Roose, "The Dynamics of a Coast with a Groin System," Proceedings of the Twelfth Conference on coastal Engineering, ASCE,pp.1001 - 1020, 1970 .
3. Borah , D. K. and A. Balloffet , " Beach Evolution Caused by Littoral Drift Barrier," Journal of Waterway, Port, Coastal and ocean Engineering, Vol. 111, No. 4, July 1985.
4. Das, M. M., " Suspended Sediment and Longshore Sediment Transport Data Review," In Proc. Int. Conf. Coastal Engng,pp. 1027 - 1048 , 1972.

5. Dean, R. G.," Heuristic Models of sand Transport in the Surf Zone, " The Institution of Engineers of Australia Sydney, pp. 208 - 213, 1973.
6. Dean, R. G., " Review of Sediment Transport Relationship and the Data Base," In proc. Workshop Coastal Sediment Transport. University of Delaware Sea Grant Program, pp. 25 - 39, 1978.
7. Dean, R. G.," Sediment Budget, principles and Applications," 20 th ICCE Dynamics of Sand Beaches, Chinese Institute of Civil and Hydraulic Engineering, Taipei, R.O.C.,Nov.1986.
8. Kamphuis, J. W.," The Coastal Mobile Bed Model," C.E. Research Report No. 76, Dept. of Civil Eng., Queens University, May, 1982.
9. Komar, P. D. and D. L. Inman, " Longshore Sand Transport on Beaches, "Journal of Geophysical Research, Vol. 75,PP. 5914 - 5927, 1970.
10. Kraus, N. C.,"Applications of a Shoreline Prediction Model,"Proc. Coastal Structures 83,ASCE,PP. 632 - 645, 1983.
11. Kraus, N. C., "Beach Change Modeling and the Coastal Planning Process,"Proc. Coastal Zone 89, ASCE,PP.553 - 567, July, 1989 .
12. Kyungduck, S., R. A. Dalrymple, "Experssion for Shoreline Advancement of Initially Plan Beach,"Journal of Waterway, Port, Coastal and Ocean Engineering, Vol. 114, No. 6, Nov. 1988.
13. Larson, M., Hanson, H., and Kraus, N.C.," Analytical Solutions of the One - line Model of Shoreline Change,"Tech. Rep. CERC - 87 - 15, U.S.Army Engr. Waterways Expt. Station, Coastal Engrg. Res. Center, Vicksburg, Miss. 72 PP., 1987.

14. Le Mahaute, B., and M. Soldate,"Mathematical Modeling of Shoreline Evoution,"Proceedings, the Sixteenth Coastal Engineering Conference, held in Hamburg, Germany,ASCE,New York, N.Y.,pp. 1163 - 1179, 1978.

15. Le Mahaute, B., and M. Soldate, "A Numerical Model for Predicting Shoreline Changes," Miscellaneous Report No. 80 - 6, Coastal Engineering Research Center, U.S. Army Corps of Engineers, Fort Belvoir, Va.,72 p., July 1980.

16. Pelanard - Considere, R. , " Essai de Theorie de L'evolution des Forme Rivages on Plages de Sable et de Galets,"4th Journess de L'Hydraulique, Energies de la Mer, Question III, Rapport No.1, pp. 289 - 298, 1954.

17. Perlin, M., " Predicting Beach Planforms in the Lee of a Breakwater," Proc. Coastal Structures 79, ASCE, PP. 792 - 808, 1979.

18. Perlin, M. and R. G. Dean , "Prediction of Beach Planforms with Littoral Controls," Proceedings of 16th Coastal Engineering Conference, ASCE, pp. 1818 - 1838, 1978.

19. Perlin, Mand R.G. Dean, "A Numerical Model to Simulate Sediment Transport in the Vicinity of Structures, "Miscellaneous Report No. 83 - 10, US Army Engineer Waterways Experiment Station, Coastal Engineering Research Center, 119 p., 1983.

Modelling Siltation at Chukpyon Harbour, Korea

B.A. O'Connor (*), H. Kim (*), K.-D. Yum (**)

(*) Civil Engineering Department, The University of Liverpool, Brownlow Street, Liverpool L69 3BX, U.K.

(**) Ocean Engineering Division, Korea Ocean Research and Development Institute, P.O. Box 29, Ansan, Seoul, 425-600, Korea

ABSTACT

The present paper describes the application of a range of computer models to describe the seasonal variation of seabed bathymetry in the vicinity of Chukpyon Harbour, which is situated on the east coast of Korea. The coastal conditions at the harbour site are dominated by waves and wave-induced currents. Consequently, wave-period-average models were used to predict waves and wave-induced-currents for three wave conditions for two wave directions. A new three-dimensional wave-period-average sediment model was then used to describe particular patterns of seabed change, which were, in turn, combined to predict seasonal changes. Use was made of a variety of field data on waves, currents, and sediment transport rates to set up and calibrate the model. Comparison of model results with field data confirmed the presence of residual sediment movement towards the harbour entrance and the location of shoal zones near the west groyne and east breakwater. The model was subsequently used to study new additions to the existing structures to control nearshore siltation.

INTRODUCTION

Considerable progress has been made over the last decade in the computer modelling of coastal processes, O'Connor [1]. Unfortunately, many of the newly-emerging models are too expensive to use for the study of long-term changes in coastal bathymetry. In addition, many models are untested in field situations. The present paper describes the application of a series of relatively simple models to study the seasonal variation of seabed bathymetry in the vicinity of Chukpyon Harbour, which is situated on the east coast of Korea, Fig. 1.

MODEL PHILOSOPHY

Initial work at Chukpyon began in 1905, but in 1963 increasing harbour siltation led to the construction of a western breakwater in an attempt to reduce the problem. Subsequently, the eastern breakwater was extended between 1968-72 in a further attempt to limit harbour dredging, but with limited success, Fig. 1. In 1986, a field study was initiated by the Pohang District Maritime and Port Authority in order to assist with both an understanding of the problem and the design of new works. Further fieldwork was undertaken in 1987, 1988 and 1989 in association with the Korean Ocean Research and Development Institute (KORDI).

The fieldwork showed that tidal effects at the site were small with the amplitude of the major tidal components (M_2, S_2, O_1, K_1) being only 4.9cm, 1.1cm, 4.3cm and 4.2cm, respectively, while wind and density-induced currents were also found to have a negligible influence on sediment movements. Wave periods were found to be in the range 4-10s and to be associated with significant wave heights up to 3m in size. The Summer and Autumn wave climates were found to be dominated by southerly waves while the Winter and Spring was dominated by northerly waves, see Fig. 2.

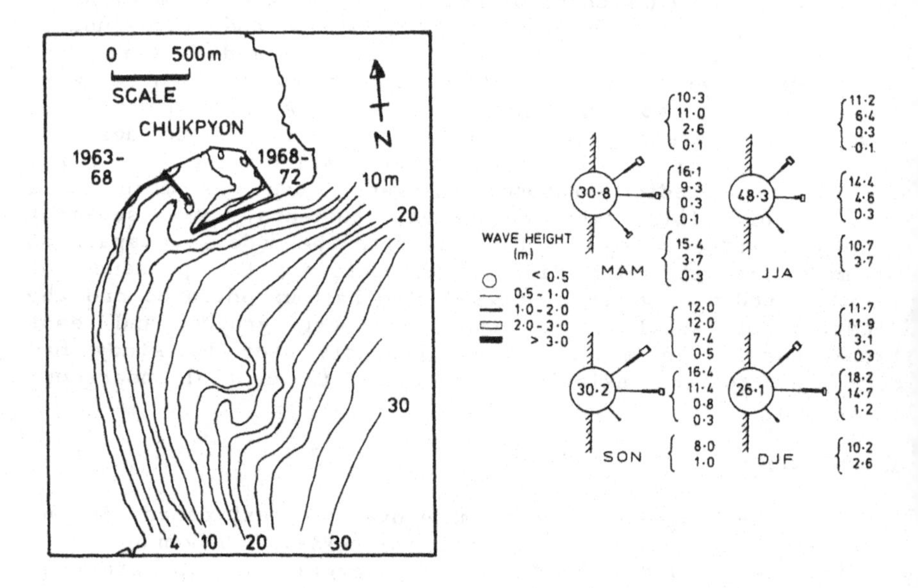

Fig. 1. Chukpyon Harbour   Fig. 2. 3-Monthly wave climate

The fieldwork suggests that the sediment transport processes at Chukpyon are likely to be dominated by waves and nearshore wave-induced currents, produced by winds from only a limited number of directions. Consequently, it is possible to use a series of simplified models to study sediment movements at the site and to build up long-term bed level changes by combining together the seabed accretion/erosion patterns produced by the dominating seasonal winds.

The individual seabed accretion/erosion patterns are obtained by using a wave propagation model to route the seasonal design wave (characterized by its significant height and peak period of the wave spectrum) inshore. The computed wave height field is then used to calculate wave radiation stresses in a wave-induced current model, assuming no interaction between the wave and current fields. The results from the wave and current models are then used to determine the movement of sediment in suspension and as bed load in a new three-dimensional sediment transport model. Finally, seabed accretion/erosion zones are determined from a sediment continuity equation. The total model system is illustrated diagrammatically in Fig. 3.

WAVE PROPAGATION MODEL

The field measurement programmes made use of wave rider buoys to determine the annual offshore wave climate. Subsequent data analysis provided details of design wave conditions for the predominant seasonal directions. These offshore conditions are routed inshore taking account of the processes of refraction, shoaling, diffraction and wave breaking using a simple finite difference solution of the equations for wave direction and wave energy flux; frictional effects being neglected.

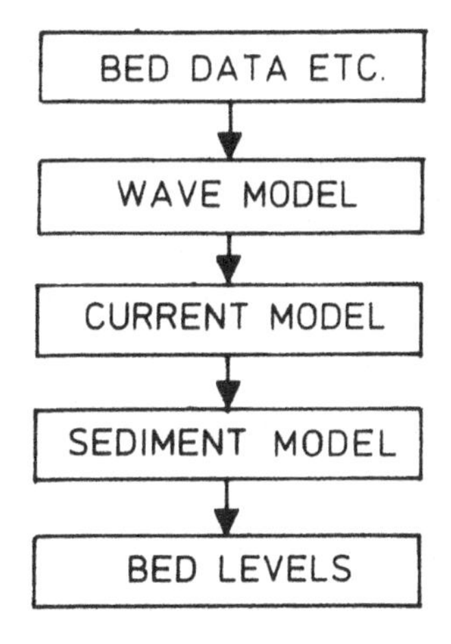

Fig. 3 Model system

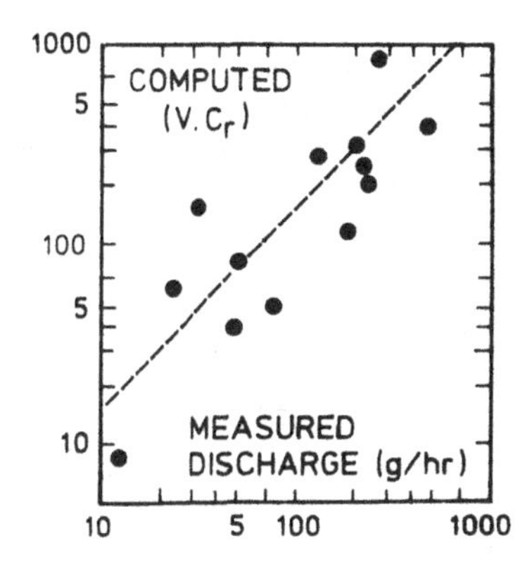

Fig. 4 Model and field sediment discharges

The model follows the approach of Ebersole et al [2]. The velocity potential (ϕ) for linear waves is expressed in complex notation as:-

$$\phi = ae^{is} \quad (1)$$

where a(x,y) is the wave amplitude; s(x,y) is the wave phase function (= $k_x x + k_y y - \sigma t$, where k_x, k_y are the x,y components of the wave number (k) and σ is the wave frequency). Substitution of equation (1) into the mild-slope form of the wave energy equation of Smith and Sprinks [4]:-

$$\frac{\partial}{\partial x}\left(cc_g \frac{\partial \phi}{\partial x}\right) + \frac{\partial}{\partial y}\left(cc_g \frac{\partial \phi}{\partial y}\right) + \sigma^2 \frac{c_g}{c} \phi = 0 \quad (2)$$

leads to Berkhoff's [3] equations for wave amplitude and direction:-

$$\frac{1}{a}\left\{\frac{\partial^2 a}{\partial x^2} + \frac{\partial^2 a}{\partial y^2} + \frac{1}{cc_g}\left(\nabla a.\nabla(cc_g)\right)\right\} + k^2 - |\nabla s|^2 = 0 \quad (3)$$

$$\nabla.(a^2 cc_g \nabla s) = 0 \quad (4)$$

where ∇ is the spatial gradient operator ($\partial/\partial x + \partial/\partial y$); x,y are longitudinal and lateral spatial cartesian co-ordinates; and c, c_g are the wave phase velocity and group velocity, respectively. Irrotationality of wave phase function is also assumed, that is:-

$$\nabla \times (\nabla s) = 0 \quad (5)$$

Equations (4) and (5) can be transformed to give two simple equations for wave direction and wave amplitude:-

$$\frac{\partial}{\partial x}(|\nabla s|\sin\theta) - \frac{\partial}{\partial y}(|\nabla s|\cos\theta) = 0 \quad (6)$$

$$\frac{\partial}{\partial x}(a^2 cc_g |\nabla s|\cos\theta) + \frac{\partial}{\partial y}(a^2 cc_g |\nabla s|\sin\theta) = 0 \quad (7)$$

where θ is the direction of wave propagation relative to the x axis.

Equations (6) and (7) are solved on a square grid of computation points using Ebersole et al's [2] simple forward marching technique with the diffraction terms in Eq. (3) omitted at grids near the sawtooth-shaped land boundaries. Wave breaking is simulated by a simple limiting depth criteria. At the open ends of breakwaters, Penney and Price's [5] diffraction theory for uniform depth is used to give wave height information in the shadow zones.

WAVE-INDUCED FLOW MODEL

Once details of the wave number and wave amplitude are known, it is possible to determine the depth-average wave-induced current pattern in the study area using the same computational grid as for the wave model.

The current model uses the standard depth-average equations of fluid momentum and continuity with the inclusion of Radiation Stress Terms, Longuet-Higgins and Stewart [6], Yoo and O'Connor [7]:-

$$\frac{\partial \eta}{\partial t} + \frac{\partial (du)}{\partial x} + \frac{\partial (dv)}{\partial y} = 0 \tag{8}$$

$$\frac{\partial u}{\partial t} + \frac{u\partial u}{\partial x} + \frac{v\partial u}{\partial y} + \frac{g\partial \eta}{\partial x} + \frac{1}{\rho d}\left[\frac{\partial S_{xx}}{\partial x} + \frac{\partial S_{xy}}{\partial y}\right] + F_x = 0 \tag{9}$$

$$\frac{\partial v}{\partial t} + \frac{u\partial v}{\partial x} + \frac{v\partial v}{\partial y} + \frac{g\partial \eta}{\partial y} + \frac{1}{\rho d}\left[\frac{\partial S_{xy}}{\partial x} + \frac{\partial S_{yy}}{\partial y}\right] + F_y = 0 \tag{10}$$

where d is the water depth, including wave set-up; u,v are depth-average velocities in the x,y co-ordinate directions; t is time; g is the acceleration due to gravity; ρ is the fluid density; η is the wave-period-average variation in mean water level; F_x, F_y are bed friction terms; S_{xx}, S_{xy}, S_{yy} are the radiation stress terms given by the equations:-

$$S_{xx} = S_{11} \cos^2\theta + S_{22}\sin^2\theta \tag{11a}$$

$$S_{xy} = S_{11} \sin\theta \cos\theta - S_{22} \cos\theta \sin\theta \tag{11b}$$

$$S_{yy} = S_{11} \sin^2\theta + S_{22} \cos^2\theta \tag{11c}$$

$$S_{11} = E(2c_g/c - 0.5); \quad S_{22} = E(c_g/c - 0.5) \tag{12}$$

with E being the wave energy per unit area of sea surface ($\rho g a^2/2$).

Lateral mixing terms are neglected in equations (9) and (10), since this leads to a cheaper model without the need for expensive extra sub-models to determine mixing coefficients. However, the effects of lateral mixing are not neglected but are introduced in a simpler way by spatial smoothing of the radiation stress terms (a linear average of the four values at neighboring points was used), together with the numerical smoothing that is present in all models (Kim [10]). Field data is used to control the degree of smoothing used in the model.

Equations (8-12) are solved using a simple explicit finite difference scheme on a space-staggered mesh; no-slip boundary conditions being used on solid boundaries with the seaward boundary being set far enough offshore to avoid reflection problems.

SEDIMENT TRANSPORT MODEL

The movement of sediment in the coastal zone is assumed to be by a combination of bed and suspended load, assuming that undertow effects can be neglected for long-term calculations. Because of the spatial non-uniformity of the sediment field near structures and the large vertical concentration gradients produced by sandy sediments, a wave-period-average (WPA), three-dimensional suspended load equation was chosen for the present study, that is:-

$$\frac{\partial c}{\partial t} + \frac{u\partial c}{\partial x} + \frac{v\partial c}{\partial y} + (w - \omega_f)\frac{\partial c}{\partial z} - \frac{\partial}{\partial x}\left(\epsilon_x \frac{\partial c}{\partial x}\right)$$

$$- \frac{\partial}{\partial y}\left(\epsilon_y \frac{\partial c}{\partial y}\right) - \frac{\partial}{\partial z}\left(\epsilon_z \frac{\partial c}{\partial z}\right) = 0 \qquad (13)$$

where c is the WPA suspended sediment concentration; u,v,w are WPA flow velocities; ω_f is a representative particle fall velocity; $\epsilon_x, \epsilon_y, \epsilon_z$ are WPA sediment diffusion coefficients in the x,y,z (vertical) co-ordinate directions, respectively.

Equation (13) is solved using the same horizontal and lateral grid system as the wave and current models. Details of the suspended sediment concentration over the flow depth at each spatial grid point in plan is provided by non-dimensionalizing the flow depth and then using an exponentially-spaced vertical grid system so that details of the large vertical concentration gradients can be resolved using the minimum number of vertical grid points.

Solution of equation (13) also requires specification of sediment concentrations at the water surface and seabed. At the surface, a no-flux condition is used, that is:-

$$\omega_f c + \epsilon_z \frac{\partial c}{\partial z} = 0 \qquad (14)$$

At the seabed, a specified WPA concentration is used related to the WPA combined wave and current shear stress, that is:-

$$c_r = A \, \overline{|\vec{\psi}_{wc}|^{1.5}} \qquad (15)$$

where c_r is determined at half a ripple height above the seabed; A is a scale factor with a value of some 6 x 10^{-6} for c_r in concentration by volume; and the overbar indicates a time-average over the wave period $({}_0\int^T (\quad) dt/T)$. ψ_{wc} is a non-dimensional form of grain Froude number given by the equation:-

$$\vec{\psi}_{wc} = \frac{(\vec{\tau}_c + \alpha \vec{\tau}_w)}{(s - 1)\, \rho g d_{50}} \qquad (16)$$

where d_{50} is a representative (50% finer) bed grain size; (s - 1) is the submerged relative density of the sediment grains ($= \rho_s/\rho - 1$ with ρ_s the density of the sediment grains); α is a scale factor with a value of some 0.2; τ_c, τ_w are the instantaneous bed shear stresses (vectors) for the current and wave fields, respectively, defined by the equations:-

$$\vec{\tau}_c = \rho \, c_{fc} \, \vec{u} |\vec{u}| \qquad (17)$$

$$\vec{\tau}_w = 0.5 \, \rho \, c_{fw} \, \vec{u}_\infty |\vec{u}_\infty| \qquad (18)$$

where c_{fc}, c_{fw} are current and wave friction factors, see O'Connor and Yoo [8], and u, u_∞ are the depth-mean current velocity and nearbed orbital wave velocity vectors, respectively.

The WPA diffusion coefficients must also be specified. Constant values are used for ϵ_x, ϵ_y with typical values of 1E-6m^2/s. The vertical coefficient is assumed to contain wave and current contributions, that is:-

$$\epsilon_z = \kappa U_{*c} z(1 - z/d) + \kappa(\overline{u}_{*wc} - U_{*c})\delta \qquad (19)$$

where κ is von Karman's constant; U_{*c} is the current shear velocity ($= \surd(|\tau_c|/\rho)$ and $\overline{u}_{*wc}$ is a combined wave/current shear velocity ($= \surd(|\tau_c + \alpha\tau_w|)/\rho$. No allowance is made for enhanced mixing in the surf zone due to breaking waves.

The combined set of equations (13 - 19) are solved using a splitting-technique to separate advective and diffusive transport. A Crank-Nicolson implicit approach is used for the vertical diffusive part while a characteristic approach is used for the advective and settling parts, see O'Connor and Nicholson [9]. Zero flux conditions are used on solid boundaries and specified concentrations, based on field data, used on inflow boundaries. The model is given a "cold-start" and run until equilibrium concentrations are obtained over the study area.

Bed level changes (z_b) are obtained by use of a total sediment continuity equation:-

$$\frac{\partial z_b}{\partial t} + \frac{1}{1 - n}\left(\frac{\partial q_x}{\partial x} + \frac{\partial q_y}{\partial y}\right) = 0 \qquad (20)$$

where n is the porosity of the deposited bed; q_x, q_y are the total bed and suspended loads, respectively.

The suspended load is determined by depth-integration of the suspended load concentrations and the wave-induced currents, assuming that the latter follow a logarithmic distribution over the flow depth, that is:-

$$u(z) = \frac{V \ln(z/z_o)}{\ln(0.368d/z_o)} \qquad (21)$$

where V is the total velocity ($\surd(u^2 + v^2)$) and z_o is the effective roughness of the seabed determined from field velocity profiles. Allowance is also made for horizontal and lateral diffusive transport.

The bed load is determined from an empirical equation based on Brown's formula (Kim [10]), which relates the sediment transport rate to the WPA grain Froude number ψ_{wc}, that is:-

$$|\vec{\phi}_b| = B\psi_c^{0.5} \; \overline{|\vec{\psi}_{wc}|^{1.5}} \qquad (22a)$$

and
$$\vec{\phi}_b = \frac{\vec{q}_b}{\omega_f d_{50}} \tag{22b}$$

where $\vec{q}_b$ is the vector of bed load transport in insitu volume (m^3) per unit width, per unit time; ψ_c is the grain Froude number of current-only motion and B is a scale factor with a typical value of 20. It is also assumed that the bed load moves in the direction of the wave-induced current vector.

APPLICATION TO CHUKPYON HARBOUR

A wide range of general data was available to test the model system. However, the period 10.6.87 to 3.11.87 was eventually selected, since bathymetric charts were available for the start and end of the period and the Harbour area was subjected to two grade B typhoons on the 16.7.87 (Thelma) and 30.8.87 (Dinah) with significant wave heights of 5.4m and 5.6m, respectively. Fortunately, continuous wave recordings were made during the period along with some measurements of sediment transport rate and suspended sediment concentration at nine points over the study area using bed load traps (ARNHEM) and horizontal drogue nets supplemented by a pump sampler on a boundary layer bed frame (Kim [10]).

Analysis of the wave records suggested that the wave climate over the study period could be divided into six representative groups composed of two predominant directions (ENE and ESE) and three wave height bands with a peak period of 7s: the largest wave heights corresponding to the typhoon conditions.

The output from the wave and current models was tested against field data at particular points and found to give a realistic representation of field conditions (Kim [10]). The field sediment data was next used to calibrate the constants in the sediment model. Quite reasonable results were obtained for both reference concentration (c_r) and vertical sediment concentration profiles, see Figs. 4 and 5.

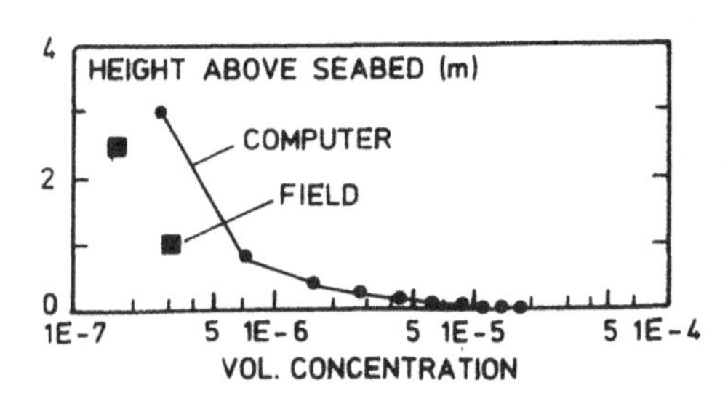

Fig. 5 Model and field concentrations

Typical wave height and wave-induced current results for the two typhoon cases are shown in Figs. 6-8. It is clear that the Eastern Breakwater provides significant protection to the harbour but that both wave directions produce strong nearshore currents which are capable of moving sediment into the harbour area.

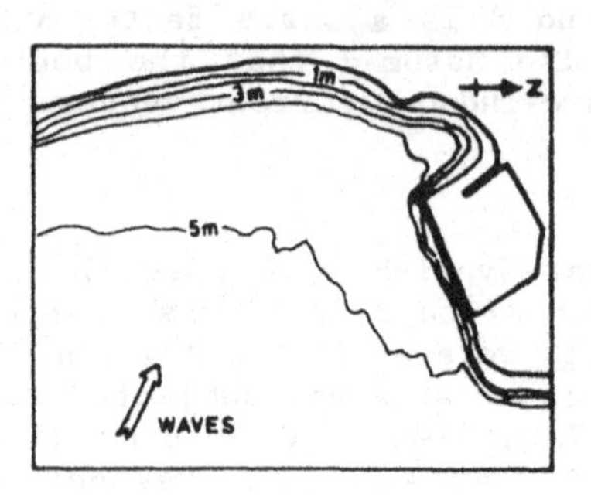

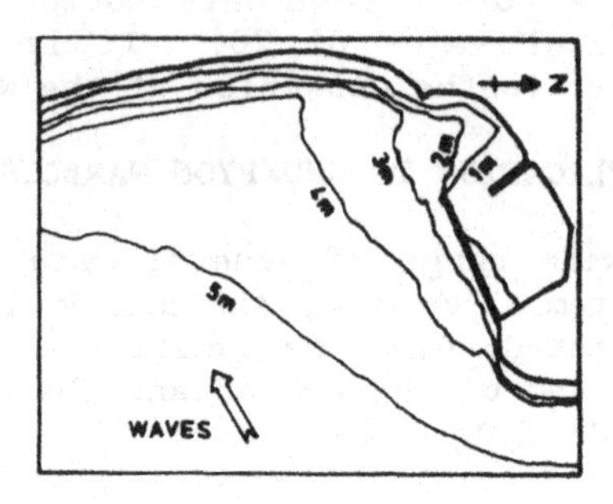

Fig. 6 Model wave heights, ESE, ENE typhoons

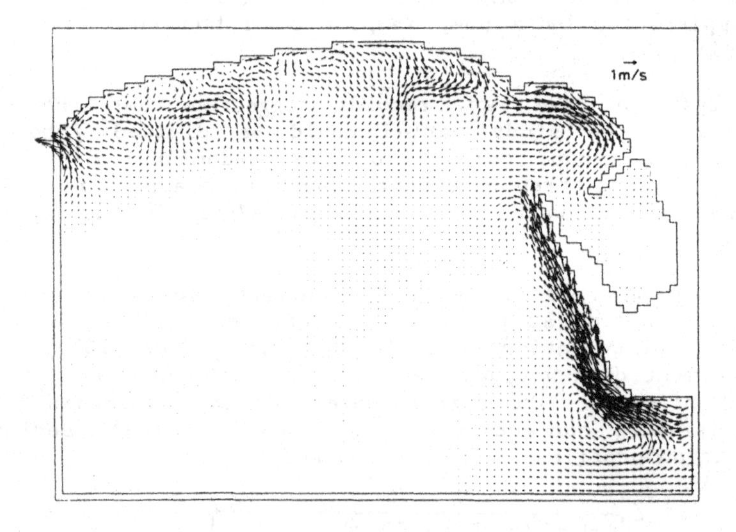

Fig. 7 ENE model currents (H_s = 5.5m)

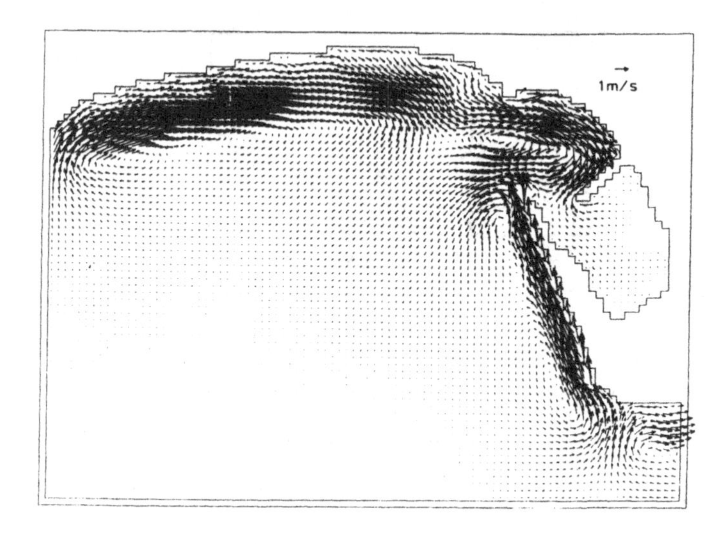

Fig. 8 ESE model currents (H_s = 5.5m)

Comparison of the sediment model bed level changes with those from the bathymetric charts confirms the effect of the wave-induced currents. Figs. 9 and 10 clearly show accretion near the Western Breakwater and off the tip of the Eastern Breakwater, although coastline accretion is less well represented. The erosion zone off the harbour entrance is also well reproduced.

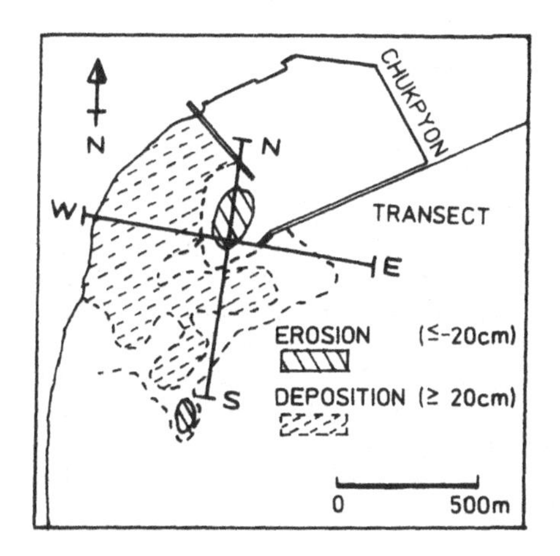

Fig. 9 Field bed level changes (117 days)

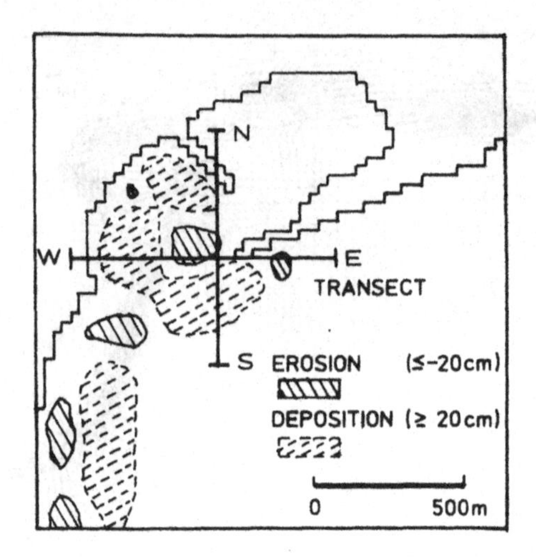

Fig. 10 Model bed level changes (117 days)

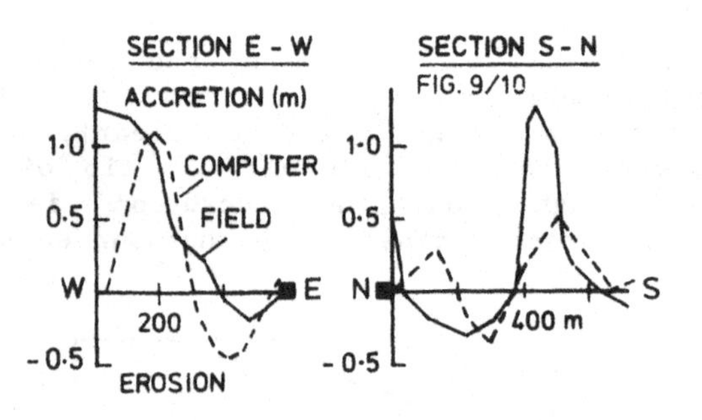

Fig. 11 Bed level changes (117 days)

Detailed bed level changes on an EW and NS transect are also shown on Fig. 11. Quite good quantitative agreement is obtained along the EW transect with less good agreement along the NS transect, although a very good location is obtained for the major shoal area. The degree of fit with field data was deemed sufficient to enable the model system to be used to study various structure modifications in order to control the nearshore siltation rates. Further work is currently in progress to improve the specification of the mixing and bed boundary conditions in the sediment model.

CONCLUSIONS

A relatively simple system of nearshore coastal models has been developed to study a particular harbour siltation problem at

Chukpyon in eastern Korea. The lack of nearshore tidal currents in the study area enables wave and current fields to be decoupled while the dominance of sediment movements by typhoon conditions has enabled the wave climate to be synthesized by a limited number of wave patterns. A new three-dimensional WPA sediment model has been interfaced with both an inshore wave prediction model and a wave-induced current model and shown to be capable of giving quite a good representation of sediment load and seabed changes in the vicinity of the harbour. The calibrated model system has proved useful in studying ways of controlling nearshore siltation.

ACKNOWLEDGEMENTS

The authors are indebted to Frances Zimmermann and Barbara Cotgreave for assistance with the preparation of the present paper. In addition, the authors are grateful for financial assistance from the Korean Government and British Council as well as the Commission of the European Directorate General for Science, Research and Development under Contract Nos. MAST-0035(C) and 0036(C).

REFERENCES

1. O'Connor, B.A. Suspended Sediment Transport in the Coastal Zone, Proc. Int. Symp. on the Transport of Suspended Sediments and its Mathematical Modelling, Florence, 1991, pp 17-63.

2. Ebersole, B.A., Cialone, M.A. and Prater, M.D. Regional Coastal Processes Numerical Modelling System: Report 1: RCPWAVE- A linear wave propagation model for engineering use, CERC Technical Report CERC-86-4, 1986.

3. Berkhoff, J.C.W. Mathematical Models for Simple Harmonic Linear Water Waves, Wave Diffraction and Refraction, Publication No. 1963, Delft Hydraulic laboratory, Delft, the Netherlands, 1976.

4. Smith, R. and Sprinks, T. Scattering of Surface Waves by a Conical Island, Journal of Fluid Mechanics, Vol. 72, pp 373-384, 1975.

5. Penney, W.G. and Price, A.T. The Diffraction Theory of Seawaves and the Shelter Afforded by Breakwaters, Phil. Trans. Roy. Soc. London, Ser. A, 244, pp 253-263, 1952.

6. Longuet-Higgins, M.S. and Stewart, T.W. Radiation Stress and Mass Transport in Gravity waves on Steady Non-uniform Currents, Journal of Fluid Mechanics, 10, pp 529-549, 1964.

7. Yoo, D.H. and O'Connor, B.A. Mathematical Modelling of Wave-induced Nearshore Circulations, in Proc. 20th Coast. Eng. Conf. (Ed. Edge, W.L.) pp 1667-1682, American Society of Civil Engineers, New York, 1986.

8. O'Connor, B.A. and Yoo, D.H. Mean Bed Friction of Combined Wave-current Flow, Coastal Engineering, 12, pp 1-21, 1988.

9. O'Connor, B.A. and Nicholson, J. A Three-dimensional Modelling of Suspended particulate Sediment Transport, Coastal Engineering, Vol. 12, pp 157-174, 1988.

10. Kim, H. Three-dimensional Sediment Transport Model, Ph.D Thesis, Dept. of Civil Engineering, University of Liverpool, 1992.

SECTION 5: POLLUTION PROBLEMS

Application of the Three-Dimensional Model to Slovenian Coastal Sea

R. Rajar

Department of Civil Engineering, University of Ljubljana, Haidrihova 28, 61000 Ljubljana, Slovenia

ABSTRACT

A three-dimensional baroclinic model based on the finite volume method is described. Some discussion about the use of two- and three-dimensional models, about turbulence modelling and about numerical diffusion is given. Two examples of model application (numerical simulation) are presented:

- Tidal circulation in the Koper bay (Slovenian Coast).
- Water exchange in Marina Koper.

INTRODUCTION

Nowadays there is no need to write about the necessity of using two and three-dimensional hydrodynamic and water quality models in surface water environmental problems. An interesting question is when is it possible to use simpler and more economic two-dimensional (2D) models and when is it necessary to apply three-dimensional (3D) models which are, in spite of high - speed computers still sometimes expensive for solving practical problems.

2D models, where the flow is "depth-averaged", can be successfully used to simulate circulation, when density differences do not play a significant role. Tidal circulation in coastal seas is typical, for example. Such models are also used for designing optimal layout of Marinas to assure sufficient flushing (Nece and Falconer [7]). Wind driven circulation can be successfully simulated by 2D models, when the computational area is large and details of the circulation near the shore are not important. 2D models can also be useful to get first approximations of some phenomena and for comparison with the final results of 3D models. In simulations of detailed flow in rivers, regarding river reclamation, or flow at or around hydraulic structures, 2D models can often successfully replace much more expensive physical models (Četina [2]).

The use of 3D models is unavoidable in all cases where the influence of density distribution cannot be neglected, be it tidal or wind-induced or purely density driven flow. Inflow of lower salinity river plumes into a coastal sea is a typical example, but often also in the simulation of tidal flows the influence of density differences cannot be neglected (Backhaus *et al.* [1]). In cases of wind-driven flows in smaller areas near the shore, the depth-averaged 2D models cannot simulate the 3D character of the flow. With wind-driven flows the density stratification causes an additional effect: an important diminishing of the shear stress between the horizontal layers, the consequence being that the surface layer circulation can differ considerably from the depth averaged circulation. Strong stratification in lakes usually also demands the use of 3D models. Though enormous effort is made to develop good 3D models, there are still some unsolved or at least not satisfactorily solved problems such as:

- Simulation of turbulent shear stresses is often not satisfactory since our understanding of turbulence phenomena is still not perfect, especially in large-scale natural conditions.
- Most numerical methods of solution are not completely satisfactory: the numerical computations are expensive and often the methods are burdened by numerical diffusion, which can render the results questionable (See Case 1.).

Most modelers agree that near the open boundaries the simulation is often not reliable. Different methods exist to replace the influences of the external region on flow inside the computational field, but sometimes only densely measured data at the open boundary can replace them. Considering all the above, there is almost always the need for verification and calibration of mathematical models. In most cases only a well planned combination of mathematical simulation and field measurements can give reliable results.

MATHEMATICAL MODELS

Two-dimensional Model

Only a brief description will be given here. As the model was used as a basis for developing the 3D model, described in the next section, some common features will be explained there. The basic equations are the depth integrated equations of motion, the advection terms are taken into account. The bottom friction is accounted for by a quadratic law. The two-equation turbulence model $k-\epsilon$ is included in the model to calculate the turbulent viscosity and shear stresses in the XZ and YZ plane (they are especially important in cases of high velocity flow in river hydraulics- Četina [2]). The finite difference method developed by Patankar [10] is used to solve the system of equations.

Three-dimensional Model

Basic Equations The model is based on the fully non-linear 3D equations of motion (Eqs. 1 to 8).

$$\frac{\partial h}{\partial t}+\frac{\partial(hu)}{\partial x}+\frac{\partial(hv)}{\partial y}+w_u-w_d=0 \tag{1}$$

$$\frac{\partial(hu)}{\partial t}+\frac{\partial(hu^2)}{\partial x}+\frac{\partial(huv)}{\partial y}+\frac{\partial(huw)}{\partial z}=-fvh-\frac{h}{\rho}\frac{\partial p}{\partial x}+$$

$$\frac{\partial}{\partial x}\left(hN_h\frac{\partial u}{\partial x}\right)+\frac{\partial}{\partial y}\left(hN_h\frac{\partial u}{\partial y}\right)+\frac{\partial}{\partial z}\left(hN_v\frac{\partial u}{\partial z}\right)+\frac{1}{\rho}\tau_{wx}-\frac{1}{\rho}\tau_{bx} \quad (2)$$

$$\frac{\partial(hv)}{\partial t}+\frac{\partial(huv)}{\partial x}+\frac{\partial(hv^2)}{\partial y}+\frac{\partial(hvw)}{\partial z}=+fuh-\frac{h}{\rho}\frac{\partial p}{\partial y}+$$

$$\frac{\partial}{\partial x}\left(hN_h\frac{\partial v}{\partial x}\right)+\frac{\partial}{\partial y}\left(hN_h\frac{\partial v}{\partial y}\right)+\frac{\partial}{\partial z}\left(hN_v\frac{\partial v}{\partial z}\right)+\frac{1}{\rho}\tau_{wy}-\frac{1}{\rho}\tau_{by} \quad (3)$$

$$\frac{\partial p}{\partial z}+\rho g=0 \quad \Rightarrow \quad p=p_a+g\int_z^H \rho\, dz \quad (4)$$

$$w_K=-\int_z^H \frac{\partial(uh)}{\partial x}+\frac{\partial(vh)}{\partial y} \quad (5)$$

$$\frac{\partial(hT)}{\partial t}+\frac{\partial(huT)}{\partial x}+\frac{\partial(hvT)}{\partial y}+\frac{\partial(hwT)}{\partial z}=$$

$$\frac{\partial}{\partial x}\left(hD_{hT}\frac{\partial T}{\partial x}\right)+\frac{\partial}{\partial y}\left(hD_{hT}\frac{\partial T}{\partial y}\right)+\frac{\partial}{\partial z}\left(hD_{vT}\frac{\partial T}{\partial z}\right) \quad (6)$$

$$\frac{\partial(hs)}{\partial t}+\frac{\partial(hus)}{\partial x}+\frac{\partial(hvs)}{\partial y}+\frac{\partial(hws)}{\partial z}=$$

$$\frac{\partial}{\partial x}\left(hD_{hs}\frac{\partial s}{\partial x}\right)+\frac{\partial}{\partial y}\left(hD_{hs}\frac{\partial s}{\partial y}\right)+\frac{\partial}{\partial z}\left(hD_{vs}\frac{\partial s}{\partial z}\right) \quad (7)$$

$$\rho=\rho(T,s) \quad (8)$$

The following symbols are used in equations (1) to (8): u, v, w - velocity components in the three space directions x, y and z, h−layer thickness, H−depth, ρ−density, p−pressure, p_a−atmospheric pressure, T−temperature, s−salinity, N_h, N_v−"horizontal" and "vertical" turbulent viscosity, D_h, D_v−dispersion coefficients, t−time, f−Coriolis parameter, τ_w−wind stress, τ_b−bottom stress. The model simulates transport and dispersion of temperature and salinity (or any other bio-chemical parameter) by the convection - diffusion equations (6) and (7). These parameters influence the density through the equation of state (8) which in turn influences the velocity field. So the model is baroclinic. The so called source/sink terms must be added to simulate the degradation or augmentation of matter due to bio-chemical reactions to obtain general "water-quality models".

Turbulence Modeling As in most models, the "eddy viscosity-diffusivity concept" is used in our 3D model. Due to the great difference of the horizontal and vertical dimensions, different "turbulent (eddy) viscosities" are used to describe the stresses in the XZ and YZ planes (N_h) and in the XY plane (N_v). The results are much more sensitive to the choice and distribution of N_v than N_h (Rajar *et al.* [12]). Therefore we use a simplified one-equation turbulence model proposed by Koutitas [4], for the calculation of N_v. A parabolic distribution of N_v with the depth is assumed, with maximum value at $0.6 \cdot H$ and zero at the bottom and at the surface. The maximum value depends on the depth and on the surface wind stress. Stable density stratification can strongly diminish the value of N_v. We use the Munk-Anderson relation with N_v and D_v being dependent on the Richardson number (Munk and Anderson [6]). Very few 3D models include a turbulence model for the horizontal eddy viscosity N_h. Usually N_h is considered constant over space and time (Cheng and Smith [3]). In most cases of circulation in coastal seas this can be acceptable. We can though write down some indications of our experience:

- In cases of high velocity flow in rivers, the spatial distribution of N_h is important (Četina [2]).
- Also in cases of inflow of larger rivers into shelf seas or lakes, when the physical phenomenon approaches a jet flow, the assumption of constant value of N_h is questionable.
- For simulation of circulation in large water bodies (in a lake) we have seen, (Rajar and Četina [11]) that constant value of N_h can give good results if at least its order of magnitude is properly estimated. When it is difficult to obtain a realistic estimate of N_h, we use our 2D model with the $k-\epsilon$ turbulence model to obtain the order of magnitude of N_h (See Case 1.).

Numerical Method The 3D model was developed from the 2D one, the basic numerical method being the same: the finite difference method of Patankar [10] (in 3D it is often cited as the "finite volume method"). "Upstream" differencing is used instead of central differencing when the Peclet number exceeds 2.0, which causes a certain amount of numerical diffusion. This problem is discussed with Case 1. Layers of the 2D model form a 3D grid, which are connected through the continuity equation (1) and through the kinematic boundary condition for the surface (5). Additional terms are included for the third dimension. Since the basic 2D model is based on the implicit method of solution, a more elaborated implicit scheme is also used to incorporate the z direction. This allows for relatively large time steps (Courant number being of the order of 10 to 40). Verification and calibration of the model was made using measurements (Rajar *et al.* [12], Rajar [13]) and satellite images in the Northern Adriatic (Rajar and Četina [14]).

CASE 1: TIDAL CIRCULATION IN KOPER BAY

Koper Bay lies at the Slovenian (Southern) coast of the Gulf of Trieste, which is located on the east side of the northernmost part of the Adriatic Sea. To determine the transport and dispersion of pollutants from different sources on the coast and to determine optimum locations of the future wastewater outlets, the general circulation in the bay was studied using a combination of measurements and mathematical modelling. Firstly only tidal flow without wind was studied, since this is the worst case for the transport and dispersion of pollutants, and since winds are weak in this region, especially in summer, when the pollution is strongest. Since not enough measurements were available to define boundary conditions along the extensive open boundary of the bay, we simulated the circulation by the so-called "nested model". First one computation for the whole Gulf of Trieste was made with the known tidal function as boundary condition. Only the $M2$ component was simulated, the range of which being 51 cm. The resulting velocity field for the mean water level during falling tide is shown in Fig.1. Some measured velocity vectors are included. The boundaries of the smaller, nested model for the Koper bay can be seen in the figure (dotted lines).

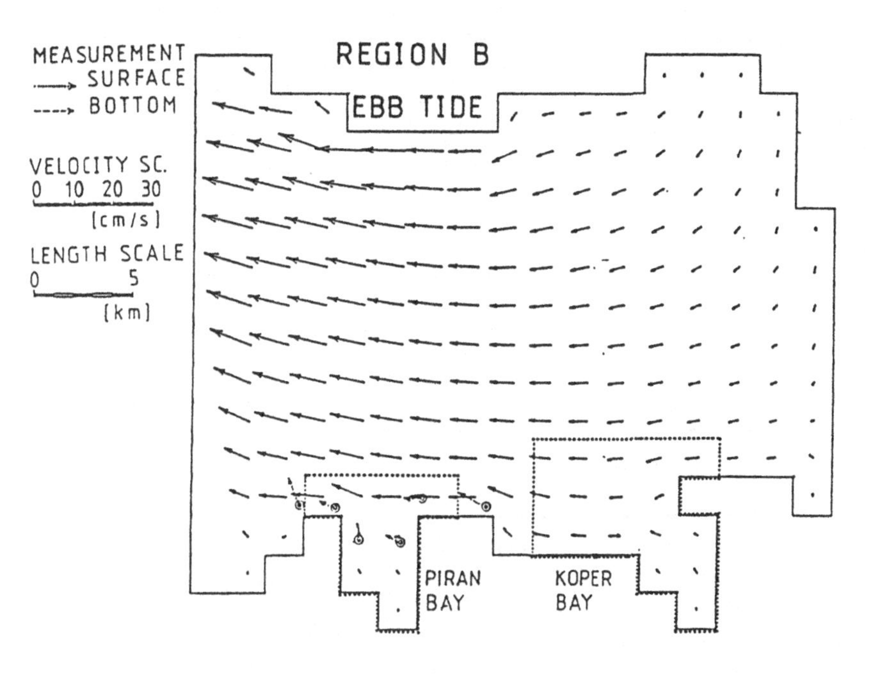

Figure 1. Tidal currents in the Gulf of Trieste during falling tide.

Velocities at the open boundaries of the smaller computational field were taken from the larger region using linear interpolation (since the grid of the smaller region is four times denser). Figs. 2a. and 2b. show tidal flow in the Koper Bay at mean water levels during rising and during falling tide. Depth-averaged measured velocities are also shown in the figures. The agreement is satisfactory. A cyclonic gyre is formed in the bay during the falling tide with very small circulation velocities. Measurements also show this pattern (Fig. 2b.). This reduces the transport and dispersion of pollutants out of the bay. As expected, this gyre was not simulated when using the coarse grid of the larger model (grid size $DX = DY = 1800\,m$). In the model of the smaller region ($DX = DY = 450\,m$) the gyre was not formed when the value of the horizontal turbulent viscosity N_h was too large (e.g. $N_h = 100\,m^2/s$). To obtain an idea of the order of magnitude of N_h we made some numerical experiments with the 2D model, which include the $k - \epsilon$ turbulence model. The computed values of N_h

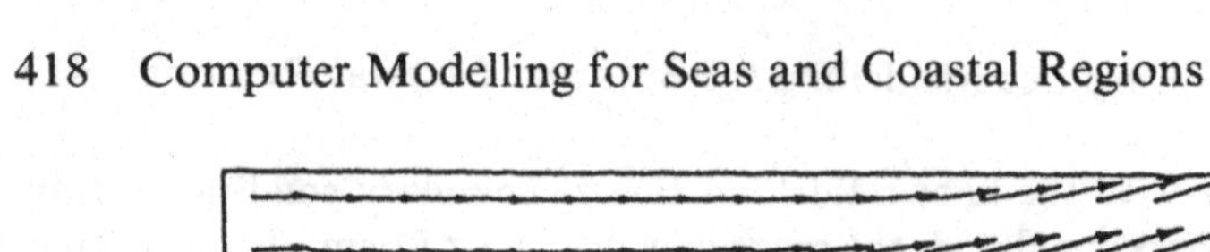

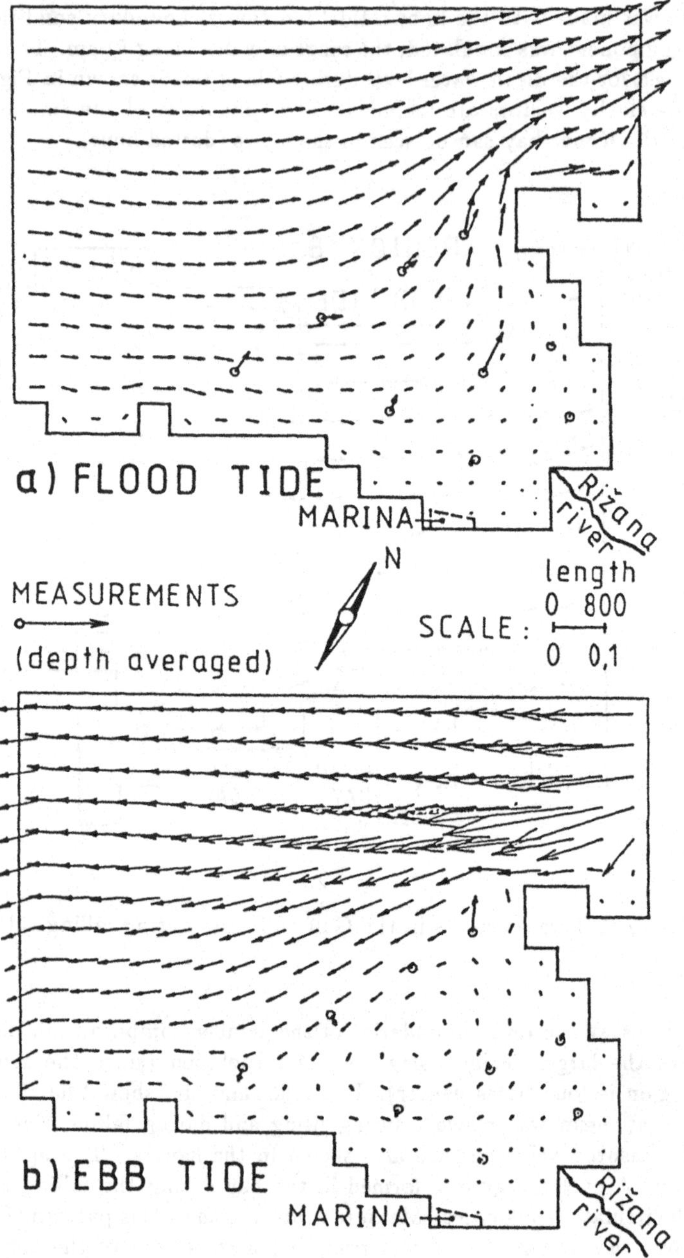

Figure 2: Tidal currents in Koper Bay during mean water level.

were between 0.01 and 0.08 m^2/s. Now the computations with different values of N_h showed that there was no difference in the circulation pattern when N_h was smaller than about $0.1\,m^2/s$. The only reason for that could be numerical diffusion. Patankar [10] gives for his numerical scheme an equation for estimation of the value of the "spurious" N_{hnum}, that is induced by the numerical diffusion:

$$N_{hnum} = \frac{v\Delta x\Delta y|\sin\alpha|}{4(\Delta y|\sin\alpha|^3 + \Delta x|\cos\alpha|^3)} \tag{9}$$

where v is the velocity magnitude and α is the angle between the velocity vector and X-axis. The maximum value of N_{hnum} is obviously obtained at $\alpha = 45°$. With an average value of $v = 0.03\,m/s$ in the inner bay, we obtain $N_{hnum} = 5\,m^2/s$. Since the angle α is generally smaller than 45° it is reasonable that total prevalence of numerical diffusion over physical diffusion occurs at about $N_h = 0.1\,m^2/s$. In our case though, the comparison with the measurements show, that the results are not yet seriously affected. But it is known that numerical diffusion affects the computation of concentration in the convection - diffusion equation much more than the velocity field (Rodi [15]).

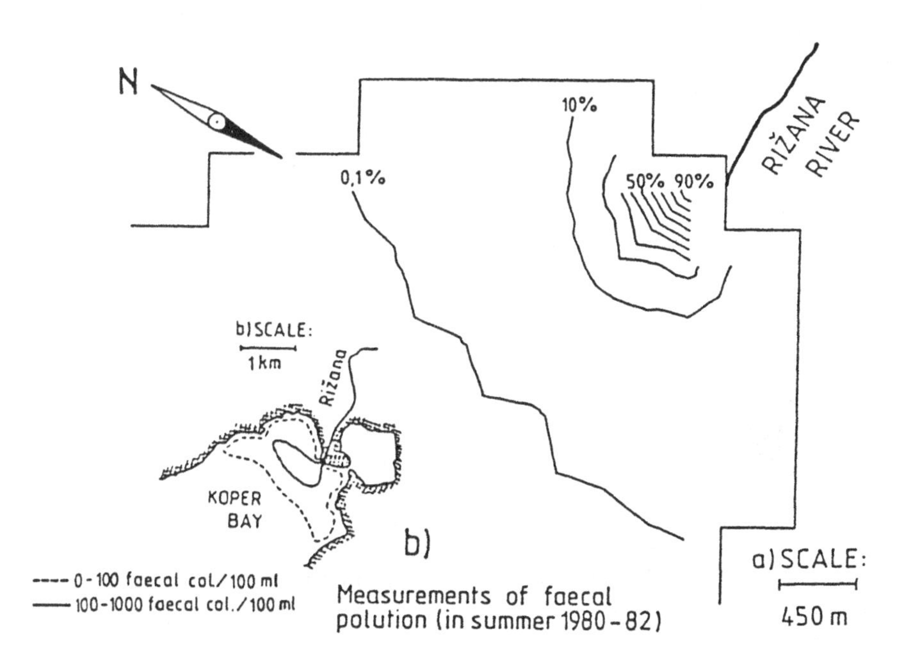

Figure 3. a) Simulation of dispersion of pollutants from Rižana river.
b) Measurements of faecal pollution (from Turk *et al* [16]).

Because of these disadvantages, we are working on including "particle tracking" techniques for the solution of the convection–diffusion equation in the finite–difference hydrodynamic model. Since the waste–water outlet (up to now only mechanical treatment) is into the Rižana river, which flows into the Koper bay, simulation of dispersion of pollutants was performed. Fig.3. shows the computed lines of equal concentration (a fictitious concentration of 100 % was assumed at the river mouth) after 45 hours of tidal simulation. A general east to north bounded flow of $v = 0.02\,m/s$ was superimposed on the tidal flow, since measurements have shown a residual current along the Slovenian coast. The only possible verification was comparison with measured dispersion of faecal pollution in the bay (Fig.3b.). The agreement is relatively good. Also the measured distribution of concentration shows an influence of a residual current along the coast.

CASE 2: WATER-EXCHANGE IN MARINA KOPER

The eastern Adriatic coast is very beautiful and with its mild weather all conditions for nautical tourism are fulfilled. During the last fifteen years many marinas have been built along the Croatian and Slovenian coast. Marina Koper is located inside the Koper Bay (see Fig.2b.). Since the tidal range is only 51 cm in this region, (the $M2$ constituent) it is difficult to assure adequate flushing of the marina by tidal circulation. The general layout is shown in Figs.4 and 5.

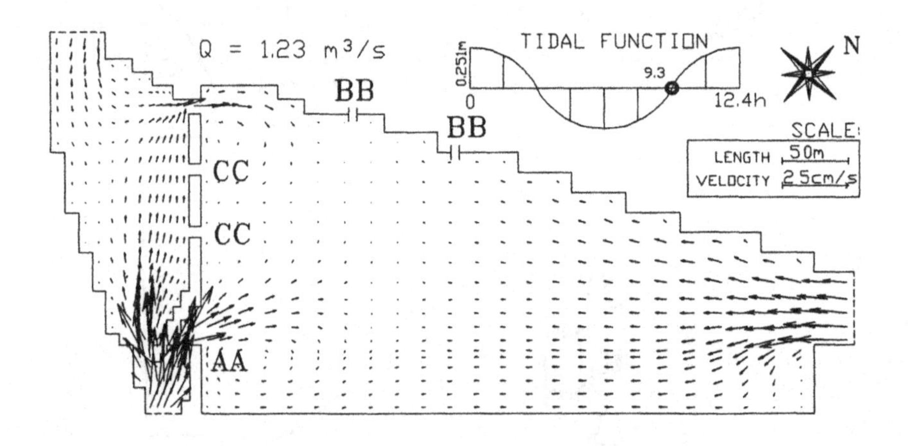

Figure 4. Tidal circulation in Marina Koper at mean water level, rising tide.

Since the location was mainly dredged, the depth in the marina is almost uniform, $H = 3\,m$. The Tidal Prism Ratio

$$\text{TPR} = \frac{vol.\,at\,high\,tide - vol.\,at\,low\,tide}{vol.\,at\,high\,tide}$$

is an approximate indicator of the tidal flushing possibilities (Nece and Falconer [7]). Here the TPR is only 0.17, which is a relatively small value, compared to e.g. some marinas in Washington state [USA] which have TPRs between 0.20 and 0.70. On the other hand in Turkey its values of only 0.05 have been noted (Ozdan [9]), but there forced flushing is used. Fortunately it will be possible to use a small river inflow to improve the marina flushing. The 2D and 3D mathematical models were used to check whether the designed marina layout allows for adequate flushing, and to propose some improvements. Tidal flow together with river inflow of $Q = 1.23\, m^3/s$ (mean annual discharge) was simulated first. The computed velocity field is shown in Fig.4. at the mean water level during rising tide. A hypothetical outside concentration with the value of 1.0 and an inside concentration of 0.0 was assumed in the computation. Fig.5. shows computed concentration inside the marina at the peak level slack water. The extent of the exchange zone can be estimated. The density differences were not taken into account in this computation, though they would undoubtedly influence the results.

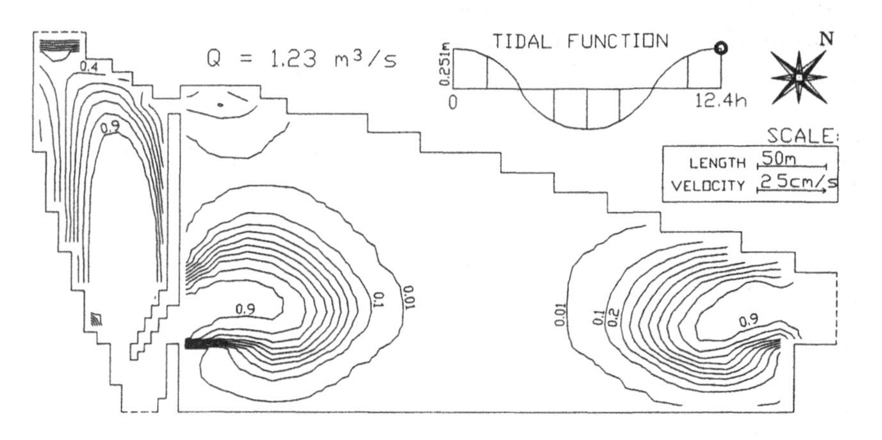

Figure 5. Marina Koper - lines of equal hypothetical concentration of peak level slack water.

The results in the two figures show, that a flow-dividing structure AA should be constructed near the inflow of the river, to enhance the flushing of the NE part of the marina. Fig.5. shows that the water exchange is not sufficient in the central part of the NE basin. Two openings in the outer breakwater (BB) are proposed and two additional openings (CC) in the wall, separating the SW and NE basins, to improve the circulation. It was strongly emphasized, that the river water entering the marina basin should be of good quality, which will be difficult to assure, especially during the hot summer season, when the river discharge is at its minimum and the marina pollution is greatest.

CONCLUSIONS

Two examples of applications of the three-dimensional model are described, showing that the model can be used as an efficient tool in solving different water-quality problems. Because of some not quite reliable steps in the modelling (simulation of turbulent shear stresses, numerical diffusion, open boundary conditions) almost always some measured data are needed.

REFERENCES

1. Backhaus, J.O., Crean, P.B. and Lee, D.K.: On the Application of a Three-Dimensional Model to the Waters Between Vancouver Island and the Mainland Coast of British Columbia and Washington State. In: Three-Dimensional Coastal Ocean Models, (Ed. Heaps, N.), pp.149-176. Am. Geophysical Union, Washington D.C. 1987.

2. Četina, M.: Some Examples of Simulation of Two-Dimensional Turbulent Free-Surface Flows. Proceedings of the X. Congress of the Yugoslav Association for Hydraulic Research, Sarajevo, 1990. (in Slovene).

3. Cheng, R.T. and Smith, P.E.: A Survey of Three-Dimensional Estuarine Models. In: Estuarine and Coastal Modelling (Ed. Spaulding, M.L.), pp. 1-15. Proceedings of the Conference Newport, Rhode Island. ASCE Publication 1990.

4. Koutitas, C.: Modelling of Three - Dimensional Wind - Induced Flows. Proc. ASCE, Hydraulic Division, pp. 1843-1865. HY 11, 1980.

5. Kuzmič, M.: Exploring the Effect of Bura over the Northern Adriatic-CZCS Imagery and a Mathematical Model Prediction. Intern. J. of Remote Sensing, No 1, 1991.

6. Munk, W.H. and Anderson, E.R.: Notes on the Theory of the Thermocline. J. of Marine Research, Vol.1, pp. 276-295. 1948.

7. Nece, R.E. and Falconer, R.A.: Hydraulic Modelling of Tidal Circulation and Flushing in Coastal Basins. Proc. Inst. Civ. Eng., Part 1,86, pp. 913-935, Oct. 1989.

8. Nece, R.E. and Layton, J.A.: Mitigating Marina Environmental Impact through Hydraulic Design. In Marinas: Planning and Feasibility. Proc. of the Int. Conference on Marinas, Southampton (Ed. Blain, W.R. and Weber, N.B.), pp. 435-450. Computational Mechanics Publications, 1989.

9. Ozdan, E.: Flushing of Marinas with Weak Tidal Motion. Marinas: Planning and Feasibility. Proc. of the Int. Conference on Marinas, Southampton (Ed. Blain, W.R. and Weber, N.B.), pp. 485-498. Computational Mechanics Publications, 1989.

10. Patankar, S.V.: Numerical Heat Transfer and Fluid Flow. Mc.Graw Hill B.C. 1980.

11. Rajar, R. and Četina, M.: Mathematical Simulation of Two-Dimensional Lake Circulation. In: HYDROSOFT 86 (Ed. Radojkovič, M., Maksimovič, C. and Brebbia, C.A.), pp. 125-134. Proc.of Int. Conf., Southampton 1986, Springer Verlag, 1986.

12. Rajar, R., Četina, M. and Tonin, V.: Influence of Linearization and of Vertical Distribution of Turbulent Viscosity on Tree-Dimensional Simulation of Currents. In: HYDROCOMP 89 (Ed. Maksimovič, C. and Radojkovič, M.), pp. 193-202. Proc. of Int. Conference, Dubrovnik 1989.

13. Rajar, R.: Three-Dimensional Modelling of Currents in the Northern Adriatic Sea. In: Proceedings of the XXIII. Congress of the IAHR, pp. C335-342. Ottawa, 1989.

14. Rajar, R. and Četina, M.: Modelling Wind-Induced Circulation and Dispersion in the Northern Adriatic. Proceedings of the XXIV. Congress of IAHR, Madrid, 1991.

15. Rodi, W.: Turbulence Models and their Application in Hydraulics. IAHR Publications, Delft, 1980.

16. Turk, V., Faganeli, J. and Malej, A.: A Look at the Pollution Problems in the Bay of Koper in Relation to the Provisional Sewage Outfall. Congress C.I.E.S.M., pp.603-608, Cannes 1892.

15. [illegible]

16. [illegible]

A Finite Element Strategy for the Analysis of the Discharge of Warm Water into Coastal Waters

M. Andreola (*), S. Bianchi (*), G. Gentile (*), A. Gurizzan (*), P. Molinaro (**)

(*) CISE Tecnologie Innovative SpA, P.O. Box 12081, Milan, Italy

(**) ENEL/CRIS, Via Ornato, 90/14 - 20162 Milan, Italy

ABSTRACT

The paper addresses two basic problems connected with the solution of the Navier Stokes equations: how to avoid oscillations in pressure when using elements not satisfying the L.B.B. condition and how to take advantage of the splitting technique in choosing a proper Δt to accurately describe convection and diffusion circumventing stability restrictions.
A presentation of the most significant results complements a short description of the capabilities of MADIAN code in analysing the physical problem relevant to the discharge of warm water from power plants together with the distinguishing aspects of the numerical method.

INTRODUCTION

The kernel of mathematical models for the study of power plants impact on the marine environment is represented by Navier-Stokes equations. Their solution requires the computation of long transients to reach stationary conditions in large zones around the discharge in domains with complex 3-D geometries.
An efficient solution of these equations is therefore crucial to perform realistic analyses with reasonable computing times.
The finite element method is attractive for its flexibility but it suffers from some drawbacks connected both to restrictions in the choice of elements for the spatial discretization and to the

unstructured sparse structure of matrices which affects the computational efficiency.
In particular, special elements should be used to avoid spurious oscillations in the solutions or filtering techniques should be applied based on the construction of "macro elements", formed by standard finite elements.
Elements based on three-linear functions for velocities and constant pressure are unstable but, despite this drawback, they are used in many codes for the simplicity of their implementation. This approximation often produces reasonable results for the velocity field while spurious oscillations in pressure are generally eliminated by crude filtering rules.
A problem connected with the use of these type of elements is the presence of rank deficiency in the matrices of the discretized model which requires some shrewdness in the solution of linear systems.
This occurrence often hampers the convergence of iterative schemes generally used, so that the computing time may become prohibitive.
Proper strategies in the simulation of long transients must be predisposed to reduce the computational effort.
The paper describes the solutions adopted in MADIAN code [1], in the attempt to get the maximum out of the choice of popular but poor elements such as those described.
MADIAN code is aimed at the analysis of the discharge of warm water from power plants into coastal waters and its present version is based on P_0/Q_0 elements for pressure and P_1/Q_1 elements for velocity, temperature and salinity. A second version using P_1-isoP_2 elements is now in progress.
After a brief description of the mathematical and numerical model of the code, the paper describes the method used to eliminate or reduce the influence of spurious oscillations and the strategy adopted to reduce the computational effort in the simulation of long transients. Numerical examples are presented to test the efficiency of the proposed choices.

OUTLINES OF THE MATHEMATICAL AND NUMERICAL MODEL

MADIAN is based on a full three dimensional transient model consisting of:

- the Reynolds averaged turbulent Navier-Stokes equations for an incompressible Newtonian fluid [2], simplified with the Boussinesq approximation [3].
- The equation of the free surface

$$\frac{\partial \eta}{\partial t} + \vec{u}_\eta \cdot \nabla \eta = w_\eta \qquad (1)$$

describing the kinematic condition where $\eta(x,y,t)$ is the difference between the actual level of the free surface and the still level z_o, u_η is the two dimensional velocity vector at the free surface and w_η is the vertical component of velocity at the free surface.

- The thermal energy equation written in terms of the variable $T^* = T-T_a$ representing the dynamic component of the thermal perturbation at study, where T_a is the natural and possibly non-uniform temperature distribution [7].
- The salinity equation and the equations for dissolved and dynamically passive substances, whose form is similar to that of the thermal energy.
- The equations of state.

The turbulence models at present implemented in the code are:

- the simplified model with constant eddy viscosity diffusivity;
- the mixing length model.

These models are modified to take into account the effects of buoyancy [5].

As for the boundary conditions references can be found in [7].

The equations previously described are solved with a fractional step method [8]. According to this technique the operators are splitt so that terms related to convection are separated from those describing diffusion-propagation. Coriolis forces, which depend on the unknown themselves are associated with the convective part, while the remaining source terms are joined to the diffusion-propagation part.

The convective equations are solved by means of a second order explicit Taylor-Galerkin method [7],[9]. The initial field for this phase is the solution of the previous time step and the boundary conditions used to advance in time are those given on inflow boundaries.

The diffusive equations are solved through a one step implicit method with initial conditions given by the solution of the convective equations in the same time interval.

The boundary conditions associated to this step are those specified for the global equations. To perform the spatial discretization isoparametric linear elements for velocity, temperature, salinity and concentration are used, while an elementwise constant pressure is considered.

The diffusion-propagation phase of the Navier-Stokes equations requires the solution of a generalized Stokes problem whose discrete form is:

$$\begin{vmatrix} A & B \\ -B^T & 0 \end{vmatrix} \begin{vmatrix} U \\ P \end{vmatrix} = \begin{vmatrix} f_1 \\ f_2 \end{vmatrix} \tag{2}$$

Where U an P are vectors of nodal velocities and pressures respectively, A is a symmetric positive definite matrix and B is a rectangular matrix.
Problem (2) is reduced to the solution of the linear system

$$\begin{aligned} &QP = s \\ &Q = B^T A^{-1} B \;; \quad s = B^T A^{-1} f_1 + f_2 \end{aligned} \tag{3}$$

The computation of nodal velocities is obtained by the first equation of system (2), once P is known.
Equation (3) is solved by means of the conjugate gradient method.

SOLUTION OF THE GENERALIZED STOKES PROBLEM

Equation (3) is singular because matrix Q has always a zero eigenvalue corresponding to the constant pressure mode. This rank deficiency can be easily eliminated because pressure is defined but for a constant. Therefore any constraint might be imposed and in general a zero mean value pressure is considered.
However the use of P_0/Q_0 - P_1/Q_1, elements can produce a further zero eigenvalue in matrix Q when a regular spatial mesh is considered [10]. This second source of singularity can be eliminated only through the computation of the related eigenvector, that is the spurious mode, what is not in general practicable. It can be shown that a simple choice of the initial guess to start the conjugate gradient method can allow the computation of the minimum norm solution of eq. (3).
Let us consider the eigenvalue problem, associated with matrix Q

$$\begin{cases} Q\xi = \xi\Lambda \\ Q\eta = 0 \end{cases} \tag{4}$$

where ξ are the modes relevant to zero eigenvalues. It must be noticed that solutions of eq. (3) only exist if the known vector does not contain contributions from ξ modes. This is normally true because of the form of vector s.

Any iterative scheme of the form

$$P^{k+1} = F(P^K, QP^K, s) \tag{5}$$

will converge to the minimum norm solution of eq. (3) only if P^{k+1} does not contain contributions from ξ modes. Otherwise the numerical scheme may diverge. This implies that P^k, and therefore the initial guess P, must lie in the subspace spanned by η modes. A simple choice of the initial guess which verifies this conditions is

$$P^o = 0 \tag{6}$$

Choosing this value as the initial time condition, the computation of pressure at time $t=t_{n+1}$ can then be performed by assuming as initial guess the value of P computed at time $t=t_n$, because the computational rule guarantees that

$$\xi^T P \,|_{t=t_n} = 0$$

This may be true for the whole transient unless the coefficients of the equations vary with time. In this case, of course, to get rid of ξ modes it's necessary to start from a null pressure at each time step. Even when coefficients are constant, because of round off errors, components of $P|_{t=tn}$, lying in the subspace spanned by ξ modes, may be introduced; thus it is advisable even in this case to start computations from null pressure every fixed number of time steps.

APPLICATION OF THE FRACTIONAL STEP METHOD

Let us consider the following convective-diffusive equation

$$\begin{cases} \dfrac{\partial U}{\partial t} = CU + DU + S \\ U(t_n) = U_n \end{cases} \tag{7}$$

where C and D are the convective and diffusive operators, respectively, S is a source term eventually depending from the solution U and from suitable boundary conditions.
The fractional step method implies the solution of the following two problems to compute the solution at

$t=t_{n+1}$:

$$\begin{cases} \dfrac{\partial \tilde{U}}{\partial t} = C\tilde{U} \\ \tilde{U}(t_n) = U_n \end{cases} \qquad t_n \le t \le t_{n+1} \tag{8}$$

for this step the only boundary conditions used are those on the inflow surface. The second problem to be solved is

$$\begin{cases} \dfrac{\partial \overline{U}}{\partial t} = D\overline{U}+S \\ \overline{U}(t_n) = \tilde{U}(t_n+1) \end{cases} \qquad t_n \le t \le t_{n+1} \tag{9}$$

Global boundary conditions are applied.
The solution U at $t=t_{n+1}$ is approximated by $\overline{U}(t_{n+1})$.
The classical formulation of the method solves the previous problems with discrete approximations based on the use of the same time step $\Delta t=t_{n+1}-t_n$ both for eq. (8) and eq. (9). This choice has the drawback that Δt must take the minimum value required for the accurate solution both of advective and diffusive equations, taking also into account stability requirements connected to the schemes used for time approximations. However this restriction seems not to be required by the basic principle of the fractional step approach and has been avoided in MADIAN code.
As described previously, eq. (8) is solved by the explicit second order Taylor Galerkin scheme whose stability condition requires that the Courant number is less than $\sqrt{3}/3$. This limitation is not necessary for the solution of eq. (9) which is performed by an implicit unconditionally stable method. In practical problems the integration time step Δt is chosen on the basis of accuracy considerations relevant to the diffusive eq. (9), while $\tilde{U}(t_{n+1})$ is obtained by integrating eq. (8) for $t_n \le t \le t_{n+1}$ with m time steps of length $\Delta\tau=\Delta t/m$.

NUMERICAL EXPERIMENTS

To test the effectiveness of the strategy used both to get rid of C.B. modes and to overcome restrictions on Δt for the integration of the Stokes problem, due to the explicit integration of the convective term, some examples are reported. More precisely, the first and second case are relevant to the problem of dealing with spurious modes for pressure; the third case shows how it is possible, while retaining the Courant condition for convection, to produce accurate results with a larger Δt for the Stokes part.

Creeping flow - Stokes equation

The first case studied is that of a creeping flow (Re << 1) and it's a pure Stokes problem: the domain and the boundary conditions are those represented in fig. 1. The analytical solution is linear in velocity and constant in pressure (u= x, v= -y, p= cost). The solution has been computed on a regular mesh of 4x4x1 hexahedra ($\Delta x = \Delta y = 0.25$ m) starting from a null velocity and constant pressure up to t= .05 s with $\Delta t = 10^{-2}$ s. The computed velocity field is exact while, as resulting from fig. 2, the pressure suffers from spurious C.B. modes when the computation starts from a non zero pressure (values oscillate among .49 and .73). These spurious modes are filtered when the initial pressure is zero being the computed results identical to the analytical ones, see again fig. 2.

Creeping flow - Navier Stokes equation

This second case is still that of a creeping flow with same boundary condition as before but now convection is present. The analytical solution has the same flow field (u= x, v= -y) but pressure is now quadratic ($p = 1/2\ \rho_o\ (x^2 + y^2) + cost$).

The computational grid is now of 20x20x1 hexahedra ($\Delta x = \Delta y = .05$ m) and $\Delta t = 10^{-3}$ s. The pressure computed from a null value is free from C.B. modes and, where there are enough grid points, accurately represents the analytical solution, see fig. 3.

Flow in a converging channel

This third example, known as the Hamel problem [11], concerns the study of a flow in a converging channel represented in fig. 4. The analytical solution consists of a self-similar purely radial velocity field with a boundary layer at the stationary walls and the pressure has a radial behaviour proportional to $1/r^2$.

The asymptotic exact solution at high Reynolds numbers is the following:

$$\begin{cases} u_r(r,\theta) &= \dfrac{\nu Re}{\alpha}\dfrac{\lambda}{r} \\ p(r,\theta) &= \dfrac{\rho\nu^2 Re}{\alpha r^2}\left(2\lambda - \dfrac{Re}{2\alpha} - 1\right) + constant \end{cases}$$

where

$$\lambda = 3Th^2\left[\sqrt{\frac{-\alpha Re}{2}(1-\frac{\theta}{\alpha})} + Th^{-1}\sqrt{\frac{2}{3}}\right] - 2$$

and the Reynolds number is defined by

$$Re = \frac{\alpha\, u_{rM}\, \Gamma}{\nu}$$

where u_{rM} is the velocity on the symmetry axis.
For symmetry reasons just half domain ($0.25 \leq r \leq 4$, $0 \leq \theta \leq \alpha$) has been considered and the calculation mesh was obtained by dividing the domain into 390 hexahedra and 868 nodes (13 elements in the θ direction, 30 elements in the r direction and only 1 in the z direction). At the inlet and outlet boundary the exact solution is imposed for velocity. The problem has been solved starting from the analytical solution of the Stokes part both for velocity and pressure with a time step $\Delta t = 10^{-4}$s (in order to meet the stability condition of the Taylor-Galerkin method). The calculation has been carried on up to t = 4 s solving at each time step one convection and one Stokes problem. The computational cost for each time step on the Alliant FX80, with three processors in parallel, is of 14.5 s and because of the large number of steps the total CPU time is very high ($\simeq$ 160 h). The results are very accurate as can be verified from fig. 5, 6, 7, but computational costs are prohibitive.
To overcome this problem experiments have been carried on using different time increments for the convection and the Stokes phase. This strategy has been pushed up to a situation in which 1000 time steps for convection were computed against 1 time step for the Stokes problem, thus the resulting global Δt is 10^{-1} s.
At the final time, t = 4s, the results are still accurate enough for velocity, see fig. 8, 9, and much less for pressure, see fig. 10, but now the computing time is 54^h, representing a saving of time of about 2/3 ($\simeq$ 66%). To verify whether the solution thus obtained lies in the same space as the exact one, a new transient has been computed starting from this solution and using now equal Δt for the convection and the Stokes problem ($\Delta t = 10^{-4}$ s). After 300 time steps the results compared well with those obtained before. In particular the computed pressure reached the waited analytical behaviour.
Thus it can be said that the strategy experimented has allowed to use an explicit scheme for convection with a time step comparable to that of an implicit scheme

with a subsequent saving of computational time.

CONCLUSIONS

The strategies described in the paper to overcome difficulties connected both to the use of elements not satisfying the L.B.B. condition and to the need of affording the simulation of long transients have been proven efficacious as shown in the test cases presented. The new version of the MADIAN code with P1-isoP2 elements will represent a further step towards a substantial decrease of the computational time for real applications.

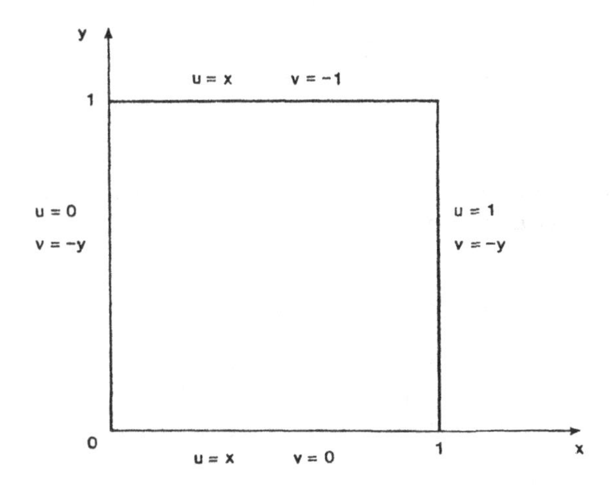

Fig.1 Domain of computation and boundary conditions for the first and second test case.

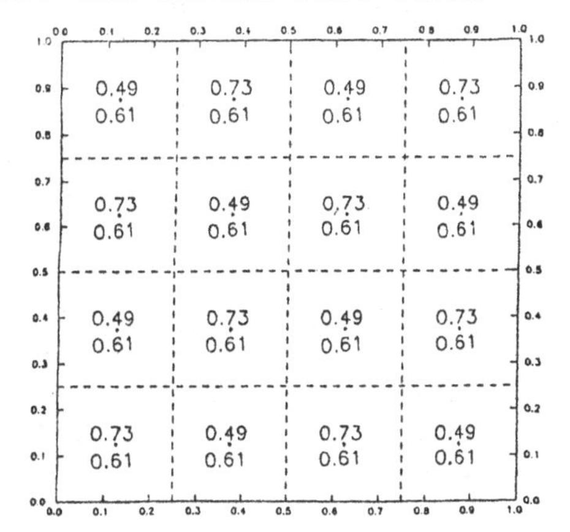

Fig. 2 Pressure field for the first test case (upper values: pressure with spurious C.B. modes; lower values: pressure after filtering).

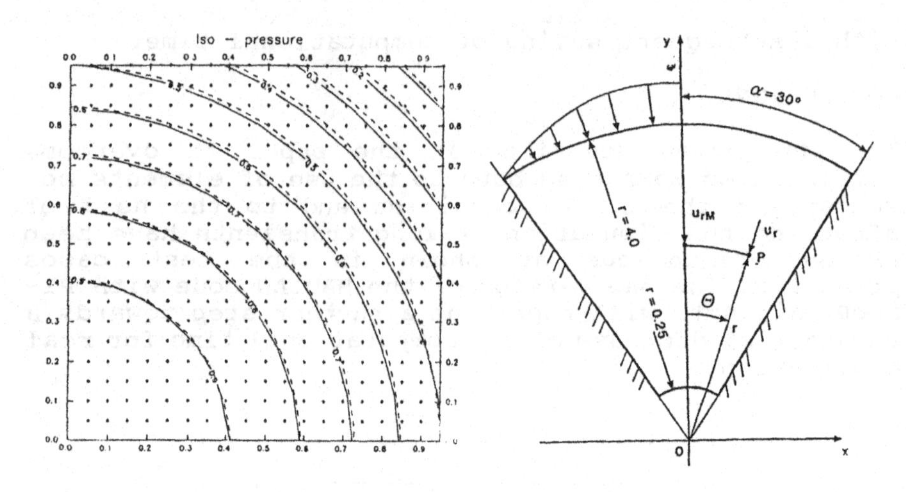

Fig.3 Pressure field for the second test case.

Fig.4 Converging 2-D channel for the Hamel problem.

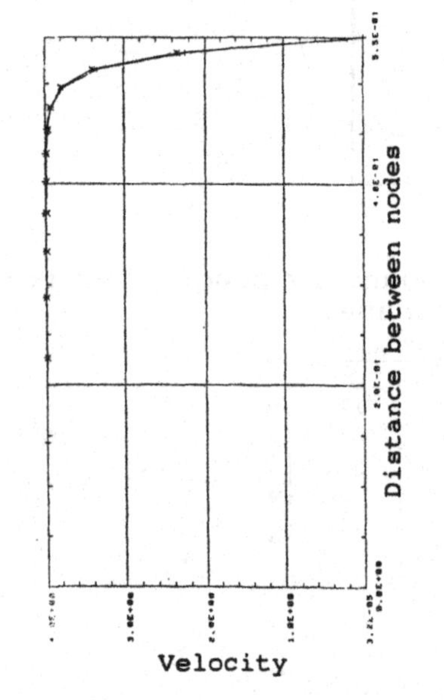

Fig. 5 Calculated and analytical radial velocity (t=4.s, r=1.043).

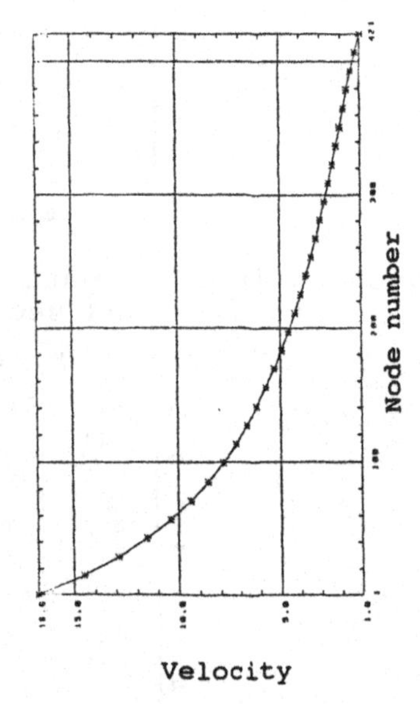

Fig.6 Calculated and analytical radial velocity (t= 4.s, θ=0°).

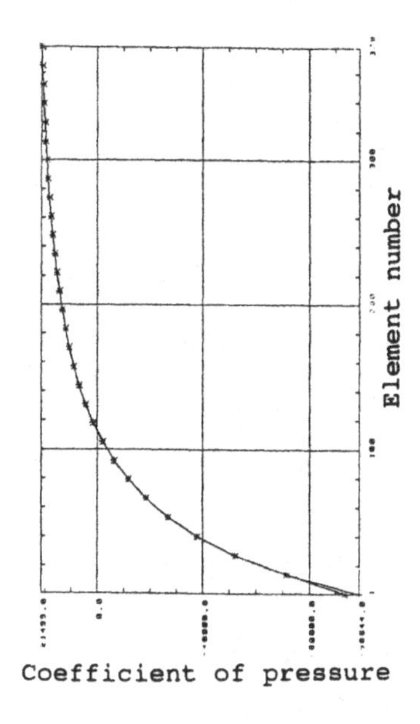

Fig.7 Calculated and analytical radial pressure (t=4.s, θ=0°).

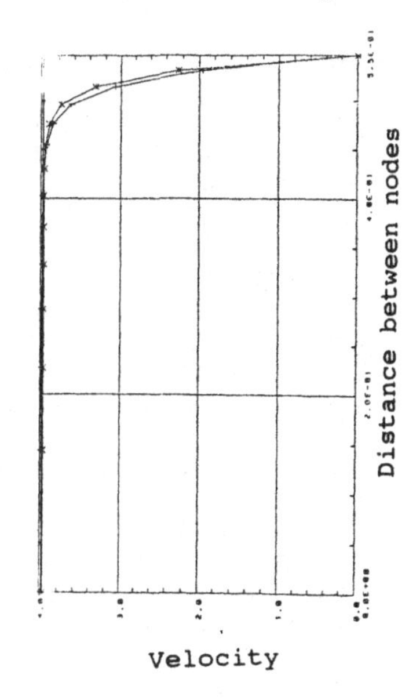

Fig.8 Calculated and analytical radial velocity (t=4.s, r=1.043).

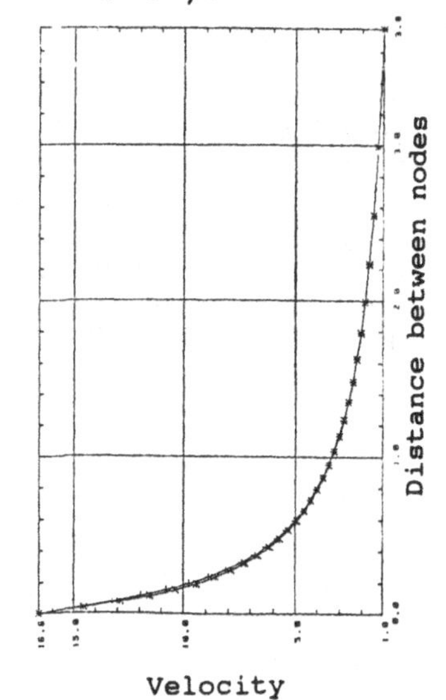

Fig. 9 Calculated and analytical radial velocity (t=4.s, θ=0°).

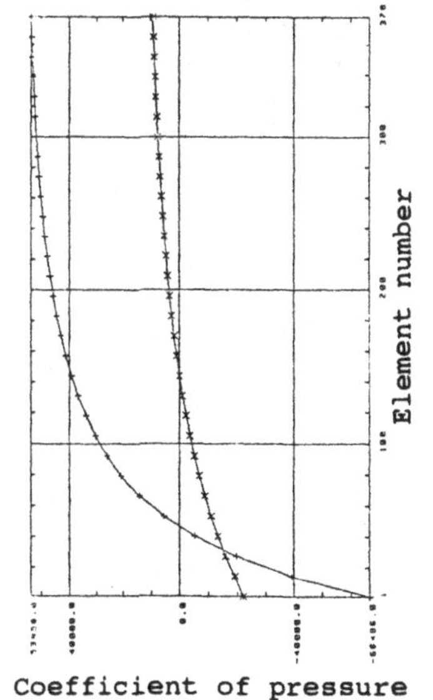

Fig.10 Calculated and analytical radial pressure (t=4.s, θ=0°).

REFERENCES

1. Andreola, M. and Gurizzan, A. MADIAN: un programma di calcolo per l'analisi di scarichi termici nelle acque costiere italiane. Prima versione CISE Report 4938, 1988.
2. Bird, R.B., Steward, W.E. and Lightfoot, E.N. Transport Phenomena, John Wiley and Sons, London and New York, 1960.
3. Gray, D.D. and Giorgini, A. The Validity of the Boussinesq Approximation for Liquids and Gases, Int. J. Heat Mass Transfer, vol. 19, pp. 545-551, 1976.
4. Gill, A.E. Atmosphere - Ocean Dynamics, Academic Press Inc., London, 1982.
5. Rodi, W. Turbulence Models and their Application in Hydraulics - A State of the Art Review, IAHR publication, Delft, The Netherlands, 1980.
6. Orlob, G.T. Mathematical Modelling of Water Quality: Streams, Lakes and Reservoirs, John Wiley and Sons, London and New York, 1983.
7. Andreola, M. et al. MADIAN: A 3-D Finite Element Code for Environmental Studies, Proc. of Hydrosoft 90, Lowell (MASS.), 3-5 April 1990.
8. Marchuk, G.I. Methods of Numerical Mathematics, Springer-Verlag, New York, 1982.
9. Donea, J. A Taylor-Galerkin Method for Convective Transport Problems, Int. J. Num. Eng. (20) 1984, pp. 101-119.
10. Sani, R.L., Gresho, P.M., and Lee R.L. The cause and cure of the spurious pressures generated by certain F.E.M. solutions of the incompressible Navier-Stokes equations: Part 1, Int. J. Num. Meth. in Fluids, vol. 1, 17-43, 1981.
11. Gartling, D.K., Nickell, R.E. and Tanner R.I. A finite element convergence stugy for accelerating flow problems, Int. J. Num. Meths. Eng., vol. 11, 1155-1174, 1977.

Quantitative Analysis of Diffusion Area of River Effluent with Satellite Remote Sensing

S. Onishi, H. Kawai

Department of Civil Engineering, Science University of Tokyo, Noda City, Chiba, Japan

ABSTRACT

Satellite remote sensing has been accepted generally in hydraulic study as an useful technique of flow visualization. In this paper, we present a method to numerically analyze the behavior of the river effluent with satellite remote sensing data. The method to find the effluent axis and estimate its diffusion area in the sea is presented and the effluent issued from the Shingu river is analyzed with Landsat TM-data as an example.

INTRODUCTION

In the present age, it seems that social and economic developments conflict with the preservation of the natural environment. In Japan, most human activities are centered on the riversides and coastlines. The water environment of the river and the sea is strongly interconnected, and at the time of planning water utilization, hydraulic engineers are required to pertinently preestimate the effects of the river effluent on the sea water environment. In-field survey works must be carried out to understand natural conditions of water quality and currents in the concerned water region. From an idealistic viewpoint, it is desirable to observe or measure instantaneously the whole aspects of the effluent in the field. But the ideal goal must be hindered by two serious obstacles, that is, the huge spatial scale of the related hydraulic phenomena and the difficulty in distinguishing the effluent distribution in the sea. The current patterns drawn with the values of current velocity, each of which is measured at different time, must be illusive. None of indices used generally to represent the water quality such as BOD, COD, salinity, temperature, turbidity, SS, conductivity and others can be used as a medium for visualization of the current in the field. However, applications of satellite remote sensing may provide us help to overcome these difficulties, since the satellite can take an extensive view of the water area periodically and visualize indirectly the current distribution through difference in reflection energy of electromagnetic waves at the water surface.

Although hydraulic engineers have been rather conservative in introducing the satellite remote sensing in their work, several studies on visualization of oceanic currents and river effluent have been reported in Japan. For examples, Tanaka and Ogihara[1] investigated the effluent issued from the Isikari river, and Onishi and Baba [2], [3], the effluents discharged into the Sea of Japan, and Sawamoto and Murakoshi [4], those into the Pacific Ocean. But in these studies, the satellite remote sensing has been mainly used as a tool of techniques for flow visualization, without introducing the expected applicability of the digital data. Considering the above, in this paper we discuss methods to analyze quantitatively the river effluent behavior, using the digital data transmitted from Landsat.

LANDSAT DATA

Orbit and sensors

The Landsats 1 through 5 were launched between 1972 and 1984 and now the Landsat 5 circles the Earth every 98.9 minutes in a nearly polar orbit, at the altitude of 705 kilometers. It can view the same area every 16 days at the same local time. The Landsat 5 loads the Multi-Spectral Scanner (MSS) and the Thematic Mapper (TM). The picture element area sampled by the MSS is about 80 meters square in the four spectral bands. The TM is more sensitive than the MSS, producing sharper and more revealing images with greater detail resolution. It measures in the seven bands, which are presented in Table 1. In this paper, the data collected by the TM are used.

Space resolution

The data of the bands 1 to 5 and 7 can image an area of 30 meters square. That of the band 6 resolve an area of 120 meters square. The diffusion area of the effluent is issued from the Shingu river,

Table 1 Range of wave length of Landsat TM

Band	Spectrum	Wavelength	Resolution	Index
1	Visible Blue	0.45 − 0.52(μm)	30 (m)	Turbidity
2	Visible Green	0.52 − 0.60	30	
3	Visible Red	0.63 − 0.69	30	
4	Nearly Infrared	0.76 − 0.90	30	Chlorophyll
5	Infrared	1.55 − 1.75	30	-
6	Thermal	10.40 − 12.50	120	Temperature
7	Infrared	2.08 − 2.35	30	-

which will be studied as an example, is presumed as 10 to 500 square kilometers, which corresponds to 11,000 to 560,000 pixels on the TM image. Therefore, with the TM data we can analyze satisfactorily the diffusion areas of this order and less than it.

Radiometric Resolution

The intensity of radiation gathered by the sensors on the satellite is converted into the digital form known as CCTs (Computer Compatible Tapes), which permit large volumes of data to be processed quickly by computers. The conversion to the digital data is based on separating the radiation intensity into binary increments. The radiometric resolution of the TM data is 256 gradations, which correspond to 8 bits in the computer. In this paper, we call each gradation as CCT count. For example, the CCT count of darkest reflectance is 0, and that of the brightest is 255. The CCT count in the water do not distribute entirely over all gradations and our previous study [5] indicates that the gadation ranges are broader in visible bands than that in infrared bands.

Visible band

The CCT counts in the visible wavelength range, namely the TM-bands 1 to 3, relate to the water color, which may vary with salinity, color and concentration of suspended solid particles as well as chemical and biological matters in the water. Engman [6] reported that the logarithms of the CCT count is proportional to the water turbidity which is larger than approximately 60 (mg/l). But in the sea where the transparency is very high degree, e.g. the tropical sea with coral reeves, the CCT count may be connected with the color at the sea bed or the depth of the sea.

Nearly infrared band

The previous investigations indicate that chlorophyll contained in algae and plankton, which is one of the important parameters in the preestimation of the river effluent effects on the sea water environment, relate well to the CCT count of this range. With respect to these, Engman and Gurney[6] and Khorram[7] developed the regression equations between the water quality parameters and the mean radiance value of different Landsat MSS-bands. But, these cannot be applied in the case that the reflection of electromagentic wave at the water surface plus pass radiance in the atmosphere are superior to the scattering intensity in the water.

Moderate infrared bands

Few studies have been reported, relating to application of those bands on the analysis of hydraulic phenomena and water quality.

THE SHINGU RIVER

The Shingu river runs through the Kii peninsula and flows into the Pacific Ocean (Figure 1). Its basin area is approximately 2360 (km^2) and yearly mean flow of 156 (m^3/s) has been recorded at the Ouga station approximately 10(km) upstream of the river mouth. In the up-stream region, several dam reservoirs have been operating in the purpose of flood control and power generation. Recently deterioration of water quality in the reservoirs, growing with the development of human activities have aroused important social interest, and it has been anxious about that the water quality deterioration of the up-stream reservoirs may influence in the water environment at the downstream river region as well as sea.

RIVER EFFLUENT PATTERNS

Plates 1(a), (b) show the images of the Landsat TM-band 2 and band 6 observed on April 19, 1988. On that day, a discharge of 397 (m^3/s) was recorded at the Ouga station. The CCT count of the visible range is proportional to the water turbidity. The Plate 1(a) suggests that the effluent axis estimated through the turbidity measurements must show deflecting to the right hand side of the river mouth. However Plate 1(b), the thermal image of the effluent, shows that the axis seems to extend almost on the straight, or even to deflect to the left side of the river mouth. Such diversity in the axis direction suggest that either of the two observed axes is affected by factors other than the river effluent itself. To uncover it, let us view around the river mouth more closely. Plates 2(a),(b) show thermal infrared images surveyed from airplanes on October 5, 1979 and November 6,1979. One can see that the river effluents flow out through the narrow opening between the sandbar and a solid river bank into the the Pacific Ocean. These suggest that the distribution of river effluent estimated through the turbidity shown in Plate 1(a) is affected by the sandbar, that is, the sand particles washed out from the sandbar are entrained into the right side of the effluent, which can result in the appearance of axis deflecting to the right hand side of the river mouth.

Next, Plates 3(a),(b) show the images of the Landsat TM-band 2 and band 6 observed on September 26,1988, when a typhoon had passed through this district a day before and the maximum rate of discharge of 1,234 (m^3/s) was recorded at the Ouga station. In this case, both the flow patterns estimated through the band 2 (water turbidity) and the band 6 (water temperature) are similar, differing from those shown in Plates 2(a),(b). Under the flood condition, the water in the river channel upstream of its mouth contain so enormous volume of solid particles that the influence of the sandbar at the downstream end shall become negligible.

ESTIMATION OF EFFLUENT AXIS

Introduction

In the above, we discussed the behavior of the river effluent in qualitative manner. But viewed from scientific angle, the quantitative approach is indispensable. Ideally speaking, it is desirable to investigate the entire sea area occupied by the effluent. However in general its works are considered to be too enormous. In fact, the diffusive feature of the effluent can be effectively estimated by distributions of water salinity and temperature along the jet axis. The distribution of water salinity and temperature along the axis is indirectly estimated through the CCT counts. The Landsat data saved in floppy disk for the personal computer which is employed in the present study consists of 512 pixels x 400 lines elements. Taking account of the accuracy practically required for the estimation of the effluent axis, we may thin out the data at certain interval instead of using the total data elements and transform in the mesh

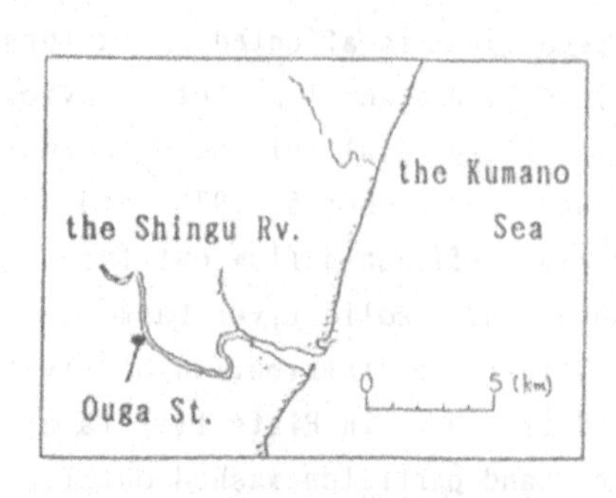

Fig. 1 The Shigu river

O P
(Po, Lo)
L

Fig. 2 P-L coordinate system

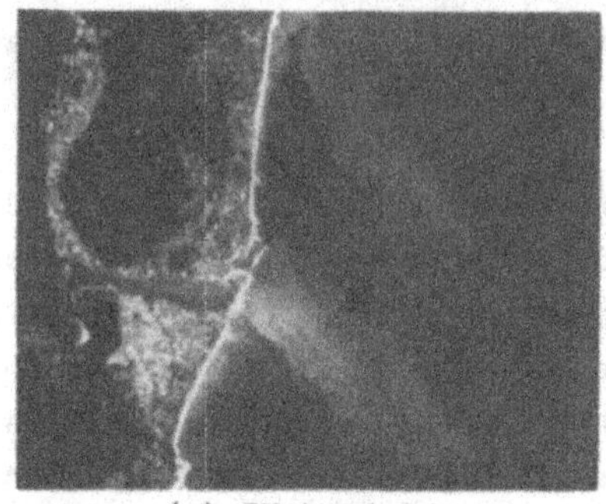

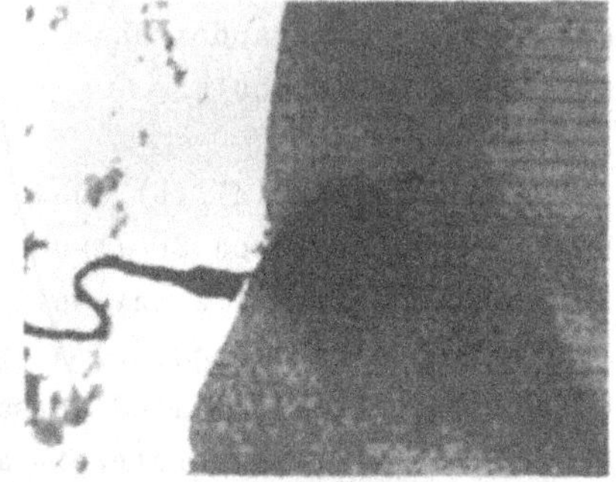

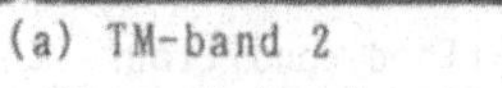

(a) TM-band 2 (b) TM-band 6

Plate 1 Landsat image observed on Apr. 19, 1988

(a) October 5, 1979 (b) November 06, 1979

Plate 2 Thermal image observed by airplanes

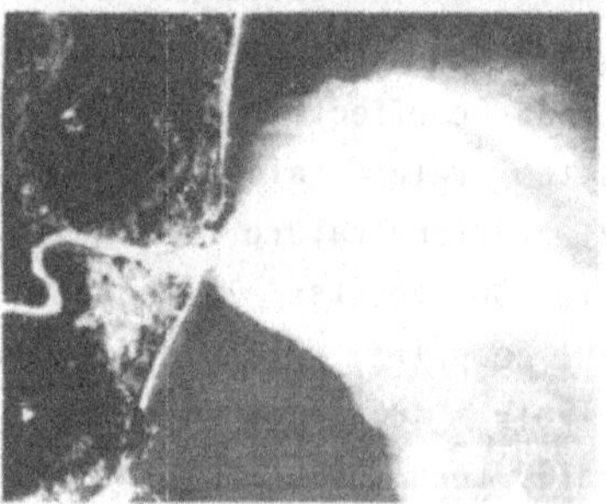

(a) TM-band 2 (b) TM-band 6

Plate 3 Landsat image observed on Sep. 26, 1988

data which is composed of 200 to 1,000 data elements. We will use the P − L coordinates system as shown in Figure 2.

CCT count distribution of cross-section

According to the jet theory, it can be assumed that the distribution of the CCT counts along a line normal to the effluent axis may be represented by a normal curve, provided that no close current exist. But practically the diffusion area of the river effluent shall be usually deformed from the normal distribution owing to physical factors such as existence of coastal current, neighboring rivers, Coriolis force and so on. Figure 3(a),(b) show the distribution of the CCT counts along the lines normal to the effluent axes estimated with the band 2 of April 19,1988 and September 29,1998. In the former, the distribution curve has a peak, against it, that in the latter is not so. When the curve has a peak, we can estimate easily the effluent axis by tracing those peak points in several sections normal to the L-axis. But in the case that the distribution curve has not a peak, it is difficult to identify the effluent axis.

Weight least squares method

Hence the weighted least square method is more useful. Let us assume that the effluent axis is a curve of third degree represented by the following equation.

$$L_{*k} = a_1 P_{*k} + a_2 P_{*k}^2 + a_3 P_{*k}^3 \quad (1)$$

where P_{*k} and L_{*k} are defined as follows.

$$P_{*k} = P_k - P_0, \quad L_{*k} = L_k - L_0 \quad (2)$$

where (P_0, L_0) is the coordinate of the river mouth. The deviation of any element mesh from the curve of the Equation (1), δ_k and the sum of deviation square multiplied by proper weight function, S are presented respectively, as follows.

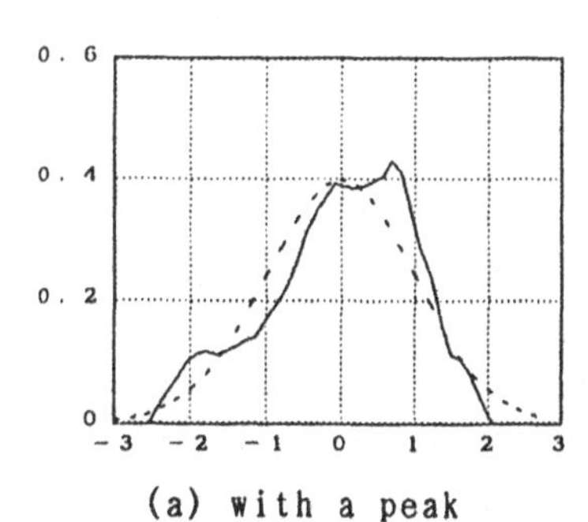

(a) with a peak

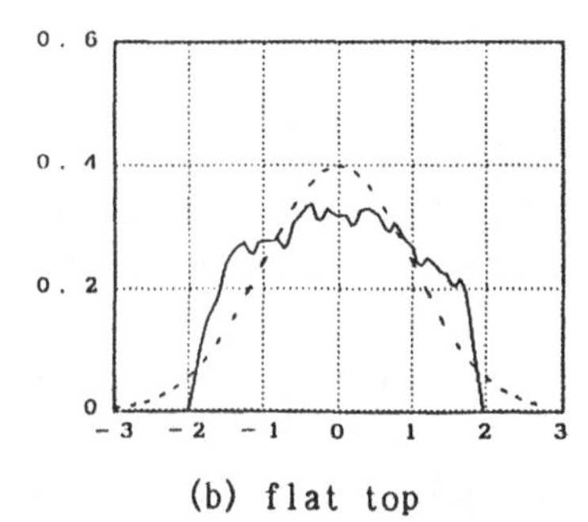

(b) flat top

Figure 3 CCT distribution in sections normal to the effluent axis

$$\delta_k = L_{*k} - (a_1 P_{*k} + a_2 P_{*k}^2 + a_3 P_{*k}^3) \quad (3)$$

$$S = \Sigma W_k \delta_k^2 \quad (4)$$

where, W_k is the weight function. Then under the condition that S becomes minimum, we yield the normal equations as following.

$$\sum_k \begin{bmatrix} W_k L_{*k} \\ W_k P_{*k} L_{*k} \\ W_k P_{*k}^2 L_{*k} \end{bmatrix} = \sum_k \begin{bmatrix} W_k P_{*k}^2 & W_k P_{*k}^3 & W_k P_{*k}^4 \\ W_k P_{*k}^3 & W_k P_{*k}^4 & W_k P_{*k}^5 \\ W_k P_{*k}^4 & W_k P_{*k}^5 & W_k P_{*k}^6 \end{bmatrix} \begin{bmatrix} a_1 \\ a_2 \\ a_3 \end{bmatrix} \quad (5)$$

These equations were led for the effluent toward the P-axis, namely for the effluent flows eastward or westward. For that toward the L-axis, we must replace P and L each other in the above equations. To eliminate the influences exerted by neighboring sources other than the river effluent to be studied, we introduce the weight functions W_k in Equation (4), which depends upon the CCT count, U_k at the k-th mesh element. Considering that the CCT count in the visible band ranges, is usually smaller at the water area than that on the land, and is approximately proportional to the water turbidity, we define the weight function W_k as follows.

$$W_k = \begin{cases} 0 & (U_k < U_{Bnd},\ U_{Mth} < U_k) \\ (U_k - U_{Bnd})^{\beta} & (U_{Bnd} \leq U_k \leq U_{Mth}) \end{cases} \quad (6)$$

where, U_{Mth} = the maximum value of the visible CCT count in the water area near the river mouth, U_{Bnd} = that at the boundary of the diffusion area, and β = constant (usually equal to 1). The CCT count in the infrared range at the offing is generally larger than that at the river mouth. Therefore, for the infrared image we define the weight function as follows.

$$W_k = \begin{cases} 0 & (U_k < U_{Mth},\ U_{Bnd} < U_k) \\ (U_{Bnd} - U_k)^{\beta} & (U_{Mth} \leq U_k \leq U_{Bnd}) \end{cases} \quad (7)$$

But in the case that the effluent temperature at the river mouth is higher than that in its surrounding sea water, the equation (6) for the visible range shall be used.

EXAMPLE OF ANALYSIS

Analysis of effluent axis

With the proposed method, we calculated the horizontal CCT count distributions of the TM-band 2 and 6 of April 19, 1988, June 06, 1988, September 25, 1988, May 08, 1989 and August 28, 1989. The results are shown in Figures 4 to 8. On the figures, the estimated effluent axes are shown as well. The discharges at the river mouth on each date are presented in the figures. From these figures we can see that when the effluent discharge is large, both the axes estimated with

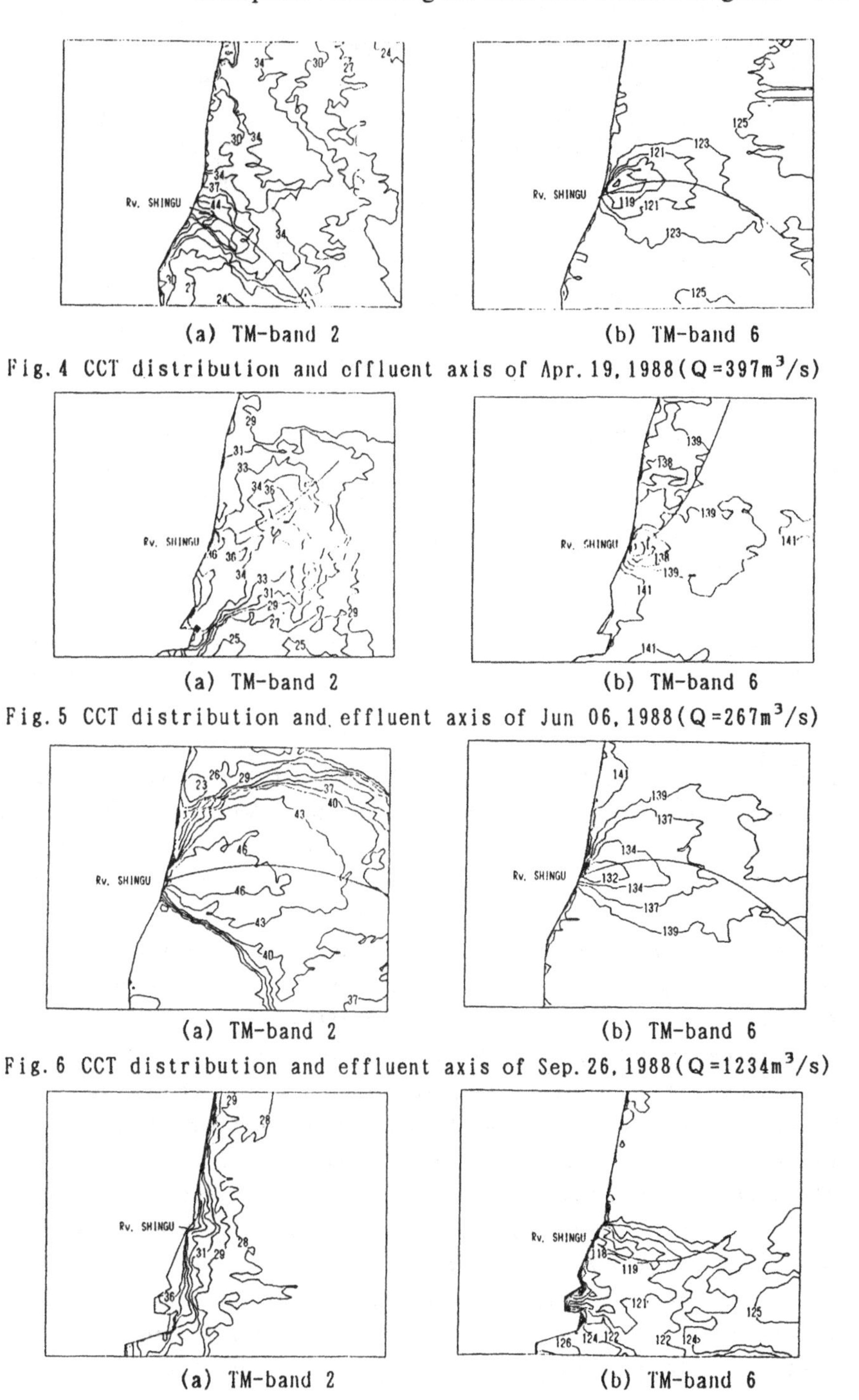

(a) TM-band 2 (b) TM-band 6

Fig. 4 CCT distribution and effluent axis of Apr. 19, 1988 ($Q = 397m^3/s$)

(a) TM-band 2 (b) TM-band 6

Fig. 5 CCT distribution and effluent axis of Jun 06, 1988 ($Q = 267m^3/s$)

(a) TM-band 2 (b) TM-band 6

Fig. 6 CCT distribution and effluent axis of Sep. 26, 1988 ($Q = 1234m^3/s$)

(a) TM-band 2 (b) TM-band 6

Fig. 7 CCT distribution and effluent axis of May 08, 1989 ($Q = 131m^3/s$)

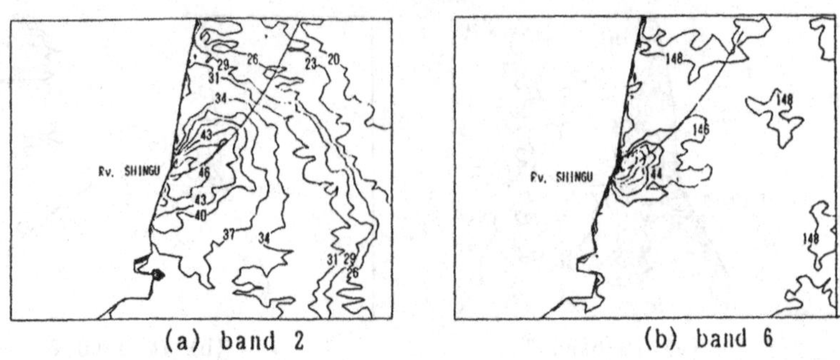

(a) band 2 (b) band 6

Fig. 8 CCT distribution and effluent axis of Aug. 28, 1989 ($Q = 1970m^3/s$)

the TM-band 2 (turbidity) and band 6 (temperature) show coincidence well. But in the case that the discharge is small, they are unlike each other. Furthermore when the discharge is extremely small, the effluented water moves keeping touch with the coast line and then it becomes hard to regress its axis (Figure 7). Indeed, under such conditions, the effects of sand drift and other materials supplied to the sea area from sources other than the river will become dominant in the distribution of CCT counts especially in the visible range.

Estimation of diffusion distance

Next, we calculate the CCT counts along the effluent axis. To remove influence which may be brought by the sun elevation and atmospheric condition into CCT count, we use the relative CCT count, U_{REL} defined by the following formula, instead of using the absolute value.

$$U_{REL} = (U_k - U_{off}) / (U_{Mth} - U_{off}) \tag{8}$$

where, U_{Mth} and U_{off} are the CCT counts at the river mouth and a point sufficiently far from the river mouth respectively, then U_k is the CCT count at any point on the effluent axis. U_{REL} is a dimensionless quantity, which is equal to 1 at the river mouth and 0 at the offing. Figures 9 (a), (b) show calculated examples of the relative CCT count distribution along the effluent axis. In the Figure 9(a), the U_{REL} curve has a peak between 8 (km) and 10 (km) distance offing from the mouth, which might be resulted from tide and unsteady features of river flow. In the figure 9(b) the relative CCT count of band 2 keeps approximately constant by the point 5(km) offshore, and then begins to converge to 0 at far offshore. But the relative CCT count of band 6 (water temperature) decreases exponentially toward offshore in the all cases investigated. An other noteworthy thing is that the curves of band 6 converge to 0 more rapidly than those of band 2 (water turbidity).

Estimation of diffusion area

Influenced area of the effluent can be estimated by counting the total number of the pixels occupied by its effluent, and it may be considered that the lower value of the relative CCT count is assumed as the limit confining the effluent, the better accuracy of estimating the influenced area. But if the limit is too low, distinguishing the diffusion area of effluent from its surrounding sea area is difficult. Therefore, in the present study we consider that the influenced area of effluent is confined with a boundary line, beyond which the relative CCT count becomes equal to or less than 0.1. The influenced area for each case calculated with the TM data is presented together with the discharge. Figure 10 indicates relation between the discharge, Q (m^3/s) and the influenced area, S (km^2), which can be represented by the following formulas.

for the band 2 : $S = 0.0296\ Q^{1.18}$ (9a)

for the band 6 : $S = 0.0483\ Q^{1.03}$ (9b)

These are in agreement with results reported by Tanaka, Sawamoto, and et. al. These results suggest that the area estimated through the water temperature shall be smaller than that through the turbidity, because of heat exchange at the water surface.

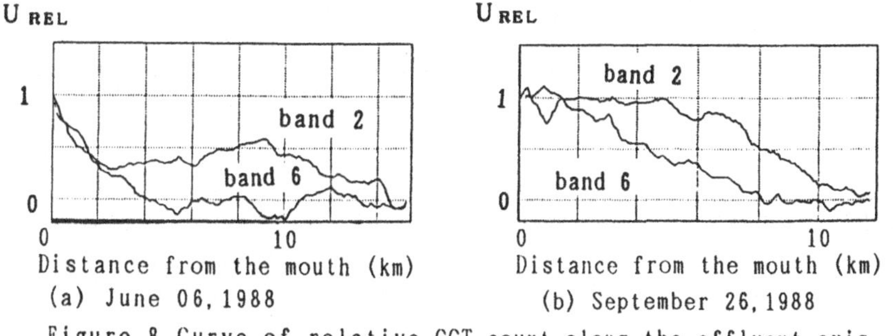

(a) June 06, 1988 (b) September 26, 1988

Figure 8 Curve of relative CCT count along the effluent axis

TANAKA and OGIHARA
SAWAMOTO and MURAKOSHI
Authors

Q	band 2	band 6
397	34	29
267	324	13
1,234	92	70
131	10	50
1,970	309	121

Fig.7 Relation between discharge Q and diffusion area S

COCLUDING REMARKS

The satellite remote sensing have been mainly used as a tool of technique for flow visualization. But the data obtained through the satellite is transmitted to users in the numerical forms and these have essentially quantitative nature. In the above, we discussed the river effluent axes in quantitative manner, and proposed the least square method weighted with the CCT count to detect the effluent axis in the sea. Applying this method to the effluent from the Shingu river, we discussed that both the axes estimated through water turbidity and temperature show coincidence. But when the discharge is small, they are unlike each other.

Secondly, with the Landsat data we discussed the diffusion area of the effluent. We consider that the influenced area of the effluent is confined with a boundary line, beyond which the relative CCT counts become equal to or less than 0.1, and show that the different influenced area will be obtained by depending upon the traced medium, in effect the water turbidity or temperature.

REFERENCES

1. Tanaka, S. and Ogihara, K : Study on Dispersion of Water Effluented from River by Landsat MSS Data using Personal Computer, JRS, Vol. 5, No. 1, pp. 69-75, 1985.
2. Onishi, S. and Baba, K. : Study of Hydraulic Behaviors of River Effluent at Coriolis Force Dominating Field by Remote Sensing, Jour. of Hydroscience and Hydraulic Eng., Vol. 5, No. 1, pp. 39-48, 1987.
3. Onishi, S. : Hydraulic Engineering and Remote Sensing, Proc. of JSCE, No. 393/II-9, pp. 9-19, 1988.
4. Sawamoto, M. and Murakoshi, J. : Study on River Effluents in Suruga Bay and from the Tenryu River, Proc. of Coastal Eng., Vol. 32, pp. 767-771, 1985.
5. Onishi, S. and Kawai, H. : Estimation of Effluent Diffusion through Satellite Remote Sensing with Personal Computer, Proc. of Hydraulic Eng., Vol. 35, pp. 173-178, 1991.
6. Engman, E. T. and Gurney, R. J. : Water Quality, Remote Sensing in Hydrology (published by CHAPMAN AND HALL), pp. 175-192, (1985).
7. Khorram, S. : Development of water Quality Models applicable throughout the Entire Sanfransisco Bay and Delta, Photogra. Eng. and Remote Sensing, pp. 53-62, 1985.

Modelling of Tidal Flow and Transport Processes: A Case Study in the Tejo Estuary

L. Portela, L. Cancino, R. Neves

Department of Mechanical Engineering, Instituto Superior Técnico, P-1096 Lisboa Codex, Portugal

ABSTRACT

A two-dimensional horizontal hydrodynamic model, a Lagrangian transport model and an Eulerian transport model, based on the finite-difference method, are presented. Application of these models to the simulation of tidal flow and transport processes in the Tejo estuary (Portugal) is described. The results of tidal flow simulation show encouraging agreement with the available field data. The results of transport processes simulation indicate variability due to the complex morphology of the Tejo estuary. The tidal prism passing through the inlet channel provides a generally efficient mixing of estuarine waters, but the dispersion of pollutants in the inner channels near industrial sites seems much less efficient. This should have significant implications for environmental management.

INTRODUCTION

The Tejo estuary, in Portugal, is surrounded by some densely populated and industrialized areas. The effects of human activities on the estuarine system, notably the impact of domestic and industrial wastes on water quality, are a source of concern for the authorities and the general public. The achievement of rational management of the Tejo estuary requires the understanding of basic physical, chemical and biological processes and the development of predictive methods for environmental management. Numerical models should play an increasingly important role.

This paper presents the first results of the numerical simulation of tidal circulation and pollutant transport in the Tejo estuary, performed at DEM-IST. Emphasis is placed on the elucidation of transport processes, using both Lagrangian and Eulerian methods.

DESCRIPTION OF THE MODELS

Hydrodynamic model

General features MOHID is a two-dimensional (2D), depth integrated, hydrodynamic model. The governing equations are the Shallow Water Equations for a vertically homogeneous fluid with constant density. The Shallow Water Equations are solved by finite differences, using an Alternating Direction Implicit (ADI) method. The 2D model can be connected to one-dimensional models (1D) and this option is often convenient (e.g. to prescribe river side boundary conditions beyond the limit of tidal propagation). MOHID also includes modules for residual flow computation and for Lagrangian particle tracking [1].

Governing equations The governing equations of the 2D, depth integrated, model are the Shallow Water Equations for mass and momentum conservation:

$$\frac{\partial z}{\partial t} + \frac{\partial (H u_1)}{\partial x_1} + \frac{\partial (H u_2)}{\partial x_2} = 0 \tag{1}$$

$$\frac{\partial (H u_1)}{\partial t} + \frac{\partial (H u_1 u_1)}{\partial x_1} + \frac{\partial (H u_1 u_2)}{\partial x_2} - H f u_2 = - g H \frac{\partial z}{\partial x_1} + H \varepsilon \frac{\partial^2 u_1}{\partial x_1^2} + H \varepsilon \frac{\partial^2 u_1}{\partial x_2^2} - \frac{g n^2}{H^{1/3}} |u| u_1 \tag{2a}$$

$$\frac{\partial (H u_2)}{\partial t} + \frac{\partial (H u_1 u_2)}{\partial x_1} + \frac{\partial (H u_2 u_2)}{\partial x_2} + H f u_1 = - g H \frac{\partial z}{\partial x_2} + H \varepsilon \frac{\partial^2 u_2}{\partial x_1^2} + H \varepsilon \frac{\partial^2 u_2}{\partial x_2^2} - \frac{g n^2}{H^{1/3}} |u| u_2 \tag{2b}$$

where t = time, x_1, x_2 = cartesian co-ordinates in the horizontal plane, z = water surface elevation above mean water level, H = total depth of flow from water surface to the bottom, u_1, u_2 = depth averaged velocity components in the x- and y-directions, f = Coriolis parameter, g = acceleration due to gravity, ε = depth averaged eddy viscosity, and n = Manning's roughness coefficient.

The corresponding equations of the 1D model, obtained by integration over the cross-section, are the de Saint Venant Equations:

$$\frac{\partial A}{\partial t} + \frac{\partial Q}{\partial x} = 0 \tag{3}$$

$$\frac{\partial Q}{\partial t} + \frac{\partial}{\partial x}\left(\frac{Q^2}{A}\right) = - g A \frac{\partial z}{\partial x} - \frac{g n^2}{A H^{4/3}} |Q| Q \tag{4}$$

where Q = flow rate, and A = cross-sectional area.

Numerical methods The partial differential equations are solved by the finite-difference method on a staggered grid. The 2D model uses a semi-implicit ADI method, which is essentially second-order accurate both in space and time. The proposed scheme leads to 6 finite-difference equations at each time-step. ADI schemes solving 6 algebraic equations have a superior ability to deal with moving boundaries, an important feature in estuaries with large intertidal areas. At each half time-step, the inversion of a tridiagonal matrix is performed using the Thomas algorithm [2].

Lagrangian transport model

General features A MOHID module allows the computation of Lagrangian advective transport, using particles and the velocity field computed by the Eulerian hydrodynamic model. Two possibilities are available: pathlines to visualize the movement of a single particle through the flow (i.e. a given particle at a number of instants); streaklines to visualize the position of particles continuously released from a prescribed source (i.e. a number of particles at a given instant).

The proposed scheme is very simple. Advection is based on a Lagrangian approach. A linear interpolation in space is performed to determine the velocity of each particle within a grid cell. No interpolation is performed in time, the computation being fully explicit. Dispersion is simulated by a random walk method.

Eulerian transport model

General features MOTAD is a 2D, depth integrated, Eulerian transport model, based on the velocity and elevation fields computed by the hydrodynamic model. Transport is described by the depth integrated advection-diffusion equation. This equation is solved by finite differences, using the QUICK scheme for the advection terms and central differencing for the dispersion terms. The mass balance equation is applicable to passive (i.e. the same depth averaged density of the water column) and conservative (i.e. it does not include a source-sink term) scalars.

Governing equation The governing equation of the 2D, depth integrated, model is the advection-diffusion equation, using the Eulerian formulation:

$$\frac{\partial HC}{\partial t}+\frac{\partial HCu_1}{\partial x_1}+\frac{\partial HCu_2}{\partial x_2}=\frac{\partial}{\partial x_1}(HD_1\frac{\partial C}{\partial x_1})+\frac{\partial}{\partial x_2}(HD_2\frac{\partial C}{\partial x_2}) \quad (5)$$

where C = depth averaged concentration of conservative scalar, and D_1, D_2= dispersion coefficients in the x- and y- directions (the crossed terms of the dispersion tensor are neglected).

The dispersion coefficients comprise the effects of turbulent diffusion, dispersion due to shear effect and additional dispersion due to sub-grid scale processes. To estimate the dispersion coefficients, several empirical formulas have been proposed, generally involving grid spacing, time step and average velocities.

Numerical methods The advection-diffusion equation is solved by finite differences, using for advection the explicit QUICK scheme, which is third-order accurate in space, and for diffusion explicit central differencing, which is second-order accurate.

The difficulties associated with the differencing of the non-linear advection terms are known. Central differences do not respect the transportive property, which stipulates that the effect of a perturbation is advected only in the direction of the velocity [3], and produce unphysical oscillations. Upwind differences possess the transportive property but introduce large numerical diffusion. The higher-order QUICK scheme, described by Leonard [4], is a compromise between central and upwind differencing, the leading truncation error being only a fourth-order dissipation. The quadratic upstream interpolation avoids most of the wiggles and the damping problems, although QUICK lacks the boundedness of upwind.

TIDAL PROCESSES SIMULATION

Study area

The Tejo estuary (Fig. 1) has a surface area of 320 km^2, of which about 40% is intertidal area. It extends for about 80 km, from Muge (tidal limit) to S. Julião da Barra (mouth). The tidal range limits are 1 and 4 m. The average freshwater input is 300 m^3/s, with wide seasonal and annual variations. The salinity distribution is well-mixed during spring tides, partially-mixed during neap tides. Estuarine waters are polluted by important domestic and industrial wastewaters, namely from the Greater Lisbon.

Application of the models

Hydrodynamic model The application of the 2D hydrodynamic model to the Tejo estuary uses a 200 m grid spacing and a 60 s time step. This 2D model is coupled to two 1D models for the tidal stretches of rivers Tejo and Sorraia. The Tejo 1D model extends to Muge, for about 30 km, and the Sorraia 1D model extends to Porto Alto, for about 5 km, using a variable space step (100 to 600 m). The Manning coefficient was set constant for the whole 2D and 1D model area: n=0.02 $s/m^{1/3}$. The eddy viscosity coefficient was also set constant for the 2D model: ε=5 m^2/s.

On the ocean side, open boundary conditions were specified by the surface elevation corresponding to the M_2 tidal constituent. Due to unavailable field data and for greater expediency, it was decided to simulate the mean tidal flow using the average amplitude of the M_2 component at Cascais (0.97 m) and a constant phase along the sea boundary. On the river side, constant inflows of 250 and 30 m^3/s

were imposed for the 1D models of rivers Tejo and Sorraia, respectively.

Lagrangian transport model Particle tracking was used to simulate contaminant transport. Transport was described by advection only, dispersion being neglected. The interpolation of the depth averaged velocities calculated by the hydrodynamic model for an average tide provided the velocities for each particle.

Particles were released at six locations: two small tributaries in the northern part of the estuary, rivers Jamor (symbol: diamond) and Trancão (symbol: square), which are important sources of domestic pollution; three major industrial sites, the shipyard near Almada (symbol: triangle oriented to the right), the chemical plant near Barreiro (symbol: triangle oriented to the top), both in the southern part of the estuary, and the chloralkali plant near Póvoa (symbol: triangle oriented to the left), in its northern part, which are important sources of toxic substances; and in the center of the estuary (symbol: circle), to simulate the accidental release of dangerous substances from a ship.

Eulerian transport model The Eulerian transport model uses the same grid and the half-time step of the hydrodynamic model, 30 s. The horizontal diffusion coefficients adopted, taking into account the shear dispersion effects, were set constant in space and time: $D_1=D_2=5.0\ m^2/s$. To focus on short-term predictions, the transport of a instantaneous discharge with a Gaussian distribution located in the center of the estuary was simulated.

Results and discussion

Tidal flow simulation Results of the average M_2 tidal flow are presented (Fig. 1, 2, 3 and 4). The flow pattern is in good qualitative agreement with the available measurements.

The first results of the hydrodynamic model have been validated comparing the tidal amplitudes and phases computed by the model with the amplitudes and phases resulting from observations in eight measuring stations. A full description of the validation results is beyond the scope of this paper. However, it is worth mentioning that the simulation of tidal flow in the Tejo estuary has achieved good accuracy in what amplitudes are concerned, but also that moderate phase discrepancies were observed. The overall results are considered encouraging.

Residual currents simulation Residual currents were obtained calculating the average velocity field and the average flow field through a tidal cycle from an Eulerian point of view. Results of the residual velocity field for different river flow rates are presented (Figs. 5 and 6). The hatched area indicates that the results are unreliable near the ocean open boundary, due to the imposition of a constant amplitude and phase of the M_2 tidal constituent.

The residual velocity field for average river flow rates (Fig. 5) reveals two large eddies, both clockwise, near Almada and in the central part of the estuary. Outside the estuary, the results suggest that the residual flow is strong, due to the large tidal prism and the narrow outflow channel. Increasing the river flow rates tenfold (Fig. 6), a situation that is not unusual during winter, the main flow of freshwater through the estuary becomes apparent. River water flows seaward mainly through the central Cala das Barcas, but is deflected to the left margin, arriving to the channel, near Almada, mainly through the south.

Lagrangian transport simulation Contaminant transport simulation, using Lagrangian methods, clarifies some of the effects of the residual currents described above. Results of streaklines are presented (Figs. 7, 8, 9 and 10).

It seems that some general considerations may be put forward: first, the narrow deep outflow channel has a significant effect in the dispersion of the particles; second, the residual eddies outside the estuary promote the removal of particles from the estuary; third, the residual eddies inside the estuary may promote, or prevent, the dispersion of the particles; fourth, and lastly, the dispersion of particles in the inner channels is much less effective.

Eulerian transport simulation A Gaussian concentration distribution (Fig. 11) illustrates the simulation of transport, including horizontal diffusion, using Eulerian methods. Concentration isolines refer to 10% and 1% of the initial peak value. After 12 hours, the maximum concentration is reduced to 17% of the initial peak value (Fig. 12).

CONCLUSIONS

Hydrodynamic and transport modelling of the Tejo estuary has demonstrated that the accuracy and stability of the selected numerical tools is quite satisfactory. The use of simple models, using both Lagrangian and Eulerian methods, has provided a good initial insight into the short-term transport processes.

The simulation of tidal processes reveals heterogeneities due to the complex morphology of the Tejo estuary. The tidal prism provides a generally efficient mixing of estuarine waters, but the dispersion of pollutants in the inner channels near industrial sites (e.g. the northern Cala do Norte) seems quite different. Water quality management should focus on these areas.

ACKNOWLEDGMENTS

The first authors gratefully acknowledge the financial support of Programa CIÊNCIA - JNICT (BD/144/90-IG and BD/BIC/M/57/90-IG).

REFERENCES

1. Portela, L. I. and Neves, R. J. J. MOHID - 2D Hydrodynamic Model. A Short Manual. *CTAMFUTL/IST Tech. Rep.*, Lisboa, 1991.

2. Neves, R. J. J. Flow Process Modelling in a Salt Marsh, in *Computer Modelling in Ocean Engineering* (Ed. Schrefler, B. A. and Zienkiewicz, O. C.), pp. 303-310, Proceedings of an Int. Conf. on Computer Modelling in Ocean Engineering, Venice, Italy, 1988. A. A. Balkema, Rotterdam, 1988.

3. Roache, P. J. *Computational Fluid Dynamics*. Hermosa, Albuquerque, 1977.

4. Leonard, B. P. A stable and accurate convective modelling procedure based on quadratic upstream interpolation. *Comput. Methods Appl. Mech. Eng.*, *19*, 59-98, 1979.

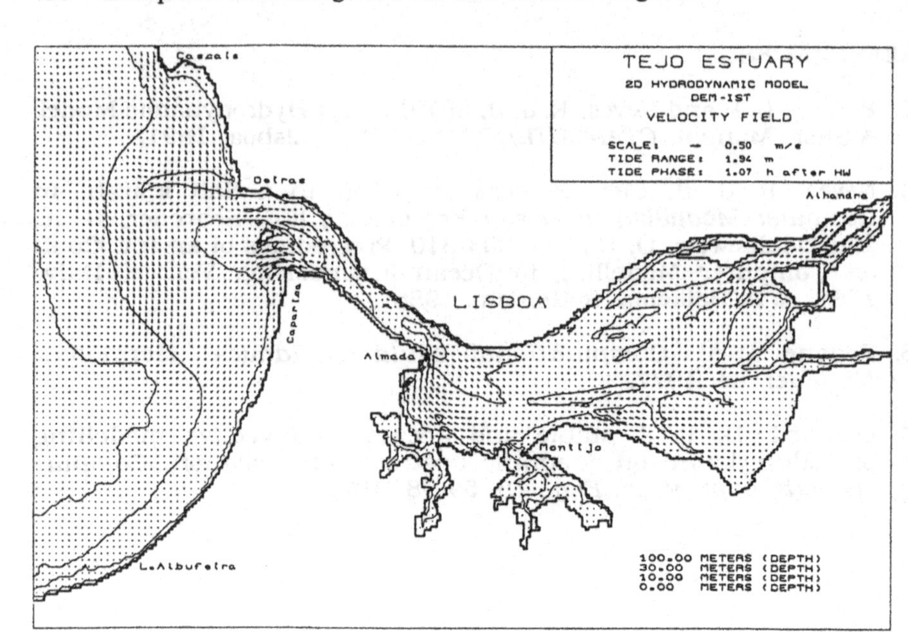

Fig. 1. Computed velocity field for the M_2 tidal constituent, 1 h 07 m after high water at Cascais.

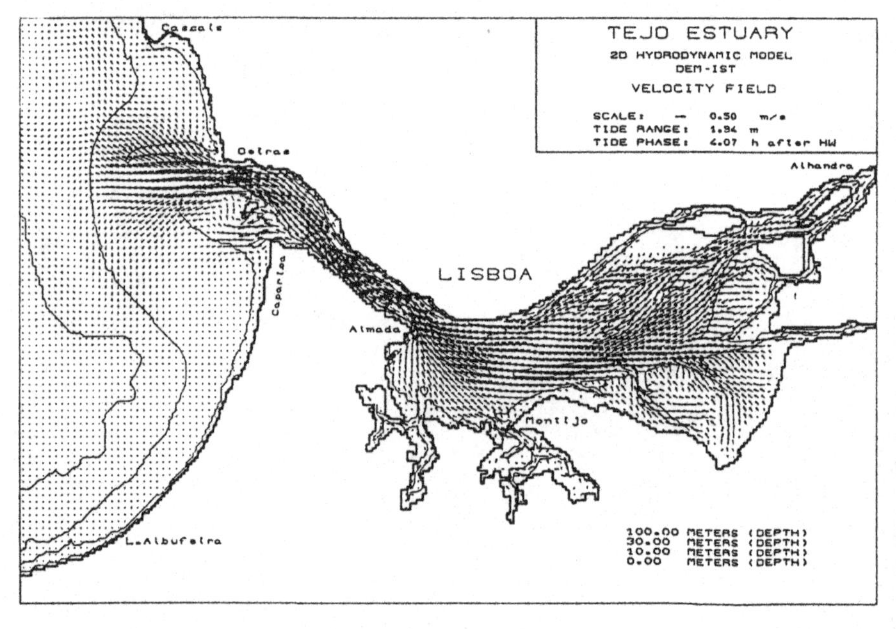

Fig. 2. Computed velocity field for the M_2 tidal constituent, 4 h 07 m after high water at Cascais.

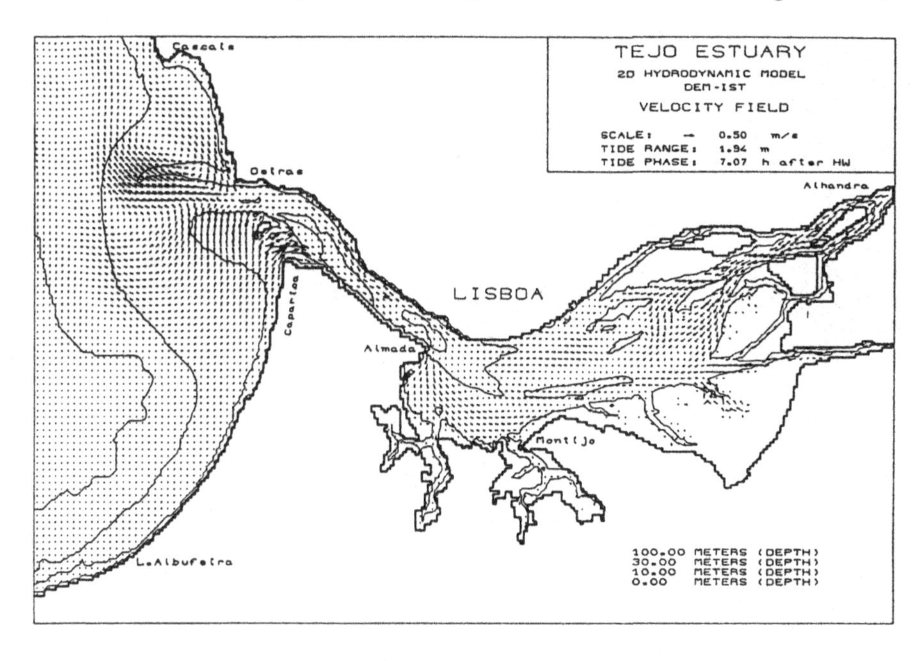

Fig. 3. Computed velocity field for the M_2 tidal constituent, 7 h 07 m after high water at Cascais.

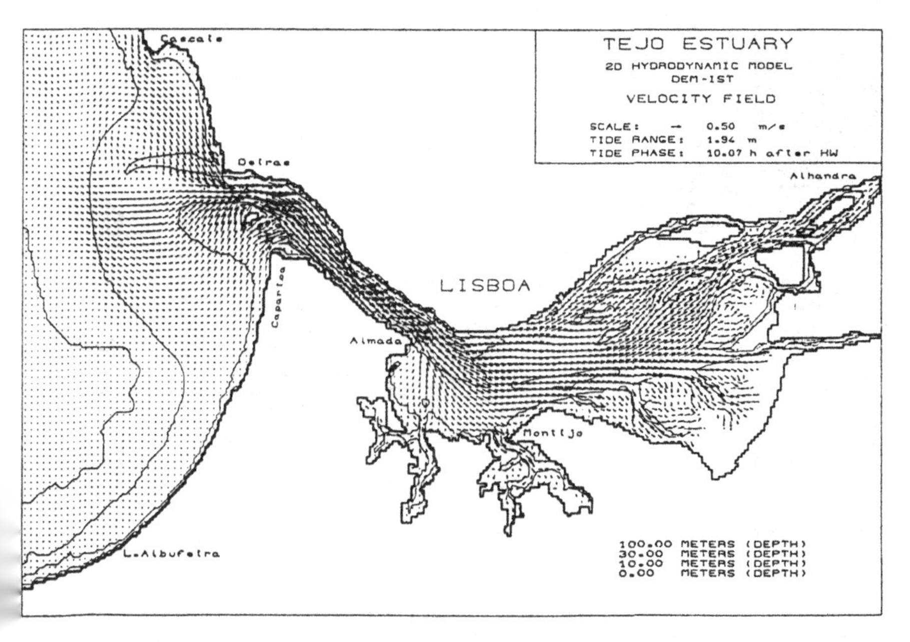

Fig. 4. Computed velocity field for the M_2 tidal constituent, 10 h 07 m after high water at Cascais.

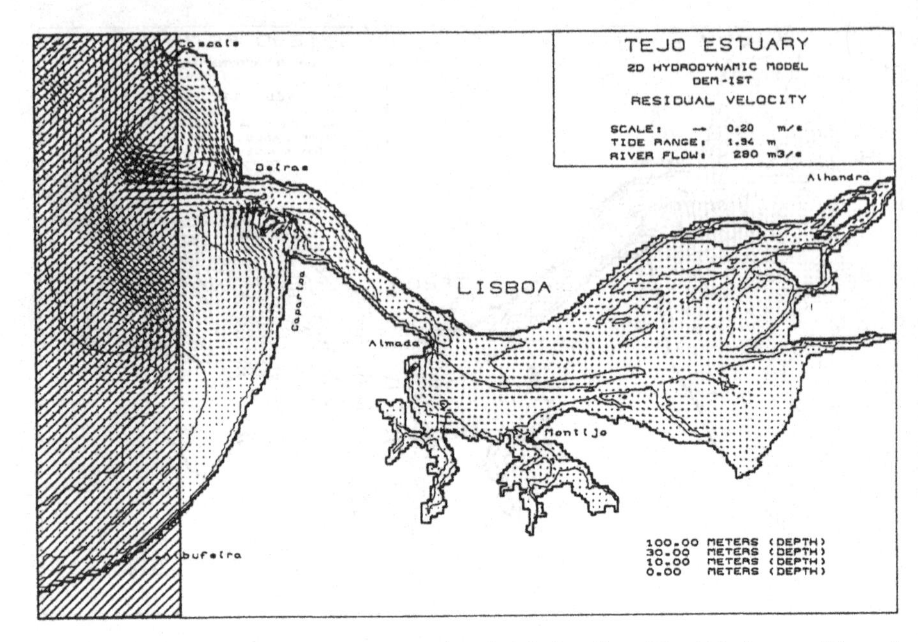

Fig. 5. Computed residual velocity field for the M_2 tidal constituent and a total river flow of 280 m^3/s.

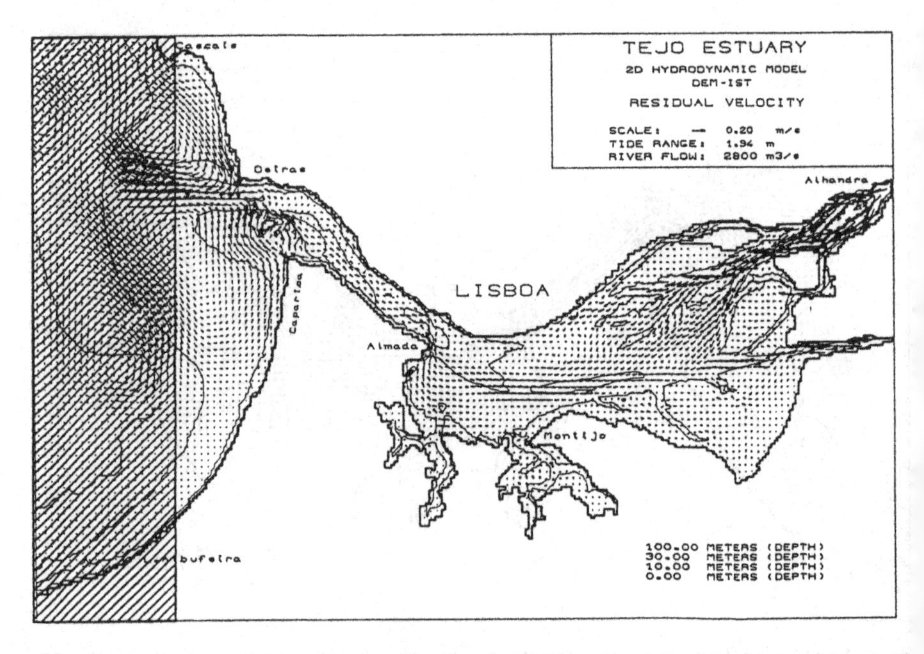

Fig. 6. Computed residual velocity field for the M_2 tidal constituent and a total river flow of 2800 m3/s.

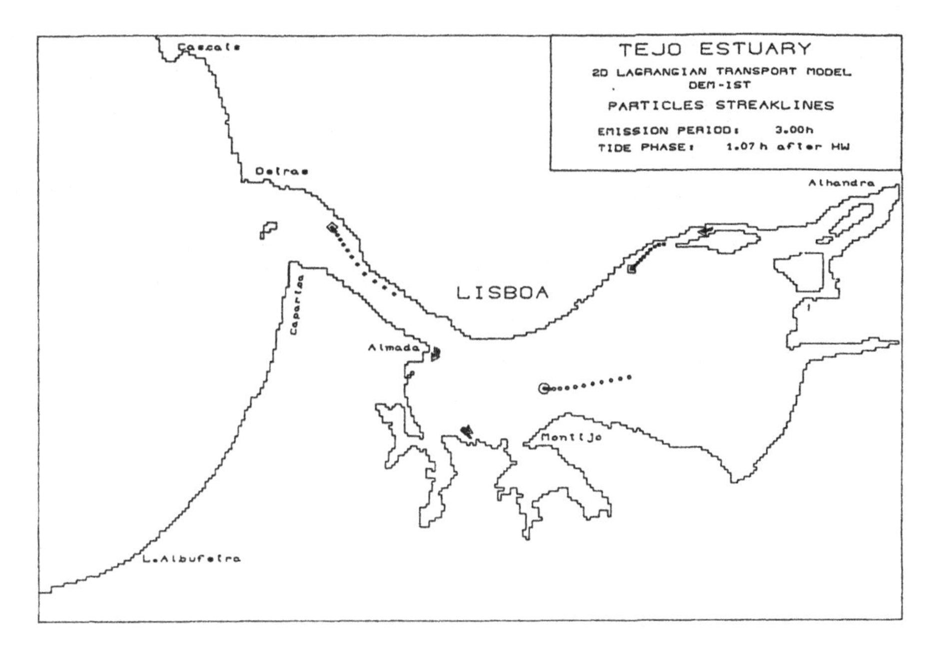

Fig. 7. Computed streaklines after 3 hours of emission, 1 h 07 m after high water at Cascais.

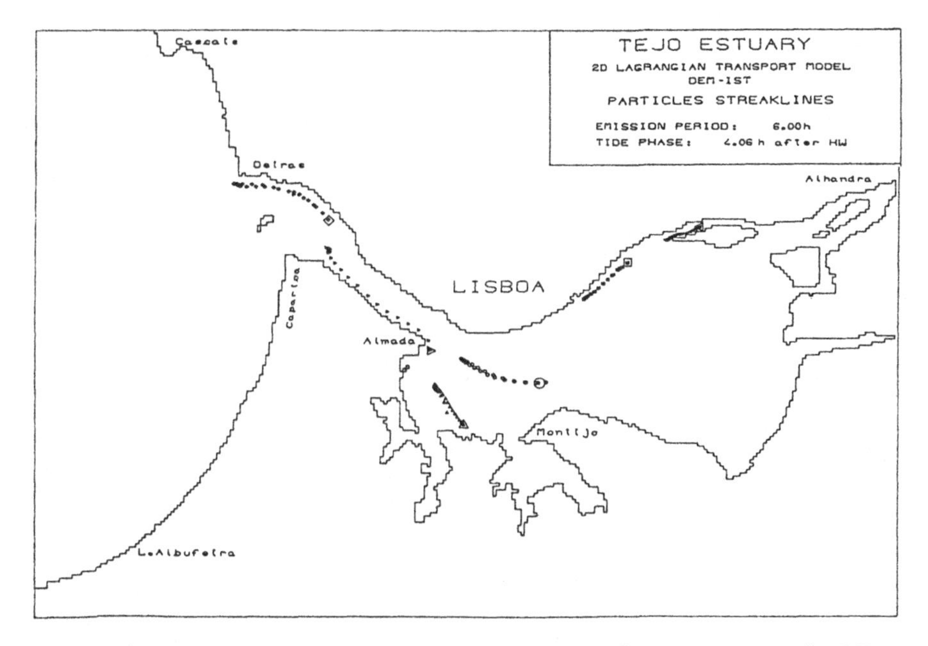

Fig. 8. Computed streaklines after 6 hours of emission, 4 h 07 m after high water at Cascais.

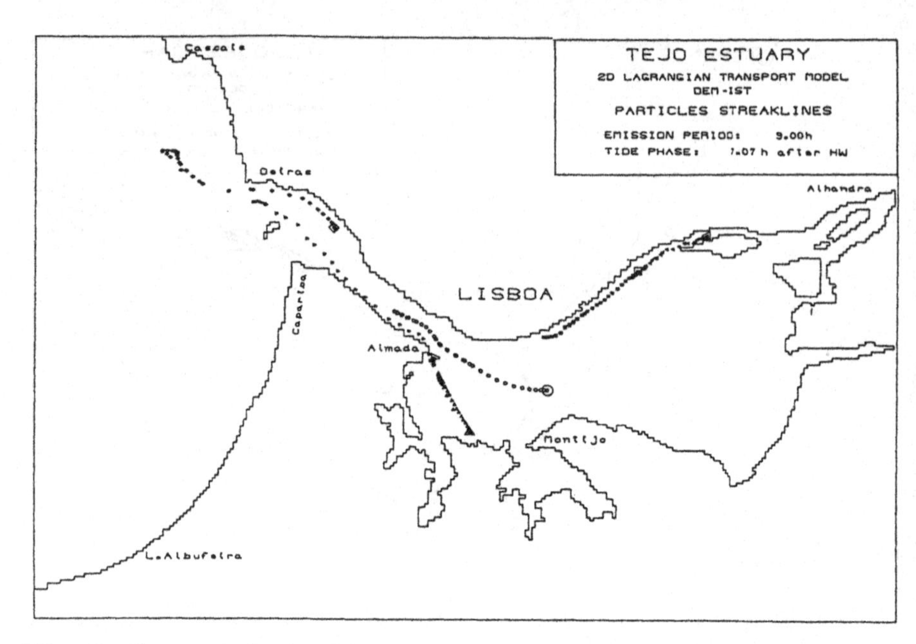

Fig. 9. Computed streaklines after 9 hours of emission, 7 h 07 m after high water at Cascais.

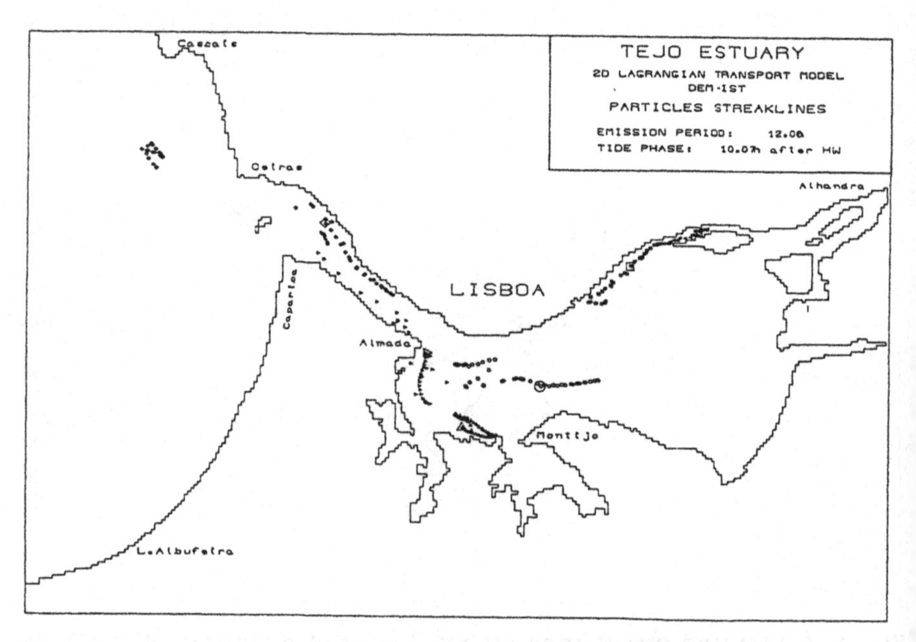

Fig. 10. Computed streaklines after 12 hours of emission, 10 h 07 m after high water at Cascais.

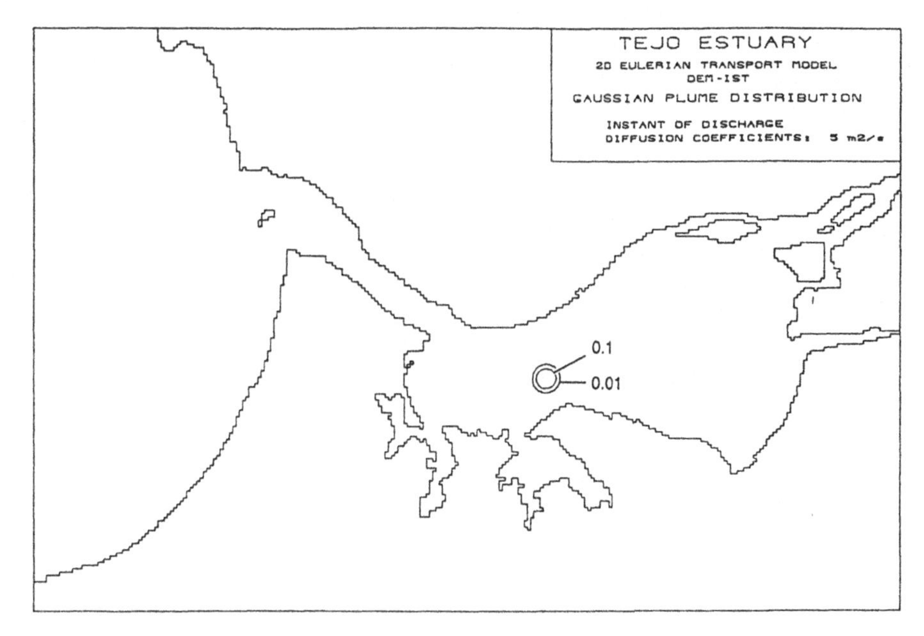

Fig. 11. Gaussian plume distribution at the instant of discharge, 10 h 32 m after high water at Cascais.

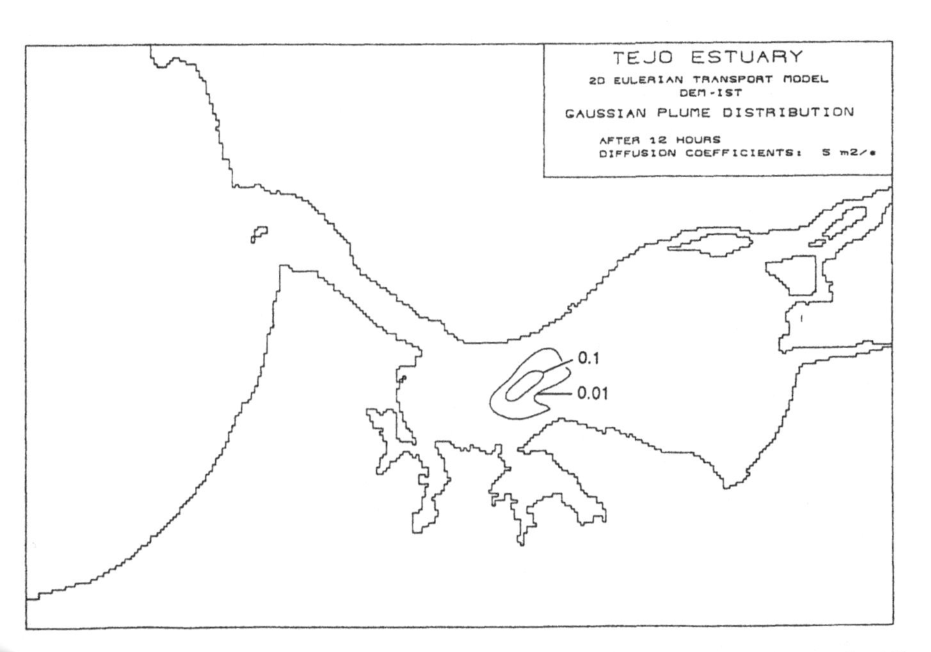

Fig. 12. Gaussian plume distribution 12 hours after emission, 10 h 07 m after high water at Cascais, for D_1=D_2=5 m^2/s.

Effluent Transport on the Solent by the Boundary Element Dual Reciprocity Method

P.W. Partridge, C.A. Brebbia
Wessex Institute of Technology, Ashurst Lodge, Ashurst, Southampton, SO4 2AA, U.K.

Abstract

The Dual Reciprocity Method is a well established technique for taking domain integrals to the boundary in Boundary Element analysis. It has been successfully applied to Diffusion Convection analysis on rectangular channels in a series of recent papers. Here the method is applied to effluent dispersion in the eastern part of the Solent, between Southampton and Portsmouth.

1 Introduction

Since its origins in 1978, the Boundary Element Method (BEM) has become a powerful tool for engineering analysis. Many problems, particularly linear ones, can be solved more efficiently and accurately using boundary elements than with other numerical techniques.

The application of BEM to more general non-linear and time dependent problems, including the Diffusion Convection equation, was hampered for a while by the need to define internal cells in order to compute domain integrals, which, whilst not introducing any new unknowns, made the method cumbersome to use.

The most powerful and versatile technique which has appeared to date for handling domain integrals without the need to define internal cells is the Dual Reciprocity Method (DRM), which was introduced by Nardini and Brebbia in 1982, [1] and has been recently generalized to a wide range of engineering problems, [2]. The method is straightforward to apply and uses simple fundamental solutions to operate on part of the equation, any remaining terms are taken to the boundary by a special transformation.

The Diffusion Convection or Transport Equation (1) has a wide range of applications in engineering ranging from heat transfer to pollution problems. The equation can be written as

$$D_x \frac{\partial^2 u}{\partial x^2} + D_y \frac{\partial^2 u}{\partial y^2} - v_x \frac{\partial u}{\partial x} - v_y \frac{\partial u}{\partial y} - ku + \phi - \frac{\partial u}{\partial t} = 0 \qquad (1)$$

Where u is concentration of a substance, temperature etc, D_x and D_y are dispersion coefficients, v_x and v_y are velocities, k is a decay parameter and ϕ is a source term. The velocities are functions of space and time, the other parameters are usually considered to be constant, though they may also be functions of space. The velocities may be obtained using a circulation model, the remaining parameters should be measured in the field.

The transport equation can be studied in BEM analysis using different fundamental solutions. Here a straightforward approach to solving equation (1) using the fundamental

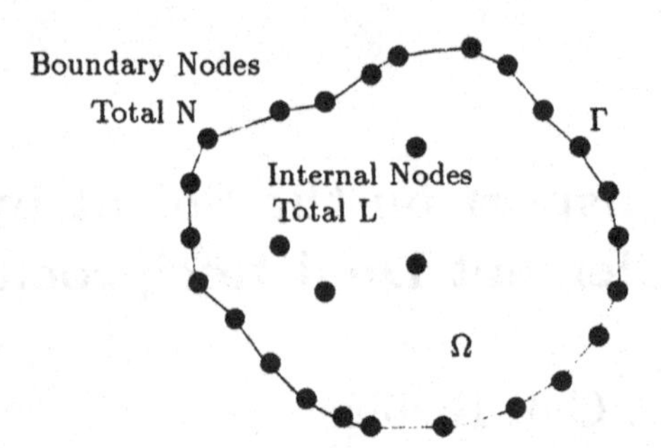

Figure 1: Boundary and Internal Nodes

solution to the Laplace Equation will be used, the domain integrals generated being taken to the boundary with the Dual Reciprocity method. This technique has been successfully applied to the solution of the diffusion convection equation on rectangular channels in references [2,3,4].

This approach, in addition to employing the simplest possible fundamental solution, allows any of the coefficients to have variable values without introducing special techniques. The equations are solved in time using a familiar time marching scheme.

The Solent is an important and very busy waterway flowing between the Isle of Wight and the English mainland. Here the eastern arm of the Solent will be studied, at its northern end lies the City of Southampton, and at the southern end the City of Portsmouth.

The region between these two cities is a development area, and as a consequence the paper studies the effect of a hypothetical new sewage outfall west of Portsmouth. This outfall and its impact on water quality has been the subject of a study using the finite element method, [5].

Here a similar study will be attempted using the Boundary Element method.

2 Fundamentals of the Dual Reciprocity Method

The DRM technique is described in [2] for a series of general problems, here only its application to equation (1) will be given in detail.

DRM may be used to take to the boundary any terms not dealt with by the fundamental solution, and can be employed in conjunction with any of the fundamental solutions available in BEM analysis. In the case of the fundamental solution to the Laplace equation all terms in (1) except those in the Laplacian need to be transferred to the right hand side to form an equation of the type

$$\nabla^2 u = b(x, y, u, t) \tag{2}$$

In order to take the integrals corresponding to the right hand side b to the boundary, the following approximation is proposed:-

$$b_i = \sum_{j=1}^{N+L} \alpha_j f_{ij}. \tag{3}$$

where b_i is the value of the function b at node i. The f_{ij} are approximating functions and the α_j unknown coefficients. The approximation is done at $(N + L)$ nodes, (N boundary nodes and L internal nodes, see figure 1). These nodes are called the DRM collocation points. The use of internal nodes is optional in most cases, however they have been shown to improve the accuracy of the solution, [2]. Internal nodes are usually defined at places where the solution is required on the domain, but notice that their definition is essential in cases where homogeneous boundary conditions are used as in this case the problem cannot accurately be described with only boundary nodes.

The functions f are defined by

$$\nabla^2 \hat{u} = f \tag{4}$$

where $\hat{u}$ is a particular solution. The type of functions f and $\hat{u}$ used in DRM will be considered shortly.

Combining (2), (3) and (4) one obtains

$$\nabla^2 u = \sum_{j=1}^{N+L} (\nabla^2 \hat{u}_j)\alpha_j. \tag{5}$$

Applying the boundary element technique in its usual form, equation (5) is multiplied by the fundamental solution and integrated over Ω, *i.e.*

$$\int_\Omega (\nabla^2 u)u^* d\Omega = \sum_{j=1}^{N+L} \alpha_j \int_\Omega (\nabla^2 \hat{u}_j)u^* d\Omega. \tag{6}$$

Integrating by parts produces

$$c_i u_i - \int_\Gamma u_i^* \left(\frac{\partial u}{\partial n}\right) d\Gamma + \int_\Gamma q_i^* u d\Gamma = \sum_{j=1}^{N+L} \left\{ \alpha_j \left(\int_\Gamma \hat{u}_j q_i^* d\Gamma - \int_\Gamma \hat{q}_j u_i^* d\Gamma + c_i \hat{u}_{ij} \right) \right\}. \tag{7}$$

In equation (7) $q = \frac{\partial u}{\partial n}$, $q^* = \frac{\partial u^*}{\partial n}$ and $\hat{q} = \frac{\partial \hat{u}}{\partial n}$.

After discretization this equation becomes

$$c_i u_i + \sum_{k=1}^{N} H_{ik} u_k - \sum_{k=1}^{N} G_{ik} q_k = \sum_{j=1}^{N+L} \left\{ \alpha_j \left(\sum_{k=1}^{N} H_{ik} \hat{u}_{kj} - \sum_{k=1}^{N} G_{ik} \hat{q}_{kj} + c_i \hat{u}_{ij} \right) \right\} \tag{8}$$

where H_{ik} and G_{ik} are the usual resultants of integration over the boundary elements, [2,6,7]. In (8), i are the source nodes, k the elements and j the DRM collocation points. Note that $\hat{q}$ is not defined for $i > N$ while $\hat{u}$ is defined for all points i. Equation (8) is written for each of the $(N + L)$ nodes i. The resulting matrix equation has N values of u and N values of q on the boundary and L values of u at interior nodes and can be written as follows;

$$\mathbf{Hu} - \mathbf{Gq} = (\mathbf{H\hat{U}} - \mathbf{G\hat{Q}})\boldsymbol{\alpha} = \mathbf{d} \tag{9}$$

The terms in c_i are incorporated onto the diagonal of $\mathbf{H}$. The vector $\boldsymbol{\alpha}$ in (9) can be calculated specializing equation (3) at all $(N + L)$ nodes, *i.e.*

$$\mathbf{b} = \mathbf{F}\boldsymbol{\alpha}$$

$$\boldsymbol{\alpha} = \mathbf{F}^{-1}\mathbf{b} \tag{10}$$

thus (9) becomes

$$\mathbf{Hu} - \mathbf{Gq} = (\mathbf{H}\hat{\mathbf{U}} - \mathbf{G}\hat{\mathbf{Q}})\mathbf{F}^{-1}\mathbf{b} = \mathbf{d} \tag{11}$$

The matrices $\hat{\mathbf{U}}$, $\hat{\mathbf{Q}}$ and $\mathbf{F}^{-1}$ are all known once f is defined.
The approximating function f is usually taken as [1,2]

$$f = 1 + r \tag{12}$$

which from (4), results in,

$$\hat{u} = \frac{r^2}{4} + \frac{r^3}{9} \tag{13}$$

where r is the distance function which appears in the BEM fundamental solution.
In the isotropic case, $D_x = D_y$, the vector $\mathbf{b}$ is set up using the nodal values of the terms transferred to the right hand side, *i.e.*

$$\mathbf{b} = \frac{1}{D_x}\left\{v_x\frac{\partial u}{\partial x} + v_y\frac{\partial u}{\partial y} + ku - \phi + \frac{\partial u}{\partial t}\right\}_{\text{nodes}} \tag{14}$$

In the orthotropic case the first two terms in (1) are written as

$$D_x\frac{\partial^2 u}{\partial x^2} + D_y\frac{\partial^2 u}{\partial y^2} = D_x\frac{\partial^2 u}{\partial x^2} + D_x\frac{\partial^2 u}{\partial y^2} + (D_y - D_x)\frac{\partial^2 u}{\partial y^2} \tag{15}$$

Equation (14) becomes

$$\mathbf{b} = \frac{1}{D_x}\left\{v_x\frac{\partial u}{\partial x} + v_y\frac{\partial u}{\partial y} + ku - \phi + \frac{\partial u}{\partial t}\right\}_{\text{nodes}} - \left\{\frac{(D_y - D_x)}{D_x}\frac{\partial^2 u}{\partial y^2}\right\}_{\text{nodes}} \tag{16}$$

Thus in the orthotropic case there is an additional term in the vector $\mathbf{b}$ for which nodal values must be obtained. This approach avoids the use of a different anisotropic fundamental solution: The additional term in (16) will simply be zero in the isotropic case.

The DRM expansion of each of the terms in equation (16) is considered below

2.1 The Decay term

This term is

$$b_1 = \frac{k}{D_x}u \tag{17}$$

in the case of k constant the term is simply

$$\mathbf{b}_1 = \frac{k}{D_x}\mathbf{u} \tag{18}$$

where $\mathbf{u}$ is a vector of nodal values of u. In the case of k varying with space then a diagonal matrix $\mathbf{K}$ needs to be defined, the non-zero terms of which are the nodal values of k/D_x and equation (18) becomes

$$\mathbf{b}_1 = \mathbf{K}\mathbf{u} \tag{19}$$

If k varies in time, this matrix will be updated at each timestep.

2.2 Convective terms

Convective terms can be easily accommodated in DRM. Consider the term in v_x

$$b_2 = \frac{v_x}{D_x}\frac{\partial u}{\partial x} \tag{20}$$

A mechanism must be established to relate the nodal values of u to the nodal values of its derivative, $\partial \mathbf{u}/\partial x$. At this point it should be remembered that the basic approximation of the DRM technique is

$$\mathbf{b} = \mathbf{F}\boldsymbol{\alpha} \tag{21}$$

A similar equation may be written for u

$$\mathbf{u} = \mathbf{F}\boldsymbol{\beta} \tag{22}$$

where $\boldsymbol{\beta} \neq \boldsymbol{\alpha}$. Differentiating equation (22) produces

$$\frac{\partial \mathbf{u}}{\partial x} = \frac{\partial \mathbf{F}}{\partial x}\boldsymbol{\beta} \tag{23}$$

Rewriting equation (22) as $\boldsymbol{\beta} = \mathbf{F}^{-1}\mathbf{u}$, then (23) becomes

$$\frac{\partial \mathbf{u}}{\partial x} = \frac{\partial \mathbf{F}}{\partial x}\mathbf{F}^{-1}\mathbf{u} \tag{24}$$

Thus in the case of v_x constant one can write the convective term in v_x as

$$\mathbf{b}_2 = \frac{v_x}{D_x}\frac{\partial \mathbf{F}}{\partial x}\mathbf{F}^{-1}\mathbf{u} \tag{25}$$

If v_x varies in space one defines a diagonal matrix $\mathbf{V}_x$ the non-zero terms of which are the nodal values of v_x/D_x. Once more, if these values are time dependent this matrix must be updated at each timestep. Equation (25) becomes

$$\mathbf{b}_2 = \mathbf{V}_x\frac{\partial \mathbf{F}}{\partial x}\mathbf{F}^{-1}\mathbf{u} \tag{26}$$

The terms in the $\partial \mathbf{F}/\partial x$ matrix are obtained by differentiating the f expansion, equation (12).

A similar expression can be written for the term in v_y in (16).

$$\mathbf{b}_3 = \mathbf{V}_y\frac{\partial \mathbf{F}}{\partial y}\mathbf{F}^{-1}\mathbf{u} \tag{27}$$

2.3 Time Derivative

The time derivative term is given by

$$b_4 = \frac{1}{D_x}\frac{\partial u}{\partial t} \tag{28}$$

Using a simple finite-difference representation for the time derivative one obtains

$$\frac{\partial u}{\partial t} = \frac{1}{\Delta t}(u_1 - u_0) \tag{29}$$

where u_1 are the nodal values of u at time $t = \Delta t$ and u_0 are the values at time $t = 0$. Thus

$$\mathbf{b}_4 = \frac{1}{\Delta t}\frac{1}{D_x}(\mathbf{u}_1 - \mathbf{u}_0) \tag{30}$$

2.4 Source term

Nodal values of the source term are included directly,

$$\mathbf{b}_5 = -\frac{1}{D_x}\phi \tag{31}$$

ϕ may be a constant or a known function of space.

2.5 Additional term for orthotropic case

Differentiating equation (23) one obtains

$$\frac{\partial^2 \mathbf{u}}{\partial x^2} = \frac{\partial^2 \mathbf{F}}{\partial x^2}\beta \tag{32}$$

Rewriting equation (22) as $\beta = \mathbf{F}^{-1}\mathbf{u}$, then (32) becomes

$$\frac{\partial^2 \mathbf{u}}{\partial x^2} = \frac{\partial^2 \mathbf{F}}{\partial x^2}\mathbf{F}^{-1}\mathbf{u} \tag{33}$$

Similarly

$$\frac{\partial^2 \mathbf{u}}{\partial y^2} = \frac{\partial^2 \mathbf{F}}{\partial y^2}\mathbf{F}^{-1}\mathbf{u} \tag{34}$$

Thus the additional term in the orthotropic case is

$$\mathbf{b}_6 = \frac{(D_y - D_x)}{D_x}\frac{\partial^2 \mathbf{F}}{\partial y^2}\mathbf{F}^{-1}\mathbf{u} \tag{35}$$

It should however be noted that in the case of expressions (33) and (34) the approximating function (12), $f = 1 + r$ cannot be used, as its second derivative is singular. For this reason, for the additional orthotropic term, the approximating function $f = 1 + r^3$ is employed instead of (12).

Collecting together expressions $\mathbf{b}_1$ to $\mathbf{b}_6$ one obtains

$$\mathbf{b} = \left\{\mathbf{V}_x\frac{\partial \mathbf{F}}{\partial x}\mathbf{F}^{-1} + \mathbf{V}_y\frac{\partial \mathbf{F}}{\partial y}\mathbf{F}^{-1} + \mathbf{K} - \frac{(D_y - D_x)}{D_x}\frac{\partial^2 \mathbf{F}}{\partial y^2}\mathbf{F}^{-1}\right\}\mathbf{u}$$

$$+\frac{1}{D_x\Delta t}(\mathbf{u}_1 - \mathbf{u}_0) - \phi \tag{36}$$

Defining

$$\mathbf{S} = (\mathbf{H}\hat{\mathbf{U}} - \mathbf{G}\hat{\mathbf{Q}})\mathbf{F}^{-1} \tag{37}$$

equation (11) becomes

$$\mathbf{Hu} - \mathbf{Gq} = \mathbf{S}\left\{\mathbf{V}_x\frac{\partial \mathbf{F}}{\partial x}\mathbf{F}^{-1} + \mathbf{V}_y\frac{\partial \mathbf{F}}{\partial y}\mathbf{F}^{-1} + \mathbf{K} - \frac{(D_y - D_x)}{D_x}\frac{\partial^2 \mathbf{F}}{\partial y^2}\mathbf{F}^{-1}\right\}\mathbf{u}+$$

$$+\frac{1}{D_x\Delta t}\mathbf{S}(\mathbf{u}_1 - \mathbf{u}_0) - \frac{1}{D_x}\mathbf{S}\phi \tag{38}$$

If one now defines

$$\mathbf{T} = \mathbf{S}\left\{\mathbf{V}_x\frac{\partial \mathbf{F}}{\partial x}\mathbf{F}^{-1} + \mathbf{V}_y\frac{\partial \mathbf{F}}{\partial y}\mathbf{F}^{-1} - \frac{(D_y - D_x)}{D_x}\frac{\partial^2 \mathbf{F}}{\partial y^2}\mathbf{F}^{-1} + \mathbf{K}\right\} \tag{39}$$

Then the following expression is obtained

$$\mathbf{Hu} - \mathbf{Gq} = \mathbf{Tu} + \frac{1}{D_x \Delta t}\mathbf{S}\left(\mathbf{u}_1 - \mathbf{u}_0\right) - \frac{1}{D_x}\mathbf{S}\phi \tag{40}$$

A linear variation of u and q within each timestep may now be defined, *i.e.*

$$u = (1 - \theta_u)u_0 + \theta_u u_1$$

$$q = (1 - \theta_q)q_0 + \theta_q q_1 \tag{41}$$

where θ_u and θ_q take values between 0 and 1. It has been found that $\theta_u = 0.5$ and $\theta_q = 1$ produce good accuracy, and for this case (40) becomes

$$\left(\mathbf{H} - \mathbf{T} - \frac{2}{D_x \Delta t}\mathbf{S}\right)\mathbf{u}_1 - 2\mathbf{G}\mathbf{q}_1 = \left(\mathbf{T} - \mathbf{H} - \frac{2}{D_x \Delta t}\mathbf{S}\right)\mathbf{u}_0 - \frac{2}{D_x}\mathbf{S}\phi \tag{42}$$

Equation (42) is reordered according to which values of u and q are known at time t_1.

$$\mathbf{Ax} = \mathbf{p}_1 + \mathbf{p}_2 \tag{43}$$

If u_1 is unknown,the column of $\mathbf{H} - \mathbf{T} - 2/(D_x\Delta t)\mathbf{S}$ is retained on the left hand side, in the same column of matrix $\mathbf{A}$, otherwise it is multiplied by the known value and transferring to the right hand side, where it is added to the vector $\mathbf{p}_1$. If q_1 is unknown the column of $-2\mathbf{G}$ is retained in $\mathbf{A}$. In addition to terms transferred from the lhs, $\mathbf{p}_1$ also contains $-2/D_x\mathbf{S}\phi$. $\mathbf{p}_2$ contains the remaining terms from (42), *i.e.*

$$\mathbf{p}_2 = (\mathbf{T} - \mathbf{H} - \frac{2}{(D_x \Delta t)}\mathbf{S})\mathbf{u}_0 \tag{44}$$

In equation (43) the vector $\mathbf{x}$ contains N unknown boundary values of u_1 and q_1, and the L unknown values of u_1 at internal points.

Letting $u_0 \rightarrow u_1$ a new timestep may be considered, this process being repeated until the total time desired is reached.

3 Application to Solent Problem

The model was applied to the study of the dispersion of effluent in the eastern part of the Solent, between Southampton and Portsmouth, in the south of England, see figure 2.

This problem has been studied using finite element analysis in reference [5]. For the study in question, velocity data for the Solent over the 13 hour spring tidal cycle was available from a field study.

As such data was not available in the present case, simplified velocity data was used, based on a study of the Tidal Streams Atlas, [8]. Velocities were generated using a sine function, having maximum values in mid-stream, diminishing to zero on the land boundaries.

The remaining data was taken from reference [5], *i.e.*

$$D_{\text{longitudinal}} = 5m^2/sec$$

$$D_{\text{transverse}} = 0.05m^2/sec$$

$$k = 0.4 \times 10^{-5}/sec \tag{45}$$

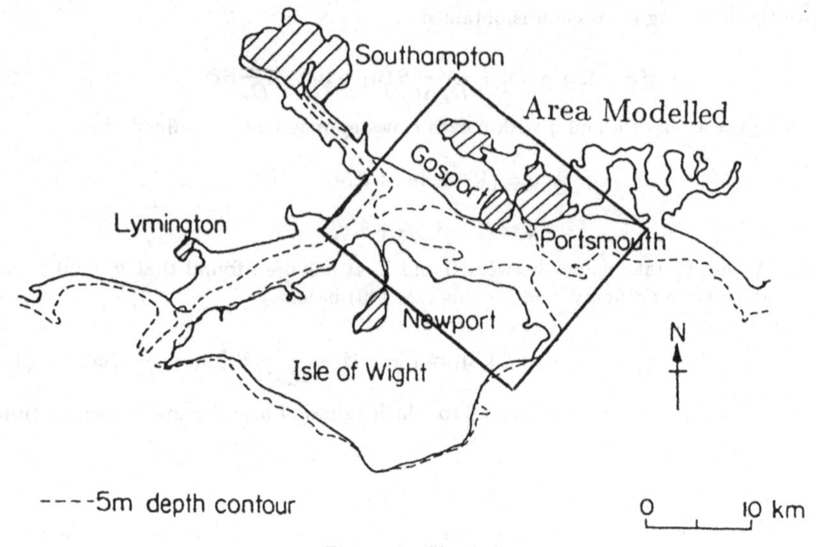

Figure 2: The Solent

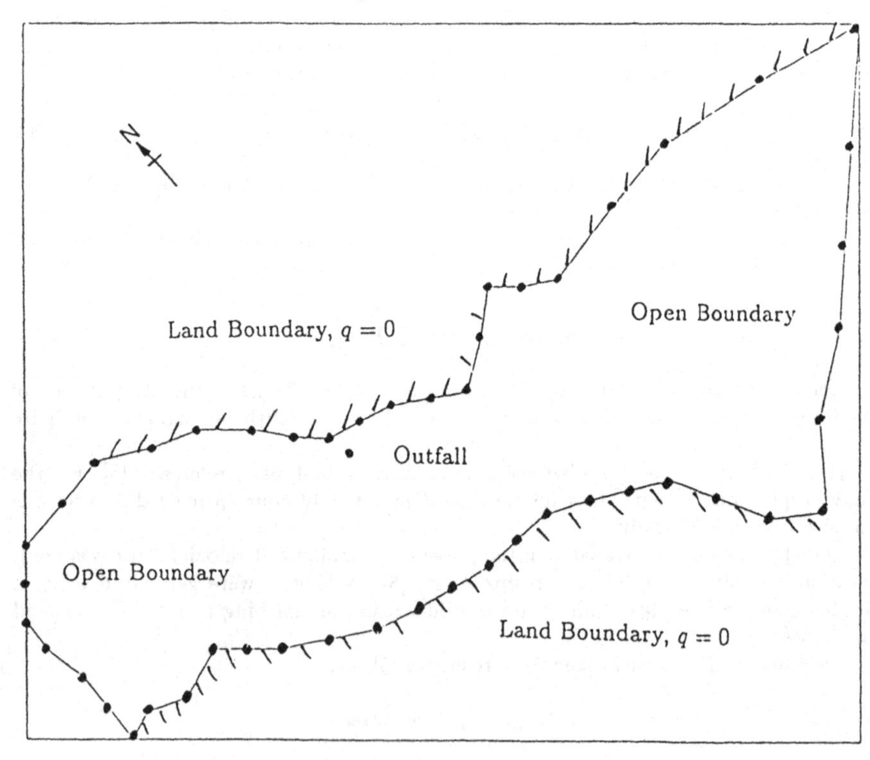

Figure 3: Boundary Discretization and Proposed Outfall

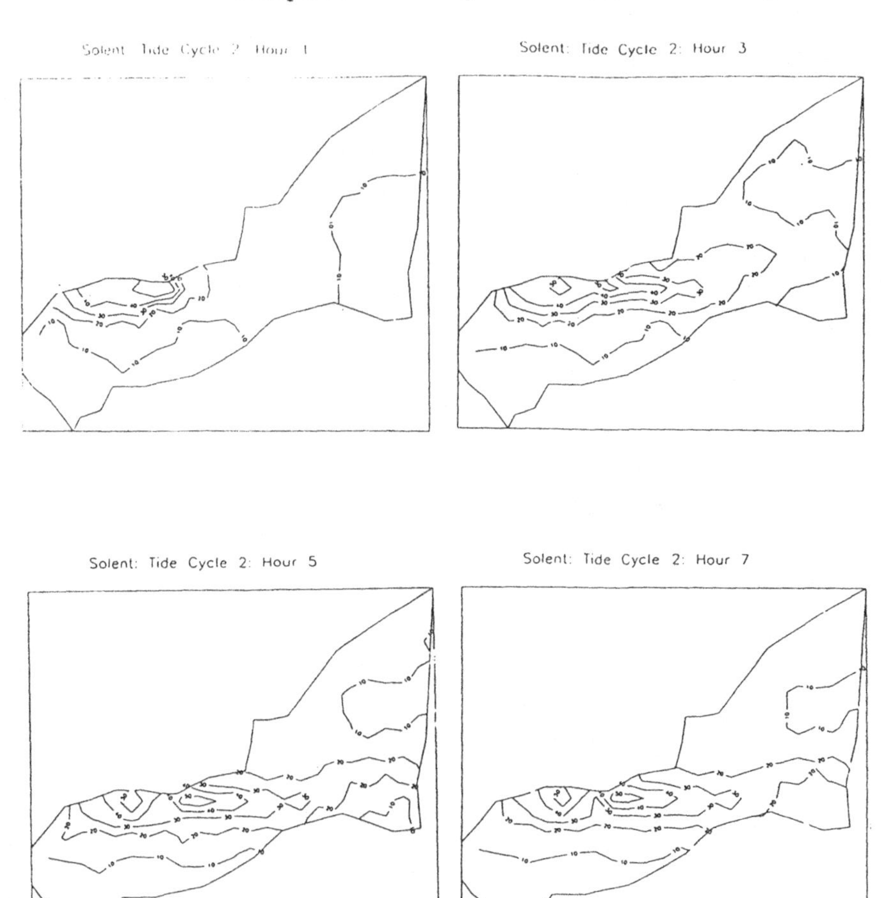

Figure 4: Boundary Element results

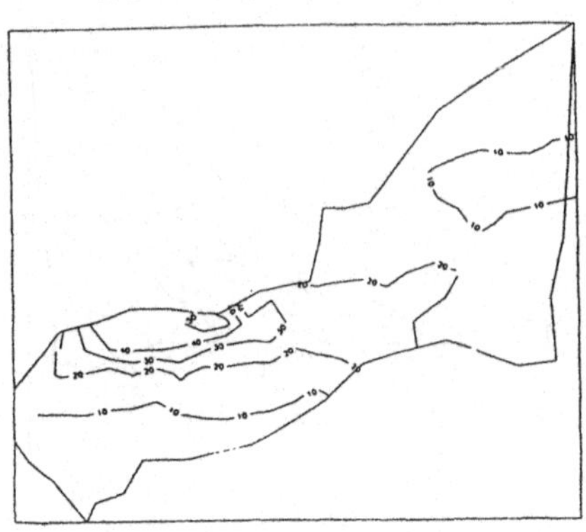
Solent: Tide Cycle 2: Hour 9

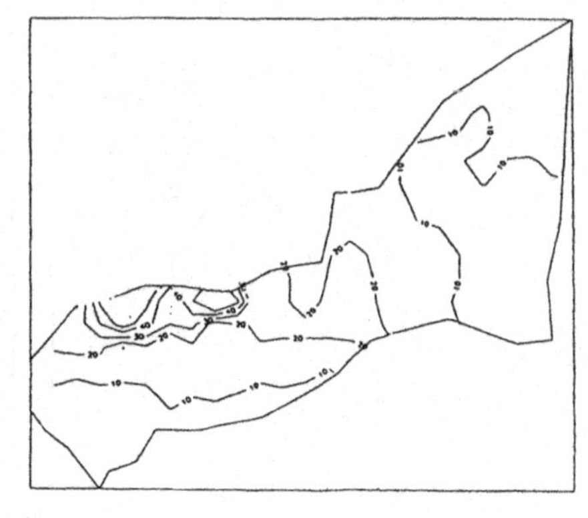
Solent: Tide Cycle 2: Hour 11

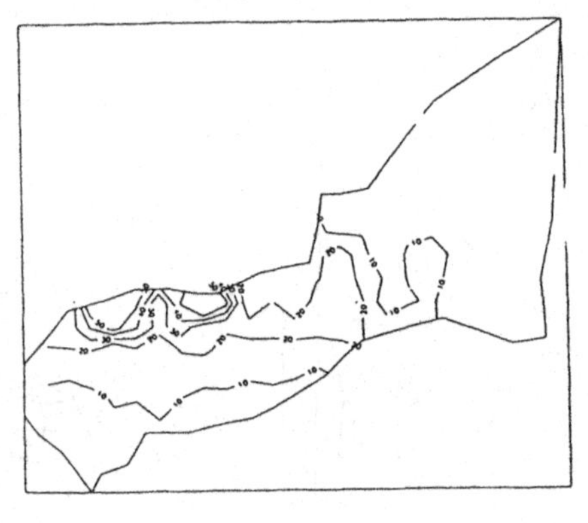
Solent: Tide Cycle 2: Hour 13

The outfall was considered at the position shown in figure 3. The discharge is given in [5] as 6000 lt/sec with a colliform density of 1.67×10^6 per 100ml, constant over the tidal cycle.

The eastern arm of the Solent is approximately 20km long and 5km wide. The boundary element discretization consisted of 47 linear elements and nodes. 52 internal nodes were also used distributed in a regular grid. At the land boundaries the condition $q = 0$ was employed. On the open sea boundaries, the correct boundary condition is $\partial^2 u/\partial n^2 = 0$, [9]. This boundary condition was approximated using a finite difference type expansion

$$q = \frac{u_a - u_b}{\Delta x_n} \tag{46}$$

for the open sea boundary nodes and those immediately adjacent. In [5] $u = 0$ was specified on these boundaries.

Results are shown in figure 4 for the second tidal cycle.

Concentrations are given in units of 1000 per 100ml.

The results were compared with the finite element solutions given in reference [5] and they tend to agree in a quantitative manner. Direct comparison is not possible however in view of the difference in velocity data, and, more important, the different treatment of the open sea boundary condition.

4 Conclusions

In this paper the boundary element method has been applied to model the dispersion convection equation using the Laplace fundamental solution and taking the remaining terms to the boundary by means of the Dual Reciprocity Method. The procedure is straightforward to implement and has the advantage that variable values of the coefficients may be easily handled.

The model has been applied to the study of effluent concentrations on the eastern arm of the Solent, for which finite element results are available, [5]. Though the results obtained should be considered as preliminary, given the simplified velocity data, the pattern produced by the BEM solution is similar to that obtained using finite elements.

5 References

1. Nardini, D. and Brebbia, C. A.
A New Approach to Free Vibration Analysis using Boundary Elements, in *Boundary Element Methods in Engineering*, Ed C. A. Brebbia, Springer-Verlag, Berlin, 1982.

2. Partridge, P. W., Brebbia, C. A. and Wrobel L. C.
The Dual Reciprocity Boundary Element Method, Computational Mechanics Publications and Elsevier, Southampton and London, 1991.

3. Partridge, P. W. and Brebbia, C. A.
The Dual Reciprocity Boundary Element Method for the Diffusion Convection Equation, IX International Conference on Computational Methods in Water Resources, Denver, Colorado, June 1992.

4. De Figueiredo, D. B. and Wrobel, L. C.
A Boundary Element Analysis of Transient Convection Diffusion Problems in *BEM XII* Eds. M. Tanaka, C. A. Brebbia and T Honma, Vol 1, Computational Mechanics Publications and Springer Verlag, 1990.

5. Connor, J. J. and Brebbia, C. A.
Finite Element Techniques for Fluid Flow, Newnes-Butterworths, London and Boston, 1977.

6. Brebbia, C. A. and Dominguez, J.
Boundary Elements: An Introductory Course, Computational Mechanics Publications and McGraw Hill, Southampton, 2nd edition, 1992.

7. Brebbia, C. A., Telles, J. and Wrobel, L.
Boundary Element Techniques: Theory and Applications In Engineering. Springer-Verlag, Berlin and N. York, 1984.

8. Tidal Streams Atlas, The Solent and Adjacent Waters, HMSO, 1962.

9. Koutitas, C. G,
Mathematical Models in Coastal Engineering, Pentech Press, London, 1988.

SECTION 6: COMPUTATIONAL TECHNIQUES

Use of Parallel Computing in Modelling Environmental Phenomena

L. Brusa

CISE Tecnologie Innovative SpA, P.O. Box 12081, 20134 Milan, Italy

ABSTRACT

The paper discusses key issues in parallel computing when applied to numerical simulations requiring different types of computations and handling of large sets of data with problem dependent structures. These features are typical of many environmental analyses based on the solution of partial differential equations in domains with complex 3-D geometries. The considered aspects concern hardware choice, algorithms and software development methodologies.

INTRODUCTION

The success of predicting the behaviour of natural or incidental phenomena and their impact on the environment is mainly based both on the use of accurate mathematical models and on the availability of software tools able to efficiently perform the relevant numerical simulations. The introduction of supercomputers in the market has opened new possibilities to the scientific community but the effective use of parallelism is not straightforward, requiring a careful examination of many aspects related both to hardware and software. The new hardware improvements allow present codes to be pushed only a limited amount beyond their current capabilities. Additional capabilities, required by the increased accuracy of mathematical models, can be obtained only through an application software able of efficiently exploiting the architectural features of the new computers.
A first aspect to be considered is therefore the choice of a proper hardware, capable of matching both

the inherent parallelism of the considered problem and the general requirements of the analyses to be performed.
A second aspect concerns the choice of numerical algorithms, which must be tailored to the chosen hardware and able of efficiently solving the algebraic problems representing the kernel of any numerical simulation.
However these two basic problems do not cover all the issues related to parallel computing because the efficiency and usability of codes also depends on the application of proper software development methodologies. These techniques should guarantee the full exploitation of compiler capabilities and an acceptable portability of software, nearly leading to obtaining the same performance, on machines with similar architectures.
Parallel software technology is relatively young and in many cases based on heuristic methodologies so that information in this field is poor. The aim of the paper is to discuss these problems on the basis of classical and recent literature (see references [1]÷[3]) and of the experience gained in the solution of fluid-dynamic problems for environmental applications, mainly discretized with finite element methods.
Some numerical results related to the use of a conjugate gradient method applied to the solution of the Stokes problem are also presented.

DRIVERS OF COMPUTATION IN ENVIRONMENTAL PROBLEMS

Many environmental phenomena are described by mathematical models based on complex non linear time dependent differential equations which must be solved in large 3-D domains. Typical examples of these problems are the studies of different types of pollution in the sea waters which often require the solution of Navier-Stokes equations coupled with turbulence models and diffusive-convective equations describing the behaviour of pollutants.
The solution of these kind of equations is a traditional driver of computation for which parallelism is often mandatory. In fact a satisfactory understanding of the physical problem requires the use of more and more sophisticated models which, in turn, require accurate numerical simulations.
The equations are generally discretized in space by means of finite difference or finite element methods and the subsequent use of time stepping discrete algorithms reduces the problem to the solution of

large and sparse systems of non linear equations. Therefore the possibility of tackling complex problems in this field mainly rely on the use of efficient linear equations solvers and on the ability to manage large sets of data organized according to the spatial pattern of a grid. Finite difference methods produce matrices with regular structure which favour vectorization and parallelization of the solution algorithm. However complex geometries are difficult to be described by this method and local mesh refinements produce a useless increase of grid points.
These drawbacks are avoided by using finite element methods but the relevant matrices have unstructured sparsity which, because of indirect addressing, hampers vectorization.
Parallelization can be fully exploited in the computation of element matrices but this task requires a small percentage of the global computing time, which mainly concerns the solution of large sparse linear systems.
The goal should be therefore the development of finite element codes able to reach performances comparable to those obtained with finite difference schemes.
In what follows, the remarks mainly refer to the development of software based on the finite element approach even though many issues are of general type and may be valid also for different problems.

HARDWARE CHOICE

Forecasts of the future parallel computing scenario indicate that highly parallel systems with distributed memories will be the predominant architectures. However it is not clear when these machines will be able to run industrial applications. Massive parallelism is used today for a narrow class of problems with special features (for example graphics or image processing) but at present these machines are not sufficiently equipped with tools, general purpose software and libraries necessary to tackle many industrial problems. The present situation sees research efforts mainly directed towards massive parallelism while the major part of today computations is performed on shared memory vector machines with a limited number of processors. Unfortunately this does not imply that these last machines are easy to use and that all the problems in the field of "coarse grain" parallelism have been solved.
The success of vector multiprocessors mainly lies in

the fact that this approach to parallelism is soft and based on "evolution" rather than "revolution". These machines are now considered as "traditional supercomputers" and allow a formal migration of old application software with no trouble, that is of most concern for industry.

Management recognizes that the existing software is a huge investment and therefore it must be preserved as much as possible. Moreover old software may partially benefit from automatic optimizations provided by the supercomputer compilers.

For these reasons shared memory multiprocessors seem today to be the most appropriate choice for those numerical simulations requiring different types of computations and handling large set of data with problem dependent structures. This happens, for example, when solving partial differential equations in domains with complex geometries. Thus in the following the attention will be focused only to this class of machines.

A fundamental question is to decide whether old software must be restructured to benefit from supercomputer architectures and at what extent.

Many restructured codes can take great advantages of vectorization reaching speedups of 20-30 against values of 2-3 given by automatic optimization.

As far as parallelism is concerned, one might object to the fact that a small number of CPUs, say 2, can at most double the performance and this speedup might not be enough to counterbalance the effort made in changing the code.

If the CPUs are properly managed by the operating system, the job stream throughput can be nearly doubled without any change in the computer programs and single users benefit from a reduction of the elapsed time. This is true, but we must be careful in interpreting the issues of evolution to avoid the risk to miss the train of technological development. Evolution is going towards machines with high degree of parallelism and also traditional supercomputers begin to have an increasing number of CPUs. When the number of CPUs becomes greater than 4 the operating system may fail in optimizing the management of resources and jobs begin to hamper themselves in the attempt to access memory or I/O ports, with consequent loss of performance.

In this case a well organized use of resources working in parallel to increase the performance of a single job seems to be a more efficient way to exploit the potentiality of the machines.

This means that the codes must be redesigned and managers must be prepared to provide long term plans on the basis at realistic evaluation of the main problems to be faced and of the cost of software restructuring.

KEY ISSUES IN PARALLEL COMPUTING: ALGORITHMS AND SOFTWARE DEVELOPMENT METHODOLOGIES.

Algorithm restructuring is in general the first step in software optimization and often copes with two opposite requirements: the need to obtain high performances through the development of machine oriented software and the necessity of preserving as much as possible existing codes, because of the huge investments required for their production. Many computer programs, including finite element codes, verify the so called "80-20 rule", that is a large part of computations (may be 80%) is performed in a small part of the code (may be 20% of the global statements) which is the one concerning the numerical solution of algebraic problems.
Therefore, by replacing the modules based on sequential algorithm with new ones optimized to exploit the architectural features of the machine the existing application software is only partially changed. But, because the algorithm kernel of the codes requires the major part of the global computing time , the performance can be substantially increased. This restructuring methodology is particularly efficient for vector computers but may present difficulties for parallel machines because in some cases local parallelization requires the change of data structure which are created in other sections of the code. Thus software revision must be enlarged. It must be noticed that a poor design of data organization may destroy the efficiency of a well optimized algorithm with unnecessary overheads due to data transfer, waiting times or I/O operations.
One of the main problems in designing parallel algorithms is the lack of realistic models to a priori evaluate the performance.
For scalar computers the number of arithmetic operations is a meaningful parameter which can help the numerical analyst in comparing the computational work relevant to different methods. This parameter, together with considerations concerning stability and accuracy, was successfully used in the past to choose the most appropriate numerical method for the solution of a given problem. Of course the machine architecture was not important for these evaluations. When parallelism comes into play other aspects must be considered: algorithm granularity, load balancing, communication and synchronization overheads, optimal

use of hyerachical memories, etc.
All these items are difficult to be taken into account in a theoretical model because they have different weights for different machines, even if machines belong to the same class of computers. Consequently numerical analysts choose algorithms on the basis of generic parallel features but only at implementation level realistic information on computational efficiency can be obtained.
It is difficult to define general methodologies for the efficient development of parallel algorithms. However two requirements seem to be important: the use of libraries and the handling of data in blocks.
In the phase of architecture-dependent implementation of algorithms the risk is to push optimizations too far with the consequence that portability might be destroyed. To reduce this drawback standard libraries such BLAS should be used as much as possible.
By using basic functions of libraries as bricks to construct algorithms, portability and performance can be enhanced. In fact BLAS is installed in every shared memory parallel computer and is optimized to exploit the architectural features of each particular machine.
At present there are not many libraries that can be really considered as standard optimized products but work is in progress in this field. In any case this work only concerns low level or middle level libraries dealing with basic operations on vectors and matrices or with the solution of algebraic problems for dense matrices. Optimized high level libraries for the solution of algebraic problems with large sparse matrices at present do not exist. If such libraries did exist, they could be used by application software programmers for the computational kernels of codes and this could help the development of a new generation of programs for parallel computers. Standard libraries optimized for different computers could guarantee portability and efficiency, at least within certain limits.
The use of algorithms structured to handle data in blocks is one of the basic issue for implementation of coarse grain parallelism. The dimension of blocks should be defined so as to optimize granularity/synchronization overheads ratios, the distribution of the computational loads among the processor and the use of local or hierarchical memories. The balancing of all this factors is not easy and is based on experience and heuristic methodologies.
A theoretical aspect should also be considered. Accuracy and stability of block algorithms is not

always well studied and few results are available even for traditional numerical methods.
Some remarks can be done about the solution of linear systems, which are the computational core of many numerical simulations.
Direct methods are generally more efficient than iterative schemes for small-size problems, but they cannot be used for large sparse systems especially for the great amount of storage required by the factorized matrix. In these cases the use of iterative methods is mandatory. For symmetric positive definite matrices the conjugate gradient method is one of the most popular schemes. Its application for finite element problems is straight forward. The matrix-vector product step simply requires the products of element matrices by subvectors (which can be concurrently performed) and the assembly of the result in a global vector. The use of linear tetrahedra might be the key to increase the performances of the finite element codes so to reach levels comparable to those obtained with finite difference programs [4]. In fact the computational rules for element matrices are very simple so that matrices can be recomputed when necessary without the necessity of being stored, as in the finite difference method. The assembly of 8 tetrahedra to form a tetrahedra with 10 nodes, as for the relevant parabolic element, can increase the granularity of the element matrix-subvector products, with a consequent reduction of the synchronization overhead. Moreover BLAS routines can be used to efficiently perform the matrix-vector products.

PERFORMANCES OF CONJUGATE GRADIENT METHOD APPLIED IN THE SOLUTION OF STOKES PROBLEM: EXPERIMENTAL RESULTS.

As an example of the gain in performance that can be obtained by restructuring algorithms for parallel computers, this section reports some results relevant to the solution of 3D Navier Stokes equations performed by the finite element MADIAN code [5]. The numerical scheme is based on a splitting technique so that at each time step pure convective equations and Stokes equations are separately solved. At present $P_0/Q_0-P_1/Q_1$ elements for pressure and velocity respectively are available. The development of a new version of the program based on P_1-iso P_2 elements is now in progress. The discretized Stokes problem has the form:

$$\begin{vmatrix} A & B \\ -B^T & 0 \end{vmatrix} \begin{vmatrix} V \\ P \end{vmatrix} = \begin{vmatrix} f_1 \\ f_2 \end{vmatrix} \qquad (1)$$

where V and P are the vectors of nodal velocities and pressures. Matrix A is symmetric positive definite and has the form

$$A = M+\Delta tK \qquad (2)$$

where Δt is the time step, M is the mass matrix and K is obtained by the discretization of the operator applied to velocity in the momentum equation.
Problem (1) is reduced to the solution of the linear system:

$$(B^T A^{-1}B)P = B^TA^{-1}f_1+f_2 \qquad (3)$$

System (3) is solved by means of the conjugate gradient method which is used also for the solution of linear systems of the form:

$$AV = b \qquad (4)$$

required at each iteration.
The results here quoted refer to the solution of system (4) for a problem discretized with 7160 linear hexahedra and 22044 unknown velocities. The following two cases are considered.
For case 1 Δt is equal to 0.1 s and the viscosity coefficient (varying in space) has a mean value of 10^{-6} m^2s^{-1}, so that the mass matrix is dominant.
In case 2 the weight of the stiffness matrix K is now more important because the mean value of the viscosity coefficient is 10^{-3} m^2s^{-1} and Δt is equal to 1000s.
For these cases system (4) has been solved by means of the conjugate gradient method preconditioned with two well known schemes: SSOR and incomplete Cholesky factorization. For this last method the subroutine MA31A of Harwell Library for sparse matrices was used. The subroutine has been only partially optimized by introducing directives to force vectorization and parallelization in some DO-loops. The conjugate gradient method based on the SSOR algorithm has been implemented without the need of

forming the global matrix by assembling the element matrices. Moreover the product of element matrices by subvectors is concurrently performed without the use of indirect addressing. The computations have been performed on ALLIANT FX/80 with 3 processors and Table 1 quotes for each method the number of iterations (NIT), the total CPU time (t_{TOT}), the times required at each iteration to perform the matrix-vector product (t_{MULT}) and to solve the preconditioning linear system (t_{PREC}). For the incomplete Cholesky factorization also the time necessary to obtain the incomplete factors is shown (t_{FACT}).

Table 1. Comparison of SSOR and incomplete Cholesky factorization preconditioners. CPU times in seconds on ALLIANT FX/80 with 3 processors. Speedups in brakets.

		CASE 1	CASE 2
SSOR	NIT	8	13
	t_{TOT}	81.88 (2.14)	122.0 (2.21)
	t_{MULT}	1.82 (2.53)	1.80 (2.60)
	t_{PREC}	7.46 (2.17)	7.0 (2.0)
INC. CHOL. FACT.	NIT	4	Convergence not reached after 300 iterations
	t_{TOT}	65.5 (1.32)	
	t_{MULT}	4.95 (1.0)	
	t_{PREC}	11.2 (1.46)	
	t_{FACT}	1055 (1.92)	1.055 (1.92)

The results show that incomplete Cholesky factorization is more efficient than SSOR for case 1, with gains in the CPU time of 20%. However the efficiency of the method greatly decreases when the condition number of the matrix increases and for case 2 convergence has not yet been reached after 300 iterations. Moreover subroutine MA31A, originally designed for scalar computers, scarcely benefits from parallelism. On the contrary the conjugate gradient method preconditioned with SSOR scheme is efficient both for cases 1 and 2 and its implementation takes advantage from the parallel architecture of the computer, being the loss of performance with respect to a theoretical speedup of 3 of about 30%.

CONCLUSIONS

There is no doubt that supercomputing is the only means to overcome the barriers established by the huge amount of computations involved in the solution of many scientific problems. The technological developments in this field are impressive and now we can choose among many different architectures which cover a wide range of costs and performance requirements.
In spite of these facilities offered by the market, the effective introduction of parallelism in real life is not straightforward. The paper has discussed difficulties and possible solutions connected with parallel software development, with special reference to problems characterized by large sets of unstructured data and intensive algorithmic computations, such as those relevant to the analysis of environmental phenomena. The awareness of these problems is important for a better understanding of benefits and drawbacks of parallelism and for a correct approach to the future way of performing scientific computing.

REFERENCES

1. Hogan, D.W., Jensen, J.C. and Cornish, M. "The cost of processing power: the process, the programmes and the processor", in High Speed Computer and Algorithm Organization (Ed. Kuck D.J., Lawrie D.M. and Sameh A.M.), pp. 371 to 377, *Proceedings of the Symposium on High Speed Computer and Algorithm Organization,* University of Illinois, 1977

2. Hyman, J.M. "Future directions in large scale scientific computing", in Large Scale Scientific Computation (ed. Parter S.V.), pp. 51 to 83, *Proceedings of a Conference Conducted by the Mathematics Research Centre of the University of Wisconsin*, Madison, 1983

3. Neves, K.W. and Kowalik, J.S. "Supercomputing: key issues and challenges", in *Supercomputing* (Ed. Kowalik J.S.), pp. 3 to 39, Nato Asi Series, vol. F62, Springer-Verlag Berlin Heidelberg, 1990.

4. Hervonet, J.M. and Péchon, P. "Modélisation numérique des écoulements à surface libre" *La Houille Blanche*, n°2, pp. 93-107, 1991

5. Andreola, M., Gurizzan, A., and Molinaro P., "MADIAN: a 3-D finite element code for environmental studies", in Hydraulic Engineering Software Applications (Ed. Blain W.R. and Ouazar D.) pp. 375 to 387, *Proceedings of the Third International Conference on Hydraulic Engineering Software*, Massachusetts, 1990.

REFERENCES

1. Hoga[illegible] D.W. [illegible] Gentleman, [illegible] and Cornish, M. "The [illegible] processing power: the process, the [illegible] and the [illegible]", In High Speed Computer and Algorithm Organization (Ed. Kuck [illegible] and Sameh A.H.), pp. 271 [illegible] Proceedings of the Symposium on High Speed Computer and Algorithm Organization, University of Illinois, 1977.

2. [illegible] "[illegible] in large scale [illegible] computing [illegible]" pp. [illegible] Proceedings of a Conference Conducted by [illegible] Research Centre [illegible] 1985.

3. [illegible]

4. [illegible] and [illegible] "[illegible] for [illegible] surface [illegible]" [illegible] pp. [illegible] 1981.

5. [illegible] Blake [illegible] Proceedings [illegible] Conference [illegible] Engineering [illegible] Massachusetts.

TELEMAC-3D: A Finite Element Code to Solve 3D Free Surface Flow Problems

J.M. Janin, F. Lepeintre, P. Péchon

Laboratoire National d'Hydraulique, Electricité de France, Chatou, France

ABSTRACT : This paper describes the algorithm of TELEMAC-3D, a code being developed to solve 3D free surface flow problems. Thanks to the use of element by element techniques, the computation takes full advantage of vectorization. Some first results are presented.

1 INTRODUCTION

Improvements in computations have made numerical codes more and more useful in Coastal Engineering. Although it would be illusory to think that small scale models will be obsolete shortly, being able to use an efficient numerical code is an asset for many studies.

L.N.H (Laboratoire National d'Hydraulique of EDF) has a code based on finite difference methods to solve 3D flows with a free surface (Donnars (1989)). However, that code obviously cannot include the "user friendly" possibilities allowed by recent developments in finite element methods, e.g. local refining of the mesh, without increasing dramatically the computation costs. Therefore, using the modern numerical techniques, L.N.H. has decided to develop a new numerical code, called TELEMAC-3D, to solve free surface flow problems. In particular, that code will be useful for problems involving pollution, thermal dilution or suspended sediment transport.

The main demands for this code are the use of finite element techniques and a low computing cost. Those requirements may seem contradictory, yet, it is now possible to meet both thanks to E.B.E. (Element By Element) techniques (Hughes (1987)).

The physical problems that we want to solve allow us to assume that the pressure is hydrostatic. As it will be explained further down, that hypothesis leads us to compute the free surface by solving shallow water like equations by means of a 2D code, TELEMAC-2D, also developed at L.N.H (Hervouet (1991)). The buoyancy effects are taken into account through the Boussinesq's approximation.

The development of the code is under way at the moment. In this paper, we intend to present the algorithm of the code by underlining what we think makes it original along with the first results obtained on test-cases.

2 AN OUTLINE OF TELEMAC-3D

TELEMAC-3D solves the Navier-Stokes equations with a free surface boundary condition and the advection-diffusion equations of the temperature, the salinity and any other needed variables (Daubert (1989)). Physical phenomena which affect the flow are included. They are for instance the influence of the temperature and the salinity on the density, the wind stress on the free surface, the heat exchange with the atmosphere and the Coriolis force. Variations of the density due to the temperature or the salinity are taken into account in the momentum equations via the Boussinesq's approximation.

The equations to be solved are :

$$\frac{du}{dt} = -\frac{1}{\rho_0}\frac{\partial p}{\partial x} + \frac{\partial}{\partial x}(\nu_x\frac{\partial u}{\partial x}) + \frac{\partial}{\partial y}(\nu_y\frac{\partial u}{\partial y}) + \frac{\partial}{\partial z}(\nu_z\frac{\partial u}{\partial z}) + f_u \tag{1}$$

$$\frac{dv}{dt} = -\frac{1}{\rho_0}\frac{\partial p}{\partial y} + \frac{\partial}{\partial x}(\nu_x\frac{\partial v}{\partial x}) + \frac{\partial}{\partial y}(\nu_y\frac{\partial v}{\partial y}) + \frac{\partial}{\partial z}(\nu_z\frac{\partial v}{\partial z}) + f_v \tag{2}$$

$$\frac{\partial u}{\partial x} + \frac{\partial v}{\partial y} + \frac{\partial w}{\partial z} = 0 \tag{3}$$

$$p = \rho_0 g(S-z) + \rho_0 g\int_z^S \frac{\Delta\rho}{\rho_0}\,dz \tag{4}$$

$$\frac{dT}{dt} = \frac{\partial}{\partial x}(\kappa_{xT}\frac{\partial T}{\partial x}) + \frac{\partial}{\partial y}(\kappa_{yT}\frac{\partial T}{\partial y}) + \frac{\partial}{\partial z}(\kappa_{zT}\frac{\partial T}{\partial z}) + f_T \tag{5}$$

with

t,x,y,z	time and space coordinates
u,v,w	velocity components
p	pressure
ρ_0	density of reference
$\Delta\rho$	density variation
ν_x, ν_y, ν_z	eddy viscosity tensor (spherical and anisotropic)
S	free surface level
T	temperature (or salinity or any other wanted variable)
$\kappa_{xT}, \kappa_{yT}, \kappa_{zT}$	diffusity tensor (spherical and anisotropic)
f_u, f_v, f_T	source terms for u,v and T
g	gravity

We assume that the pressure is hydrostatic. Such a hypothesis is in fact valid for most of the problems that we want to solve with TELEMAC-3D. It allows us to compute the free surface by solving equations similar to the shallow-water equations. This point is developed under 3.5. However, what should be underlined here is that the computation of those equations is done by a shallow-water code, TELEMAC-2D (Hervouet (1991)).

For turbulence, TELEMAC-3D offers a choice between two zero equation models and one two equation model. The eddy viscosity is either constant or computed by a mixing length model or a k-ε model.

3 THE ALGORITHM

Given the solution at $t=t^n$, we compute the solution at $t^{n+1}=t^n + \Delta t$. The numerical method solves the equations by means of a decomposition in fractional steps (i.e. we split the operators). Each numerical operator can this way be treated by an adequate method. The resolution is achieved in three steps : an advection step, a diffusion step and a free surface-continuity-pressure step. The time derivatives are thus written as follows :

$$\frac{\partial f}{\partial t} = \frac{f^{n+1} - f_{diff}}{\Delta t} + \frac{f_{diff} - f_{adv}}{\Delta t} + \frac{f_{adv} - f^n}{\Delta t} \qquad (6)$$

with

f^{n+1}	solution at t^{n+1}
f_{diff}	result of the diffusion step
f_{adv}	result of the advection step
f^n	solution at t^n

Before detailing those steps, we would like to describe the space discretization that we use and how we deal with the movement of the free surface.

3.1 *The space discretization*

The domain to be studied is limited by a bottom defined by $z=Z_f(x,y)$, a free surface defined by $z=S(x,y)$ and on the sides by a vertical cylinder.

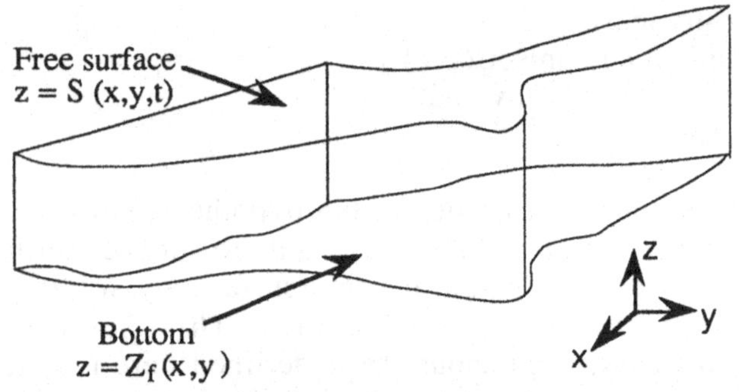

Fig 1 : domain of computation

The space discretization that we have adopted is a finite element discretization in prisms. The quadrangular sides of the prisms are vertical. The interpolation is so that it is linear when restricted to the triangles or the quadrangles of the prism.

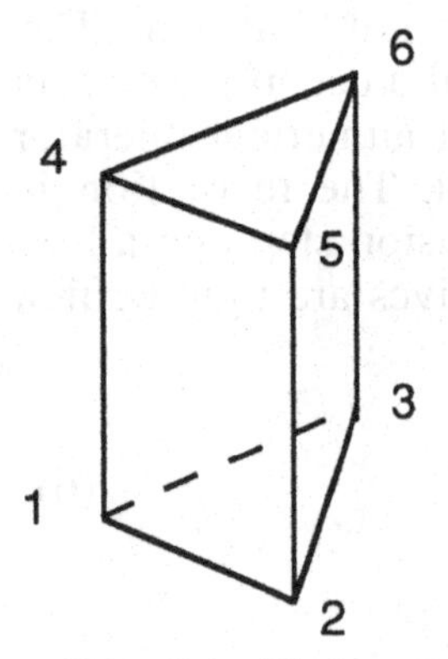

Fig 2 : Finite element used for the space discretization (local numbering is mentioned)

That way, the horizontal 2D projection of the mesh is made of triangles, one of the types of finite element used by TELEMAC-2D. Note that since TELEMAC-2D can also use quadrangles, we could have taken the brick as a finite element for TELEMAC-3D. Nevertheless, it is easier to mesh complicated domains with triangles than with quadrangles.

To mesh the 3D domain, we just need to mesh the 2D horizontal domain and then to duplicate it along the vertical. That is for every point M(x,y) of the 2D mesh, we define several points N(x,y,z) for which :

$$z = Z_f(x,y) + \theta\ (\ S(x,y,t) - Z_f(x,y)\ )\ \text{with}\ 0. \leq \theta \leq 1. \qquad (7)$$

3.2 *The movement of the free surface*

Due to the movement of the free surface, the 3D mesh moves with time. The coordinate z of any point of the mesh depends on time. Yet, it is possible to work on a mesh independent of time with a change of variables. One classical change of variables consists of switching from z to z^* according to the σ–transformation :

$$z^* = \bar{S}\frac{z - Z_f(x,y)}{S(x,y,t) - Z_f(x,y)} \qquad (8)$$

where $\bar{S}$ is a given positive constant. The corresponding mesh will be called afterwards the σ–mesh. It is made of points independent of time $N^*(x,y,z^*=\theta\ \bar{S}\ )$.

Solving equations (1) to (5) expressed in that new set of coordinates (x,y,z^*,t) is a way to deal with the free surface movement. We have chosen that option for the advection step. For the diffusion step, some terms of the equations become complicated and, numerically speaking, require special attention. Thus, those troublesome diffusion terms are very often neglected for the sake of computational efficiency. So, preferring not to drop any terms, we have prefered to make another approximation, which is to solve the diffusion step in the real mesh at $t=t^n$. To be treated correctly after the advection step, the diffusion step should be solved in the real mesh at $t=t^{n+1}$ or in the σ-mesh. As for the continuity equation, it is solved in the σ-mesh. Let us detail what are the implications of the σ-mesh for the advection and the continuity equations.

In the new set of coordinates (x,y,z^*,t), the time derivative $\frac{d}{dt}a(x,y,z,t)$ of any function a becomes :

$$\left(\frac{\partial a}{\partial t}\right)_{x,y,z^*} + u\left(\frac{\partial a}{\partial x}\right)_{y,z^*,t} + v\left(\frac{\partial a}{\partial y}\right)_{x,z^*,t} + w^*\left(\frac{\partial a}{\partial z^*}\right)_{x,y,t} = 0 \tag{9}$$

where

$$w^* = \frac{dz^*}{dt} = \left(\frac{\partial z^*}{\partial t}\right)_{x,y,z} + u\left(\frac{\partial z^*}{\partial x}\right)_{y,z,t} + v\left(\frac{\partial z^*}{\partial y}\right)_{x,z,t} + w\left(\frac{\partial z^*}{\partial z}\right)_{x,y,t} \tag{10}$$

Therefore, by defining a new advection velocity $\overrightarrow{u^*}(u,v,w^*)$, the advection equation remains the same.

Since we assume that the pressure is hydrostatic, the vertical velocity w is only needed throughout a time step during the advection step. And what is in fact required is w^*. Therefore, it is advantageous to compute directly w^* by solving the continuity equation in the σ-mesh. Expressed in the set of coordinates (x,y,z^*,t), the continuity equation reads :

$$\left(\frac{\partial h}{\partial t}\right)_{x,y,z^*} + \left(\frac{\partial(hu)}{\partial x}\right)_{y,z^*,t} + \left(\frac{\partial(hv)}{\partial y}\right)_{x,z^*,t} + h\left(\frac{\partial w^*}{\partial z^*}\right)_{x,y,t} = 0 \tag{11}$$

The vertical velocity, w^*, is known at the bottom and at the free surface, where it is either equal to zero in case of a wall boundary (impermeability condition) or to any set value in case of an inlet or an outlet of water at the bottom or at the free surface. The latter case is encountered for instance in the study of the thermal plume of some nuclear power plants.

Equation (11) integrated along the vertical between 0 and z^* and between z^* and $\overline{S}$ yields :

$$z^*\left(\frac{\partial h}{\partial t}\right)_{x,y,z^*} + \int_0^{z^*}\left[\left(\frac{\partial(hu)}{\partial x}\right)_{y,z^*,t} + \left(\frac{\partial(hv)}{\partial y}\right)_{x,z^*,t}\right]dz^* + h\left[w^*(z^*) - w^*(0)\right] = 0 \tag{12}$$

and

$$\left(z^*-\overline{S}\right)\left(\frac{\partial h}{\partial t}\right)_{x,y,z^*} + \int_{\overline{S}}^{z^*}\left[\left(\frac{\partial(hu)}{\partial x}\right)_{y,z^*,t} + \left(\frac{\partial(hv)}{\partial y}\right)_{x,z^*,t}\right]dz^* + h\left[w^*(z^*) - w^*(\overline{S})\right] = 0 \tag{13}$$

Recall that $w^*(0)$ and $w^*(\bar{S})$ are known. By combining (12) and (13), we obtain equation (14) which satisfies both boundary conditions at $z^*=0$ and $z^*=\bar{S}$:

$$h\,\bar{S}\,w^*(z^*) = h\left(\bar{S}-z^*\right)w^*(0) + h\,z^*\,w^*(\bar{S}) + \tag{14}$$

$$z^*\int_0^{\bar{S}}\left[\left(\frac{\partial(hu)}{\partial x}\right)_{y,z^*,t} + \left(\frac{\partial(hv)}{\partial y}\right)_{x,z^*,t}\right]dz^* - \bar{S}\int_0^{z^*}\left[\left(\frac{\partial(hu)}{\partial x}\right)_{y,z^*,t} + \left(\frac{\partial(hv)}{\partial y}\right)_{x,z^*,t}\right]dz^*$$

Note also that no time discretisation is required to solve (14) contrary to (12) or (13).

Apart from the advantage of dealing correctly with the free surface movement, the σ-mesh allows fast and simple computations due to its regularity.

3.3 *The advection step*

The solution is computed by means of a characteristic curve method in the σ-mesh. Characteristic curves are evaluated by a Runge-Kutta method with an explicit velocity field (i.e. taken at $t=t^n$). The interpolation of the advected variable is linear at the foot of the characteristic.

It is well known that such methods generate a lot of damping. However, this can be greatly reduced by refining the mesh at the cost of increasing the computing time. But, because those methods are unconditionally stable, it is possible to take a large time step which outbalances the effect of mesh refining. Note also that the cost of computation is not increased when there are many variables to advect (the characteristic curve is the same for all the variables).

3.4 *The diffusion step*

The diffusion is solved in the real mesh at $t=t^n$. Computation is achieved through a finite element formulation.

The two main points in that step are the computation of the matrices; here a mass and a diffusion matrix, and the resolution of the matrix system.

Recall that we take advantage of E.B.E. methods. A matrix is stored element by element. In fact, in our case, as initiated by the development of TELEMAC-2D, the diagonal of a matrix is assembled and only the extra diagonal elements are stored element by element. On vectorial computers, such a data structure allows the vectorisation of the matrix vector product.

Elements of matrices are computed exactly. The value of every element is explicitly written in the program in the form of a polynomial, thanks to the use of a symbolic calculus software REDUCE.

To be precise, the terms of the diffusion matrix are not exactly computed. They are of the form :

$$\int_{prism} \frac{\partial \varphi_i \partial \varphi_j}{\partial x \; \partial x} dP$$

with φ_i and φ_j basis functions at points i and j of the mesh.

In order to simplify the computation, we assume that the prisms have horizontal triangles. A prism without horizontal triangles is thus replaced by a prism of the same volume with horizontal triangles. This is due to the fact that the jacobian of the mapping between the prism and the reference prism appears at the denominator in the computation of the integral. That jacobian is constant only if the triangles of the prism are horizontal. Thus, the exact expression of the integral contains a logarithm most of the time, which would impede the computing efficiency. The consequence of that approximation is illustrated in figure 3. A passive effluent is distributed linearly along the vertical in a close flume with a steep bump on the bottom. The distribution of the effuent should remain the same, since diffusion has no effect in such a case. This is what is found by TELEMAC-3D, except for some wiggles on the isolines of the effluent above the bump. Since those wiggles are very small, the approximation is valid.

Eventually, that step leads to the resolution of a matrix system AX=B where A is a symmetric definite positive matrix. The system may thus be solved by means of a conjugate gradient algorithm, which is well suited to our type of matrix storage because it requires only matrix-vector products. To diminish the number of iterations of the conjugate gradient algorithm, preconditioning of the matrix A is useful. We currently use two kinds of preconditioning, scaling preconditioning and Crout preconditioning as described in Hughes (1987). The latter is in fact the most efficient one and can be compared to Choleski preconditioning for assembled matrices. On one of our test cases described below, the wind test case, Crout preconditioner reduced the computing cost of the diffusion step by as much as 30% when compared to diagonal scaling.

3.5 *The free surface-continuity-pressure step*

At this point, the advection and diffusion operators have been treated. So, by integrating along the vertical (between the bottom and the free surface) the momemtum and continuity equations, we obtain the classical shallow water equations without the advective and diffusive terms. Recall that this is possible because we suppose that the pressure is hydrostatic. These equations read :

$$\frac{\partial h}{\partial t} + \frac{\partial(\bar{u}h)}{\partial x} + \frac{\partial(\bar{v}h)}{\partial y} + G = 0 \qquad (9)$$

$$\frac{\bar{u}^{n+1} - \overline{u_{diff}}}{\Delta t} + g\frac{\partial h}{\partial x} = -g\frac{\partial Z_f}{\partial x} + F_x \qquad (10)$$

$$\frac{\bar{v}^{n+1} - \overline{v_{diff}}}{\Delta t} + g\frac{\partial h}{\partial y} = -g\frac{\partial Z_f}{\partial y} + F_y \qquad (11)$$

with

$h = S - Z_f$	water height
G	source term due to any inlet or oulet of water at the bottom or at the free surface.
$\bar{u}^{n+1}$	mean velocity along x at t^{n+1}
$\bar{v}^{n+1}$	mean velocity along x at t^{n+1}
$\overline{u_{diff}}$	velocity u_{diff} (diffusion result) integrated along vertical
$\overline{v_{diff}}$	velocity v_{diff} (diffusion result) integrated along vertical
F_x, F_y	buoyancy terms

Mean velocities are computed according to the formula :

$$\bar{u} = \frac{1}{(S-Z_f)}\int_{Z_f}^{S} u\,dz$$

The source term G reads :

$$G = \frac{h}{S}\left[w^*(\bar{S}) - w^*(0)\right]$$

Buoyancy terms are given by :

$$F_x = g\int_{Z_f}^{S}\frac{\partial}{\partial x}\int_{z}^{S}\frac{\Delta\rho}{\rho_0}dz\,dz' \; , \; F_y = g\int_{Z_f}^{S}\frac{\partial}{\partial y}\int_{z}^{S}\frac{\Delta\rho}{\rho_0}dz\,dz'$$

These 2D integrated equations are solved by TELEMAC-2D. TELEMAC-2D solves the 2D shallow water equations by means of finite element methods. One of the finite elements available is the triangle. That code has been written by also taking advantage of E.B.E. methods (that is non assembly of matrices, vectorizable matrix vector product...). Its computing cost is between 0.05 and 0.1 s per time step per thousand points. It is currently used by L.N.H. for real applications in ocean and river engineering. More information about TELEMAC-2D can be obtained in Galland. Note that we do not use the full potientality of TELEMAC-2D, because we do not have diffusive and advective terms in our integrated equations.
TELEMAC-2D yields the free surface S at t^{n+1}, which allows us to compute the pressure p^{n+1} and the horizontal components of the velocity u^{n+1} and v^{n+1}. w^{n+1} is then known by solving the continuity equation.

3.6 *Boundary conditions*

Boundary conditions are roughly and numerically speaking of three types : Dirichlet, Neumann and degree of freedom.

Dirichlet boundary conditions are taken into account in the three steps of the algorithm. The equation to solve at the boundary is replaced by the required equality. Neumann boundary conditions appear naturally in the finite element formulation of the diffusion equation. Hence, they are treated in the diffusion step.The last type is the easiest to deal with since there is nothing to change in the corresponding equation. Many classic boundary conditions enter in the classification listed above. For instance, wind, wall and bottom frictions are Neumann boundary conditions. Values imposed at the inlet of a domain may be considered of the Dirichlet type.

However, there are other boundary conditions which are not of any of the previous types like wall boundary conditions. In the case of TELEMAC-3D, wall boundary conditions undergo a special treatment in the last step. First, they appear naturally in the finite element formulation of TELEMAC-2D, when the mean velocities $\bar{u}^{n+1}$ and $\bar{v}^{n+1}$ are computed. Then, they are again taken into account when u^{n+1} and v^{n+1} are computed, by imposing $\vec{u}.\vec{n}=0$ ($\vec{u}$ velocity vector and $\vec{n}$ outward normal vector) on the side walls if any. Note that on the side walls w^{n+1} is not affected by a wall boundary condition since the side walls are vertical. The wall boundary condition at the bottom is satisfied by solving equation (13).

4 RESULTS

TELEMAC-3D is run on a CRAY YMP which allows full vectorization of many E.B.E. subroutines. Its computing cost varies between 0.1s and 0.15s per time step per thousand points.

It is developed according to a quality procedure, which requires a complete validation. Planned test cases cover the range of applications of the code. Here, we present two test cases along with the first results of an environmental application.

4.1 *Ekman test case*

It consists of computing in an ocean with a flat bottom, the velocity vertical profile due to the actions of the wind and the Coriolis force. The depth of the ocean is finite and there is no slip on bottom. There is an analytical solution to that problem to which numerical results may be compared. That test case is particularly useful to test the quality of the diffusion step and the computation of the horizontal components of the velocity in the last step of the algorithm. Indeed, the veering of the velocity vector along the vertical is due to the combined effect of friction and the Coriolis force, both taken into account in the diffusion step. The test case is done with a wind velocity of 32m/s, a Coriolis parameter of 1.1 10^{-4} s^{-1}, a viscosity of 0.1 $m^2.s^{-1}$ and four different depths of water 13m, 33m, 67m and 160m. The computation domain is a square of side 2km meshed in 2D with 488 points. There are, along the vertical, either 9 planes (for water depths of 13m and 33m) or 18 planes (for water depths of 67m and 160m). The computed velocities are in very good agreement with the analytic solution (figures 4, 5 and 6).

4.2 *Quarter circle test-case*

That test-case has been extensively studied by Lynch (1991). It consists of the computation of a tide entering a quarter circle. It is more complicated than the Ekman test-case since it involves a varying bathymetry and movement of the free surface. Yet, there is still an analytic solution allowing comparisons with numerical results. With such a test-case, we can judge of the quality of the diffusion and the free surface-continuity-pressure steps, except for the computation of w which remains very low. The advection step is of no importance here.

The tidal wave enters the domain through the outer quarter circle (see figure 7). The three other boundaries are walls. The friction on the bottom is supposed infinite (no slip condition). The horizontal viscosity coefficients are equal to zero and the vertical viscosity coefficient is given by : $\nu_z = 0.1\ \omega h^2$

The 2D mesh is made of 322 points. There are 11 planes along the vertical. The bathymetry is dependent upon the distance from the origin according to :

$$Z_f = -\ 3.048 \left(\frac{r}{60{,}960}\right)^2$$

The computations of TELEMAC-3D are in very good agreement with the analytic solutions for the free surface (figure 8) as well as for the velocity vertical profiles (figure 9).

4.3 *Dilution of a fresh water discharge in the salted lagoon of Berre*

The aim is to study dilution of the fresh water of the River Durance in the salted water of the lake of Berre connected to the Mediterranean sea through the channel of Caronte.

The currents in the lagoon are induced by 3 phenomena :

- The tide which enters the lagoon through the flume of Caronte.
- The wind which creates significant currents in the lagoon. It generally blows from north to south.
- The buoyancy effects due to the fresh water discharge of the Durance river.

The heterogeneity of both the currents and the salinity requires a 3D code like TELEMAC-3D. Figure 10 shows a first simulation without wind. Currents and stratification are represented.

5 CONCLUSION

The use of finite elements is no longer synonym of important computation costs, thanks to E.B.E. techniques. This is good news since finite element methods provide the user (and the programmer) with a convenient environment, as we have been able to experience it with the codes TELEMAC-2D and TELEMAC-3D.

REFERENCES

Daubert, O., Gest, B., Lepeintre, F. (1989)
Code 3D à surface libre MITHRIDATE - Equations de base - Note d'avancement
EDF report HE-41/89.38 (1989)

Donnars, P., Manoha, B. (1989)
Three dimensional numerical modelling of thermal impact for Gravelines nuclear power plant. Proc 23rd IAHR Congress. Ottawa. August 1989

Hervouet, J.M. (1991)
TELEMAC, a fully vectorized finite element software for shallow water equations.
Proc second international conference on Computer Methods and Water Resources. October 7-11 1991. Rabat. Morocco.

Hughes, T.J.R., Ferencz, R.M., Hallquist, J.O. (1987)
Large-scale vectorized implicit calculations in solid mechanics on a Cray XMP/48 utilizing EBE preconditioned conjugate gradients. *Computer Methods in Applied Mechanics and Engineering* **61**:215-248

Lynch, D.R., Werner, F.E. (1991)
'Three dimensional hydrodynamics on finite elements. Part II : Non-linear time-stepping model', *Int. j. numer. methods fluids*, **12**, 507-533 (1991).

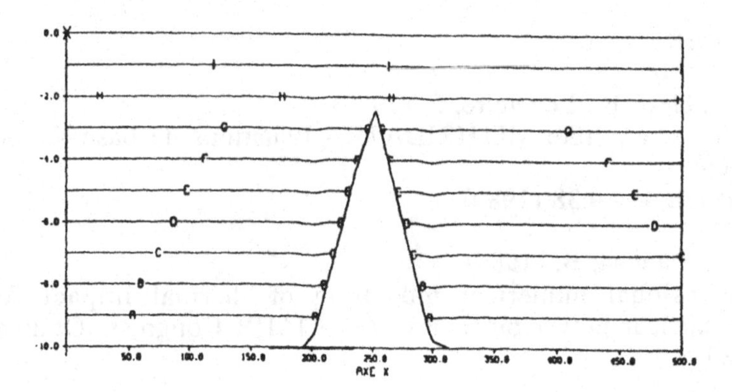

Figure 3 : absence of diffusion of a passive effluent with a linear distribution in a closed flume (A=2., B=4., ..., I=18)

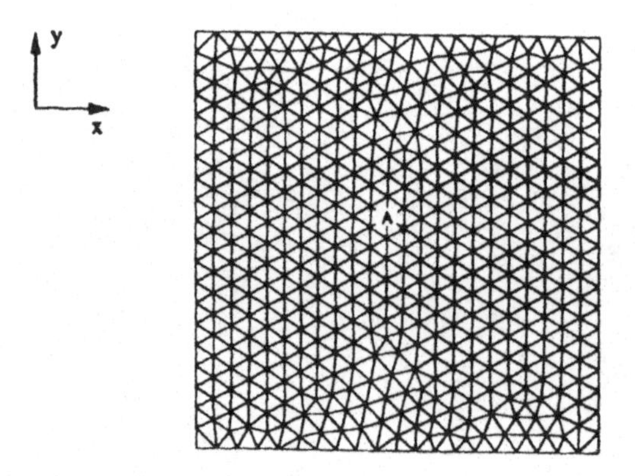

Figure 4 : 2D mesh for the Ekman test-case (488 points, 894 triangles). Velocity vertical profiles of figures 5 and 6 were taken at point A.

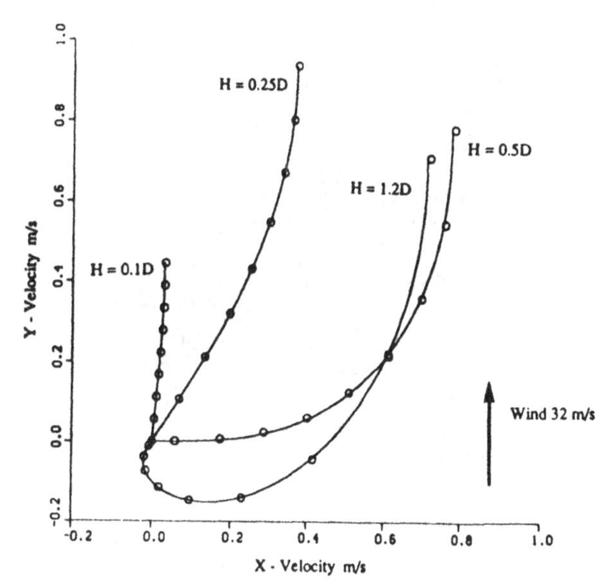

Figure 5 : computed velocity vertcal profiles (circles) compared to analytic profiles (full lines) at point A of figure 4. H is the depth of water (13m, 33m, 67m and 160m). D is the depth of frictional influence = 133.7m.

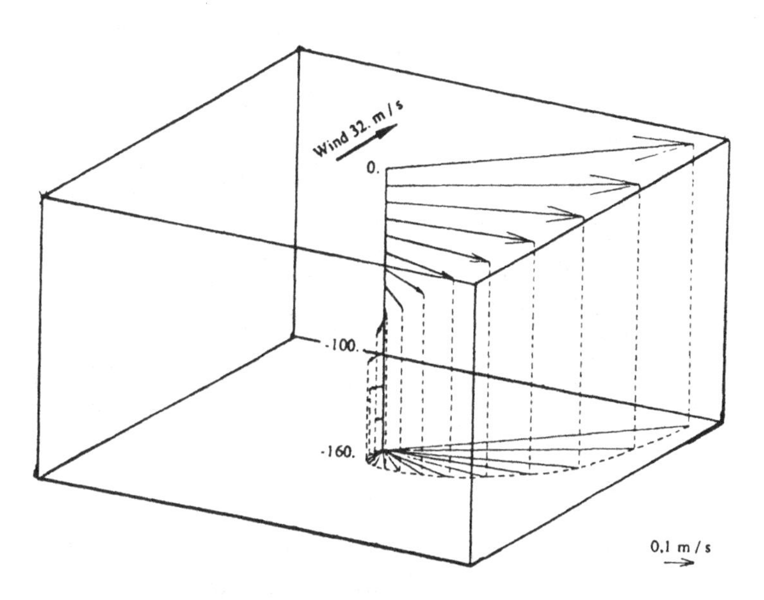

Figure 6 : perspective of the velocity vertical profile.

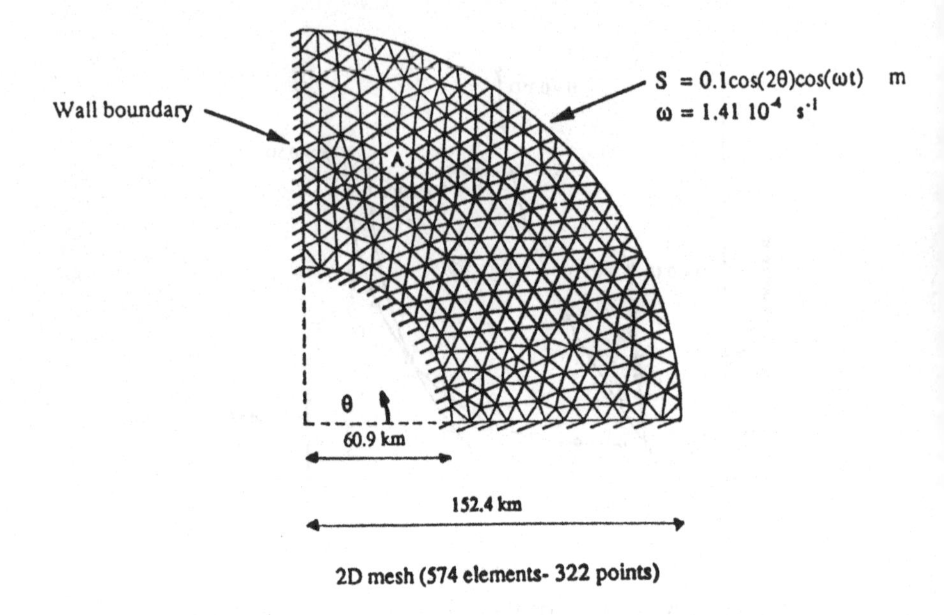

Figure 7

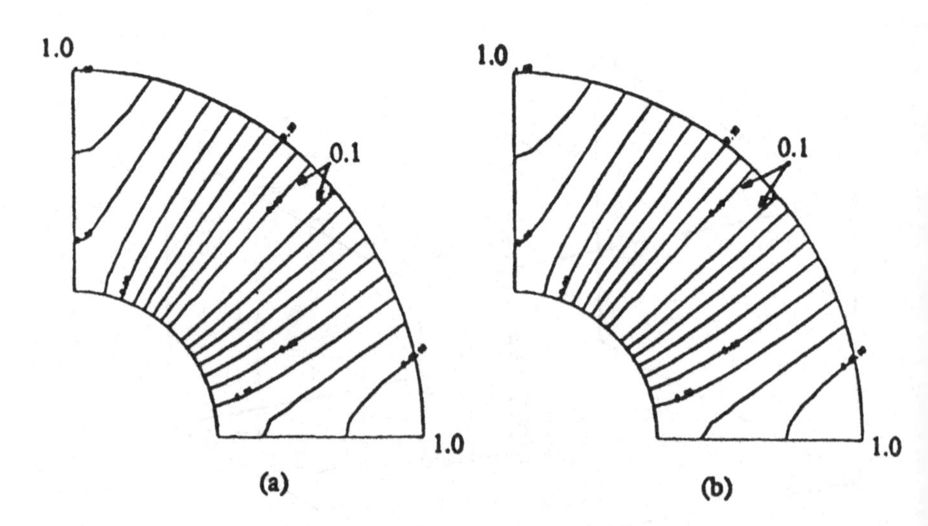

Figure 8 : Tidal amplitudes
(a) Normalized numerical result
(b) Normalized analytic result

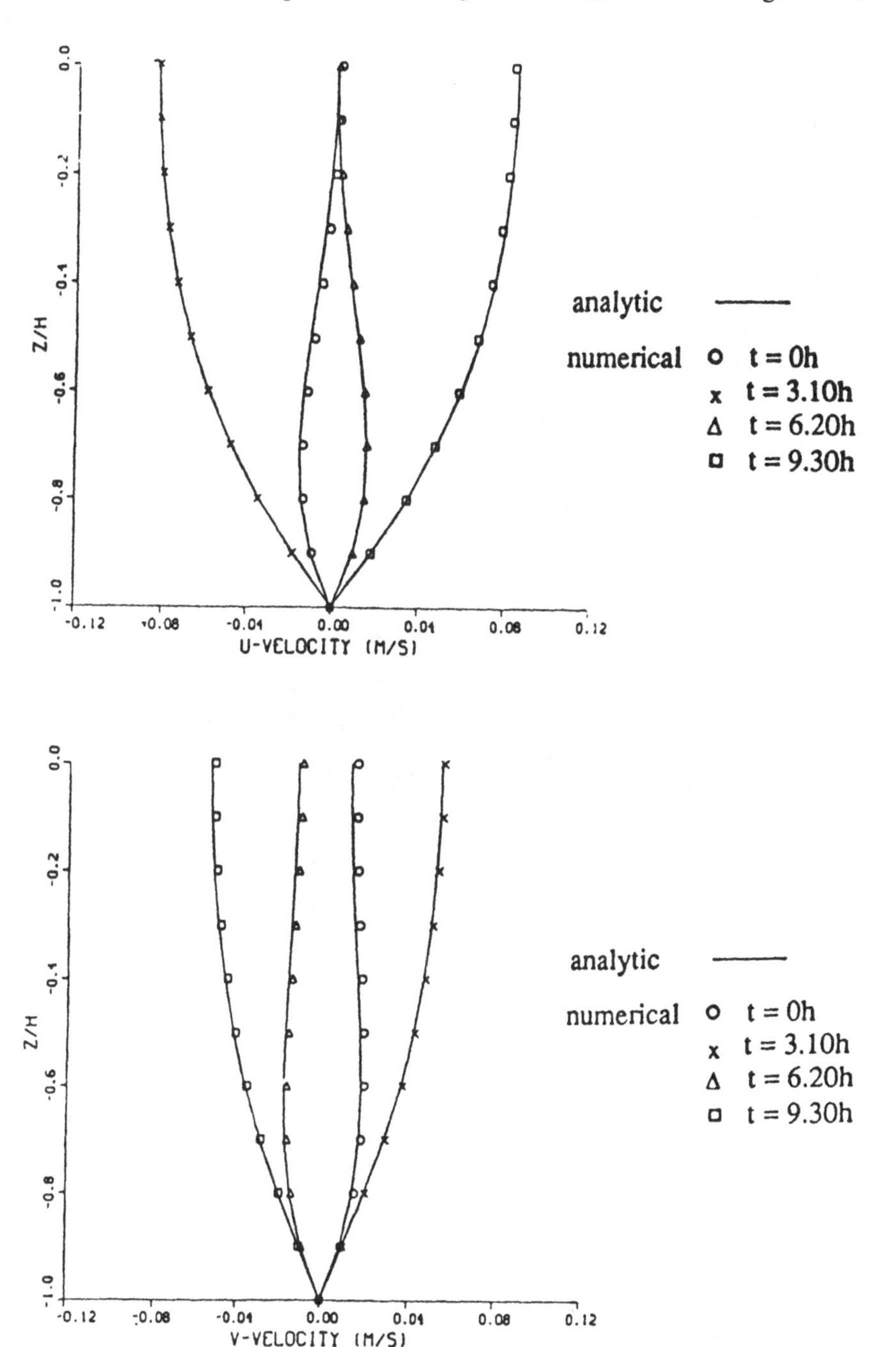

Figure 9 : comparison of analytic and numerical vertical velocity profiles for u and v at four points in time during a tide. The vertical is at point A of the 2D mesh (see figure 7)

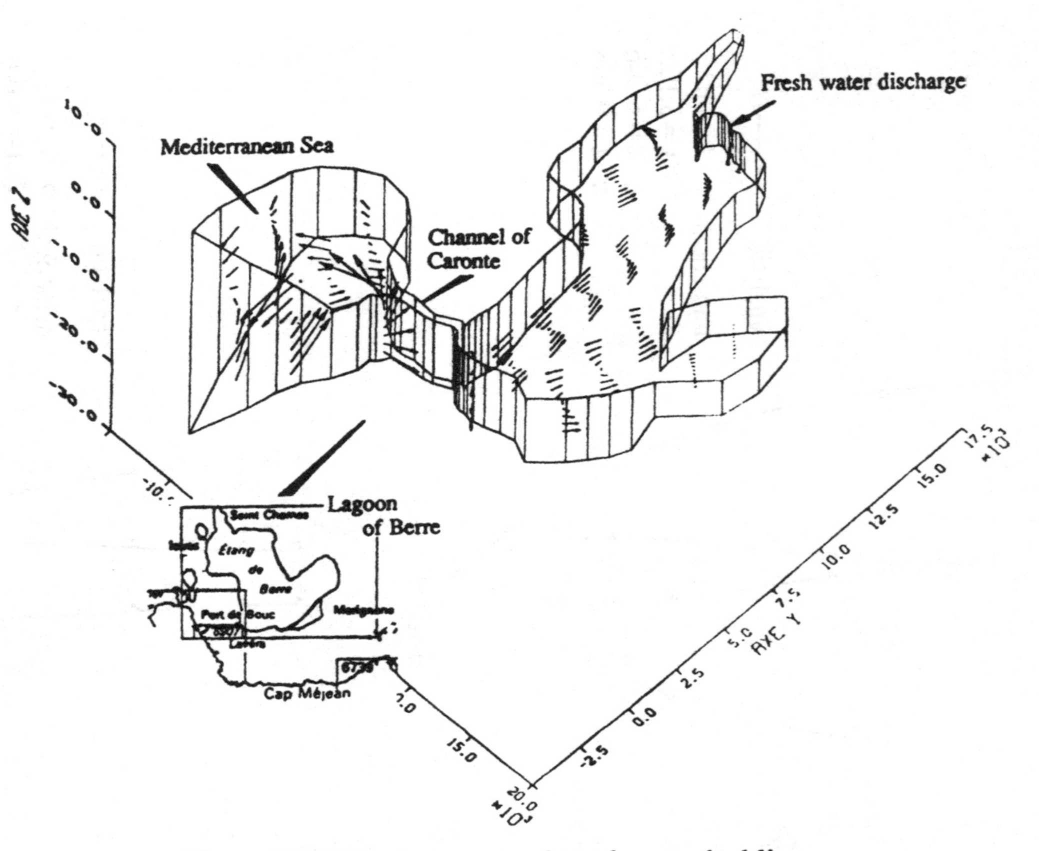

Figure 10 : Velocity vectors along few vertical lines

Dropped Object Risk Assessment

D.W. Begg, D. Fox
Brighton and Portsmouth Polytechnic, U.K.

ABSTRACT

As a part of any subsea field development proposal there is a need to consider the potential risk of damage to the production system arising from equipment accidentally dropped from vessels overhead. This paper describes an analysis methodology in which extensive use is made of computer simulation of fully directional dropped object trajectories due to the major environmental inputs (Wave, Wind Induced Current and Tidal Current). Potential dropped objects, their physical characteristics and drop frequencies are identified. Many drops from each surface drop zone, under environmental conditions generated to match the directional distributions are used to develop seabed impact footprints for each object and each drop zone. Cumulative impact frequencies over a subsea grid element give impact scatter diagrams, impact contour maps and 3-D impact probability density functions. The simulations are for the first time fully directional allowing a detailed consideration of overhead hazards in the conceptual stage of a seabed facility layout.

INTRODUCTION

The evaluation of these risks is established by means of determining the number of objects which may be dropped during the design life of the field and predicting their distribution over the seabed area adjacent to the proposed satellite well sites. The development of the field should have regard for the drop zones and associated high risk areas beneath floating production facility FPF or work-over vessel port and starboard pedestal cranes as well as the moonpool.

OBJECTIVES

The overhead hazard analysis aims to:

i) Demonstrate the probable magnitude of the risk of dropped objects over the proposed subsea facilities.

ii) Assess the results and bring to the operators attention areas where the envisaged impact frequencies could give cause for concern suggesting if possible means of risk reduction.

iii) Provide Probability Contour Plots which will enable alternative lower risk arrangements to be investigated.

BASIS OF CALCULATIONS

From information provided by an operator the proposed location of the seabed facilities at the satellite well location and similarly the proposed location of the vessel's cranes over the equipment are known. All contour plots can thus be transposed to a co-ordinate system with respect to the crane under consideration to allow superimposing of equipment onto them or, vice-versa, using them as overlays to amend arrangement drawings.

Environmental Conditions.

Previous analyses have used the probable maximum values for prevailing environmental forces which will act on an object dropped during the course of field operation. For the purpose of the simulations used by this current technique, fully directional environmental data have been obtained from the Marex Technology Limited and the Marine Information Advisory Service [1]; enabling a true probabilistic approach rather than an extreme value approach.

Rig Facilities.

With regard to the overhead hazard appraisal, of prime interest is the location and sweep area of the rig cranes used for supply boat cargo transfers. The static drop zone is assumed to extend to the full outboard crane sweep as equipment drops may occur at any point between lift-off and landing. From details of the rig the vessel outline and the extent of this 'drop map' can be defined in terms of a 2m grid of points in the form of an array.

When moored on location the rig will be subject to environmental forces which will possibly result in a horizontal offset as well as other surge and or sway motions. Prediction of the magnitude of this effect would add a second order effect to the displacements currently estimated. However the way in which the results are presented allows the effects of this offset to be evaluated. Future rig monitoring programmes could be utilised to gain more comprehensive information on vessel motion under a range of environmental inputs for direct evaluation during the computer simulations. In the absence of any other data a vessel offset of up to 6% of the sea depth should be allowed for, which in this case could be as great as 9.0 m.

Intervention Tasks.

The cranes are assumed to contribute to the hazard of dropped objects primarily by supply boat operations, the transfer of equipment and supplies being carried out during work-over periods; Wireline equipment handling, Coil tubing equipment handling and Wireline Coil tubing operations should also be considered based on previous experience.

Drop Frequencies.
The failure rates noted as design values are arrived at from a collation of experience and both published and confidential information.

Handling operations are assigned a 0.002 (for all Crane operations) or 0.005 (Moonpool) failure rate, dependent on their means of rigging or connection. Improvements to standard lifting techniques with the provision of lifting eyes where possible can bring about reductions in these risks to 0.001 and 0.003 respectively

Dropped Object Information.
The final basic information input for the study comprises information relating to the specific objects which may be dropped during the course of normal field operations and including workover and future drilling operations.

A listing of some of the objects included in an appraisal is shown in Table 1. below. Parameters for each object, including weight, dimensions, projected areas and terminal velocity constant 'T_{vk}', were compiled to provide input for the analysis. Drag coefficients associated with these objects have generally been assumed to fall within the range 1.0 to 1.5 for all irregular objects [2]; with cylindrical casing the actual value being determined at the time of simulation.

Study Methodology.
The objective of the methodology is to estimate the total numbers of dropped object impacts per square metre of seabed area over the analysis period. The results being presented as scatter diagrams or contour maps able to be superimposed on layout plans of the seabed facilities.

The methodology was applied to surface activities relating to rig operation and comprised the following procedural steps:

i) Identify potential dropped objects, their physical characteristics and drop frequencies.

ii) Examine environmental conditions and develop magnitude probability distributions by direction for waves, wind and current, where previously only extreme omni-directional conditions have been assumed to prevail.

iii) For each object, simulate a large number of drops from each surface drop footprint, under environmental conditions generated to match the directional distributions calculated in ii).

iv) From iii) develop seabed impact footprints for each object and each drop zone.

v) Cumulate the impact frequencies in each impact grid element to give impact contour maps, 3-D surface impact graphs and impact scatter diagrams.

Generation of Impact Probability Distributions

An object will fall, if dropped, into a seabed impact footprint for the appropriate surface drop zone. The surface drop zones considered under this study are those associated with crane operations and Moonpool intervention tasks. The seabed impact footprint for any particular object has been broken down into a rectangular grid of constant element width and constant element breadth (2m). The trajectory of an object as it descends throughout the sea depth can be estimated from analysis of the forces acting on it during the time it takes to fall at its terminal velocity. It should be noted that during this study incremental depth calculations as small as 0.001m have been investigated to ascertain the most appropriate calculation step size. The three main forces considered to affect the motion of the object are:

1. Current

2. Wave

3. Wind Drift Current

The location of the points of impact thus determined give rise to Seabed impact footprints for each object under each environmental input. It has been assumed that no correlation exists between the direction of Tide and Wave/Wind under normal operating conditions.

Using this approach for each object dropped from within an appropriate surface drop zone a seafloor "CUMULATIVE IMPACT CONTOUR MAP" indicating the number of impacts for a given total can be generated. These matrices can then be summed for all the objects in a particular damage group related by their lift frequency and if necessary differing failure rates to give the probabilities for each of the three areas already discussed. Sub-sea equipment can then be superimposed upon these probability distributions to allow assessment of total mean expected probabilities. This is obtained by summing the probabilities over the number of 2m squares covered by the equipment allowing for the mean object size as a buffer zone.

DESCRIPTION OF MODEL

The simulation routines for object trajectory and thus seabed impact location assume that we need only consider Drag and Inertia forces [2] and that the object enters the water at a position vertically beneath the point from which it was dropped at its terminal velocity. The object proceeds to fall through the depth of the sea to the seabed at this terminal velocity, which may vary from drop to drop, dependant upon the value of the drag coefficient which is allowed to take values randomly between 1.0 and 1.5. During this free fall it acted upon by three environmental inputs; Wave, Tidal Current and Wind induced Current: which may be directionally related. These three individual displacement vectors are then combined to give an impact distribution for a particular object. For the purpose of simulation the most complex effect is that due of wave force variation with depth. Both Airy wave theory and Stokes Non-Linear wave theory have been investigated.

Wave Theory

Solutions developed for standing and progressive small amplitude water waves [3] provide the basis for applications to numerous problems of engineering interest. The water particle kinematics and the pressure field within the waves are directly related to the calculation of forces on bodies.

Linear Waves

The horizontal velocity under the wave is given by Airey theory.

Nonlinear Waves

The water waves that have been discussed above are assumed to be small amplitude waves, which satisfy linearised forms of the kinematic and dynamic free surface boundary conditions. We have seen during simulation with both linear and non linear wave theory that the linear wave theory has given results almost exactly the same as non linear wave theory for all objects tested. Extension of the linear theory to a second-order Stokes theory is presented for completeness. The desire being to use a water wave theory to best satisfy the demands of the simulation where the requirement is to estimate probable impact locations rather than provide an exact solution . The simulation calculations proved both wave theories to be sufficiently accurate with the finite incremental depth change during object trajectory calculation being of far greater importance to seabed impact location. In the ensuing calculations a depth increment of 0.1m was chosen for iterative calculation.

Higher order terms have been shown to be of negligible importance to seabed impact location determination also according to the MAREX environmental report these terms have zero coefficients for the waves of this measured waves in the location of current Northern North Sea developments.

Wave Direction and Height Probabilities

Rather than use omni-directional wave data; the directional frequency tables provided in the environmental report, have been used to provide a look-up table for use during simulation computation.

Tidal Current Speed and Direction Probabilities

Tidal current simulation lead to by far the greatest magnitudes of object displacement, as the force is applied in the same direction over the full depth of the sea(3). The data supplied by the company related to extreme currents which is not applicable for simulation purposes. It was therefore necessary to obtain data from adjacent sites of current meter deployment available through the Marine Information Advisory Service (MIAS). The data from the three scatter diagrams for these locations have been summed to give a look-up table of 'Current Direction and Speed Probabilities'.

The current profile over the lower 25% of the depth has been approximated by a power law.

Wind Drift Current Speed and Direction Probabilities

The simulation of this current is very similar to tidal current except that with the tidal current the object is accelerating whereas with wind drift the object is decelerating with depth (4). Again it was necessary to obtain the cumulative probability distribution of winds by direction to allow realistic simulation.

The local wind induced current speed at the surface is assumed to be 3% of these hourly mean values. The associated current profile due to local wind being approximated by a logarithmic relationship.

RESULTS OF SIMULATIONS

Computer simulation of the above effects is carried out for each of the types of object for a sufficiently large number of drops to give a smooth representation of the number of impacts per square metre even at the extremes of displacement.

Wave Force Trajectory Component

The wave characteristics vary in accordance with the look-up table already discussed whilst the object may enter the sea surface at a random phase of this wave. It is of interest to note that there appears to be no relationship between wave height and object displacement from the drop centre. The highest waves did not always produce the largest displacement. Also that the effect of wave forces is relatively omni-directional possibly due to the derived nature of the supplied data. An improvement in the simulation would be the use of measured data relating to wave height and direction.

Tidal Current Force Trajectory Component

The current characteristics varying in accordance with the look-up table already discussed. This force is predominant in determining the object displacement. The effect of tidal current forces is not omni-directional with the impact distributions showing a distinct skew. The shape of the frequency distribution function is definitely object dependent again illustrating the benefits to be gained from this type of simulation.

Wind Drift Force Trajectory

The wind characteristics varying in accordance with the look-up table already discussed. This effect is of least importance in the determination of object displacement but due to the probability of its coincidence with wave direction it can not be neglected. The effect of wind drift current forces is almost omni-directional with the impact distributions showing a possible discontinuity in the measured data. This would be improved by finer resolution of wind direction measurements. The impact frequency distribution for wind drift current alone would give rise to displacements less than 1.0m for relatively high wind speeds.

Combined Object Impact Distributions

The results of the individual simulations of the effects of Wave, Tidal Current and Wind are combined at run time, fig.1., for each of the objects to give an estimate of the impact distribution for a specific object dropped under normally occurring environmental conditions. The 8 directions for wind and wave and the 18 directions for current are distributed again at run time to give a quasi full field effect in the form of scatter diagrams, fig.2. Routines then sum the impact frequencies for a 2m grid size over a 60m x 60m square centred on the drop origin. This can be displayed in one of two ways.

1. As a 2-D plot for a particular object showing the object number the maximum number of impacts in any one 2m x2m square and the total number of combined simulated drops. With grid squares coloured to represent the relative frequency of impact occurrence. Up to 15 different levels being distinguishable on the screen.

2. As a 3-D surface is found by regression analysis of x,y,z i.e. Eastings, Northings, Frequency of Occurrence data output from the previous routines. This can also allow smoothing of the data if required as well as contour plotting and output of impact frequencies over the 5184 grid squares.

PROBABILITY PLOTS

Superimposition of the main items of subsea equipment onto the 10 year impact probability maps allows direct reduction of the mean impact risk over this period for all objects. Additional probabilities may be calculated from the contour plots provided by counting the number of 2m squares covered by the equipment under consideration allowing for a suitable buffer zone of say 2-4m

CONCLUSIONS

By far the greatest risk to subsea equipment is that associated with cargo handling by FPF or work-over vessel cranes as well as moonpool activities whist directly over the well heads. Risk reduction can only be achieved by running all pipelines and umbilicals outwith the crane drop zone or using a combination of beneficial vessel offset and single crane operation.

It is noted that whilst prevailing environmental forces may appear essentially omnidirectional in nature the ability of the simulation to model the measured environmental data provides the opportunity for risk reduction by re-arranging the proposed facilities.

It is also noted that the analysis methodology used being fully directional with respect to environmental inputs allows a detailed consideration of overhead hazards in the conceptual stage of seabed facility layouts, which may be relevant for possible future field developments. The requirement for true 'Time at Level' type data has been apparent throughout this study. To improve this simulation technique a review of system specification of surface and subsea monitoring should be undertaken to provide details of fully directional environmental data together with the associated vessel response as well as time series monitoring of all lifting operations.

REFERENCES

1. 'Environmental Conditions for the Alwyn North Extension', Marex confidential report.

2. 'Dynamics of Marine Structures', Report UR8 Ciria Underwater Engineering Group, June 1977.

3. Dean and Dalrymple 'Wave Water Mechanics for Engineers and Scientists' , 1984.

Object Number	Description	Length (m)	Width (m)	Height (m)	Mass (kg)
1	10Full Height Cont.	3.048	3.048	1.8288	2000
2	10Half Height Cont.	3.048	3.048	0.9144	1420
3	20FH	6.096	6.096	1.8288	3000
4	20HH	6.096	6.096	0.9144	2640
6	10FH	3.048	3.048	1.8288	4000
7	10FH	3.048	3.048	1.8288	5000
8	10FH	3.048	3.048	1.8288	6000
9	10FH	3.048	3.048	1.8288	7000
10	10FH	3.048	3.048	1.8288	8000
12	10FH	3.048	3.048	1.8288	10000
13	10FH	3.048	3.048	1.8288	12000
14	10HH	3.048	3.048	0.9144	2420
15	10HH	3.048	3.048	0.9144	3420
16	10HH	3.048	3.048	0.9144	4420
18	10HH	3.048	3.048	0.9144	6420
19	10HH	3.048	3.048	0.9144	7420
21	10HH	3.048	3.048	0.9144	9420
22	10HH	3.048	3.048	0.9144	10420
23	10HH	3.048	3.048	0.9144	11420
24	20FH	6.096	6.096	1.8288	8000
26	20FH	6.096	6.096	1.8288	10000
27	20FH	6.096	6.096	1.8288	11000
28	20FH	6.096	6.096	1.8288	13000
29	20HH	6.096	6.096	0.9144	3640
30	20HH	6.096	6.096	0.9144	6640
31	Riser Joint	15.24	15.24	0.168	2000
32	LMRP	3.05	3.05	2.59	7824
33	TRT	3.658	3.658	2.59	12193
34	Stress Joint	6.096	6.096	0.61	4064
35	SLF	10.06	10.06	2.13	7824
36	LUBE	9.754	9.754	0.61	1524
37	Blow Out Preventer	7.036	7.036	2.35	181488
38	CASING	18	18	0.762	9000
39	CASING	12	12	0.473	1560
40	CASING	12	12	0.346	1200
41	CASING	12	12	0.273	1056
43	CASING	12	12	0.178	564
44	CASING	12	12	0.127	420
45	CASING	9	9	0.241	2970
46	CASING	9	9	0.203	1980
48	STABIL	3	3	0.45	667
49	STABIL	2.4	2.4	0.343	625
51	STABIL	2	2	0.215	700
52	DRILST	9	9	0.127	288

TABLE 1. Dropped Object Data

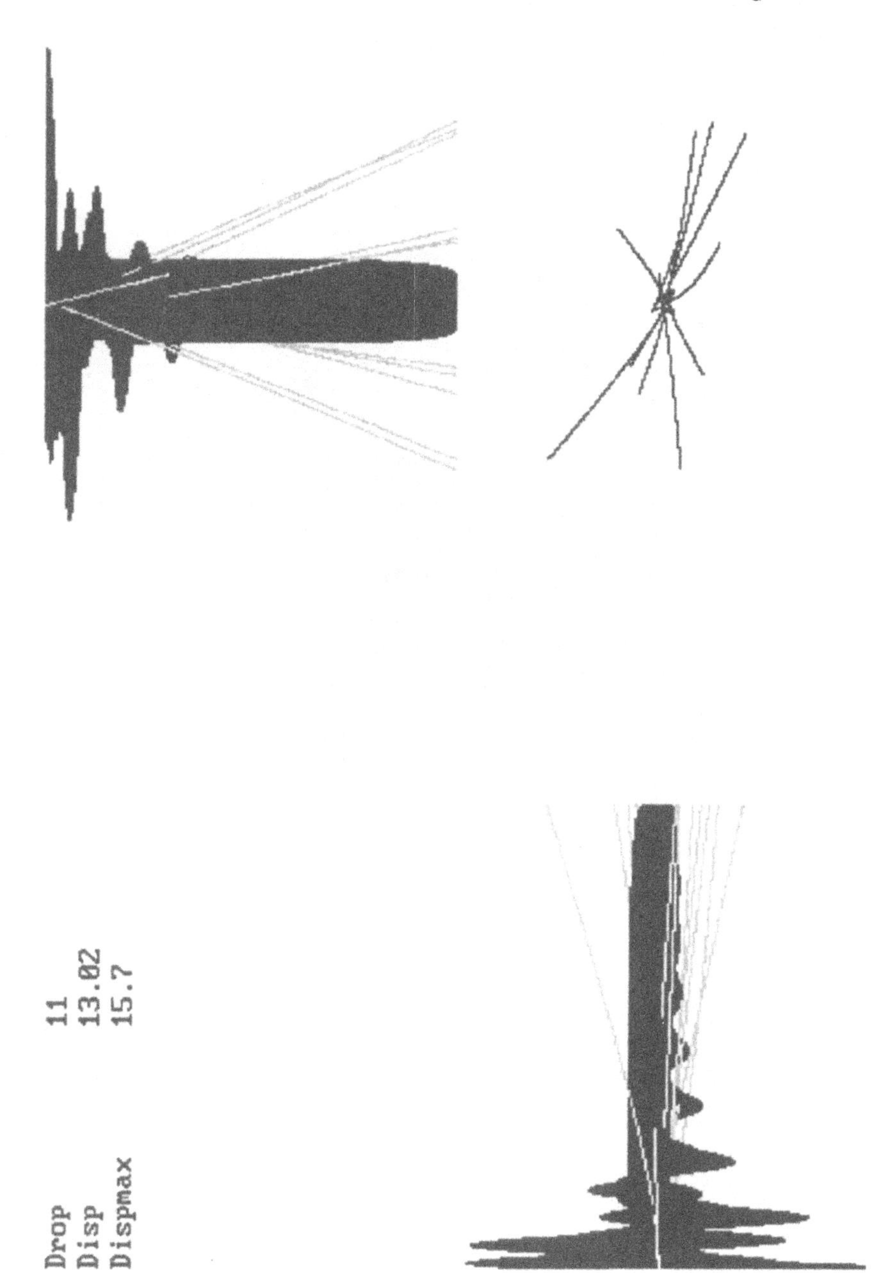

Figure 1. Typical Simulation showing Dropped Object Trajectories in both plan and elevations.

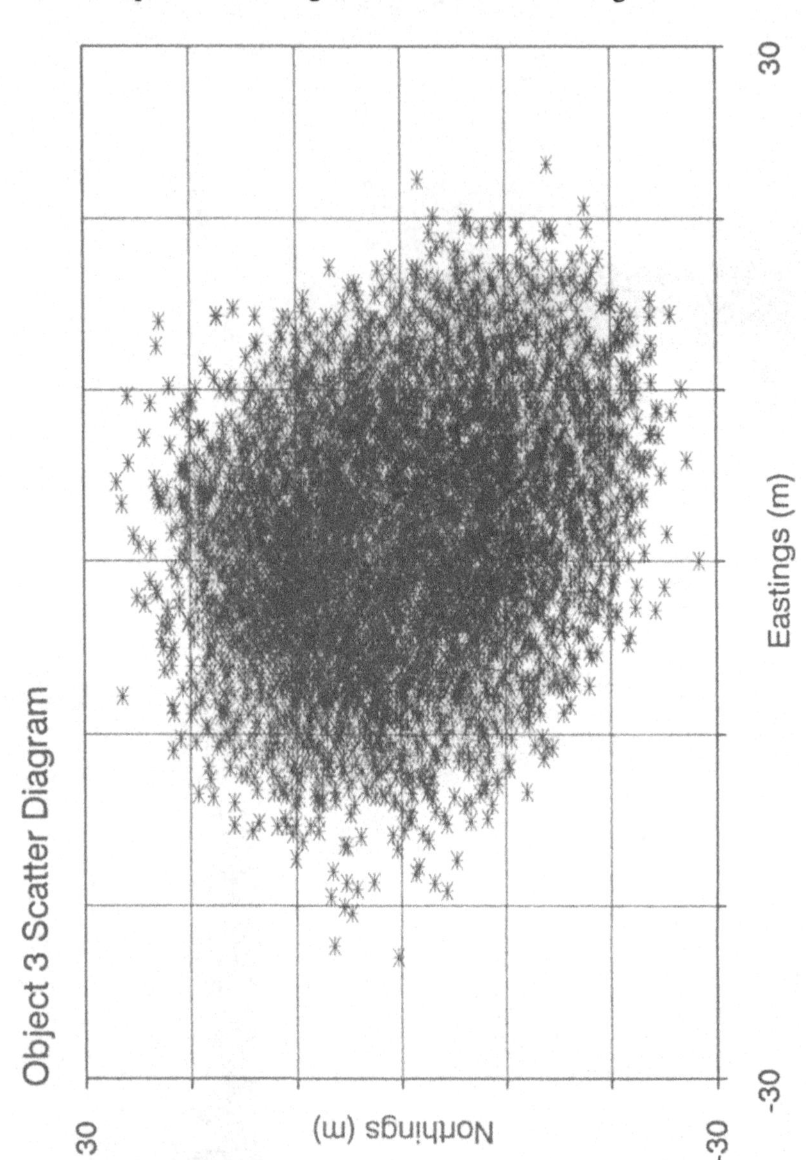

Figure 2. Seabed Scatter Diagram for Typical Object

SEDCO PORT CRANE

71 m x 71 m

0 E 0 N

436 drop zones.

Sedco Starboard Crane

71 m x 71 m

0 E 0 N

426 drop zones.

Figure 3. Crane Surface Drop Zones.

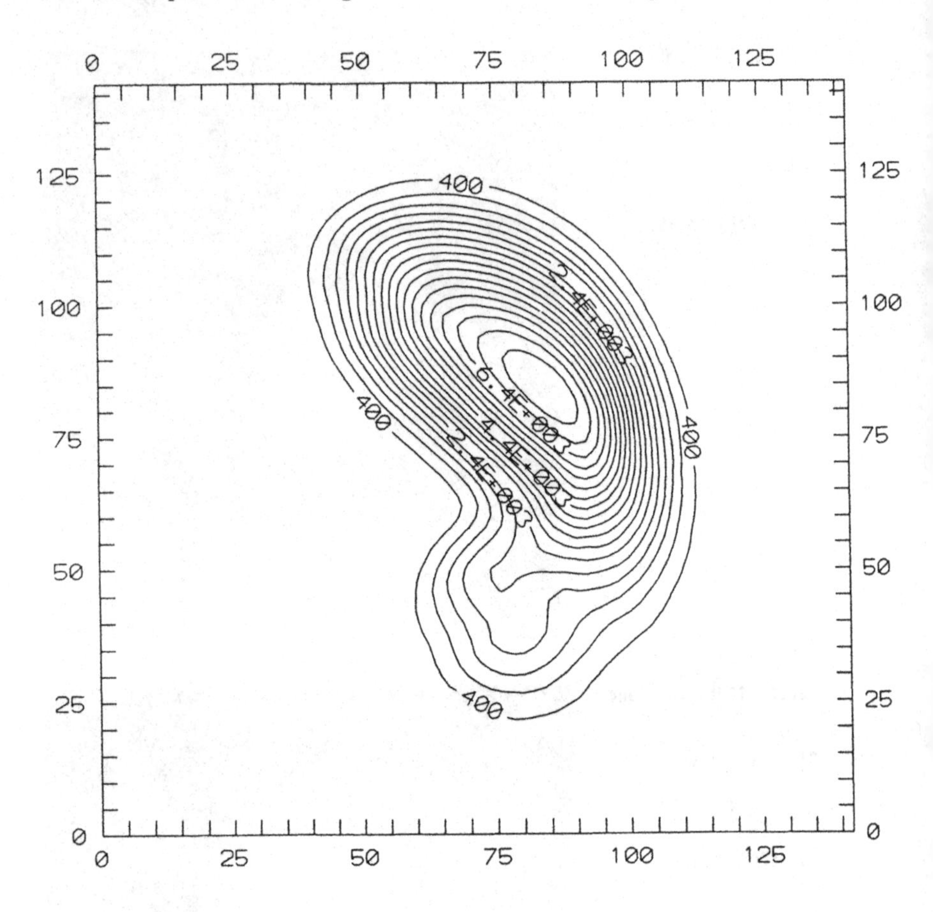

Figure 4. Starboard Crane Mean Expected Probability Distribution

AUTHORS' INDEX

Printed in the United States
By Bookmasters